Sequences II

Renato Capocelli   Alfredo De Santis
Ugo Vaccaro
Editors

Sequences II

Methods in Communication, Security, and Computer Science

With 56 Illustrations

Springer-Verlag

New York Berlin Heidelberg London Paris
Tokyo Hong Kong Barcelona Budapest

Renato Capocelli (deceased)
formerly with:
Department of Mathematics
University of Rome "La Sapienza"
Rome
Italy

Alfredo De Santis
Ugo Vaccaro
Dipartimento di Informatica ed
Applicazioni
Universitá di Salerno
84081 Baronissi (SA)
Italy

Library of Congress Cataloging-in-Publication Data
Sequences II (1991: Positano, Italy)
    Sequences II / Renato Capocelli, Alfredo De Santis, Ugo Vaccaro,
editors.
        p. cm.
    Includes bibliographical references and index.
    ISBN-13: 978-1-4613-9325-2   e-ISBN-13: 978-1-4613-9323-8
    DOI: 10.1007/978-1-4613-9323-8
    1. Sequences (Mathematics) --Congresses. I. Capocelli, Renato M.
II. De Santis, Alfredo. III. Vaccaro, Ugo. IV. Title. V. Title:
Sequences two.
QA292.S45 1991
515.24--dc20   92-32461

Printed on acid-free paper.

Production managed by Dimitry L. Loseff; manufacturing supervised by Jacqui Ashri.
Camera-ready copy prepared by the editors.

9 8 7 6 5 4 3 2 1

Renato M. Capocelli

(May 3, 1940 – April 8, 1992)

In Memoriam

Shortly after the editing of the papers presented to the Workshop *Sequences '91: Methods in Communication, Security, and Computer Science* was completed, Renato M. Capocelli died at the age of 52. We dedicate this volume to his memory as a sign of everlasting gratitude, appreciation, and love.

Salerno, May 10, 1992
A.D.S.
U.V.

Preface

This volume contains all papers presented at the workshop "Sequences '91: Methods in Communication, Security and Computer Science," which was held Monday, June 17, through Friday, June 21, 1991, at the Hotel Covo dei Saraceni, Positano, Italy.

The event was sponsored by the Dipartimento di Informatica ed Applicazioni of the University of Salerno and by the Dipartimento di Matematica of the University of Rome.

We wish to express our warmest thanks to the members of the program Committee: Professor B. Bose, Professor S. Even, Professor Z. Galil, Professor A. Lempel, Professor J. Massey, Professor D. Perrin, and Professor J. Storer. Furthermore, Professor Luisa Gargano provided effective, ceaseless help both during the organization of the workshop and during the preparation of this volume. Finally, we would like to express our sincere gratitude to all participants of the Workshop.

R.M.C.
A.D.S.
U.V.
Salerno, December 1991

Contents

Computer Science

Security

Automata and Combinatorics on Words

Contributors

Dave Bayer
Barnard College
Columbia University
New York, NY USA

Ian F. Blake
Department of Electrical and
Computer Engineering
University of Waterloo
Waterloo, Ontario CANADA

Mario Blaum
IBM Research Division
Almaden Research Center
San Jose, CA USA

Bella Bose
Department of Computer Science
Oregon State University
Corvalis, OR USA

Dany Breslauer
Department of Computer Science
Columbia University
New York, NY USA

Jehoshua Bruck
IBM Research Division
Almaden Research Center
San Jose, CA USA

Andrei Z. Broder
DEC Systems Research Center
Palo Alto, CA USA

Renato M. Capocelli (deceased)
Dipartimento di Matematica
Universitá di Roma "La Sapienza"
Rome ITALY

Benny Chor
Department of Computer Science
Technion
Haifa ISRAEL

Gérard D. Cohen
Ecole Nationale Supérieure des
Télécommunications
Paris FRANCE

Martin Cohn
Computer Science Department
Brandeis University
Waltham, MA USA

Claude Crépeau
Laboratoire de Recherche en
Informatique
Université Paris-Sud
Orsay FRANCE

Maxime Crochemore
LITP, Institut Blaise Pascal
Paris FRANCE

Paul Cull
Department of Computer Science
Oregon State University
Corvalis, OR USA

Larry J. Cummings
Faculty of Mathematics
University of Waterloo
Waterloo, Ontario CANADA

Steve Cunningham
Department of Computer Science
California State University
Stanislaus, CA USA

Aldo de Luca
Dipartimento di Matematica
Universitá di Roma "La Sapienza"
& Instituto di Cibernetica del CNR
Rome/Naples ITALY

Alfredo De Santis
Dipartimento di Informatica
Universitá di Salerno
Baronissi ITALY

Yvo Desmedt
Department of Electrical Engineering
and Computer Science
University of Wisconsin-Milwaukee
Milwaukee, WI USA

Michele Elia
Dipartimento di Elettronica
Politecnico di Torino
Torino ITALY

David Eppstein
Department of Computer Science
University of California
Irvine, CA USA

Shimon Even
Computer Science Department
Technion
Haifa ISRAEL

Yair Frankel
Department of Electrical Engineering
and Computer Science
University of Wisconsin-Milwaukee
Milwaukee, WI USA

Matthew Franklin
Computer Science Department
Columbia University
New York, NY USA

George H. Freeman
Department of Electrical and
Computer Engineering
University of Waterloo
Waterloo, Ontario CANADA

Zvi Galil
Department of Computer Science
Columbia University
New York, NY USA
& Department of Computer Science
Tel-Aviv University
Tel-Aviv ISRAEL

Luisa Gargano
Dipartimento di Informatica
Universitá di Salerno
Baronissi ITALY

Raffaele Giancarlo
AT&T Laboratories
Murray Hill, NJ USA
& Dipartimento di Matematica
Universitá di Palermo
Palermo ITALY

Roberto Grossi
Dipartimento di Informatica
Universitá di Pisa
Pisa, ITALY

Dan Gusfield
Computer Science Division
University of California
Davis, CA USA

Stuart Haber
Bellcore
Morristown, NJ USA

Jim Halloway ·
Department of Computer Science
Oregon State University
Corvalis, OR USA

Tom Head
Mathematical Sciences
State University of New York
Binghamton, NY USA

Giuseppe F. Italiano
Department of Computer Science
Columbia University
New York, NY USA
& Dipartimento di Informatica e
Sistemistica
Universitá di Roma "La Sapienza"
Rome ITALY

János Körner
IASI – CNR
Rome ITALY

Giles Lachaud
Equipe ATI, CIRM
Marseille-Luminy FRANCE

Gad M. Landau
Department of Computer Science
Polytechnic University
Brooklyn, NY USA

Abraham Lempel
Hewlet-Packard Laboratories
Palo Alto, CA USA

Ami Litman
Bellcore
Morristown, NJ USA

Simon N. Litsyn
Department of Electrical
Engineering - Systems
Tel-Aviv University
Ramat-Aviv ISRAEL

Fabrizio Luccio
Dipartimento di Informatica
Universitá de Pisa
Pisa ITALY

James L. Massey
Signal and Information Processing
Laboratory
Swiss Federal Institute of Technology
Zürich SWITZERLAND

Harold F. Mattson, Jr.
School of Computer & Information
Science
Center for Science and Technology
Syracuse, NY USA

Neri Merhav
Department of Electrical Engineering
Technion
Haifa ISRAEL

Thomas Meittelholzer
Signal and Information Processing
Laboratory
Swiss Federal Institute of Technology
Zürich SWITZERLAND

Rosanna Montalbano
Dipartimento di Matematica
Universitá di Palermo
Palermo ITALY

Alon Orlitsky
AT&T Bell Laboratories
Murray Hill, NJ USA

Rafail Ostrovsky
MIT Laboratory for Computer
Science
Cambridge, MA USA

Linda Pagli
Dipartimento di Informatica
Universitá di Pisa
Pisa ITALY

Dominique Perrin
LITP, Institut Blaise Pascal
Université Paris
Paris FRANCE

Michael O. Rabin
Aiken Computation Laboratory
Harvard University
Cambridge, MA USA
& The Institute of Mathematics
Hebrew University of Jerusalem
Jerusalem ISRAEL

John H. Reif
Computer Science Department
Duke University
Durham, NC USA

Antonio Restivo
Dipartimento di Matematica
Universitá di Palermo
Palermo ITALY

Wojciech Rytter
Institute of Informatics
Warsaw University
Warsaw POLAND

Miklós Sántha
Laboratoire de Recherche en

Informatique
Université Paris-Sud
Orsay FRANCE

Baruch Schieber
IBM Research Division
T.J. Watson Research Center
Yorktown, NY USA

Gadiel Seroussi
Hewlet-Packard Laboratories
Palo Alto, CA USA

Andrea Sgarro
Dipartimento di Matematica e
Informatica
Universitá di Udine
Udine ITALY
& Dipartimento di Scienze
Matematiche
Universitá di Trieste
Trieste ITALY

Netta Shani
Department of Computer Science
Technion
Haifa ISRAEL

Jacques Stern
GRECC, DMI
École Normale Supérieure
FRANCE

James A. Storer
Computer Science Department
Brandeis University
Waltham, MA USA

W. Scott Stornetta
Bellcore
Morristown, NJ USA

Peter R. Stubley
Department of Electrical and

Computer Engineering
University of Waterloo
Waterloo, Ontario CANADA

Esko Ukkonen
Department of Computer Science
University of Helsinki
Helsinki FINLAND

Ugo Vaccaro
Dipartimento di Informatica
Universitá di Salerno
Baronissi ITALY

Stefano Varricchio
Dipartimento di Matematica
Universitá degli Studi dell'Aquilla
L'Aquilla ITALY
& LITP
Institut Blaise Pascal
Université Paris
Paris FRANCE

Ramarathnam Venkatesan
Bellcore
Morristown, NJ USA

Andreas Weber
Fachbereich Informatik
J.W. Goethe-Universität
Frankfurt am Main GERMANY

Moti Young
IBM Research Division
T.J. Watson Research Center
Yorktown Heights, NY USA

Jacob Ziv
Department of Electrical Engineering
Technion
Haifa ISRAEL

Communication

On the enumeration of dyadic distributions [1]

Ian F. Blake, George H. Freeman and Peter R. Stubley
Department of Electrical and Computer Engineering
University of Waterloo, Waterloo, Ontario Canada N2L 3G1

I. Introduction.

The problem of interest is the number of solutions to the diophantine equation

$$\sum_{i=1}^{n} \frac{1}{2^{x_i}} = 1 \ , \quad 0 \le x_1 \le x_2 \le \cdots \le x_n \ , \quad x_i \epsilon Z^+ \tag{1}$$

The equation is the case of equality in the Kraft-McMillan inequality for uniquely decipherable codes with length distribution $\{x_i\}$, on an alphabet of size two. A probability distribution of the form $\{1/2^{x_i} \ i = 1, 2, \cdots, n\}$ will be referred to as dyadic and the relationship between dyadic distributions and uniquely decipherable code trees is immediate.

All trees considered here are binary and a tree will be called complete if it represents a uniquely decipherable code whose word lengths satisfy (1). They are pictured as progressing left-to-right from a root node of degree 2 (no parent) to leaf nodes of degree 1 (no children). All other nodes are of degree 3. The level (or depth) of a node is the length of its unique path from the root. Each leaf in the complete binary tree corresponds to a codeword in the uniquely decipherable code. A leaf at level x_i will be thought of both as representing a codeword having length x_i in the code and an element 2^{-x_i} in a dyadic distribution. Equation (1) is satisfied for the distribution under an appropriate ordering of the n leaf nodes. Two complete binary trees yielding the same distribution will be called equivalent. In that case each node is distinguishable only by its level l and may appear as either child of any node at

[1]This work supported by Natural Science and Engineering Research Council Grants A7382 and A6658

3

level $l - 1$ having children. The canonical form of such a tree is viewed as having each node in its uppermost possible position. At each level then all nonleaf nodes are above leaf nodes. A complete tree in canonical form will be referred to as a Huffman tree and the one-to-one correspondence between dyadic distributions and Huffman trees is clear.

The purpose of this note is to discuss the literature on the enumeration problem for such trees or distributions. The next section introduces the motivation for consideration of the problem. The interesting approaches to this problem in the literature are considered in section III. Section IV gives a generating function approach to the problem communicated to the authors by David Jackson [6].

II. Motivation of the problem.

The application of interest is to data compression algorithms and it will be useful to think of the trees in the following manner. Consider a semi-infinite binary string. The bits determine a path through the tree where at a given node a zero indicates taking the upper branch and a one the lower branch. Any semi-infinite binary string contains a unique prefix representing a path from the root to a leaf. The complete tree parses the string.

The motivation for considering the problem of (1) lies in the notion of dual tree coding [4], a data compression technique that is briefly explained as follows. Consider an i.i.d. binary source where the probability of a zero is p. Given a complete binary tree with n leaves, the source induces a probability distribution on the tree which parses source sequences and the tree is referred to as a parse tree. The Huffman code is then constructed for this induced distribution - to each leaf of the the parse tree is associated a binary code word. Alternatively one can think of a reversed tree (root on the right), the code tree (a Huffman tree) also with n leaves that codes the leaves of the parse tree. This formulation of the problem is similar in spirit to the well known Lempel-Ziv algorithm. Notice that it includes ordinary Huffman coding, where the code tree is of fixed length (fixed to variable length) and Tunstall (variable-to-fixed length). From this formulation an adaptive Tunstall algorithm has recently been formulated [9] that performs very well.

If $P = \{p_i\}$ is the distribution induced on the parse tree by the source $\{P(0) = p\}$ and $Q = \{q_i = 1/2^{x_i}\}$ is the distribution induced on the corresponding Huffman tree (if the length of the codeword in the Huffman tree for the probability p_i in the parse tree is x_i, the induced probability is $1/2^{x_i}$), define the divergence between the two distributions as

$$
\begin{aligned}
D(P, Q) &= \sum_i p_i \log \frac{p_i}{q_i} \\
&= \sum_i p_i \log p_i - \sum_i p_i \log q_i
\end{aligned}
$$

4

$$= -H(P) + E(C),$$

where $H(P)$ is the entropy of the parse tree distribution and $E(C)$ is the expected length of the code tree words. Thus the divergence in this context is the redundancy of the Huffman code. If $E(P)$ is the average length of the parsed sequences then [7] $H(P) = H(S)E(P)$. Consequently the compression ratio of the dual tree coding scheme is given by

$$\rho = \frac{E(C)}{E(P)}$$
$$= \frac{D(P,Q) + H(S)E(P)}{E(P)}$$
$$= \frac{D(P,Q)}{E(P)} + H(S).$$

In order to minimize the compression ratio one has to minimize $D(P,Q)/E(P)$. The problem of interest then is, for a given iid source with probability of zero p, and for a given code size n, determine the parse tree that minimizes the compression ratio. As n becomes large, $E(P)$ will become large and choosing the parse tree to minimize the divergence should lead to a useful compression algorithm.

The dual tree coding algorithms considered attempt to minimize the divergence between the two induced distributions and the problem seems complex. Consider the simpler problem: given a distribution $Q = \{q_i\}$ of size n, determine the tree with n leaves that induces the distribution $P = \{p_i\}$ that minimizes $D(P,Q)$. A suboptimal algorithm has recently been proposed [8] and it has been noted that a difficulty in determining the optimal tree is that a tree giving a very small $D(P,Q)$ may be topologically dramatically different from the optimal tree. Thus conventional tree growth algorithms seem of limited use in determining the optimal tree.

The motivation for the problem was then, conceptually, for each Huffman code tree (and its induced distribution) determine the parse tree that minimizes $D(P,Q)$. Such an approach runs through all possible Huffman trees with n leaves and their number is of interest. The algorithms developed did not in fact require this enumeration but the problem remained of interest. In particular, the asymptotic behaviour of enumeration is sought.

III. Approaches to the enumeration problem.

For small numbers of leaves there are simple recursions that quickly yield the numbers required. Following Golomb [5] let $\tau(n,k)$ denote the number of coding trees with n leaves

5

of which k pairs are at the deepest level. The recursion

$$\tau(n,k) = \sum_{j \geq \lfloor (k+1)/2 \rfloor} \tau(n-k,j) \tag{2}$$

reflects the fact that from a tree with $n-k$ leaves and j pairs of leaves at the deepest level, and thus $2j$ leaves at this level, $j \geq \lfloor (k+1)/2 \rfloor$ one can grow on k of these leaves to yield a tree with n leaves and k pairs at the deepest level. Finally

$$\tau(n) = \sum_k \tau(n,k)$$

and it is a straight forward matter to use this recursion to determine $\tau(n)$ for any reasonable value of n, although the magnitudes grow quickly. The approach is simple but appears difficult to draw asymptotic information from.

In earlier work Even and Lempel [3] give an algorithm for generating all code trees of size n that leads to the same recursion. It proceeds by defining proper words of length n as follows. A binary n-tuple $b = (b_1, b_2, \cdots, b_n)$, $b_i \epsilon \{0,1\}$, of weight u can be expressed as $(0^{r_0} 10^{r_1} 1 \cdots 0^{r_{u-1}} 10^{r_u})$ where 0^a represents a run of zeros of length a, and $\sum_i r_i = n - u$. The n-tuple b is called proper if it satisfies the conditions (i) $r_0 = 0$ (the sequences start with a one) and (ii) $r_{i+1} \leq 2r_i + 1$, $i = 0, 1, \cdots, u-1$. To each such proper n-tuple corresponds a coding tree constructed recursively as follows. $T(0)$ corresponds to the two leaf tree. To construct $T(i)$ from $T(i-1)$, $i = 1, 2, \cdots, n$ we use the rules: if $b_i = 1$ grow the uppermost leaf at the deepest level of $T(i-1)$; if $b_i = 0$ grow the uppermost leaf at the second deepest level of $T(i-1)$. Condition (ii) on the integer set r_i guarantees the availability of leaves to grow on at the required level. An algorithm is given that generates all proper words of length n and hence all complete trees, up to equivalence, with n leaves. It is further shown that if $f(n,l)$ is the number of proper words terminating in 10^l the function satisfies the recursion

$$f(n,l) = \sum_{k=\lceil (l-1)/2 \rceil} f(n-l-1,k).$$

It is easily recognized that $\tau(n,k) = f(n-2,k-1)$ and the recursion of (2) results.

An earlier work by Bende [1] followed a slightly different path. Consider a dyadic distribution with n elements $\{1/2^{x_i}, i = 1, 2, \cdots, n, x_1 \leq x_2 \leq \cdots \leq x_n, x_i$ a positive integer, $\sum_i 1/2^{x_i} = 1\}$, or

$$2^{x_n - x_1} + 2^{x_n - x_2} + \cdots + 2^{x_n - x_{n-1}} + 1 = 2^{x_n}$$

or alternatively

$$2^{x_n - x_1} + 2^{x_n - x_2} + \cdots + 2^{x_n - x_{n-1}} = 2^{x_n} - 1.$$

6

Define $A(n, r)$ to be the number of distinct positive integer solution sets $\{m_1, m_2, \cdots, m_n\}$ to the equation

$$r = 2^{m_1} + \cdots + 2^{m_n}.$$

From (2) it follows that

$$\tau(n) = \sum_{s \geq 1} A(n - 1, 2^s - 1).$$

It follows readily that if

$$F(z, x) = 1 + \sum_{r=1}^{\infty} \sum_{n=1}^{\infty} A(n, r) z^n x^r$$

is the bivariate generating function for the $A(n, r)$, then

$$F(z, x) = \prod_{k=0}^{\infty} \left(\frac{1}{1 - zx^{2^k}}\right) = \prod_{k=0}^{\infty} (1 + zx^{2^k} + z^2 x^{2 \cdot 2^k} + \cdots). \tag{3}$$

It follows that

$$F(z, x)(1 - zx) = F(z, x^2)$$

from which the following recursions derive:

$$A(n, 2k) = A(n - 1, 2k - 1) + A(n, k)$$

and

$$A(n, 2k + 1) = A(n - 1, 2k).$$

Our interest in (3) is only for r of the form $2^s - 1$ and the generating function is difficult to use to determine the asymptotic behaviour. Notice that

$$A(n, 2^{n-1}) = A(n - 1, 2^{n-1} - 1) + A(n, 2^{n-2})$$

$$A(n, 2^{n-2}) = A(n - 1, 2^{n-2} - 1) + A(n, 2^{n-3}), \quad \text{etc}$$

and it follows that

$$A(n, 2^{n-1}) = \sum_{i \geq 1} A(n - 1, 2^{n-i} - 1) = \tau(n)$$

an interesting result in itself.

Another method of generating all solutions to the diophantine equation (1) is noted in Boyd [2] (credited to Bende). From a given solution $(x_1, x_2, \cdots, x_n)$, $0 \leq x_1 \leq x_2 \cdots \leq x_n$, x_i a positive integer, if $x_n = m$ the equation can be expressed as

$$2^m = k_0 2^0 + \cdots + k_m 2^m, \quad \sum_i k_i = n$$

7

and note there are k_i leaves at depth $m-i$ and that $k_m = 0$ if $n > 1$. From this representation one obtains k_0 representations of 2^{m+1}, namely

$$2^{m+1} = (2i)2^0 + (k_0 - i)2^1 + k_1 2^2 + \cdots + k_m 2^{m+1} , \quad i = 1, 2, \cdots, k_0$$

and since $2i + (k_0 - i) + \cdots + k_m = n + i$ these representations correspond to growing i leaves at depth m of the previous tree to give i pairs of leaves at the new greatest depth, $m + 1$. . The description leads to the construction of a graph A in the following manner. A node reached by an edge labeled k in A is connected to $2k$ nodes at the next depth labeled $\{1, 2, \cdots, 2k\}$, a process described by the production rule $k \to \{1, 2, \cdots, 2k\}$. A node is assigned the label $n + i$ if the previous node had label n and the node is reached by an edge with label i. The process is started from a root node with label 1 and an edge with label 1 to a node with label 2. The required $\tau(n)$ is the number of nodes in A labeled n.

From this approach Boyd was able to derive the asymptotics of $\tau(n)$ and showed in fact that

$$\tau(n) \approx \alpha \lambda^n + O(\mu^n) , \quad 0 \le \mu < \lambda$$

where $\alpha = 0.14185325 \pm 6 \cdot 10^{-8}$ and

$$1.794147176 < \lambda < 1.794147186.$$

IV. The generating function approach.

Jackson [6] has communicated to the authors a direct generating function technique for the $\tau(n)$ and his method is outlined here. While more involved than the method of Bende, the technique itself is of interest. It is hoped to use the development to further investigate the asymptotic properties of $\tau(n)$.

Consider the equivalent problem to (1) of determining the number of positive integer m-tuples $\{k_1, k_2, \cdots, k_m\}$ such that for $n > 0$

$$(i) \ \sum_{j=1}^{i-1} \frac{k_j}{2^j} \le 1 \quad (ii) \ \sum_{j=1}^{m} \frac{k_j}{2^j} = 1 \quad (iii) \ \sum_{j=1}^{m} k_j = n.$$

The corresponding tree would have k_j leaves at depth j. Equivalently one could define p_j to be the number of nodes at depth j that are grown on and seek the integer m-tuple $\{p_0, p_1, \cdots, p_{m-1}\}$ where

$$\sum_{j=1}^{i-1} \frac{k_j}{2^j} = 1 - \frac{p_{i-1}}{2^{i-1}}, \quad i = 1, 2, \cdots, m.$$

8

Thus

$$k_i = 2p_{i-1} - p_i , \quad i = 1, 2, \cdots, m-1$$
$$p_0 = 1, , \quad p_m = 0 , \quad k_m = 2p_{m-1}.$$

The problem thus reduces to determining the number of integer vectors $\{p_0, p_1, \cdots, p_m\}$ such that $p_i \leq 2p_{i-1}$, $p_0 = 1$, $p_m = 0$. Now

$$\sum_{i=1}^{m-1} k_i = 2\sum_{i-1}^{m-1} p_{i-1} - \sum_{i=1}^{m-1} p_i$$

or

$$n - k_m = 2\sum_{i-1}^{m-1} p_{i-1} - \sum_{i-1}^{m-1} p_i$$

or

$$n = 2\sum_{i=1}^{m} p_{i-1} - \sum_{i=1}^{m-1} p_i = 2\sum_{i=0}^{m-1} p_i - \sum_{i=1}^{m-1} p_i = 2p_0 + \sum_{i=1}^{m-1} p_i$$

or

$$n = 2 + \sum_{i=1}^{m} p_i \quad \text{since} \quad p_0 = 1, \quad p_m = 0.$$

Define the function

$$F_m(x) = \sum_{p_1=0}^{2} x^{p_1} \sum_{p_2=0}^{2p_1} x^{p_2} \cdots \sum_{p_{m-1}}^{2p_{m-2}} x^{p_{m-1}}$$

and it is clear that the coefficient of x^{n-2} in $F_m(x)$, $[x^{n-2}]F_m(x)$, is the number of trees with n leaves of depth at most m.

The key steps in the development are described. Define the commuting variables $u_j \to u^j$, $x_j \to x^j$ and define the generating function

$$\Gamma_m(u, x) = \sum_{p_0=0}^{\infty} u_{p_0} \sum_{p_1=0}^{2p_0} x_{p_1} \cdots \sum_{p_{m-1}=0}^{2p_{m-2}} x_{p_{m-1}}$$

and further define

$$\Gamma(t, u, x) = \sum_{m=1}^{\infty} t^m \Gamma_m(u, x).$$

Define the set of indice pairs:

$$\pi_1 = \{(i, j) | 0 \leq j \leq 2i, \; i, j \geq 0\}$$

9

and let $A = \mathcal{I}(\pi_\infty)$ be the incidence matrix of π_1 and let B be such that $A + B = J$, the all ones matrix. Thus B is the incidence matrix of the indice pairs:

$$\pi_2 = \{(i,j)|j > 2i \geq 0,\ i,j \geq 0\},\quad B = \mathcal{I}(\pi_\in)$$

Let $U = \mathrm{diag}(u_0, u_1, u_2, \cdots)$, $X = \mathrm{diag}(x_0, x_1, x_2, \cdots)$. Then it is readily established that

$$\Gamma(t, u, x) = 1 + t\mathrm{tr}(U(I - tAX)^{-1}J).$$

A little manipulation then shows that

$$\Gamma(t, u, x) = 1 + \frac{t\mathrm{tr}(U(I + tBX)^{-1}J)}{1 - t\mathrm{tr}(X(I + tBX)^{-1}J)}$$

The evaluation of the various matrix terms are straightforward and it yields:

$$\Gamma(t, u, x) = 1 + \frac{\sum_{k=0}^{\infty}(-1)^k \frac{t^{k+1}}{1 - ux^{2^{k+1}-2}} \prod_{j=1}^{k} \frac{x^{2^j-1}}{1 - x^{2^j-1}}}{1 + \sum_{k=0}^{\infty} \frac{(-1)^{k+1}t^{k+1}}{1 - x^{2^{k+1}-1}} \prod_{j=1}^{k} \frac{x^{2^j-1}}{1 - x^{2^j-1}}}$$

Define $H(t, x) = [u]\Gamma(t, u, x)$ from which we obtain:

$$H(t, x) = \frac{\sum_{k=0}^{\infty}(-1)^k t^{k+1} x^{2^{k+1}-2} \prod_{j=1}^{k} \frac{x^{2^j-1}}{1 - x^{2^j-1}}}{1 + \sum_{k=0}^{\infty}(-1)^{k+1} \frac{t^{k+1}}{1 - x^{2^{k+1}-1}} \prod_{j=1}^{k} \frac{x^{2^j-1}}{1 - x^{2^j-1}}}$$

then $F_m(x) = [t^m]H(t, x)$, the coefficient of t^m in $H(t, x)$ and the number of trees of depth at most m with n leaves is $[x^{n-2}]F_m(x)$.

It appears very difficult to obtain the singularities of this generating function, and thence to determine the asymptotic behaviour.

V. Comment.

This note has dealt with the enumeration of complete binary trees but it is clear that the extension to r-ary alphabets is immediate. Boyd [2] has observed that if $\tau_r(n)$ is the number of complete trees for the r-ary case then

$$\tau_r(n) \approx \alpha_r \lambda_r{}^n$$

where λ_r tends to 2 as r tends to infinity, perhaps a surprising result as to how slow the growth is with alphabet size. The technique of Boyd [2] to determine the λ_r's is an approximate one and it would be of interest to determine a more straightforward technique to investigate the asymptotics. In particular it would be of interest to use the generating function developed to determine these growth parameters more directly.

References.

[1] S. Bende, The diophantine equation $\sum_{i=1}^{n} 2^{-z_i} = 1$ and its connection with graph theory and information theory (Hungarian), *Mat. Lapok.* vol. 18 (1967), 323-3227.

[2] D.W. Boyd, The asymptotic number of solutions of a diophantine equation from coding theory, *J. Comb. Theory (A)* vol. 18 (1975), 210-215.

[3] S. Even and A. Lempel, Generation and enumeration of all solutions of the characteristic sum condition, *Information and Control* vol.21 (1972), 476-482.

[4] G.H. Freeman, Asymptotic convergence of dual-tree entropy codes, *Proc. Data Compression Conference*, Snowbird, Utah, April, 1991, 208-217.

[5] S.W. Golomb, Sources which maximize the choice of a Huffman coding tree, *Information and Control* vol. 45 (1980), 263-272.

[6] D. Jackson, personal communication.

[7] F. Jelinek and K.S. Schneider, On variable-length-to-block coding, *IEEE Trans. Inform. Theory* vol. IT-18 (1972), 765-774.

[8] Peter R. Stubley and Ian F. Blake, On a discrete probability distribution matching problem, submitted.

[9] Peter R. Stubley and Ian F. Blake, An adaptive Tunstall coding algorithm, submitted.

Detection of Skew in a Sequence of Subsets

Mario Blaum and Jehoshua Bruck
IBM Research Division
Almaden Research Center
San Jose, CA 95120

Abstract

Let C be a set whose elements are subsets of $\{1, 2, \ldots, n\}$. We call the set C a code and its elements codewords. Assume that we transmit a sequence $\underline{c_1}, \underline{c_2}, \ldots, \underline{c_m}, \ldots$, where each $\underline{c_i}$ is a codeword in C with its elements in any order and the receiver receives a sequence $\underline{c'_1}, \underline{c'_2}, \ldots, \underline{c'_m}, \ldots$, where each $\underline{c'_i}$ is equal to $\underline{c_i}$ with perhaps some skew from the previous and following codewords. The idea is to decode correctly the codewords when there is no skew between consecutive codewords, and to detect the occurrence of skew below a certain threshold.

In this paper, we define the concept of skew and give necessary and sufficient conditions for codes detecting skew not exceeding a certain threshold. We present some code constructions and discuss their optimality. Finally, we mention possible applications for parallel asynchronous communications.

1 Introduction

Consider a communication channel that consists of several sub-channels transmitting simultaneously. As an example of this scheme consider a board with several chips where the sub-channels represent wires connecting between the chips and differences in the lengths of the wires might result in asynchronous reception. Namely, we would like to transmit a binary vector of length n using n parallel channels/wires. Every wire can carry only one bit of information. Each wire represents a coordinate of the vector to be transmitted. The propagation delay in the wires varies. The problem is to find an efficient communication scheme that will be delay-insensitive.

Clearly, this problem is very common and arises in every system that incorporates transmission of information over parallel lines. Currently, there are two approaches for solving it in practice:

1. There is a clock that is shared by both the transmitter and the receiver and the state of the wire at the time of the clock represents the corresponding bit of information.

This is a synchronous type of communication (which is not always feasible due to the difficulties in clock distribution and the fact that the transmitter might be part of an asynchronous system).

2. Asynchronous type of communications. Here the idea is to send one vector at a time and have a handshake mechanism. Namely, the transmitter sends the following vector only after getting an acknowledgment that the current vector was completely received by the receiver.

A natural question with regard to the asynchronous type of communication is: how does the receiver know that the reception is complete? This problem was studied by Verhoeff [9]. He describes the forgoing physical model as a scheme in which the sender communicates with the receiver via parallel tracks by rolling marbles (that correspond to a logical 1) in the tracks. Although the marbles are sent in parallel, the channels are asynchronous. This means that marbles are received randomly and at different instants.

Before presenting Verhoeff's result we introduce some notation. Let us represent the tracks with the numbers $1, 2, \ldots, n$. After the m-th marble has arrived, the receiver obtains a sequence $X_m = x_1, x_2, \ldots, x_m$, where $1 \leq x_i \leq n$, the number x_i meaning that the i-th marble was received at the x_i-th track. The set $\{x_1, x_2, \ldots, x_m\}$ is the support (i.e., the set of non-zero coordinates) of a vector and determines uniquely a binary vector. From now on, the sequence $X_m = x_1, x_2, \ldots, x_m$ will denote either the sequence as defined above, or a binary vector as defined by its support. Also, X may denote either a vector or its support. This abuse of notation should be clear from the context.

For example, let a vector $X = 0110$ and a vector $Y = 0100$. In the language of sets we have $X = \{2, 3\}$ and $Y = \{2\}$. Clearly, when the receiver gets a marble in track number 2, it is not clear whether he just received Y or he should wait to get a marble in track number 3 (this will correspond to receiving X).

In [9], the following problem has been studied: assuming that a vector X is transmitted, once reception has been completed, the receiver acknowledges receipt of the message. The next message is sent by the sender only after the receipt of the acknowledgement. The problem is finding a code $\mathcal{C}$ whose elements are messages such that the receiver can identify when transmission has been completed. It is easy to see, as proved in [9], that the codes having the right property are the so called unordered codes, i.e., all its elements are unordered vectors (we say that two binary vectors are unordered when their supports are unordered as sets—one set is not a subset of the other).

One of the disadvantages of using the asynchronous type of communication is the fact that the channel is not fully utilized. Namely, there is at most one vector in the wires at any given time. This becomes very critical when the transmission rates are getting higher and lines are getting longer.

In this paper, we present a novel scheme that enables a pipelined utilization of the channel. In addition, our scheme has the important feature of not using a handshake (acknowledgement) mechanism for each transmission. Hence, there is no need of communication between receiver and sender, unless a problem has been detected by the receiver.

13

We note here that if one is ready to pay in performance, then a possible strategy, if acknowledgment of messages is not allowed, is that the sender will wait long enough between messages. So, if the sender sends a codeword X followed by a codeword Y, it will be very unlikely that a marble from Y will arrive before the reception of X has been completed. With this scheme, we can again use unordered codes as in [9].

The purpose of this paper is to study parallel asynchronous pipelined communication without acknowledgement. The main difficulty in this scheme is that a certain number of marbles from the second message might arrive before reception of the current message has been completed, a condition that we call *skew*.

It turns out that skew should be defined using two parameters. Assume that we transmit a vector X followed by a vector Y. In general, since X is transmitted first, the marbles from X will tend to arrive before the marbles from Y. Also, if a marble in Y arrives before the transmission of X has been completed, it is very likely that few marbles remain in X. Let us call t_1 the maximum number of marbles that may remain in X when a marble from Y arrives. Also, we do not expect too many marbles from Y to arrive before the transmission of X has been completed. Let us call t_2 the maximum number of marbles from Y that may arrive before the completion of X.

Our approach to dealing with skew is to use coding theory methodology and to try to identify the properties of a family of vectors (a code) that can handle the skew. In this paper, we study the case in which we want to detect that skew has occurred, and then invoke a protocol that will halt transmission and allow for retransmission. Codes detecting skew are called *skew-detecting* codes. Formally:

Definition 1.1 We say that a code C is (t_1, t_2)-skew-detecting if, for any pair of codewords $X, Y \in C$ such that codeword X is transmitted followed by codeword Y, and the skew between X and Y is limited by the following two conditions:

1. at most t_1 marbles may still be missing from X when a marble from Y arrives; and

2. at most t_2 marbles from Y may arrive before all the marbles from X have arrived;

then C will correctly decode X when there is no skew between X and Y, and will detect at a certain point the presence of skew provided it does not exceed the t_1 and t_2 constraints.

In some applications, we might want to go further and not only detect, but also correct the skew, allowing for continuous operation. Codes capable of correcting skew are called *skew-tolerant* codes. However, skew-tolerant codes are more complicated than skew-detecting codes and are not considered here [3].

We illustrate Definition 1.1 with an example.

Example 1.1 Consider the following code: $C = \{X, Y\}$, where $X = 10000$ and $Y = 01111$. Claim: Code C is (3,3)-skew-detecting. In effect, assume that X is transmitted first followed by Y. If the first marble arrives in track 1, then we conclude that X has been transmitted.

However, if the first marble does not arrive in track 1, it does not necessarily mean that Y has been transmitted, since up to 3 marbles from Y may arrive before X. So, if marbles from Y arrive before reception of X has been completed, then at least one marble will remain in Y when the marble in track one arrives. Hence, we will never receive neither X nor Y, and the decoder detects this situation.

Something similar occurs when Y is transmitted followed by X.

Although Example 1.1 is very simple, the reader is urged to comprehend it, since the general case involves a similar reasoning. The necessary and sufficient conditions for a code to be (t_1, t_2)-skew-detecting, to be given in the next section, will allow to explain immediately why the code in Example 1.1 is (3,3)-skew-detecting.

The paper is organized as follows. In Section 2, we prove the characterization theorem for (t_1, t_2)-skew-detecting codes and present an algorithm for detection of skew. In Section 3, we use the characterization theorem to construct efficient (t_1, t_2)-skew-detecting codes. In Section 4, we address the issue of the optimality of the codes obtained in Section 3.

2   Necessary and sufficient conditions and decoding for (t_1, t_2)-skew-detecting codes

In this section, we study (t_1, t_2)-skew-detecting codes as given by Definition 1.1. We give a characterization in terms of distance between codewords, starting by necessary conditions and then proving that these conditions are also sufficient. The sufficient conditions are proven by providing a decoding algorithm, and showing that the decoding algorithm correctly decodes a codeword when there is no skew, and detects the presence of skew when this skew does not exceed the (t_1, t_2) constraints.

We start by giving some notation.

Given two binary vectors X and Y of length n, we denote by $N(X, Y)$ the number of coordinates in which X is 1 and Y is 0 [2]. For example, if $X = 10110$ and $Y = 00101$, we have $N(X, Y) = 2$ and $N(Y, X) = 1$. Notice that $N(X, Y) + N(Y, X) = d_H(X, Y)$, where d_H denotes Hamming distance.

The following theorem gives necessary conditions for a code to be (t_1, t_2)-skew-detecting.

Theorem 2.1 Let a code C be (t_1, t_2)-detecting, and let X and Y be codewords in C. Let $t = \min\{t_1, t_2\}$ and let $T = \max\{t_1, t_2\}$. Then, at least one of the following two conditions occurs:

(a) $\min\{N(X, Y), N(Y, X)\} \geq t + 1$

or

(b) $\min\{N(X, Y), N(Y, X)\} \geq 1$ and $\max\{N(X, Y), N(Y, X)\} \geq T + 1$.

Proof: Assume that for a given (t_1, t_2)-skew-detecting code C the conditions are not satisfied. Namely, there exist X and $Y \in C$ such that

15

$$\min\{N(X,Y),N(Y,X)\} \le t \tag{1}$$

and

$$\min\{N(X,Y),N(Y,X)\} = 0 \quad \text{or} \quad \max\{N(X,Y),N(Y,X)\} \le T. \tag{2}$$

If $N(X,Y) = 0$, then $X \subseteq Y$. But then the code would be unable to distinguish, when X is transmitted followed by Y, which codeword was sent first, either X or Y. So, conditions (1) and (2) become

$$1 \le \min\{N(X,Y),N(Y,X)\} \le t \quad \text{and} \quad \max\{N(X,Y),N(Y,X)\} \le T. \tag{3}$$

Without loss of generality, assume that $N(X,Y) \le t_1$ and $N(Y,X) \le t_2$.
If Y is transmitted followed by X, it might well happen that the marbles in $X \cap Y$ arrive first, then the marbles in $Y - X$, then the marbles in $X - Y$ and finally the marbles in $X \cap Y$. However, we may receive the same sequence if X is transmitted followed by Y, since $|X - Y| = N(X,Y) \le t_1$, hence marbles from Y may start to arrive before reception of X is completed. Also, since $|Y - X| = N(Y,X) \le t_2$, the marbles in $Y - X$ may arrive before the marbles in $X - Y$. Hence, reception of the sequence $X \cap Y, Y - X, X - Y, X \cap Y$ may correspond both to a transmission of Y followed by X with correct reception or to a transmission of X followed by Y with some skew at reception that does not violate the (t_1, t_2)-constraints. This contradicts the hypothesis that the code is (t_1, t_2)-skew-detecting. $\square$

Theorem 2.1 states necessary conditions for a code to be (t_1, t_2)-skew-detecting. It turns out that these conditions are also sufficient, as stated in the next theorem.

Theorem 2.2 Let t_1 and t_2 be positive integers with $t = \min\{t_1, t_2\}$ and $T = \max\{t_1, t_2\}$. Let $\mathcal{C}$ be a code such that, for any $X, Y \in \mathcal{C}$, at least one of the following two conditions occurs:
(a) $\min\{N(X,Y),N(Y,X)\} \ge t + 1$
or
(b) $\min\{N(X,Y),N(Y,X)\} \ge 1$ and $\max\{N(X,Y),N(Y,X)\} \ge T + 1$.
Then, code $\mathcal{C}$ is (t_1, t_2)-skew-detecting.

The proof is based on the following Decoding Algorithm:

Algorithm 2.1 (Decoding Algorithm for (t_1, t_2)-skew-detecting codes) Set the initial conditions as $X_0 = \emptyset$ and $m \leftarrow 1$. Then:

START: Input the new arrival x_m.
      If $x_m \in X_{m-1}$, then detect an error and stop.
      Else, let $X_m = x_1, x_2, \ldots, x_m$.
            If $X_m \in \mathcal{C}$, then output X_m, reset $m \leftarrow 1$ and go to **START**.
            Else, let $m \leftarrow m + 1$ and go to **START**.

16

We now prove Theorem 2.2 by showing that Decoding Algorithm 2.1 will correctly decode any codeword X when no skew with a second message Y has occurred, and will detect the occurrence of skew not exceeding the (t_1, t_2) constraints.

Proof: Assume that $X \in C$ has been transmitted and we have received $X_m = x_1, x_2, \ldots, x_m$. If $X_m = X$, then the algorithm will correctly decode X. So, it remains to show that if $X_m \neq X$, then $X_m \notin C$.
If $X_m \subset X$ or $X \subset X_m$, $X_m \neq X$, then $X_m \notin C$ since the code is unordered.
So, assume that X and X_m are unordered and $X_m \in C$. According to the t_1 and t_2 constraints, at most t_1 marbles remain in X and at most t_2 marbles are in X_m and not in X; this means, $N(X, X_m) \leq t_1$ and $N(X_m, X) \leq t_2$, contradicting conditions (a) and (b) in the theorem. Since the algorithm will always decode correctly when the marbles of X are received before a marble from a second codeword arrives, and will never output a codeword when this condition is not verified, then the erroneous reception will be at some point detected. Hence C is (t_1, t_2)-skew-detecting. □

Notice that when skew not exceeding the (t_1, t_2) constraints has occurred, this situation will be detected soon, i.e., when a repeated arrival is detected. In a worst case situation, reception of n marbles will produce the whole set $\{1, 2, \ldots, n\}$, so the error will be detected when the next marble arrives.
The following two corollaries are clear from the necessary and sufficient conditions.

Corollary 2.1 A code is (t_1, t_2)-skew-detecting if and only if it is also (t_2, t_1)-skew-detecting. Moreover, the decoding algorithm is the same for the two pairs of conditions.

Corollary 2.2 A code C is (t, t)-skew-detecting if and only if, for any $X, Y \in C$,

$$\min\{N(X, Y), N(Y, X)\} \geq 1 \quad \text{and} \quad \max\{N(X, Y), N(Y, X)\} \geq t + 1.$$

Next we illustrate the necessary and sufficient conditions and the decoding algorithm for (t_1, t_2)-skew-detecting codes with an example.

Example 2.1 Let $C = \{X, Y, Z\}$, where

$$
\begin{aligned}
X &= 00110 &\leftrightarrow \{3, 4\} \\
Y &= 01101 &\leftrightarrow \{2, 3, 5\} \\
Z &= 10001 &\leftrightarrow \{1, 5\}.
\end{aligned}
$$

Notice that $N(X, Y) = 1$ and $N(Y, X) = 2$, $N(X, Z) = 1$ and $N(Z, X) = 2$, and $N(Y, Z) = 2$ and $N(Z, Y) = 1$. According to Corollary 2.2, code C is (1,1)-skew-detecting.
Assume now that the sender transmits the following sequence: X, Y, Z, which corresponds to 3,4,2,3,5,1,5, but the receiver obtains the sequence

$$3 \ 4 \ 5 \ 2 \ 1 \ 3 \ 5.$$

We can observe that skew has occurred between Y and Z: the 5th arrival, in track 1, corresponds to Z, and the 6th arrival, in track 3, corresponds to Y. Since the skew between Y and Z does not violate the $(1,1)$ constraints, it should be detected by the decoding algorithm. So, let us apply Decoding Algorithm 2.1 to the received sequence.

Step 1: $\{3\}$ is not in the code.
Step 2: $\{3,4\} = X$ is in the code. Accept it and reset.
Step 3: $\{5\}$ is not in the code.
Step 4: $\{5,2\}$ is not in the code.
Step 5: $\{5,2,1\}$ is not in the code.
Step 6: $\{5,2,1,3\}$ is not in the code.
Step 7: $5 \in \{5,2,1,3\}$. Repeated arrival. Skew error detected!

Let us complete the example by showing that if the $(1,1)$ constraints are broken, skew may be undetected. This example also illustrates the proofs of Theorems 2.1 and 2.2.
Assume, as before, that X, Y, Z is sent but the sequence

$$3 \ 4 \ 5 \ 1 \ 2 \ 3 \ 5$$

is received. We observe that the fourth arrival occurs in track 1, which corresponds to Z. When this occurs, there are two marbles left in Y, 2 and 3. Hence, $t_1 = 1$ has been broken. Notice that the received sequence corresponds to a skew-free reception of X, Z, Y. Hence, this skew cannot be detected.

In the next sections we discuss actual constructions of codes that are (t_1, t_2)-skew-detecting as well as optimality issues.

3 Constructions of (t_1, t_2)-skew-detecting codes

In this section, we give two constructions of (t_1, t_2)-skew-detecting codes.
The first construction involves a family of codes well known in literature: the so called t-error correcting/all unidirectional error detecting codes [2, 4, 5, 8]. A t-error correcting/all unidirectional error detecting (EC/AUED) code satisfies condition (a) [8], which is sufficient for (t_1, t_2)-skew-detection when the skew does not exceed the (t_1, t_2) constraints, and $t = \min\{t_1, t_2\}$. We state this fact in the next proposition.

Proposition 3.1 Let t_1 and t_2 be positive integers and $t = \min\{t_1, t_2\}$. Let $\mathcal{C}$ be a t-EC/AUED code. Then $\mathcal{C}$ is (t_1, t_2)-skew-detecting.

An efficient way of constructing a t-EC/AUED is as follows: first encode the information bits using a t-error correcting code. Then append a tail giving the code property (a). Efficient tail matrices may be found in [2, 4, 5].

The second family of codes that we consider are the so called error correcting unordered (ECU) codes. A t-ECU code is a code that can correct t errors and any two codewords are unordered. Formally:

Definition 3.1 We say that a code C is error correcting unordered (ECU) with minimum distance d if, for any $X, Y \in C$,

1. $d_H(X, Y) = N(X, Y) + N(Y, X) \geq d$.

2. $\min\{N(X, Y), N(Y, X)\} \geq 1$.

The connection between ECU codes and (t_1, t_2)-skew-detecting codes is given by the following lemma:

Lemma 3.1 Let t_1 and t_2 be positive integers and $t = \min\{t_1, t_2\}$. Let C be an ECU with minimum distance $\geq t_1 + t_2 + 1$. Then C is (t_1, t_2)-skew-detecting.

Proof: Let $t = \min\{t_1, t_2\}$ and $T = \max\{t_1, t_2\}$. Let $X, Y \in C$, and assume that condition (a) is violated, say, $N(X, Y) \leq t$. The codewords are unordered, and also,

$$N(Y, X) = d_H(X, Y) - N(X, Y) \geq t_1 + t_2 + 1 - t = T + 1.$$

Hence, X and Y satisfy condition (b), proving that the code is (t_1, t_2)-skew-detecting. $\quad\square$

In the particular case in which $t_1 = t_2 = t$, an ECU code with minimum distance $2t + 1$ gives a (t, t)-skew-detecting code.

Next, we describe a method to construct systematic ECU codes. The construction is in fact a generalization of the well known Berger construction [1].

Construction 3.1 Assume that we want to construct an ECU code C with minimum distance d and dimension k. Choose an $[n', k, d]$ error correcting (EC) code C'. Let $\underline{u}$ be an information vector of length k. Then proceed as follows:

1. Encode $\underline{u}$ into a vector $\underline{v} \in C'$.

2. Let j be the Hamming weight of $\underline{v}$. Then append to $\underline{v}$ the complement of the binary representation of $\lfloor j/d \rfloor$.

The code obtained with this encoding procedure is ECU with minimum distance d.

Before proving that the code is ECU, we observe the following:

19

1. The code $\mathcal{C}$ has length $n' + \lceil \log_2 \lceil (n'+1)/d \rceil \rceil$.

2. The Berger construction corresponds to the special case in which $\mathcal{C}'$ is the $[k, k, 1]$ code.

3. The code $\mathcal{C}$ is systematic if the code $\mathcal{C}'$ is systematic.

4. We may sometimes make the construction more efficient when the all-1 vector is in $\mathcal{C}'$ by taking a coset of this code. The construction is analogous but we have less than $n' + 1$ different weights in the coset [5].

5. For a table with the best error correcting codes, see [10].

Lemma 3.2 The code $\mathcal{C}$ obtained in Construction 3.1 is ECU with minimum distance d.

Proof: It is clear that the minimum distance is d. Assume that we have two codewords $\underline{u}$ and $\underline{v}$ in $\mathcal{C}'$ with weights i and j respectively, where $i \le j$. Notice that $N(\underline{v}, \underline{u}) > 0$. Let $\underline{t_u}$ and $\underline{t_v}$ be the tails when we encode using Construction 3.1. We will prove that $N\left((\underline{u}, \underline{t_u}), (\underline{v}, \underline{t_v})\right) > 0$.
We have two possibilities: either $\lfloor i/d \rfloor = \lfloor j/d \rfloor$ or $\lfloor i/d \rfloor \ne \lfloor j/d \rfloor$.
If $\lfloor i/d \rfloor = \lfloor j/d \rfloor$, then $j - i \le d - 1$. If $\underline{u} \subseteq \underline{v}$, then $d_H(\underline{u}, \underline{v}) = j - i \le d - 1$, a contradiction. So, in particular, $(\underline{u}, \underline{t_u})$ and $(\underline{v}, \underline{t_v})$ are unordered.
If $\lfloor i/d \rfloor \ne \lfloor j/d \rfloor$, then, in particular, $\lfloor i/d \rfloor < \lfloor j/d \rfloor$. According to Construction 3.1, $\underline{t_u}$ as a binary number is larger than $\underline{t_v}$ as a binary number. This means, $N(\underline{t_u}, \underline{t_v}) > 0$. Since we had that $N(\underline{v}, \underline{u}) > 0$, it follows that $(\underline{u}, \underline{t_u})$ and $(\underline{v}, \underline{t_v})$ are unordered. $\square$

Example 3.1 Assume that we want to construct a $(1, 2)$-skew-detecting code of dimension k. The first approach is to encode the k information bits into an $[n', k, 3]$ Hamming code. We then append a tail in such a way that the code becomes 1-EC/AUED.
Take for instance $k = 57$. We first add 6 redundant bits in order to encode the information into a $[63, 57, 3]$ Hamming code. Using the table in [5], we see that we have to add 9 bits to make the code 1-EC/AUED (and hence, $(1,2)$-skew-detecting). This gives a total of 15 redundant bits.
The second approach is to use Construction 3.1 to obtain an ECU code with minimum distance 4. We first have to encode into a $[64, 57, 4]$ extended Hamming code. Then, to make the code unordered, we have to add a tail of length $\lceil \log_2 \lceil 65/4 \rceil \rceil = 5$ bits. This gives a total of 12 redundant bits, so, for $k = 57$, the second method is more efficient than the first. If we take a coset of this code, the weight distribution goes from 1 to 63, so we have 63 different weights. Now, $\lceil \log_2 \lceil 63/4 \rceil \rceil = 4$ bits, so we save one bit in the total redundancy.

We note here that condition (b) in Theorem 2.2 can be satisfied by encoding first into an asymmetric error correcting code and then appending a tail to make it unordered.
In the next section, we deal with the issues of optimality of ECU codes.

4 Optimal Error Correcting Unordered Codes

In the previous section we have presented two general constructions of codes that meet the necessary and sufficient conditions. The second construction is based on error correcting codes to which a tail is added in such a way that the code is unordered.

In this section, we prove the optimality of Construction 3.1 for the extended Hamming codes with minimum distance 4 and for certain BCH codes.

We consider the optimality of Construction 3.1 in the following sense: the tail added to the error correcting code has minimal length, i.e., it is impossible to find a shorter tail making the code unordered.

We begin by defining the concept of a *chain* of vectors.

Definition 4.1 A set of binary vectors $\{V_1, V_2, \ldots, V_m\}$ is a chain of length m if any two vectors in the set are ordered.

The idea in proving the optimality of our constructions is to exhibit a long enough chain of codewords in the error correcting code. The following lemma gives the key.

Lemma 4.1 Let $\{C_1, C_2, \ldots, C_m\}$ be a chain of vectors, each being a codeword in a given code $\mathcal{C}$. Then the length of the a tail that we have to add to $\mathcal{C}$ to make it unordered is at least $\lceil \log_2 m \rceil$ bits.

Proof: Since all the codewords in the chain are ordered, we need to have a different tail for every one of them to make them unordered. Hence, we need at least m different tails. □

We prove the optimality of some of our constructions by exhibiting chains of length $\lceil n/d \rceil + 1$ in an $[n, k, d]$ code. First we prove the optimality of our construction for the extended Hamming code by exhibiting a chain of $2^{m-2}+1$ codewords in a code of length 2^m.

Proposition 4.1 The $[2^m, 2^m - m - 1, 4]$ extended Hamming code contains a chain of $2^{m-2}+1$ codewords.

Proof: The columns of the parity check matrix of a $(2^m, 2^m - m - 1, 4)$ extended Hamming code are:

$$\{(v_1, v_2, \ldots, v_m, 1)^T \; : \; (v_1, v_2, \ldots, v_m) \in \{0, 1\}^m\}.$$

Note that we can arrange the columns in the parity check matrix in pairs such that the first m bits are complementary. Namely, column $(v_1, v_2, \ldots, v_m, 1)^T$ is paired with $(\bar{v}_1, \bar{v}_2, \ldots, \bar{v}_m, 1)^T$. Hence, the sum of a pair of columns in this arrangement gives the vector $(1, 1, \ldots, 1, 0)^T$ and the sum of 2 pairs (4 columns) is the all-0 vector. We call this matrix H_m. For example,

$$H_3 = \begin{pmatrix} 0 & 1 & 0 & 1 & 0 & 1 & 0 & 1 \\ 0 & 1 & 0 & 1 & 1 & 0 & 1 & 0 \\ 0 & 1 & 1 & 0 & 0 & 1 & 1 & 0 \\ 1 & 1 & 1 & 1 & 1 & 1 & 1 & 1 \end{pmatrix}.$$

Consider the extended Hamming code that corresponds to the matrix H_m. It follows from the construction of H_m that the following set of $2^{m-2} + 1$ codewords is a chain:

$$\{(1111)^i 0^{2^m - 4i} \; : \; 0 \le i \le 2^{m-2}\}$$

where S^i, S a binary vector, is a vector obtained by concatenating S i times. □

The second result is related to BCH codes. We prove that in many cases we can exhibit chains of codewords that show the optimality of our construction. The key to exhibiting long chains is the following lemma [6].

Lemma 4.2 Consider a binary t-error-correcting BCH code defined in a standard way, i.e., as a cyclic code of length $2^m - 1$. Let a and b be two integers such that

$$a \cdot b = 2^m - 1$$

and

$$a \ge 2t + 1.$$

Then the following b polynomials correspond to codewords:

$$z_1(X) = 1 + X^b + X^{2b} + \cdots + X^{(a-1)b}$$

and for $2 \le i \le b$,

$$z_i(X) = X^{i-1} z_1(X).$$

Using this lemma we can prove the following:

Proposition 4.2 Given a t-error correcting BCH code of length $2^m - 1 = a \cdot b$ where a and b are integers, and $a \ge 2t + 1$, we can exhibit a chain of length $b + 1$.

Proof: The proof follows from lemma 4.2. The chain consists of the all-0 vector and the set of b vectors that correspond to partial sums of the polynomials from Lemma 4.2 as follows:

$$\{\sum_{i=1}^{j} z_i \; : \; 1 \le j \le b\}.$$

□

Example 4.1 Consider the case $t = 2$, namely $2t + 1 = 5$. We can exhibit a chain of $((2^m - 1)/5) + 1$ codewords in all the cases in which $2^m - 1 \equiv 0 \pmod 5$. For example, for $m = 4$ we can exhibit a chain of length 4. In general, we can exhibit a long chain whenever $m \equiv 0 \pmod 4$ (by Fermat's Theorem). Similarly, for $2t + 1 = 7$, we can exhibit a long chain for all all the cases in which $m \equiv 0 \pmod 6$.

To summarize, we proved in this section that our construction of a t-ECU code is optimal when we consider the extended Hamming codes and certain BCH codes.

5 Conclusions

We have studied a problem in parallel asynchronous communications allowing skew between consecutive messages and showed that there are codes that can detect a certain amount of skew. We found necessary and sufficient conditions for codes that can detect a predetermined amount of skew, constructed codes satisfying the necessary and sufficient conditions and studied their optimality. Finally, we provided efficient encoding and decoding algorithms.

References

[1] J. M. Berger, "A note on error detecting codes for asymmetric channels," Information and Control, Vol. 4, pp. 68-73, March 1961.

[2] M. Blaum and H. van Tilborg, "On t-Error Correcting/All Unidirectional Error Detecting Codes," IEEE Trans. on Computers, vol. C-38, pp. 1493-1501, November 1989.

[3] M. Blaum and J. Bruck, "Coding for Parallel Asynchronous Communications," IBM Research Report 8168 (75127), June 1991.

[4] F. J. H. Boinck and H. van Tilborg, "Constructions and bounds for systematic tEC/AUED codes," IEEE Trans. on Information Theory, vol. IT-36, No. 6, pp. 1381-1390, November 1990.

[5] J. Bruck and M. Blaum, "New Techniques for Constructing EC/AUED Codes," IBM Research Report, RJ 6818 (65352), May 1989, to appear in IEEE Trans. on Computers.

[6] R. H. Deng and M. A. Herro, "DC-Free Coset Codes," IEEE Trans. on Information Theory, vol. IT-34, pp. 786-792, July 1988.

[7] F. J. MacWilliams and N. J. A. Sloane, "The Theory of Error-Correcting Codes," Amsterdam, The Netherlands: North-Holland, 1977.

[8] D. K. Pradhan, "A new class of error-correcting detecting codes for fault-tolerant computer application," IEEE Trans. on Computers, vol. C-29, pp. 471-481, June 1980.

[9] T. Verhoeff "Delay-insensitive codes - an overview," Distributed Computing, 3:1-8, 1988.

[10] T. Verhoeff "An updated table of minimum distance bounds for binary linear codes," IEEE Trans. on Information Theory, vol. IT-33, pp. 665-680, Sept. 1987.

ASYMMETRIC ERROR CORRECTING CODES[*]

Bella Bose

Department of Computer Science, Oregon State university, Corvallis, OR 97331-3902

Steve Cunningham

Department of Computer Science, California State College,Stanislaus, CA 95380

Abstract: Non-linear but cyclic codes capable of correcting asymmetric errors are described. For these codes the syndromes directly give the symmetric functions of the error locations and so these codes are much easier to decode. The hardware implementation of the decoding algorithm is given. In many cases the information rate of these codes is as good as or better than the corresponding BCH codes.

I. Introduction

Error correcting and detecting codes are extensively used in improving the reliability of computer and communication systems [1]. Most known classes of binary codes [1-7] have been developed for use on binary symmetric channel, where the error probabilities 1 $\rightarrow$ 0 and 0 $\rightarrow$ 1 are equal. However, in certain systems, the observed errors are highly asymmetric and so the appropriate model may be in fact Z-channel, in which one type of error, say 0 $\rightarrow$ 1, is impossible. For example, in optical communications, photons may fade or fail to be detected (1 $\rightarrow$ 0 error) but the creation of spurious photons (0 $\rightarrow$ 1 error) is impossible [8,9]. In MNOS LSI memories, the logic 1 and 0 are represented by the presence and absence of charges respectively; in these memories the most likely errors are of asymmetric type since a charge cannot be created but leakage of charge is possible [10-12]. Furthermore, it is observed that alpha-particles cause only asymmetric errors in semiconductor memories [13]. All these systems can be modeled as Z-channel. This paper deals with the design of asymmetric error correcting codes.

It is obvious that a code capable of correcting t errors on a symmetric channel is also capable of correcting t errors in a Z-channel. However, a t-asymmetric error correcting

[*] This research is supported by NSF grant MIP-9016143.

code of length n is expected to have more number of codewords than the corresponding symmetric error code. In the past most of the researchers have concentrated their effort in this direction [11,12,14-24]. These results show that we may be able to get more codewords but 'not quite a lot'. In fact, Borden[22] has shown the following relation. Let $A(n, t)$ be the maximum number of codewords of length n capable of correcting t asymmetric errors and $S(n, t)$ be the corresponding function for symmetric coding. Then

$$S(n, t) \leq A(n, t) \leq (t+1)S(n, t) \qquad (1\text{-}1)$$

Thus, we can have only limited success in the search for higher information rate asymmetric codes. Moreover, at present not many satisfactory techniques are available (other than C-R codes [12] and Varshmov codes [15,16]) in actual code design.

Due to these facts, we will be satisfied if we could design asymmetric error correcting codes which have simpler encoding/decoding algorithms than the symmetric coding. The codes designed in Section II, to a large extent, achieve this goal. For these codes, we have shown that the syndromes directly give the symmetric functions of the error locations. Thus these codes are much simpler to decode. The information rate of these codes are better than or equal to BCH codes. However at this point, it is not clear whether there exits a simple encoding algorithm for these codes. Further works needs to be done.

In this paper, the code capabilities are analyzed under the assumption of $0 \rightarrow 1$ asymmetric errors, but they are valid (or easily modified) for $1 \rightarrow 0$ type asymmetric errors.

II Error Correcting Codes

2.1 Code Design

A t-(asymmetric) error correcting code of length n can be designed as described below.

Let $\{ \alpha_0 = 0, \alpha_1, \alpha_2, \ldots\ldots, \alpha_n \}$ be the elements of GF(n+1). Define a function F from binary n-tuples to GF(n+1) as follows.

F: $GF(2)^n \rightarrow GF(n+1)$ such that

$$F(X) = \begin{bmatrix} F_1(X) \\ F_2(X) \\ \vdots \\ F_t(X) \end{bmatrix}$$

where

$$F_1(X) = F_1((x_n, x_{n-1}, x_{n-2},..., x_1))$$

$$= \sum_{x_i=1} \alpha_i$$

$$F_2(X) = \sum_{\substack{j_1 < j_2 \\ x_{j1} = x_{j2} = 1}} \alpha_{j1} \alpha_{j2}$$

$$\tag{2-1}$$

$$\cdot$$
$$\cdot$$
$$\cdot$$

$$F_i(X) = \sum_{\substack{j_1 < j_2 < \cdots < j_i \\ x_{j1} = x_{j2} = \cdots = x_{ji} = 1}} \alpha_{j1} \alpha_{j2} \ldots \alpha_{ji}$$

$$\cdot$$
$$\cdot$$
$$\cdot$$

$$F_t(X) = \sum_{\substack{j_1 < j_2 < \cdots < j_t \\ x_{j1} = x_{j2} = \cdots = x_{jt} = 1}} \alpha_{j1} \alpha_{j2} \ldots \alpha_{jt}$$

(At this point we would like to point out that a somewhat similar method was given by Graham and Sloane in [25] for establishing a lower bound on constant weight codes of length n and distance 2t+2. However, here we are interested in developing codes for correcting t-asymmetric errors and in particular their decoding algorithms. These results are initially given by us in [26-28]. Recently we came to know that some of these results are independently given by Prof. Klove in [29]).

Example 2.1:

Let α be a root of $X^3 + X + 1$, which is a primitive polynomial. Then all elements in $GF(2^3)$ can be represented in binary as $Q = 000$, $\alpha^0 = 001$, $\alpha^1 = 010$, $\alpha^2 = 100$, $\alpha^3 = 011$, $\alpha^4 = 110$, $\alpha^5 = 111$, and $\alpha^6 = 101$. Suppose $X = 1010011$ is a 7 bit vector. For double error correction X will be mapped into

$$F(X) = \begin{bmatrix} F_1(X) \\ F_2(X) \end{bmatrix}$$

where

$$F_1(X) = F_1(1010011)$$
$$= \alpha^6 + \alpha^4 + \alpha^1 + \alpha^0$$
$$= 101 + 110 + 010 + 001$$
$$= (000)$$
$$F_2(X) = \alpha^6 \alpha^4 + \alpha^6 \alpha^1 + \alpha^6 \alpha^0 + \alpha^4 \alpha^1 + \alpha^4 \alpha^0 + \alpha^1 \alpha^0$$
$$= \alpha^6 (\alpha^4 + \alpha^1 + \alpha^0) + \alpha^4 (\alpha^1 + \alpha^0) + \alpha^1 \alpha^0$$
$$= \alpha^2 = (100) .$$

Note that F partitions 2^n tuples into $(n+1)^t$ classes $C_1, C_2, \ldots, C_{(n+1)^t}$ such that if $X, Y \in C_i$ then $F(X) = F(Y)$.

<u>Claim</u> Each of C_i, $i = 1, 2, \ldots, (n+1)^t$ is a t-error correcting code.

Proof: The proof given here also establishes the decoding algorithm for the code.

Suppose, $\forall X \in C_i$, let $F(X) = \begin{bmatrix} F_1(X) \\ F_2(X) \\ \vdots \\ F_t(X) \end{bmatrix} = \begin{bmatrix} \delta_1 \\ \delta_2 \\ \vdots \\ \delta_t \end{bmatrix}$

Let there be s $0 \to 1$ errors in $X = (x_n, x_{n-1},..., x_1)$ where $s \leq t$, and (the weights of) the error locations be $\beta_1, \beta_2,..., \beta_s$. Evaluate $F(X')$ where X' is the received word. Thus

$$F_1(X') = S_1 = \delta_1 + \sigma_1$$
$$F_2(X') = S_2 = \delta_2 + \sigma_1\delta_1 + \sigma_2$$
$$F_3(X') = S_3 = \delta_3 + \sigma_1\delta_2 + \sigma_2\delta_1 + \sigma_3$$

$$\cdot$$

(2-2)

$$\cdot$$

$$\cdot$$

$$F_s(X') = S_s = \delta_s + \sigma_1 \delta_{s-1} + \sigma_2\delta_{s-2} +...+ \sigma_s$$
$$F_{s+1}(X') = S_{s+1} = \delta_{s+1} + \sigma_1 \delta_s + \sigma_2\delta_{s-1} +...+ \sigma_s\delta_1 + \sigma_{s+1}$$

$$\cdot$$

$$\cdot$$

$$\cdot$$

$$F_t(X') = S_t = \delta_t + \sigma_1 \delta_{t-1} + \sigma_2\delta_{t-2} +... + \sigma_s\delta_{t-s} +...+ \sigma_t$$

where

$\sigma_1, \sigma_2,..., \sigma_s$ are the elementary symmetric functions corresponding to the error locations $\beta_1, \beta_2,... , \beta_s$. i.e.

$$\sigma_1 = \beta_1 + \beta_2 + \cdots + \beta_s$$
$$\sigma_2 = \prod_{i<j} \beta_i\beta_j$$

(2-3)

$$\sigma_3 = \prod_{i_1<i_2<i_3} \beta_{i1}\beta_{i2}\beta_{i3}$$

$$\sigma_s = \beta_1\beta_2 \cdots \beta_s$$

We call the S_i's in equation (2-2) as syndromes. From (2-2) we can find the values of σ_1, σ_2, ... , σ_s.

Thus the roots of the equation

$$X^s - \sigma_1 X^{s-1} + \sigma_2 X^{s-2} + ... + (-1)^s \sigma_s = 0$$

give the error locations, $\beta_1, \beta_2, \cdots , \beta_s$.

Remark 1

When the number of errors s, is strictly less than t, σ_{s+1} to σ_t will be 0 and $\sigma_s \neq 0$. Thus from syndromes we can know the number of errors.

Remark 2

Suppose C_i is chosen such that $\delta_1 = 0$, $\delta_2 = 0$, ..., $\delta_t = 0$. Then for s $0 \to 1$ errors, where $s \leq t$, the syndromes will be

$$S_1 = \sigma_1$$
$$S_2 = \sigma_2$$

$\cdot$

$\cdot$ $\qquad\qquad\qquad\qquad$ (2-4)

$\cdot$

$$S_s = \sigma_s \neq 0$$
$$S_{s+1} = 0$$

$\cdot$

$\cdot$

$\cdot$

$$S_t = 0$$

Thus, in this case the syndromes directly give the elementary symmetric function values of the error location elements.

Remark 3

As it has been proved before, each of C_i is a t-(asymmetric) error correcting code. Thus it is easy to verify that one of the C_i's contains at least $2^n / (n+1)^t$ code words. When $n = p^q - 1$, the number of code words in the proposed code is better than many BCH codes of same parameters. However, we emphasize that the main feature of the proposed code is the simplicity of the decoding algorithm and not the information rate.

From the above discussions, it is clear how to decode the code and it is formally mentioned here.

Decoding algorithm:

step1. From the received word X', find the syndromes. This is equivalent to finding F(X').

step2. From the syndromes evaluate the symmetric functions σ_1, σ_2, ..., σ_t (This step can be eliminated if we take the code C_i such that $\forall x \in C_i$, $F(X) = 0$. Even for other case, this step is not difficult).

step3. Find the roots of $X^s - \sigma_1 X^{s-1} + \sigma_2 X^{s-2} - ... + (-1)^s \sigma_s$. (This can be done using Chien search method[1-7]). The roots give the error locations.

2.2 Comparison of Decoding Complexity with BCH Codes

A t-(symmetric) error correcting BCH code C can be defined as follows.

$A(X) = (a_0 \ a_1 \ a_2 ... \ a_{n-1}) = a_0 + a_1 X + a_2 X^2 + ... + a_{n-1} X^{n-1}$ is a code word in C if and only if $A(\alpha^i) = 0$ for i = 1, 2, 3, .., 2t, where α is a primitive root of GF(2^n). For this code, the decoding process consists of 3-steps.

1) From the received word, A'(X), find the syndromes S_1, S_2,..., S_{2t}, where $S_i = A'(\alpha^i)$.

2) Calculate the symmetric functions σ_1, σ_2,..., σ_s (of error locations) from S_i's.

3) Find the roots of $X^s - \sigma_1 X^{s-1} + \sigma_2 X^{s-2} - ... + (-1)^s \sigma_s = 0$. These roots give the error locations.

Among these, steps 1 and 3 can be implemented easily whereas step 2 is the most difficult one.

For the proposed asymmetric code, calculation similar to step 2 is not needed at all (by taking $F(X) = 0$ for all X in the code). Even if we choose a code C such that $F(X) \neq 0$ for all $X \in$ C, the symmetric functions can be found in a straight forward manner. In addition, the syndromes calculation of the proposed code can be implemented in a simple way as explained in the next section. Thus the proposed codes are much easier to decode.

2.3 Syndromes Calculation Circuit Design

In order to find the ith syndrome S_i, we need to evaluate $F_i(X')$ for $i = 1, 2, ...,t$, where X' is the received word. The algorithm given below can be used to find S_i's. Let the received word $X' = (x_n, x_{n-1}, ... x_1)$.

Algorithm to find syndrome calculation:

$S_{old}(0) = 1$
for j = 1 to t do   // Initialize the syndrome registers //
   $S_{old}(j) = 0$
end;
for i = 1 to n do
if $x_i = 1$ then do
   begin
        for j = 1 to t do
           $S_{new}(j) = S_{old}(j) + \alpha^j \cdot S_{old}(j-1)$ // α is a primitive root in an
        end;   appropriate Galois Field   //
        for j = 1 to t do
           $S_{old}(j) = S_{new}(j)$
        end
   end
end

At the end of the algorithm $S_{old}(j)$ contains the syndrome S_j. The hardware implementation of the above algorithm is shown in Fig 1.

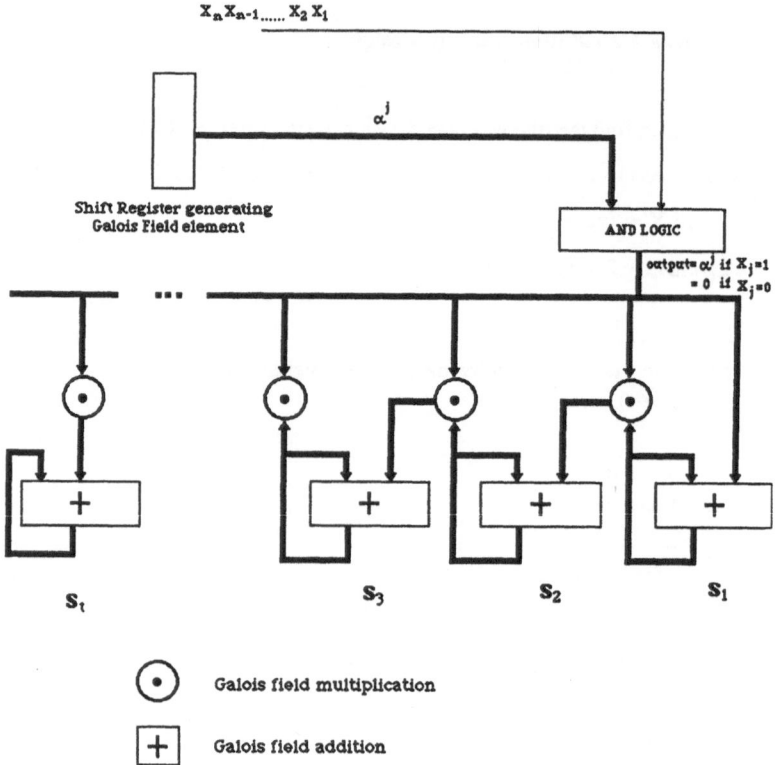

$X_n X_{n-1} ...... X_2 X_1$

α^j

Shift Register generating
Galois Field element

AND LOGIC

output= α^j if X_j=1
      = 0 if X_j=0

S_t S_3 S_2 S_1

⊙ Galois field multiplication

+ Galois field addition

Fig 1. Syndromes Calculation Circuit For The Proposed Code

III. Conclusion

In this paper we have designed asymmetric error correcting codes which have simple decoding algorithms. The information rate of these codes is as good as or better than the BCH codes. In general these codes are non-linear but cyclic (assuming that $\forall X$ in the code $F(X) = 0$). Using Newton's identities, we can also prove that the proposed t-asymmetric code is a subset of $t/2$ symmetric error correcting BCH code. An efficient design of encoding algorithm for these codes is still an open problem.

REFERENCE

1. S. Lin and D. J. Costello,Jr., Error Control Coding : Fundamentals and Applications, Prentice-Hall, Inc. Englewood Cliffs, New Jersey, 1983.

2. R. E. Blahut, Theory and Practice of Error Control Codes, Reading, MA: Addison-Wesley Publishing Comapany, California, 1983.

3. W. W. Peterson and E. J. Weldon, Jr., Error Correcting Codes, 2nd ed., MIT Press, Cambridge, Mass., 1972.

4. I. F. Blake and R.C. Mullin, The Mathematical Theory of Coding, Academic Press, New York 1975.

5. F. J. MacWilliams and N. J. A. Sloane, The Theory of Error Correcting Codes, North-Holland, Amsterdam, 1977.

6. R.J. McEliece, The Theory of Information and Coding, Addison-Wesley, Reading, Mass, 1977.

7. E. R. Berlekamp, Algebraic Coding Theory, Aegean Park Press, Laguna Hills, California, 1984.

8. J. R Pierce, "Optical channels: practical limits with photon counting," IEEE Trans. on Comm. vol. Com-26, pp. 1819-1821, Dec. 1978.

9. R. J. McEliece, "Practical codes for photon communications," IEEE Trans. on Information Theory, vol. IT-27, pp. 393-398, Jul. 1981.

10. T.R.N. Rao and E. Fujiwara, Error Control Coding for Computer Systems, Prentice-Hall, Englewood Cliffs, New Jersey, 1989.

11. T.R.N. Rao and A.S. Chawla, "Asymmetric error codes for some LSI semiconductor memories," in Proc. Annu. Southeastern Symp. on System Theory, pp. 170-171, 1975.

12. S. D. Constantin and T.R.N. Rao, "On the theory of binary asymmetric error correcting codes," Inform. and Control, 40, pp. 20-36, 1979.

13. R. J. McEliece, "The reliability of computer memories," Scientific America, vol. 252, pp. 88-95, Jan. 1985.

14. W. H. Kim and C. V. Freiman, "Single error codes for asymmetric binary channels," IRE Trans. Inform. Theory, vol. IT-5, pp.62-66, June 1959.

15. R.R. Varshamov, "A class of codes for asymmetric channels and a problem from the additive theory of numbers," IRE Trans. Inform. Theory, vol. IT-19, pp. 92-95,Jan.1973.

16. R. J. McEliece, "Comment on 'A class of codes for asymmetric channels and a problem from the additive theory of numbers'," IRE Trans. Infor. Theory, vol. IT-19, p.137, Jan. 1973.

17. R. J. McEliece and E. R. Rodemich, "The Constantin-Rao construction for binary asymmetric error-correcting codes," Inform and Cont. 44, pp. 187-196, 1980.

18. P. Delsarte and P. Piret, "Bounds and constructions for binary asymmetric error-correcting codes," IEEE Trans. Inform. Theory, pp. 125-128, Jan. 1981.

19. P. H. Delsarte and P. H. Piret, "Spectral Enumerators for certain additive error correcting codes over integer alphabets," Information and Control, 48, pp. 193-210, 1981

20. T. Helleseth and T.Klove, "On group-theoretic codes for asymmetric channels," Inform. and Cont., 49, pp. 1-9, Jan. 1981.

21. T. Klove, "Upper bounds on codes correcting asymmetric errors," IEEE Trans. Inform. Theory, vol. IT-28, pp.128-130, Jan. 1981.

22. J. M. Borden, "A low-rate bound for asymmetric error correcting codes," IEEE Trans. on Infor. Theory, vol. IT-29, pp. 600-602, Jul. 1982.

23. J. H. Weber, C. de Vroedt, and D. E. Boekee, "New upper bounds on the size of codes correcting asymmetric errors," IEEE Trans. Inform. Theory, vol. IT-33, pp. 434-437, May. 1987.

24. J. H. Weber, C. de Vroedt, and D. E. Boekee, "Bounds and constructions for binary codes of length less than 24 and asymmetric distance less than 6," IEEE Trans. Inform. Theory, vol. IT-34, pp. 1321-1331, Sep. 1988.

25. R. L. Graham and N. J. A. Sloane, "Lower bounds for constant weight codes," IEE Trans. Inform. Theory, vol. IT-26, 37-43, Jan. 1980.

26. B. Bose and S. Cunningham, 'On asymmetric error correcting codes', Tech Report, Dept. of Computer Science,Oregon State University, Corvallis, OR., 97331, March 1982.

27. S. Cunningham, 'On asymmetric error correcting codes', M.S. Thesis, Dept. of Computer Science, Oregon State University, Corvallis, OR., 97331,March1982.

28. B. Bose and S. Cunningham, 'On asymmetric error correcting codes', Inter. Symposiun on Information Theory, St. Javite, Canada, September 1983.

29. T. Klove, 'Error correcting codes for the asymmetric channel', Tech. Report, Institute for Informatikk, Thormohlensgt, 55, N-5008, Bergen, Norway, 1981 and 1982.

BINARY PERFECT WEIGHTED COVERINGS (PWC)

I. The Linear Case

Gérard D. Cohen

Ecole Nationale Supérieure des Télécommunications

46 rue Barrault, C-220-5

75634 Paris cédex 13, France

Email: cohen@inf.enst.fr

Simon N. Litsyn

Dept. of Electrical Engineering-Systems

Tel-Aviv University

Ramat-Aviv

69978, Israel

Email: litsyn@genius.tau.ac.il

H. F. Mattson, Jr.

School of Computer and Information Science

4-116 Center for Science & Technology

Syracuse, New York 13244-4100

Email: jen@SUVM.acs.syr.edu, jen@SUVM.bitnet

Abstract

This paper deals with an extension of perfect codes to fractional (or weighted) coverings. We shall derive a Lloyd theorem—a strong necessary condition of existence—and start a classification of these perfect coverings according to their diameter. We illustrate by pointing to list decoding.

1   Introduction

Most codes involved in error-correction use nearest-neighbor decoding, i.e., the output of the decoder is the nearest codeword to the received vector. There has been renewed interest lately (see, e.g., [11]) in list decoding, where the decoder output is a list with given maximal size: correct decoding now means that the actually transmitted codeword is in the list. The size of the list could be constant (see the perfect multiple

36

coverings studied in [17]) or an increasing function of the distance between the received vector and the code, so as to guarantee a given level of confidence. For example, a codeword would be given a list consisting of itself ($m_0 = 1$ in our notations), whereas vectors at distance R (the covering radius of the code) would have lists of maximal size. An application of list codes could be to spelling checking, with the code being the English vocabulary, and the "ambient" space being any combination of letters with maximal length n.

2 Notations and known special cases

We denote by F^n the vector space of binary n-tuples, by $d(\cdot\ ,\ \cdot)$ the Hamming distance, by $C[n,k,d]R$ a linear code C with length n, dimension k, minimum distance $d = d(C)$ and covering radius R [8]. In this paper we consider mostly codes with $d \geq 2$. We denote the Hamming weight of $x \in F^n$ by $|x|$.

$A(x) = (A_0(x), A_1(x) \ldots A_n(x))$ will stand for the weight-distribution of the coset $C + x, x \in F^n$; thus

$$A_i(x) := |\{c \in C : d(c,x) = i\}|.$$

Given an $(n+1)$-tuple $M = (m_0, m_1, \ldots, m_n)$ of weights, i.e., rational numbers in $[0,1]$, we define the *M-density of C* at x as

$$\theta(x) := \sum_{i=0}^{n} m_i\, A_i(x) = <M, A(x)>. \qquad (2.1)$$

We consider only *coverings*, i.e., codes C such that $\theta(x) \geq 1$ for all x.

$$C \text{ is a } \textit{perfect M-covering} \text{ if } \theta(x) = 1 \text{ for all } x. \qquad (2.2)$$

We define the *diameter* of an M-covering as

$$\delta := \max\{i : m_i \neq 0\}.$$

To avoid trivial cases, we usually assume that $m_i = 0$ for $i \geq n/2$, i.e., $\delta < n/2$. Here are the known special cases.

$$\text{Classical perfect code: } m_i = 1 \text{ for } i = 0, 1, \ldots \delta. \qquad (2.3)$$

$$\text{Perfect multiple coverings: } m_i = 1/j \text{ for } i = 0, 1, \ldots \delta \qquad (2.4)$$
$$\text{where } j \text{ is a positive integer [17, 6].}$$

$$\text{Perfect L-codes: } m_i = 1 \text{ for } i \in L \subseteq \{1, 2, \ldots \lfloor n/2 \rfloor\}. \text{ See [13] and [7].} \qquad (2.5)$$

37

3 The covering equality

For a perfect M-covering C one gets from the definition:

$$\sum_{i=0}^{n} m_i \, A_i(x) = 1 \text{ for all } x.$$

Summing over all x in F^n and permuting sums, we get

$$\sum_{i=0}^{n} m_i \sum_{x \in F^n} A_i(x) = 2^n.$$

For $i = 0$, the second sum is $|C| = 2^k$, for $i = 1$ it is $2^k n$, and so on. Hence we get the following analog of the Hamming condition.

Proposition 3.1 A covering C is a perfect M-covering if and only if

$$\sum_{i=0}^{n} m_i \binom{n}{i} = 2^{n-k} \tag{3.1}$$

□

As mentioned earlier, we want to avoid trivial solutions to (3.1) as, e.g.,

$$m_i = 1 \quad \text{for all } i \ (k = 0),$$
$$m_i = 1 \quad \text{for } 0 \leq i \leq \lfloor n/2 \rfloor \text{ for odd } n \ (k = 1).$$

In fact, we are interested in getting perfect M-coverings with small diameter. However, we shall prove in the next section a strong lower bound on δ.

We can interpret (3.1) in a geometrical way: we define a weighted sphere around any vector c in F^n by means of the function

$$\mu_c(x) := m_{d(c,x)}. \tag{3.2}$$

For $d(c, x) > \delta$, $\mu_c(x) = 0$; hence δ can be viewed as the radius of the weighted sphere, denoted by $S(c, \delta)$. Set

$$\mu(S(c, \delta)) := \sum_{x} \mu_c(x) = \sum_{i=0}^{n} m_i \binom{n}{i};$$

then (3.1) becomes

$$\mu(S(c, \delta)) = 2^{n-k}$$

so that C is a *perfect weighted covering* (PWC) of F^n.

Equation (3.2) is reminiscent of a *fuzzy membership* function, as studied, e.g., in [5].

4 A Lloyd theorem

We denote by $P_{n,i}(x)$ the Krawtchouk polynomial, for $0 \le i \le n$,

$$P_{n,i}(x) = \sum_{0 \le j \le i} (-1)^j \binom{n-x}{i-j} \binom{x}{j}. \tag{4.1}$$

We now prove

Theorem 4.1 *An $[n,k,d]$ R code C is a perfect $(m_0, m_1, \ldots, m_\delta)$-covering only if the Lloyd polynomial*

$$L(x) := \sum_{0 \le i \le \delta} m_i \, P_{n,i}(x)$$

has among its roots the s nonzero weights of $C^\perp$.

Corollary 4.1 $s \le \delta$.

Proof of the Theorem: (Adapted from [1], Chapter II, Section 1, which records A. M. Gleason's proof of the classical Lloyd theorem.) We use the group algebra $\mathcal{A}$ of all formal polynomials

$$\sum_{a \in F^n} \gamma_a X^a$$

with $\gamma_a \in Q$, the field of rational numbers.

Define

$$S := \sum_{0 \le i \le \delta} m_i \sum_{|a|=i} X^a. \tag{4.2}$$

We let the symbol C for our code also stand for the corresponding element in $\mathcal{A}$, namely,

$$C := \sum_{c \in C} X^c. \tag{4.3}$$

Then we find from Section 3 that

$$SC = \sum_{c \in C} X^c \cdot S = F^n := \sum_{a \in F^n} X^a. \tag{4.4}$$

Characters on F^n are group homomorphisms of $(F^n, +)$ into $\{1, -1\}$, the group of order 2 in $Q^\times$. All characters have the form χ_u for $u \in F^n$, where χ_u is defined as

$$\chi_u(v) = (-1)^{u \cdot v} \text{ for } u, v \in F^n.$$

We use linearity to extend χ_u to a linear functional defined on $\mathcal{A}$:

For all $Y \in \mathcal{A}$ if $Y = \sum_{a \in F^n} \gamma_a X^a$, then $\chi_u(Y) := \sum \gamma_a \chi_u(a)$.
It follows that

$$\chi_u(YZ) = \chi_u(Y)\chi_u(Z) \text{ for all } Y, Z \in \mathcal{A}.$$

It is known [1, 10] that for any $u \in F^n$, if $|u| = w$, then

$$\chi_u \left(\sum_{|a|=i} X^a \right) = P_{n,i}(w).$$

It follows that

$$\chi_u(S) = L(w).$$

From (4.4), furthermore, we see that

$$\chi_u(SC) = \chi_u(S)\chi_u(C) = 0$$

for all $u \neq 0$.

Now if $u \in C^{\perp}$, then

$$\chi_u(C) = \sum_{c \in C} (-1)^{u \cdot c} = |C| = 2^k.$$

Thus $\chi_u(S) = 0$. $\qquad\qquad\qquad\qquad\qquad\qquad\qquad\qquad\qquad\qquad\quad$ □

5   The case when $\delta = s$

By a result of Delsarte [10], (3.10), one can always choose $\delta = s$. Let us reformulate his result.

Proposition 5.1 *A code C is a perfect M-covering with $\delta(M) = s$. In that case the m_i's are uniquely determined by*

$$m_i = \alpha_i, \quad 0 \leq i \leq s,$$

where α_i is the ith coefficient in the Krawtchouk expansion of the annihilator polynomial $\alpha(x)$ of $C^{\perp}$. Here

$$\alpha(x) := 2^{n-k} \prod_{w \in W} \left(1 - \frac{x}{w}\right),$$

and W is the set of s nonzero weights of vectors in $C^{\perp}$.

5.1 Uniformly packed codes

In [2] a code is called *uniformly packed* (u.p.) if there exist rational numbers $\alpha_0, \alpha_1, \ldots \alpha_R$ such that for any x in F^n, $\sum \alpha_i A_i(x) = 1$ holds. An extensive account of u.p. codes appears in [12]. With our notations, this reads:

Proposition 5.2 *A uniformly packed code C is a perfect M-covering with $\delta(M) = R(C)$.*

In that case, $R = s = \delta$ and Proposition 5.1 applies. The reason is that $R \leq s \leq \delta$ in general. The first inequality is Delsarte's Theorem, (3.3) of [10]; the second is Corollary 4.1.

Examples of u.p. codes (see [10], Section (3.1)):

$$QR[47, 24, 11]7, \quad \text{with } M = (1, 1, 1, 1, 1/9, 1/9, 1/9, 1/9)$$
$$\text{extended } QR[48, 24, 12]8, \quad \text{with } M = (1, 1, 1, 1, 5/27, 1/9, 1/9, 1/9, 1/54).$$

5.2 Strongly uniformly packed codes

This concept is introduced in [15]: An $[n, k, d = 2e+1]R$ code C is *strongly uniformly packed* (s.u.p.) if $R \leq e+1$ and for any x such that $d(x, C) \geq e$, the following holds:

$$|B(x, e+1) \cap C| \; = \; r \; (\text{independent of } x).$$

Here $B(x, e+1)$ denotes the sphere of radius $e+1$ centered at x in F^n. Of course, if $d(x, C) \leq e-1$, then by the triangle inequality

$$|B(x, e+1) \cap C| \; = \; 1.$$

Such a code will be denoted by $SUP(n, e, r)$. We have just proved

Proposition 5.3 *An $SUP(n, e, r)$ is a perfect M-covering with $m_0 = m_1 = \cdots = m_{e-1} = 1, m_e = m_{e+1} = 1/r$.*   □

Note that a code C can be a perfect M-covering for different M's. For example, the $[23, 12, 7]3$ Golay code is an $SUP(23, 3, 6)$, hence a perfect $(1, 1, 1, 1/6, 1/6)$-covering by Proposition 5.3. On the other hand, since this code is perfect, it is also a perfect $(1, 1, 1, 1)$-covering. We saw in (5.1) a sufficient condition for the uniqueness of M.

41

5.3 The case of diameter one

If $\delta = 1$, then $R = s = 1$, and $L(x) = m_0 + m_1(n - 2x) = -2xm_1 + 2^{n-k}$. Theorem (4.1) implies that $C^\perp$ has a unique nonzero weight, namely, $x = 2^{n-k-1}/m_1$.

Since $d \geq 2$, $C^\perp$ has no coordinates identically 0. Therefore $C^\perp$ consists of $1/m_1$ copies of the simplex code [14] with $m_1 = 1/t$, for some integer t. Thus $n = t(2^{n-k}-1)$.

Now from $m_0 + m_1 n = 2^{n-k}$, we get $m_0 = 2^{n-k} - n/t = 1$ (which also follows directly from $d \geq 2$).

Proposition 5.4 *A perfect M covering with $\delta = 1$ exists iff $n = t(2^i - 1)$, $m_0 = 1$, $m_i = 1/t$ for some integer t.*

Proof. The "only if" part is proved just above. The "if part" also! Namely, take for C the dual of the code consisting of t copies of the simplex code. □

For its intrinsic interest we shall present an alternate description of these $\{1, 1/t\}$-coverings.

Definitions. Let $C[n, k, d]R$ and $C'[n', k', d']R'$ be two linear codes.

$$\text{Set} \quad \chi_c(x) = \begin{cases} 0 & \text{if } x \in C \\ 1 & \text{otherwise.} \end{cases}$$

Note that χ_c is the complement of the usual indicator function. We then extend this function to a mapping $\chi : F^{nn'} \rightarrow F^{n'}$ by setting

$$\chi(x) := (\chi_c(x_1), \chi_c(x_2), \ldots \chi_c(x_{n'}))$$

where the x_i's are in F^n, for $1 \leq i \leq n'$, and

$$x = (x_1, x_2, \ldots, x_{n'}) \text{ is their concatenation.}$$

We are now ready to define $C \otimes C'$ as follows:

$$C \otimes C' := \{z \in F^{nn'} : \chi(z) \in C'\}.$$

Remark that $C \otimes C'$ is *not linear* in general, because χ_c is not.

Proposition 5.5 *$C \otimes C'$ has length nn', minimum distance $\min\{d, d'\}$, and covering radius RR'.*

Proof. Easy. □

Proposition 5.6 *Suppose that $d(C') \geq 2$. Then $C \otimes C'$ is linear if and only if C is an $[n, n-1]$ code. In that case, $C \otimes C'$ is an $[nn', nn' - (n' - k')]$ code.*

Proof. If C is an $[n, n-1]$ code, then χ_c is linear (check!), and so is χ, and $C \otimes C' = \chi^{-1}(C')$.

Conversely, suppose C has codimension at least 2. Let a, b be in different cosets of C; then $\chi_c(a) = \chi_c(b) = \chi_c(a + b) = 1$. Let c' be a word of C' with first component $= 1$ and $x = (a, x_2, x_3, \ldots, x_{n'})$ a codeword of $C \otimes C'$ (i.e., in $\chi^{-1}(C')$). Then $y := (b, x_2, x_3 \ldots, x_{n'})$ is also in $\chi^{-1}(C')$. Now $x + y = (a + b, 0, 0 \ldots, 0)$ is not in $C \otimes C'$, since $\chi(x + y) = (1, 0, \ldots 0) \notin C'$. Hence $C \otimes C'$ is not linear.

Let us now prove the dimensional part of the proposition: consider χ', the restriction of χ to $C \otimes C'$. Since $ker\,\chi \subset C \otimes C'$, $ker\,\chi' = ker\,\chi$. But $\dim(ker\,\chi) = n' \cdot \dim C = n'(n-1)$. Combined with $Im(\chi') = C'$, this yields:

$$
\begin{aligned}
\dim C \otimes C' &= \dim(Im(\chi')) + \dim(ker(\chi')) \\
&= k' + n'(n-1) \\
&= nn' - (n' - k').
\end{aligned}
$$

$\square$

To avoid $d(C) = 1$, and hence $d(C \otimes C') = 1$, we choose for C the parity code $[n, n-1, 2]1$ which is unique with such parameters. Then $C \otimes C'$ is an $[nn', nn' - (n' - k'), 2]R'$ code.

Proposition 5.7 *Let x and x' be such that $d(x, C) = R, d(x', C') = R'$. Suppose that $A_R(x)$ and $A'_{R'}(x')$ are independent of x. Then for $C \otimes C'$ the coefficient $A_{RR'}(z)$ is the same for any z such that $d(z, C \otimes C') = RR'$, and*

$$
A_{RR'} = A_R A'_{R'}.
$$

$\square$

Corollary 5.1 *The "if" part of Proposition 5.4 (alternate proof).*

$$
\begin{array}{ll}
\text{Choose} \quad C[t, t - 1, 2]1 & A_1 = t \\
\phantom{\text{Choose} \quad} C'[2^i - 1, 2^i - i - 1, 3]1 & A'_1 = 1
\end{array}
$$

Then $C \otimes C'$ is a $[t(2^i - 1), t(2^i - 1) - i, 2]1$ with $A_1(x) = t$ for all $x \notin C \otimes C'$, i.e., a $\{1, 1/t\}$-covering.

$\square$

If we omit the condition $m_0 = 1$, i.e., we set $d = 1$, we get an extended family of PWC by adding to all code words from PWC in Corollary 5.1 all possible tails of length l. Let C be $[t(2^i - 1), t(2^i - 1) - i, 2]1$ PWC, and C'' consist of all vectors (c, x) of length $t(2^i - 1) + l$, where c belongs to C, and x is from F^l.

Proposition 5.8 C'' is a $[t(2^i-1)+l, t(2^i-1)+l-i, 1]1$ PWC, i.e., a $\{1-l/t, 1/t\}$-covering.

Proof. Linearity and parameters are trivial. Let us consider a vector $y = (y_1, y_2)$, where y_1 and y_2 are from $F^{t(2^i-1)}$ and F^l respectively. If y_1 belongs to C, then $A_0(y) = 1$ and $A_1(y) = l$ (all the code vectors of the shape (y_1, x), $d(y_2, x) = 1$). If $d(y_1, C) = 1$, then $A_0(y) = 0$, and $A_1(y) = t$ (all the code words of shape (z, y_2), $d(z, y_1) = 1$). Solving the system $m_0 + l \cdot m_1 = 1$, $t \cdot m_1 = 1$, we get the statement.

From the uniqueness of the above system we conclude

Proposition 5.9 For $\delta = 1$ all the possible PWC are described in Proposition 5.8.

Remark. In particular, setting $l = t - 1$ in Proposition 5.8 gives a complete characterization of the parameters of binary linear PMC with diameter 1, simpler than the one in [15].

6   The case of diameter 2

Now we take $\delta = 2$. By Corollary 4.1, we know that s is at most 2. We shall treat separately the two possible values of s. First, notice the obvious implication following from (2.1):

$$m_0 \neq 1 \Longrightarrow d \leq \delta.$$

Under the assumption $\delta = 2$ this becomes

$$m_0 \neq 1 \Longrightarrow d \leq 2. \tag{6.1}$$

Therefore, if the code C corrects at least one error, then $m_0 = 1$. Since $R \leq s \leq 2$, C is quasi-perfect and, in fact, (λ, μ) uniformly packed. Much is known, although not everything, about these codes [12, 10], and we shall not consider them here. Hence we assume

$$m_0 \neq 1, \quad \text{which implies} \quad d = 2$$

44

from (6.1) and our blanket assumption $d \geq 2$. In fact, we shall restrict ourselves to perfect multiple coverings (2.2); i.e., set

$$m_0 = m_1 = m_2 = 1/j.$$

From the definition in Theorem 4.1, the Lloyd polynomial $L(x)$ satisfies

$$jL(x) = 2x^2 - 2(n+1)x + 1 + n + \binom{n}{2}. \tag{6.2}$$

If we use (3.1) we may write

$$jL(x) = 2x^2 - 2(n+1)x + j2^{n-k}. \tag{6.3}$$

Since $s \geq 1$, $L(x)$ has at least one integral root. But the sum of the roots is $n+1$, so both are integral. Solving (6.1) we find that the roots of $L(x)$ are

$$\frac{1}{2}\left(n + 1 \pm \sqrt{n-1}\right) = 1 + \frac{m^2 \pm m}{2}, \tag{6.4}$$

where we have set

$$n = 1 + m^2 \tag{6.5}$$

for some integer m.

6.1   PMC with $s = 1$

Proposition 6.1 *The only perfect multiple covering code with $s = 1$, $d = 2$, and $\delta = 2$ is the $[2, 1, 2]$ code with $j = 2$.*

Proof.   Let C be an $[n, k, d]$ code satisfying the hypotheses. Since $d = 2$, $C^\perp$ has repeated coordinates but none identically zero. Therefore $C^\perp$ is, for some integer $t \geq 2$, the t-fold repetition of the simplex code [14] of type $[2^i - 1, i, 2^{i-1}]$, where $i = \dim C^\perp = n - k$. Thus $n = t(2^{n-k} - 1)$. From (6.5) we get

$$t(2^{n-k} - 1) = 1 + m^2. \tag{6.6}$$

But there are no solutions for (6.6) if $n - k \geq 2$. For let p be any prime dividing $2^{n-k} - 1$. Then -1 is a quadratic residue mod p, from (6.6). Therefore $p \equiv 1$ (mod 4). It follows that $2^{n-k} - 1 \equiv 1$ (mod 4), a contradiction.

Therefore there are no PMC with $\delta = 2$ and $s = 1$ except for $n - k = 1$. And in this case there is only the $[2, 1, 2]$ code. The reason is that with $d \geq 2$ it must be the

$[n, n-1, 2]$ code. From the definition in Section 1 it easily follows that n can be only 2. $\qquad\square$

Allowing d to be 1, we find that the only possibility for the check matrix is the t-fold repetition of $g(S_i)$ (generator matrix of a simplex code of length $2^i - 1$) with l zero-columns appended, yielding $n = t(2^i - 1) + l$. It amounts to appending all possible tails of length l to codewords described in Corollary 5.1. It is easy to check that there are 2 kinds of covering equalities (namely, vectors coinciding with, or being at distance 1 from, codewords on the first $t(2^i - 1)$ coordinates):

$$m_0 + lm_1 + \binom{t}{2}(2^i - 1)m_2 + \binom{l}{2}m_2 = 1$$
$$tm_1 + (2^{i-1} - 1)t^2 m_2 + tlm_2 = 1.$$

This implies

$$t^2 - t(2^i + 1 + 2l) + (l^2 + l + 2) = 0$$

which has discriminant

$$D = (2^i + 1)^2 + 2^{i+2}l - 8.$$

We get a PMC iff $D = x^2$ has integer solutions. For example, the values $i = 3, l = 3, t = 14$ yield the PMC $[101, 98]$ with $j = 644$. Of course, for $i = t$ we get $8l + 1 = x^2$ having all odd x as solutions.

Parameters of a series of PMC for $s = 1, \delta = 2$:

$$i = 1 \quad l = \tfrac{x^2 - 1}{8}, \qquad x \equiv \pm 1 \pmod 4;$$

$$i = 2 \quad l = \tfrac{x^2 - 1}{16} - 1, \quad x \equiv \pm 1 \pmod 8;$$

$$i = 3 \quad l = \tfrac{x^2 - 3}{32} - 2, \quad x \equiv \pm 3 \pmod{16};$$

$$i = 4 \quad l = \tfrac{x^2 - 25}{64} - 4, \quad x \equiv \pm 5 \pmod{32};$$

$$i = 5 \quad l = \tfrac{x^2 - 57}{128} - 8, \quad x \equiv \pm 21 \pmod{64};$$

$$\ldots\ldots$$

6.2 PMC with $s = 2$

We have found the following PMC codes C in this case $(d = s = \delta = 2)$.

$$\begin{array}{lll} C & & C^\perp \\ [5,1;5] & j=1 & [5,4;2,4] \\ [5,2,2] & j=2 & [5,3;2,4] \\ [5,3,2] & j=4 & [5,2;2,4] \\ [10,7,2] & j=7 & [10,3;4,7] \\ [37,32,2] & j=22 & [37,5;16,22] \\ [8282,8269,2] & j=4187 & [8282,13;4096,4187] \end{array}$$

(6.7)

The first is a classical perfect code. The notation $[n,k;w_1,w_2,\ldots]$ stands for an $[n,k]$ code in which all nonzero weights are among $w_1,w_2,\ldots$. In the above codes $C^\perp$, since $s=2$, both weights are present. All the above codes C are PMC codes.

These codes arise from the following two constructions.

Notation

$g(C)$ generator matrix of code C

S_i i-dimensional simplex code

First Construction. We construct a 2-weight code $C^\perp$ by setting $g(C^\perp)$ equal to $g(S_i); c^h$ where c is any column of $g(S_i)$. For example, the $[5,3,2]$ code for $j=4$ above has

$$g(C^\perp) = \left[\begin{array}{ccc|cc} 1 & 1 & 0 & 1 & 1 \\ 0 & 1 & 1 & 0 & 0 \end{array}\right].$$

Here $i=2$ and $h=2$. There is no loss of generality in taking c to be a unit vector. In general we have

$$g(C^\perp) \;=\; g(S_i); c^h. \tag{6.8}$$

The weights in $C^\perp$ are 2^{i-1} and $2^{i-1}+h$.

We will now calculate the values

$$D := A_0(x),\; A_1(x),\; A_2(x)$$

for the cosets of C:

Identify the cosets with the syndromes, which are columns of S_i.

(i) The code C has

$$D = 1, 0, \binom{h+1}{2}$$

since column c occurs $h+1$ times in $g(C^\perp)$.

47

(*ii*) For any column c' of $g(S_i)$ other than c,

$$D = 0, 1, \frac{2^i - 2}{2} + h,$$

since there are $(2^i - 2)/2$ vectors v of weight 2 in any coset of weight 1 in the Hamming code $S_i^{\perp}$. Column c is covered by one of those vectors v. We may replace c there by any of its h clones.

(*iii*) For column c,

$$D = 0, h + 1, \frac{2^i - 2}{2}.$$

Now the code C will be a PMC iff the sum of D is the same in all three cases:

$$j = 1 + \binom{h+1}{2} = 1 + h + \frac{2^i - 2}{2}. \tag{6.9}$$

This equation can be written

$$1 + \binom{h}{2} = 2^{i-1}. \tag{6.10}$$

All solutions of this Diophantine equation are known [16]. They exist precisely for $h = 0, 1, 2, 3, 6$, and 91.

Since h is the difference between the two weights in $C^{\perp}$, $h = m$ as defined in (6.4) and (6.5). Thus we consider

$$h = m = 0, 1, 2, 3, 6, \text{ and } 91.$$

The corresponding values of j, from (6.9), are

$$j = 1, 2, 4, 7, 22, 4187.$$

Since $i = \dim(C^{\perp})$, $i = n - k$. We may calculate i from (6.10). We get

$$n - k = 0, 1, 2, 3, 5, 13.$$

The first two cases have $s = 0$ and 1. They are nevertheless the PMC codes C shown here:

C	j	$C^{\perp}$
$[1, 1; 1]$	1	$[1, 0; 0]$
$[2, 1; 2]$	2	$[2, 1; 2]$.

The next cases are the $[5, 3, 2]$ code in our table (6.7), and the three larger codes of (6.7).

It remains to account for the $[5, 1; 5]$ code and the $[5, 2, 2]$ code.

Second Construction. The $[5, 1; 5]$ code C has $s = R = 2$. Since it is perfect, it is a PMC with $j = 1$. If we now let C_2 be a coset of C of weight 2, and define

$$C_1 := C \cup C_2,$$

we get a $[5, 2, 2]$ code C_1. This construction obviously doubles the value of j for any PMC code with $R = 2$. Thus we get the second code in (6.7).

Since C_1 has $R = 2$ as well, we may arrive at the $[5, 3, 2]$ PMC code again by applying the second construction to C_1.

Note. The first construction yields nothing if we repeat the simplex code in $C^\perp$. I.e., if

$$g(C^\perp) = t \times g(S_i); c^m$$

then the smaller weight in $C^\perp$ is

$$t \cdot 2^{i-1} = 1 + \frac{m^2 - m}{2},$$

from (6.4). The length is

$$n = t(2^i - 1) + m = 1 + m^2,$$

from (6.5). These easily imply $t = 1$. We have proved

Proposition 6.2 *The only PMC codes with $d = s = \delta = 2$ obtainable by the First Construction are those in (6.7).*

Conjecture 6.1 *We conjecture the nonexistence of PMC codes with $d = s = \delta = 2$ other than those in (6.7).*

7 List codes

Recall from the introduction that in list decoding, to every x in F^n (received vector) is attached a list of at most K candidates (transmitted codewords). Following [4, 11], we denote by (n, e, K) a code C enabling the correction of up to e errors by list decoding with maximal list size K. This is equivalent to

$$|B(x, e) \cap C| \leq K \quad \text{for all} \quad x, \tag{7.1}$$

where $B(x, e)$ denotes the sphere of radius e centered at x. A code satisfying (7.1) is also called a K-fold e-packing, and a perfect multiple covering (2.2) if equality holds in (7.1) for all x (see [17]). From (5.2), the following is immediate.

Proposition 7.1 *An $SUP(n, e, r)$ is a list code $(n, e + 1, r)$.*

Refining the definition of a list code, we denote by $\mathcal{K} = \{K_i\}$ the set of possible list sizes attached to x's, with max $K_i = K$. Here is a small table of these codes.

n	e	$\mathcal{K}$	Comments
23	4	$\{1, 6\}$	Golay code
23	5	$\{1, 22\}$	Golay code
47	7	$\{1, 9\}$	QR
48	8	$\{1, 24, 29, 34, 39, 44, 45, 49, 54\}$	Ext. QR

Acknowledgement

We are pleased to acknowledge that this problem arose in discussions with I. Honkala in Veldhoven in June, 1990.

References

[1] E. F. Assmus, Jr., H. F. Mattson, Jr., R. Turyn, "Cyclic codes," Final Report, Contract no. AF 19(604)-8516, AFCRL, April 28, 1966. Sylvania Applied Research Laboratory, Waltham, Mass. (Document no. AFCRL-66-348.)

[2] L. A. Bassalygo, G. V. Zaitsev, V. A. Zinoviev, "On Uniformly Packed Codes," *Problems of Inform. Transmission*, vol. 10, no. 1, 1974, pp. 9–14.

[3] L. A. Bassalygo, G. V. Zaitsev, V. A. Zinoviev, "Note on Uniformly Packed Codes," *Problems of Inform. Transmission*, vol. 13, no. 3, 1977, pp. 22–25.

[4] V. M. Blinovskii, "Bounds for codes in the case of list decoding of finite volume," *Problems Inform. Transmission* 22, no. 1, (1986), pp. 7–19 (English).

[5] B. Bouchon, G. Cohen, "Partitions and Fuzziness," *J. of Math. Analysis and Applications*, vol. 116, no. 1, 1986, pp. 166–183.

[6] R. F. Clayton, "Multiple Packings and Coverings in Algebraic Coding Theory," Thesis, Univ. of California, Los Angeles, 1987.

[7] G. D. Cohen, P. Frankl, "On tilings of the binary vector space," *Discrete Math.* 31, 1980, pp. 271–277.

[8] G. D. Cohen, M. Karpovsky, H. F. Mattson, Jr., and J. R. Schatz, "Covering radius—Survey and recent results," *IEEE Trans. Inform. Theory* **IT-31**, pp. 328–343, 1985.

[9] A. A. Davydov, L. M. Tombak, "Number of minimal-weight words in block codes," *Problems of Inform. Transmission*, vol. 24, no. 1, pp. 11–24, 1988.

[10] P. Delsarte, "Four Fundamental Parameters of a Code and Their Combinatorial Significance," *Information and Control*, vol. 23, no. 5, pp. 407–438, 1973.

[11] P. Elias, "Error-correcting codes for list decoding," *IEEE Trans. Inform. Theory* 37 (1991), pp. 5–12.

[12] J. M. Goethals, H. C. A. van Tilborg, "Uniformly packed codes," *Philips Res. Repts* 30, pp. 9–36, 1975.

[13] M. Karpovsky, "Weight Distribution of Translates, covering radius and perfect codes correcting errors of the given multiplicities," *IEEE Trans. Inform. Theory* IT-27, pp. 462–472, 1981.

[14] J. E. MacDonald, "Design methods for maximum minimum-distance error-correcting codes," *IBM J. Res. Devel.* vol. 4, pp. 43–57, 1960.

[15] N. V. Semakov, V. A. Zinoviev, G. V. Zaitsev, "Uniformly Packed Codes," *Problems of Inform. Transmission*, vol. 7, 1971, no. 1, pp. 38–50.

[16] Th. Skolem, P. Chowla, and D. J. Lewis, "The Diophantine equation $2^{n-2} - 7 = x^2$ and related problems," *Proc. Amer. Math. Soc.* 10 (1959), pp. 663–669.

[17] G. J. M. van Wee, G. D. Cohen, S. N. Litsyn, "A note on Perfect Multiple Coverings of Hamming Spaces," *IEEE Trans. Inform. Theory*, vol. 37, no. 3, pp. 678–682, May 1991.

Read/Write Isolated Memory

Martin Cohn
Computer Science Department
Brandeis University
Waltham MA 02254
(marty@cs.brandeis.edu)

Abstract

We find upper and lower bounds on the code rate for a binary, rewritable medium
in which adjacent locations may not both store 1's, and adjacent locations may
be not altered during a single rewriting pass. If the true rate is close to our lower
bound, then a trivially simple code comes close to optimality.

1   Introduction

A *read/write isolated memory (RWIM)* is a binary storage medium with two restrictions:

- Adjacent 1's may not appear in the recording itself.

- Adjacent locations may not be altered during a single rewriting pass.

The first restriction is typical of magnetic recording, and has recurred in optical
recording. The general case was first studied by Freiman and Wyner[FrWy] and a subcase
by Kautz[Kaut]. The second restriction has arisen more recently in the context of bar
codes[Robi] and rewritable optical discs[CoZe]. In this paper we consider the conjunction
of the two restrictions. We compute some bounds on asymptotically achievable code
rates, and mention one simple code.

The rewritable-memory problem has four variants, according to whether the encoder
and decoder, respectively, are aware or not aware of the previous state of recording. In
all four cases a rate of 0.5 bits/symbol is constructively achieved by a code that fixes
alternate symbols at '0' and simply writes or rewrites the binary input data in the non-
fixed positions. The code rate for each restriction taken alone is upper bounded by
$0.694\ldots = log_2\phi$, where ϕ is the larger root of the Fibonacci recurrence[Kaut, CoZe],
so this bound applies *a fortiori* to the conjunction of the two restrictions. The case of
greatest practical interest occurs when the encoder *is* aware of the previously recorded
state (by having somehow stored it or by reading it on the fly), but the decoder is *not*
aware of the previous state.

$$
\begin{array}{c}
00 \\
01 \\
10
\end{array}
\begin{bmatrix}
1 & 1 & 1 \\
1 & 1 & 0 \\
1 & 0 & 1
\end{bmatrix}
\qquad
\begin{array}{c}
000 \\
001 \\
010 \\
100 \\
101
\end{array}
\begin{bmatrix}
1 & 1 & 1 & 1 & 1 \\
1 & 1 & 0 & 1 & 1 \\
1 & 0 & 1 & 0 & 0 \\
1 & 1 & 0 & 1 & 1 \\
1 & 1 & 0 & 1 & 1
\end{bmatrix}
$$

$$
\begin{array}{c}
0000 \\
0001 \\
0010 \\
0100 \\
0101 \\
1000 \\
1001 \\
1010
\end{array}
\begin{bmatrix}
1 & 1 & 1 & 1 & 1 & 1 & 1 & 1 \\
1 & 1 & 0 & 1 & 1 & 1 & 1 & 0 \\
1 & 0 & 1 & 0 & 0 & 1 & 0 & 1 \\
1 & 1 & 0 & 1 & 1 & 0 & 0 & 0 \\
1 & 1 & 0 & 1 & 1 & 0 & 0 & 0 \\
1 & 1 & 1 & 0 & 0 & 1 & 1 & 1 \\
1 & 1 & 0 & 0 & 0 & 1 & 1 & 0 \\
1 & 0 & 1 & 0 & 0 & 1 & 0 & 1
\end{bmatrix}
$$

Figure 1: Code Words and Adjacency Matrices for Word Lengths 2, 3, and 4

2 Block Rewriting

We will consider the limiting rate attainable by block rewriting in which no adjacent changes are made and no adjacent 1's are ever recorded. A *k-tuple* of 0's and 1's with no adjacent 1's can describe either a k-bit recorded code word, or a k-bit pattern of non-adjacent changes. For every $k \geq 1$ it is easily seen there are f_k such k-tuples, where f_k is the kth Fibonacci number [Kaut]. [1] For example, consider the sets of the $f_2 = 3, f_3 = 5,$ and $f_4 = 8$ code words of lengths two, three, and four, each set listed in lexicographic order in Figure 1. Given each set of words, we can construct graphs, or more clearly, binary adjacency matrices, with entry 1 corresponding to each pair of codewords whose bitwise difference has no adjacent 1's. That is, the $(i,j)^{th}$ entry of A_k is 0 if and only if the ith and j^{th} code words of length k differ in at least two adjacent positions; $A_2, A_3,$ and A_4 are shown in Figure 1. In what follows we always take the sets of words to be ordered lexicographically.

In general we can partition the set W_k of f_k words into three classes:

- A subset W_k^{00} of f_{k-2} words, all with prefix 00;

- a subset $W_k^{01} = W_k^{010}$ of f_{k-3} words, all with prefix 01 (actually 010);

- a subset $W_k^1 = W_k^{10}$ of f_{k-2} words, all with prefix 10.

The union of W_k^{00} and W_k^{01} is the (ordered) subset W_k^0 of f_{k-1} words with prefix 0.

The words in W_k^0 among themselves have adjacency matrix A_{k-1} because the first bit does not change; the words in $W_k^1 = W_k^{10}$ among themselves have adjacency matrix A_{k-2} because the first two bits don't change. The adjacencies between W_k^{00} and W_k^{10} are again

---

[1] We take $f_1 = 2, f_2 = 3, f_3 = 5\ldots$, a shift of the conventional Fibonacci indexing, to simplify notation.

53

given by A_{k-2}. There are no adjacencies between W_k^{01} and W_k^{10} because such adjacencies would entail changing both the first two bits. This gives a recursive definition of the $f_k \times f_k$ adjacency matrix A_k:

$$A_{-1} = A_0 = [1]; \qquad A_1 = \begin{bmatrix} 1 & 1 \\ 1 & 1 \end{bmatrix}; \qquad A_k = \begin{bmatrix} A_{k-1} & A_{k-2} \\ & 0 \\ A_{k-2} & 0 & A_{k-2} \end{bmatrix} \qquad (1)$$

where matrices of zeros are appropriately dimensioned. An alternative, equivalent recurrence is

$$A_k = \begin{bmatrix} A_{k-2} & A_{k-3} & A_{k-2} \\ & 0 & \\ A_{k-3} & 0 & A_{k-3} & 0 \\ A_{k-2} & & 0 & A_{k-2} \end{bmatrix} \qquad (2)$$

Note that A_k is symmetric.

We would like to find, or at least to bound, the *capacity* of the associated graph and its equivalent channel. The capacity per word of length k was defined by Shannon[Shan] and shown to be $\log_2 \lambda_k^*$, where λ_k^* is defined to be the *index*[CvDoSa] of the adjacency graph, that is, the largest real root of the characteristic polynomial of the adjacency matrix A_k. The capacity per symbol is thus $\lim_{k \to \infty} (\log_2 \lambda_k^*)/k$.

3 Bounds on Channel Capacity

3.1 Lower Bound

The constructive lower bound on capacity, achieved by coding only alternate positions, is 0.5 bits/symbol. A slightly better lower bound can be found via the following theorem:[CvDoSa, p. 84]

Theorem 1 (Collatz, Sinogowitz) *Let $\bar{d}$ be the mean value of the valencies and λ^* the greatest eigenvalue of a graph G. Then $\lambda^* \geq \bar{d}$, where equality holds if and only if G is regular.*

The mean valency of a symmetric graph is one-half the average row sum of its adjacency matrix. From Equation 1 we know that the dimension of A_k satisfies the Fibonacci recurrence and hence is the Fibonacci number f_k. Let s_k be the number of 1's in A_k, that is, $s_k = \Sigma_{i,j}(A_k)_{i,j}$. From Equation 1 we know also that s_k satisfies the recurrence $s_k = s_{k-1} + 3s_{k-2}$. Thus for some constants c_1 and c_2 we have

$$f_k \asymp c_2 \left(\frac{1 + \sqrt{5}}{2} \right)^k$$

$$s_k \asymp c_1 \left(\frac{1 + \sqrt{13}}{2} \right)^k$$

and the average row sum of A_k is asymptotically s_k/f_k.

Theorem 2 *The capacity of the RWIM satisfies*

$$
\begin{aligned}
\text{Capacity} &= \lim_{k \to \infty} \frac{1}{k} \log_2 \lambda_k^* \\
&\geq \lim_{k \to \infty} \frac{1}{k} \log_2(s_k/f_k) \\
&= \log_2(1 + \sqrt{13}) - \log_2(1 + \sqrt{5}) \\
&= 0.509 \ldots
\end{aligned}
$$

It should be noted that Theorem 1 applies to graphs with no loops, but subtracting f_k from s_k would not change the capacity. Theorem 1 is actually a special case (corresponding to the trivial partition) of the *interlacing theorem*[CvDoSa, p. 19]:

Theorem 3 (Sims) *Let A be a real symmetric matrix. Given a partition into blocks of sizes $n_i > 0$, consider the corresponding blocking $A = (A_{ij})$ so that A_{ij} is an $n_i \times n_j$ block. Let e_{ij} be the sum of the entries in A_{ij} and put $B = (e_{ij}/n_i)$ (i.e., the average row sum in A_{ij}.) Then the spectrum of B is contained in the spectrum of A.*

It seems inviting to use Theorem 3 in conjunction with the decomposition of Equation 1, and indeed a single application multiplies the bound of Theorem 1 by a factor of about 3. But this constant factor disappears in the capacity computation; what is needed for a real change is the compounded effect of $\Omega(k)$ multiplicative factors.

3.2 Upper Bounds

In the case of upper bounds, again a simple argument gives a better result than a more complicated argument. A graph with n vertices and m edges has eigenvalues satisfying

$$\lambda(1) + \ldots + \lambda(n) = 0; \qquad \lambda(1)^2 + \ldots + \lambda(n)^2 = 2m.$$

Then the largest (real) eigenvalue satisfies $\lambda^* = \lambda(1) \leq \sqrt{2m(1 - 1/n)}$ [CvDoSa, p. 221.] In our case the term $1/n$ is asymptotically insignificant, and we have $\lambda_k^* \leq \sqrt{s_k}$.

Theorem 4

$$
\begin{aligned}
\text{Capacity} &\leq \lim_{k \to \infty} \frac{1}{k} \log_2 \sqrt{s_k} \\
&= \lim_{k \to \infty} \frac{1}{k} \log_2 \left(\frac{1 + \sqrt{13}}{2} \right)^{k/2} \\
&= \frac{1}{2} \log_2 \left(\frac{1 + \sqrt{13}}{2} \right) = 0.6017 \ldots
\end{aligned}
$$

Once again it is tempting to improve the bound by making use of the recursive decomposition of A_k, but once again the result is disappointing. We use a bound on the *index* of the union of two graphs with identical vertex sets, that is, the largest eigenvalue of the sum of their adjacency matrices[CvDoSa; Theorem 2.1, p. 51]

Theorem 5 (Courant-Weyl) *If B and C are real symmetric matrices of order n, and if $D = B + C$, then $\lambda_k^*(D) \leq \lambda_k^*(B) + \lambda_k^*(C)$.*

Theorem 6 $\lambda_k^* \leq \lambda_{k-1}^* + \phi\lambda_{k-2}^*$ *(where $\phi = 1.618\ldots$ is the larger Fibonacci root.)*

Proof:

Decompose A_k as in Equation 1 into the sum

$$A_k = \begin{bmatrix} \mathbf{A_{k-1}} & 0 & 0 \\ 0 & 0 & 0 \\ 0 & 0 & 0 \end{bmatrix} + \begin{bmatrix} 0 & & A_{k-2} \\ & 0 & 0 \\ A_{k-2} & 0 & A_{k-2} \end{bmatrix}$$

The largest eigenvalue of the first matrix is just λ_{k-1}^*. The connected portion of the second matrix is the Kronecker (or Cartesian) product of $\begin{bmatrix} 0 & 1 \\ 1 & 1 \end{bmatrix}$ and A_{k-2}, so its largest eigenvalue is the product of ϕ and λ_{k-2}^*. The inequality follows from Theorem 5. □

Corollary

$$\lambda_k^* \leq c\left(\frac{1 + \sqrt{1 + 4\phi}}{2}\right)^k; \quad \text{Capacity} \leq \log_2(1 + \sqrt{1 + 4\phi}) - 1 = 0.9005\ldots$$

The slack in this bound might have been anticipated from its application to the case of $\lambda_2^* = 2.414$ for which the bound gives $2 + 1.618\ldots = 3.618\ldots$ Using the more complicated matrix recurrence of Equation 2, the recurrence for λ_k^* becomes $\lambda_k^* \leq 2\lambda_{k-2}^* + \phi\lambda_{k-3}^*$, and the consequent bound on the capacity is $0.7787\ldots$, better than above, but still much worse than the simpler bound of $0.6017\ldots$ The accumulating slack introduced by successive union bounds is obviously detrimental.

4   Conclusion

4.1   Numerical Results

The following table lists the matrix dimension f_k, the dominant eigenvalue λ_k^*, and $C_k/k = \frac{1}{k}\log_2\lambda_k^*$, capacity per symbol, as functions of wordlength k.

56

k	f_k	λ_k^*	C_k/k
1	2	2.000000	1.000000
2	3	2.414215	0.635777
3	5	4.086133	0.676912
4	8	5.345956	0.604612
5	13	8.434573	0.615263
6	21	11.510559	0.587481
7	34	17.517792	0.590107
8	55	24.487541	0.576747
9	89	36.525244	0.576758
10	144	51.788146	0.569455
11	233	76.349987	0.568596
12	377	109.182967	0.564217
13	610	159.856988	0.563126
14	987	229.787737	0.560297

The tabulated rates upper-bound the capacity measured in *bits per symbol per rewrite* for blocks of length k symbols. Since blocks can not in general be freely concatenated in space (consider, e.g., $k = 1$,) the rates also upper bound the capacity in bits per symbol. These values can be also be used to compute rates asymptotically achievable under concatenation if each block is buffered from its successor by an idle position, as was done to get the constructive code of rate 0.5 from the case $k = 1$.

Notice that in this tabulation although the capacities are not strictly decreasing, the odd-indexed and even-indexed capacity sequences *do* decrease, suggesting that the dominant-eigenvalue sequence might be approximated by a weighted sum of two exponentials, for example,

$$\lambda_k^* \doteq 1.25 \times (1.451)^k - .25 \times (-1.08)^k.$$

Such an approximation in turn suggests that λ_k^* might satisfy a second-order ordinary linear difference equation, but I have no rationale for such a claim. If the approximation *is* valid, the limiting capacity would be about $\log_2 1.451 \doteq 0.537$ bits per symbol.

The true limiting code rate has been shown to lie between 0.509 and 0.561. If it should be close to 0.5, the trivial rule "code only in alternate positions" gives a very efficient code. The simplicity of this coding rule would be remarkable in view of the apparent difficulty of devising an efficient practical code for the channel obeying just the single restriction "no adjacent changes" [CoZe].

5 Acknowledgment

I am grateful to Gerard Cohen for a preprint of his paper with Zemor; to Ira Gessel who pointed me to [CvDoSa]; and to Mor Harchol who programmed and ran the tabulated computations.

6 References

[CoZe] G. Cohen and G. Zemor, "Write-Isolated Memories" *Preprint, submitted to Discrete Mathematics* (1988)

[CvDoSa] D. Cvetković, M. Doob, and H. Sachs, *Spectra of Graphs* Academic Press, Harcourt Brace Jovanovich, New York (1980)

[FrWy] C. V. Freiman and A.D. Wyner, "Optimum Block Codes for Noiseless Input-Restricted Channels" *Information and Control, 7* (1964) 398-415

[Kaut] W.H. Kautz, "Fibonacci Codes for Synchronization Control" *IEEE Transactions on Information Theory, IT-11* (April 1965) 284-292

[Robi] J.P. Robinson "An Asymmetric Error-Correcting Ternary Code" *IEEE Transaction on Information Theory, IT-24, 2* (1978) 258-261

[Shan] C.E. Shannon *The Mathematical Theory of Communication* U. of Illinois Press, Urbana (1949)

Polynomial-time construction of linear codes with almost equal weights

Gilles Lachaud

Equipe A.T.I., C.I.R.M.

Marseille-Luminy

and

Jacques Stern

G.R.E.C.C., D.M.I.

École Normale Supérieure

If C is a binary linear code with $d \leq w(x) \leq D$ for every non-zero codeword x in C, we define the *disparity* of C to be the ratio

$$r(C) = \frac{D}{d}$$

The single-weight codes are those with $r = 1$. Any single-weight binary linear code C of dimension k and length n is such that

$$k \leq 1 + \log_2(n)$$

thus, there is no infinite family of single-weight codes whose length is linearly bounded with respect to the dimension.

If $\mathbf{C} = (C_i)_{i \geq 0}$ is an infinite family of binary codes whose length goes to infinity, we let

$$r(\mathbf{C}) = \limsup_{i \to \infty} r(C_i) \qquad ; \qquad R(\mathbf{C}) = \limsup_{i \to \infty} \frac{\dim(C_i)}{\text{length}(C_i)}$$

In the present paper, we state that there are infinite families $\mathbf{C}$ of binary linear codes with polynomial complexity of construction which are such that $r(\mathbf{C})$ is arbitrarily close to 1 and $R(\mathbf{C}) > 0$ (cf. Theorem 2 for a precise statement). In fact, we first prove that there are families with weights close to $\frac{1}{2}$ (Theorem 1) and analyzing this construction, we draw conclusions on the disparity of the resulting family. The method we use is a concatenation process: the inner families are the projective Reed-Muller codes ([6, 7, 11]) and the outer families are the modular Goppa codes of Tfasman, Vlădut and Zink ([3, 4, 5, 9, 11]), constucted from the Goppa theory ([1, 2, 5])? Our result can be viewed as a consequence of the concatenation bound of Katsman, Tfasman and Vlădut (see [3, 4]).

Theorem 1 *For every q and every k $\geq$ 4, there is an infinite family of codes $(C_{i,k})_{i\geq 0}$ over $\mathbf{F}_q$, with polynomial complexity of construction, such that* length$(C_{i,k}) \to \infty$ *and*

$$R(C_{i,k}) = k\frac{q-1}{q^k - 1}\left(\frac{1}{q^{k/2} - 1} + o(1)\right)$$

and, for every $x \in C_{i,k} - \{0\}$

$$\left|\frac{\text{weight}(x)}{\text{length}(C_{i,k})} - \frac{q-1}{q}\right| \leq \left(\frac{2}{q^{k/2} - 1} + o(1)\right)$$

In plain words, this means that there is an infinite family of codes over $\mathbf{F}_q$, with polynomial complexity of construction, whose relative weights are as close to $\frac{q-1}{q}$ as we wish. Note that this is in accordance with the normal distribution obtained for some binary codes and discussed in [8, pages 282–288].

If $\mathbf{C} = (C_i)_{i\geq 0}$ is an infinite family of codes over a finite field, we let

$$\delta(\mathbf{C}) = \lim_{i\to\infty}\frac{\text{dist}(C_i)}{\text{length}(C_i)}$$

Next, we introduce the *polynomial code domain* (cf. [5, 9, 11]), which is the set of points (R, δ) in the square $[0,1] \times [0,1]$ such that there exists an infinite family of polynomially constructible q-ary codes $\mathbf{C} = (C_i)_{i\geq 0}$, with $\delta(\mathbf{C}) = \delta$ and $R(\mathbf{C}) = R$.

Corollary 1 *If k is even, there is a pair (R_k, C_k) in the polynomial-time domain with*

$$R_k \sim \frac{1}{4}\frac{q^3}{(q-1)^2}(z_k)^3 \log_q \frac{1}{z_k} \qquad \text{where} \qquad z_k = \frac{q-1}{q} - \delta_k$$

In particular, for $q = 2$, we get:

$$R_k \sim 2(z_k)^3 \log_2 \frac{1}{z_k} \qquad \text{with} \qquad z_k = \frac{1}{2} - \delta_k$$

In order to understand the proper meaning of this result, recall that it follows from the Varshamov-Gilbert lower bound (see [8, 11]) that the curve $\mathbf{R} = \mathbf{R}_{VG}(\delta)$ is included in the general code domain (defined as above but without any polynomial-time restriction), where

$$\mathbf{R}_{VG}(\delta) = 1 + \delta \log_q \left(\frac{\delta}{q-1} \right) + (1-\delta) \log_q (1-\delta)$$

Since

$$\mathbf{R}_{VG}(\delta) \sim \frac{q^2}{2(q-1) \ln q} z^2 \qquad \text{when} \qquad z \to 0$$

we notice that the Varshamov-Gilbert curve is above the points defined in the corollary.

Let us also point out that, from the Varshamov-Gilbert bound, we can infer the existence of families $\mathbf{C}$ of binary linear codes (which are not necessarily polynomial-time constructible), with disparity $r(\mathbf{C}) \leq 1 + \epsilon$ and with

$$r(\mathbf{C}) \geq \frac{1}{2 \ln 2} \epsilon^2$$

Compared to this last statement, Theorem 1 specializes as follows:

Theorem 2 *For every $\epsilon > 0$, there are families $\mathbf{C}$ of polynomially constructible binary linear codes with disparity $r(\mathbf{C}) \leq 1 + \epsilon$ and with rate*

$$r(\mathbf{C}) \geq \frac{1}{4} \epsilon^3 \log_2 \left(\frac{1}{\epsilon} \right) + O(\epsilon^3)$$

References

[1] Goppa, V.D., *Algebraico-geometric codes*, Izv. Akad. Nauk S.S.S.R., **46**(1982); Math. U.S.S.R Izvestiya, **21**(1983), 75-91.

[2] Goppa, V.D., *Geometry and codes*, Kluwer Acad. Pub., Dordrecht, 1988.

[3] Katsman, G.L., Tsfasman, M.A. and Vlădut, S.G., *Modular Curves and Codes with polynomial complexity of construction*, Problemy Peredachi Informatsii **20**(1984), 47-55; Problems of information Transmission **20**(1984), 35-42.

[4] Katsman, G.L., Tfasman, M.A. and Vlădut, S.G., *Modular Curves and Codes with a polynomial construction*, IEEE Transactions on Information Theory, **30**(1984), 353-355.

[5] Lachaud, G., *Les codes géométriques de Goppa*, Séminaire Bourbaki 1984/85, exp. 641, Astj'erisque **133-134**, 189-207.

[6] Lachaud, G., *Projective Reed-Muller codes*, Coding Theory and Applications, Proc. 2nd Int. Coll. Paris 1986, Lect. Notes in Comp. Sci. **311** (1988), 125-129.

[7] Lachaud, G., *The parameters of projective Reed-Muller codes*, Discrete Math. **81** (1990), 1-5.

[8] McWilliams, F.J., Sloane, N.J.A., *The Theory of Error-Correcting Codes*, North-Holland, Amsterdam, 1977.

[9] Manin, Yu. I., Vlădut, S.G., *Codes linéaires et Courbes Modulaires*, Itogi Nauki i Tekhniki **25**(1984), 209-257; J. Soviet Math. **30**(1985), 2611-1643.

[10] Tsfasman, M.A., Vlădut, S.G., Zink, Th., *Modular Curves, Shimura Curves and Goppa Codes better than Varshamov-Gilbert bound*, Math. Nachr. **109**(1982), 21-28.

[11] Tsfasman, M.A., Vlădut, S.G., Algebraico-geometric Codes, Kluwer Acad. Pub., Dordrecht (1991).

Welch's Bound and Sequence Sets for Code-Division Multiple-Access Systems

James L. Massey and Thomas Mittelholzer

Signal and Information Processing Laboratory
Swiss Federal Institute of Technology
CH-8092 Zürich

Abstract

Welch's bound for a set of M complex equi-energy sequences is considered as a lower bound on the sum of the squares of the magnitudes of the inner products between all pairs of these sequences. It is shown that, when the sequences are binary (±-1 valued) sequences assigned to the M users in a synchronous code-division multiple-access (S-CDMA) system, precisely such a sum determines the sum of the variances of the interuser interference seen by the individual users. It is further shown that Welch's bound, in the general case, holds with equality if and only if the array having the M sequences as rows has orthogonal and equi-energy columns. For the case of binary (±-1 valued) sequences that meet Welch's bound with equality, it is shown that the sequences are uniformly good in the sense that, when used in a S-CDMA system, the variance of the interuser interference is the same for all users. It is proved that a sequence set corresponding to a binary linear code achieves Welch's bound with equality if and only if the dual code contains no codewords of Hamming weight two. Transformations and combination of sequences sets that preserve equality in Welch's bound are given and used to illustrate the design and analysis of sequence sets for non-synchronous CDMA systems.

1 Introduction

The aim of this paper is to show the central role played by Welch's bound [1] in the synthesis and/or analysis of sequence sets or multi-sets (in which the same sequence can appear more than once) for use in code-division multiple-access (CDMA) systems.

Section 2 motivates the interest in the sum of the squares of the magnitudes of the inner products between the sequences in a sequence (multi-)set by showing that this sum determines the sum of the variances of the interuser interference experienced by the individual users in a synchronous CDMA (S-CDMA) system. Welch's bound on the sum of the magnitudes of the squares of the inner products between the sequences in a complex sequence (multi-)set is introduced in Section 3 where the apparently new necessary and sufficient condition for equality is derived. It is further shown that (multi-)sets of equi-energy sequences that achieve Welch's bound with equality enjoy an interesting "uniformly good" property that, for the S-CDMA case, implies that the variance of the interuser interference experienced by a user is the same for all users. Section 4 treats the construction of sequence sets from binary linear codes and gives the necessary and sufficient condition for such a sequence set to achieve equality in Welch's bound. In Section 5, transformations and combination of sequence (multi-)sets are considered that preserve equality in Welch's bound and that are useful in synthesizing and/or analyzing sequence (multi-)sets for use in non-synchronous CDMA systems. Section 6 contains some concluding remarks.

2 Synchronous CDMA Systems

In most spread-spectrum multiple-access systems of the CDMA type, each of the users, say user i, is assigned a binary (±1 valued) *spreading sequence* of length L, say

$$x^{(i)} = [x_1^{(i)}, x_2^{(i)}, \dots, x_L^{(i)}],$$

where L is the *spreading factor* of the spread-spectrum system. The binary (±1 valued) data sequence

$$\dots, B_{-1}^{(i)}, B_0^{(i)}, B_1^{(i)}, \dots$$

of user i is then expanded to a binary (±1 valued) sequence at L times the original data rate by using the original data symbols to control the polarity of the spreading sequence $x^{(i)}$. Each component of the expanded binary (±1 valued) data symbol

$$B_k^{(i)} x^{(i)} = [B_k^{(i)} x_1^{(i)}, B_k^{(i)} x_2^{(i)}, \dots, B_k^{(i)} x_L^{(i)}]$$

is called a *chip*. In this manner, one creates the binary (±1 valued) sequence

$$\dots, B_{-1}^{(i)} x^{(i)}, B_0^{(i)} x^{(i)}, B_1^{(i)} x^{(i)}, \dots$$

that forms the input to the modulator of user i. This modulator might, for instance, be a binary phase-shift-keyed modulator, and all users would use the same carrier frequency.

In a synchronous CDMA (S-CDMA) system, all users are in exact synchronism (relative to the receiver) in the sense that not only are their carrier frequencies and phases the same, but also their expanded data symbols are aligned in time. With the usual assumption of additive white Gaussian noise, this implies that the demodulator output for the k-th data symbol interval can be written as the L-chip sequence

$$r_k = \sum_{j=1}^{M} B_k^{(j)} x^{(j)} + n_k \tag{2.1}$$

where the L-chip noise sequence

$$n_k = [n_{k1}, n_{k2}, \dots, n_{kL}],$$

which is independent of the data symbols, is a sequence of independent Gaussian

64

random variables, each with mean 0 and variance $1/\gamma$, where γ is the signal-to-noise ratio defined as the received energy of an expanded data symbol divided by the two-sided noise power spectral density of the additive white Gaussian noise.

In a *conventional CDMA receiver*, the sequence r_k is further processed separately for each user in the manner that, say for user i, r_k is *matched-filtered* (or "correlated") with the spreading sequence $x^{(i)}$ of that user to produce the detection statistic $S_k^{(i)}$ for the data symbol $B_k^{(i)}$. Mathematically, matched filtering is just the operation of computing the inner product so that

$$S_k^{(i)} = \langle r_k, x^{(i)} \rangle = \sum_{l=1}^{L} r_{kl} \, x_l^{(i)}$$

where

$$r_k = [r_{k1}, r_{k2}, \dots, r_{kL}].$$

With the help of (2.1) and the fact that

$$\langle x^{(j)}, x^{(j)} \rangle = L \qquad (2.2)$$

for all j, the data symbol detection statistic for user i becomes

$$S_k^{(i)} = L \, B_k^{(i)} + \sum_{\substack{j=1 \\ j \neq i}}^{M} B_k^{(j)} \langle x^{(j)}, x^{(i)} \rangle + \eta_k^{(i)}, \qquad (2.3)$$

where

$$\eta_k^{(i)} = \langle n_k, x^{(i)} \rangle = \sum_{l=1}^{L} n_{kl} \, x_l^{(i)}$$

is a Gaussian random variable with mean 0 and variance L/γ that is independent of the data symbols. Because the data symbols of the M users are themselves statistically independent and each has mean 0 and variance 1, the sum

$$\xi_k^{(i)} = \sum_{\substack{j=1 \\ j \neq i}}^{M} B_k^{(j)} \langle x^{(j)}, x^{(i)} \rangle \qquad (2.4)$$

in (2.3), which represents the *interuser interference experienced by user i*, has mean 0 and variance

$$\sigma^2(i) = \sum_{\substack{j=1 \\ j \neq i}}^{M} |\langle x^{(j)}, x^{(i)} \rangle|^2. \qquad (2.5)$$

65

It is common in the analysis of CDMA systems to make the so-called *Gaussian assumption* that the interuser interference experienced by user i is a Gaussian random variable; the central-limit theorem ensures that this assumption is generally valid when the number M of users is large.

Summarizing, we have seen that, in a conventional S-CDMA system, the detection statistic for the data symbol $B_k^{(i)}$ of user i can be written as

$$S_k^{(i)} = L\,B_k^{(i)} + \xi_k^{(i)} + \eta_k^{(i)} \tag{2.6}$$

where $\xi_k^{(i)}$ and $\eta_k^{(i)}$, under the Gaussian assumption, are independent zero-mean Gaussian random variables. The noise $\eta_k^{(i)}$ has variance L/γ, while the interuser interference $\xi_k^{(i)}$ has variance given by (2.5), which we write here in the more convenient form

$$\sigma^2(i) = \sum_{j=1}^{M} |\langle x^{(j)}, x^{(i)}\rangle|^2 - L^2. \tag{2.7}$$

It follows that the sequence design problem for S-CDMA, when a conventional receiver is used and when the system is judged by the *worst interuser interference* σ^2_{wc} *experienced by any user,* can be phrased as follows:

Problem 1: Choose the binary (±1 valued) sequences $x^{(1)}, x^{(2)}, \ldots, x^{(M)}$ of length L to minimize

$$\sigma^2_{wc} = \max_i \sigma^2(i) = \max_i \sum_{j=1}^{M} |\langle x^{(j)}, x^{(i)}\rangle|^2 - L^2. \tag{2.8}$$

The optimally fair solution to Problem 1 will result from a solution, when it exists, to the following problem.

Problem 2: Choose the binary (±1 valued) sequences $x^{(1)}, x^{(2)}, \ldots, x^{(M)}$ of length L to minimize

$$\sigma^2_{TOT} = \sum_{i=1}^{M}\sum_{j=1}^{M} |\langle x^{(j)}, x^{(i)}\rangle|^2 - M\,L^2 \tag{2.9}$$

over all choices of such sequences and then (if possible) to satisfy the further condition that, for $1 \le i \le M$,

$$\sigma^2(i) = \sum_{j=1}^{M} |\langle x^{(j)}, x^{(i)}\rangle|^2 - L^2 = \frac{1}{M}\,\sigma^2_{TOT}. \tag{2.10}$$

It is primarily this second problem that we will address in the remainder of this paper. To place our later results into better perspective, we first consider here the trivial case where the interuser interference can be totally eliminated.

66

Condition for No Interuser Interference: The binary (±1 valued) length L sequences $x^{(1)}, x^{(2)}, ... , x^{(M)}$ give $\sigma^2_{TOT} = 0$ [and hence also $\sigma^2(i) = 0$ for $1 \leq i \leq M$] if and only if $x^{(1)}, x^{(2)}, ... , x^{(M)}$ are orthogonal, i.e., if and only if

$$\langle x^{(j)}, x^{(i)} \rangle = 0, \text{ all } i \neq j.$$

Proof:
$$\sum_{i=1}^{M} \sum_{j=1}^{M} |\langle x^{(j)}, x^{(i)} \rangle|^2 \geq \sum_{i=1}^{M} |\langle x^{(i)}, x^{(i)} \rangle|^2 = M L^2$$

with equality if and only if $x^{(1)}, x^{(2)}, ... , x^{(M)}$ are orthogonal. ❑

It follows immediately that $\sigma^2_{TOT} = 0$ is possible only when $M \leq L$, since there can be at most L orthogonal non-zero sequences of length L. Thus, $\sigma^2_{TOT} = 0$ when M = L is possible if and only if the $L \times L$ matrix having $x^{(1)}, x^{(2)}, ... , x^{(M)}$ as rows is a *Hadamard matrix* [2, p. 129]. Except for the trivial cases L = 1 and L = 2, Hadamard matrices exist only when L is divisible by 4; they may exist for all L divisible by 4 but this has not been proved. Hadamard matrices are known to exist whenever L is a power of 2 [2, p. 130-131]. However, the case $M \leq L$ is not of real interest in S-CDMA systems. The motivation for attaining the complete synchronization that characterizes an S-CDMA system is that this should allow the system to accomodate many more users than one can tolerate with non-synchronous CDMA.

3 Welch's Bound

The starting point for our finding solutions to Problem 2 above will be the bound on the sum of the squares of the magnitudes of inner products given by Welch in 1974 [1]. Because this bound applies generally to sequences with complex components and because the general case is as easy to treat as the special case of sequences with ±1 components, we will hereafter allow the sequences

$$x^{(i)} = [x_1^{(i)}, x_2^{(i)}, ... , x_L^{(i)}]$$

to be in C^L, the vector space of L-tuples over the complex field C with the inner product defined as

$$\langle x^{(j)}, x^{(i)} \rangle = \sum_{k=1}^{L} x_k^{(j)} x_k^{(i)*} \tag{3.1}$$

where the asterisk denotes complex conjugation.

We first derive an elementary property of squares of inner products in C^L that is the key not only to a simple derivation of Welch's bound, but also and more interestingly to a recognition of when equality holds in that bound.

Lemma 1: (Row-Column Equivalence) Let $y^{(1)}, y^{(2)}, \ldots, y^{(L)}$ denote the columns of the $M \times L$ array whose rows are the sequences $x^{(1)}, x^{(2)}, \ldots, x^{(M)}$ in C^L, then

$$\sum_{i=1}^{M} \sum_{j=1}^{M} |\langle x^{(j)}, x^{(i)} \rangle|^2 = \sum_{k=1}^{L} \sum_{l=1}^{L} |\langle y^{(l)}, y^{(k)} \rangle|^2. \tag{3.2}$$

Proof: Because $\langle x^{(i)}, x^{(j)} \rangle = \langle x^{(j)}, x^{(i)} \rangle^*$, we have

$$\sum_{i=1}^{M} \sum_{j=1}^{M} |\langle x^{(j)}, x^{(i)} \rangle|^2 = \sum_{i=1}^{M} \sum_{j=1}^{M} \langle x^{(j)}, x^{(i)} \rangle \langle x^{(i)}, x^{(j)} \rangle$$

$$= \sum_{i=1}^{M} \sum_{j=1}^{M} \sum_{k=1}^{L} x_k^{(j)} x_k^{(i)*} \sum_{l=1}^{L} x_l^{(i)} x_l^{(j)*}$$

$$= \sum_{k=1}^{L} \sum_{l=1}^{L} \sum_{i=1}^{M} x_l^{(i)} x_k^{(i)*} \sum_{j=1}^{M} x_k^{(j)} x_l^{(j)*}$$

$$= \sum_{k=1}^{L} \sum_{l=1}^{L} \langle y^{(l)}, y^{(k)} \rangle \langle y^{(k)}, y^{(l)} \rangle$$

$$= \sum_{k=1}^{L} \sum_{l=1}^{L} |\langle y^{(l)}, y^{(k)} \rangle|^2,$$

where we have used the fact that the column vector $y^{(k)}$ is just

$$y^{(k)} = (x_k^{(1)}, x_k^{(2)}, \ldots, x_k^{(M)}). \quad \square$$

We will also have need for the following simple result.

Lemma 2: If $a_1, a_2, \ldots, a_L$ are real numbers, then

$$\sum_{k=1}^{L} (a_k)^2 \geq \frac{1}{L} \left(\sum_{k=1}^{L} a_k \right)^2 \tag{3.3}$$

with equality if and only if $a_1 = a_2 = \ldots = a_L$.

Proof: Consider a random variable X that takes on the value a_k with probability $1/L$ for $1 \leq k \leq L$. Because the square function x^2 is strictly convex-$\cup$ on the whole real line, it follows from Jensen's inequality that

$$E[X^2] = \sum_{k=1}^{L} \frac{1}{L}(a_k)^2 \geq E[X]^2 = (\sum_{k=1}^{L} \frac{1}{L} a_k)^2$$

with equality if and only if $a_1 = a_2 = ... = a_L$. Multiplying both sides of this inequality by L gives the lemma. □

We are now ready for the proof of Welch's bound.

Welch's Bound: If $x^{(1)}, x^{(2)}, ... , x^{(M)}$ are sequences in C^L and all have the same "energy" L, i. e., if

$$\| x^{(i)} \|^2 = \langle x^{(i)}, x^{(i)} \rangle = L \tag{3.4}$$

for $1 \leq i \leq M$, then

$$\sum_{i=1}^{M} \sum_{j=1}^{M} |\langle x^{(j)}, x^{(i)} \rangle|^2 \geq M^2 L \tag{3.5}$$

with equality if and only if the *columns* $y^{(1)}, y^{(2)}, ... , y^{(L)}$ of the $M \times L$ array whose rows are $x^{(1)}, x^{(2)}, ... , x^{(M)}$ are *orthogonal* and all columns have the same energy, i.e.,

$$\| y^{(k)} \|^2 = M \tag{3.6}$$

for $1 \leq k \leq L$.

Remark: Welch's bound was originally stated, and is usually treated, as a lower bound on the maximum value of $|\langle x^{(j)}, x^{(i)} \rangle|$ for $i \neq j$. This form of the bound is easily obtained from (3.5), but it seems to us that Welch's bound is more fundamentally a bound on the sum of the squares of the magnitudes of the inner products between the sequences. The condition for equality in (3.5) appears not to have been given previously.

Proof:

$$\sum_{k=1}^{L} \sum_{l=1}^{L} |\langle y^{(l)}, y^{(k)} \rangle|^2 \geq \sum_{k=1}^{L} |\langle y^{(k)}, y^{(k)} \rangle|^2 = \sum_{k=1}^{L} (\| y^{(k)} \|^2)^2$$

with equality if and only if $y^{(1)}, y^{(2)}, ... , y^{(L)}$ are orthogonal. But Lemma 2 now gives

$$\sum_{k=1}^{L} (\| y^{(k)} \|^2)^2 \geq \frac{1}{L}(\sum_{k=1}^{L} \| y^{(k)} \|^2)^2 = \frac{1}{L}(\sum_{k=1}^{L} \sum_{i=1}^{M} |x_k^{(i)}|^2)^2 = \frac{1}{L}(\sum_{i=1}^{M} \sum_{k=1}^{L} |x_k^{(i)}|^2)^2$$

$$= \frac{1}{L}(\sum_{i=1}^{M} \| x^{(i)} \|^2)^2 = \frac{1}{L}(M L)^2 = M^2 L$$

where equality holds if and only if $\| y^{(k)} \|^2$ has the same value for $1 \leq k \leq L$ (and hence

69

this value must be M.) ◻

We now show a rather surprising consequence of the situation when equality holds in (3.5), i.e., when the *columns* $y^{(1)}, y^{(2)}, ... , y^{(L)}$ of the $M \times L$ array whose rows are the equi-energy sequences $x^{(1)}, x^{(2)}, ... , x^{(M)}$ are *orthogonal*.

Proposition 1: (The Uniformly-Good Property) If $x^{(1)}, x^{(2)}, ... , x^{(M)}$ are sequences in C^L such that $\| x^{(i)} \|^2 = L$ for $1 \le i \le M$ and such that equality holds in (3.5), then

$$\sum_{j=1}^{M} |\langle x^{(j)}, x^{(i)} \rangle|^2 = M L$$

for $1 \le i \le M$.

Proof:
$$\sum_{j=1}^{M} |\langle x^{(j)}, x^{(i)} \rangle|^2 = \sum_{j=1}^{M} \langle x^{(j)}, x^{(i)} \rangle \langle x^{(i)}, x^{(j)} \rangle = \sum_{j=1}^{M} \sum_{k=1}^{L} x_k^{(j)} x_k^{(i)*} \sum_{l=1}^{L} x_l^{(i)} x_l^{(j)*}$$

$$= \sum_{k=1}^{L} \sum_{l=1}^{L} x_k^{(i)*} x_l^{(i)} \sum_{j=1}^{M} x_k^{(j)} x_l^{(j)*} = \sum_{k=1}^{L} \sum_{l=1}^{L} x_k^{(i)*} x_l^{(i)} \langle y^{(k)}, y^{(l)} \rangle.$$

But equality in (3.5) implies $\langle y^{(k)}, y^{(l)} \rangle = 0$ for all $k \ne l$ and $\langle y^{(k)}, y^{(l)} \rangle = M$ for $k = l$ so that

$$\sum_{j=1}^{M} |\langle x^{(j)}, x^{(i)} \rangle|^2 = \sum_{k=1}^{L} x_k^{(i)*} x_k^{(i)} M = \| x^{(i)} \|^2 M = L M. ◻$$

We next consider the special case when all components of all sequences have unit magnitude, i.e., when $|x_k^{(i)}| = 1$ for $1 \le i \le M$ and $1 \le k < L$. [We note that this special case includes the case of interest for CDMA systems where the sequences have ± 1 components.] For this special case, equalities (3.4) and (3.6) are automatically fulfilled so that Welch's bound simplifies as follows.

Welch's Bound for Sequences with Unit-Amplitude Components: If $x^{(1)}, x^{(2)}, ... , x^{(M)}$ are sequences in C^L such that $|x_k^{(i)}| = 1$ for $1 \le i \le M$ and $1 \le k < L$, then

$$\sum_{i=1}^{M} \sum_{j=1}^{M} |\langle x^{(j)}, x^{(i)} \rangle|^2 \ge M^2 L \tag{3.7}$$

with equality if and only if the *columns* $y^{(1)}, y^{(2)}, ... , y^{(L)}$ of the $M \times L$ array whose rows are $x^{(1)}, x^{(2)}, ... , x^{(M)}$ are *orthogonal*. Moreover (by Proposition 1), if equality holds in (3.7), then

$$\sum_{j=1}^{M} |\langle x^{(j)}, x^{(i)} \rangle|^2 = M L \tag{3.8}$$

for $1 \le i \le M$.

4 Optimum S-CDMA Sequence Sets from Linear Codes

We will call the binary (±1 valued) sequences $x^{(1)}$, $x^{(2)}$, ... , $x^{(M)}$ of length L a *Welch-Bound-Equality* (WBE) sequence set (or sequence multi-set if these M sequences are not all distinct) if equality holds in (3.7) or, equivalently from (2.9), if

$$\sigma^2_{TOT} = M L (M - L). \tag{4.1}$$

It follows further from Proposition 1 that, for a WBE sequence (multi-)set, the variance of the interuser interference experienced by user i is

$$\sigma^2(i) = L (M - L) \tag{4.2}$$

for $1 \le i \le M$. In other words, *WBE sequence (multi-)sets are optimal for S-CDMA systems in that they provide a solution to Problem 2* that was formulated in Section 2. In this section, we will give a simple, but powerful, construction of WBE sequence sets based on linear error-correcting codes. First, we note that the number M of sequences in a WBE sequence (multi-)set must be even, because the parity of $\langle y^{(k)}, y^{(l)} \rangle$ equals the parity of M and hence cannot vanish for $k \ne l$ unless M is even (where here and hereafter we exclude the trivial case where L = 1). Moreover, because L non-zero vectors can be orthogonal only if their length M is at least L, it follows that equality in (3.7) is possible only when $M \ge L$. Thus, for a given L, we see that it will generally be easier to satisfy the orthogonality condition on $y^{(1)}$, $y^{(2)}$, ... ,$y^{(L)}$ as M becomes larger. Thus, the minimization of the variance of interuser interference when $M \ge L$ is quite the reverse of the problem of complete elimination of interuser interference that was discussed at the end of Section 2.

With a binary (±1 valued) sequence $x = [x_1, x_2, ... , x_L]$, we can and will associate a binary (GF(2) valued) sequence $b = [b_1, b_2, ... ,b_L]$ where b_k is 0 or 1 according as x_k is +1 or -1, respectively. In this manner, we can and will associate to any set of vectors in $GF(2)^L$ a corresponding set of binary (±1 valued) sequences of length L. For ease of later reference, we note here that if x and x' are any two binary (±1 valued) sequences of length L and if b and b' are the corresponding vectors in $GF(2)^L$, then

$$\langle x', x \rangle = \sum_{k=1}^{L} x_k' x_k = L - 2 d(b', b) \tag{4.3}$$

where d(.,.) denotes the *Hamming distance* between the indicated vectors, i.e., the number of components in which these vectors differ.

We recall that a binary (L, K) *linear code* V is just a K-dimensional subspace of $GF(2)^L$ considered as a vector space of dimension L over the finite field GF(2) and that such a code contains 2^K codewords [2, p. 40]. We recall also that the *dual code* $V^\perp$ is the set of all $b' = [b_1', b_2', ... ,b_L']$ in $GF(2)^L$ such that $b_1 b_1' + b_2 b_2' + ... + b_L b_L' = 0$ for all $b = [b_1, b_2, ... ,b_L]$ in V, and that this dual code is a binary linear (L, L - K) code [2, p. 44].

We now characterize completely the WBE sequence sets corresponding to linear codes.

71

Proposition 2: The binary (±1 valued) sequence set corresponding to the binary linear code V is a WBE sequence set if and only if the *dual code* $V^\perp$ contains no codewords of Hamming weight two, i. e., with exactly two non-zero components.

Proof: Consider any positions k and l, $1 \le k < l \le L$. We will determine the condition such that columns $y^{(k)}$ and $y^{(l)}$ of the array with rows $x^{(1)}, x^{(2)}, \dots , x^{(M)}$ are *not* orthogonal when this sequence set corresponds to the binary linear code V. Now the subset of codewords $b = [b_1, b_2, \dots ,b_L]$ in V such that $b_k = b_l$ or, equivalently, such that $b_k + b_l = 0$ is a subspace U of V, where we note that the codeword $0 = [0, 0, \dots ,0]$ is always in this subspace U. If $U \ne V$, then this subspace has a single coset in V distinct from itself, namely the set of all codewords $b = [b_1, b_2, \dots ,b_L]$ such that $b_k \ne b_l$ or, equivalently, such that $b_k + b_l = 1$. Thus, if $U \ne V$, $y^{(k)}$ and $y^{(l)}$ will disagree in exactly half of their components and hence $\langle y^{(k)}, y^{(l)} \rangle = 0$. But if $U = V$, then $y^{(k)}$ and $y^{(l)}$ will agree in all their components and hence $\langle y^{(k)}, y^{(l)} \rangle = M \ne 0$. Thus $y^{(k)}$ and $y^{(l)}$ will not be orthogonal for all $k \ne l$ if and only if, for some $k \ne l$, $b_k + b_l = 0$ in all codewords **b**. But this latter condition is just the condition that the dual code contains the weight two vector whose 1's are in positions k and l. □

The interesting WBE sequence sets specified by Proposition 2 are those of the following corollary, whose truth follows from the fact that the *minimum distance* of a linear code is equal to the minimum Hamming weight of its non-0 codewords [2, p. 41].

Corollary to Proposition 2: The binary (±1 valued) sequence set corresponding to the binary linear code V is a WBE sequence set if the minimum distance $d^\perp$ of the dual code $V^\perp$ is at least three.

The dual code $V^\perp$ of a linear code V contains a word of Hamming weight one whose 1 is in position k if and only if $b_k = 0$ in all codewords $b = [b_1, b_2, \dots ,b_L]$ of V, i.e., if and only if the k-th component of the codewords in V is *idle*. If $V^\perp$ contains no weight two codewords, then, because $V^\perp$ is also a linear code, it can contain at most one codeword of Hamming weight one. Moreover, deleting the corresponding idle component from all codewords of V will then give a linear code V' whose dual code $V'^\perp$ contains no codewords with Hamming weights one or two, and thus, unless the dual code is the linear dual code $V'^\perp$ has minimum distance $d^\perp \ge 3$. It follows that *the Corollary to Proposition 2 actually gives all the linear codes corresponding to WBE sequence sets except for the trivial generalization to linear codes obtained by inserting exactly one idle component into all the codewords of one of the former codes.*

Recall from the discussion in Section 2 that the i-th user in a CDMA system will transmit in every expanded data symbol period either his spreading sequence $+x^{(i)}$ or its negative $-x^{(i)}$ according as his corresponding data bit is +1 or -1, respectively. Thus, it is often desired that no sequence in a CDMA sequence set be the negative of another sequence, i. e., that $x^{(i)} \ne -x^{(j)}$ for all $i \ne j$ or, equivalently for the corresponding sequences in $GF(2)^L$, $b^{(i)} \ne b^{(j)} + 1$ for all $i \ne j$ where $1 = [1, 1, \dots , 1]$ is the all-one vector. We will call a sequence set $x^{(1)}, x^{(2)}, \dots , x^{(M)}$ with the property that $x^{(i)} \ne -x^{(j)}$

for all i ≠ j a *unipolar* sequence set. The following characterization is immediate.
Characterization of Unipolar Sequence Sets Corresponding to Linear Codes: The binary (±1 valued) sequence set correponding to a binary linear code V is unipolar if and only if the all-one word $\mathbf{1}$ = [1, 1, ... , 1] is *not* a codeword in V.

We illustrate the ideas of this section with two simple examples.

Example 1: Let **v** be a binary *maximal-length sequence* (or *m-sequence* or *pseudo-noise sequence*) of length $L = 2^m - 1$ where $m \geq 2$ [2, p. 222]. Let T denote the left cyclic shift operator. Then $\mathbf{0}$, **v**, T(**v**), ... , T^{L-1}(**v**) are the codewords in a binary linear code V (for which $\mathbf{1}$ is not a codeword) with minimum distance $d = 2^{m-1}$ whose dual code $V^{\perp}$ is a Hamming code with minimum distance $d^{\perp} = 3$ [2, p. 223]. It follows from the Corollary that the binary (±1 valued) sequence set corresponding to V is a unipolar WBE sequence set of M = L + 1 sequences. It follows further from (4.2) that, when this sequence set is used in an S-CDMA system, the interuser interference experienced by user i has variance

$$\sigma^2(i) = L \tag{4.4}$$

for $1 \leq i \leq M$. [We note that the results in this simple example could also have been obtained by conventional arguments. The code V is an equidistant code in the sense that the Hamming distance between any of its two codewords is the same [2, p. 223], namely 2^{m-1}. This implies from (4.3) that

$$\langle x^{(j)}, x^{(i)} \rangle = L - 2 \cdot 2^{m-1} = -1 \tag{4.5}$$

for all i ≠ j, which then with the aid of the definition (2.5) gives (4.4).]

Example 2: Let α be a primitive element of the finite field $GF(2^4)$ [2, p. 158] and let V be the (L = 15, K = 6) Bose-Chaudhuri-Hocquenghem (BCH) code such that $\alpha^0 = 1$, α, and α^3 are zeroes of the generator polynomial g(X) of this code [2, p. 271]. This BCH code has minimum distance d = 6 [2, Appendix D] and the all-one vector $\mathbf{1}$ is not a codeword in this code. The dual code $V^{\perp}$ is a (15, 9) cyclic code with minimum distance $d^{\perp} = 3$ [2, Appendix D]. It follows from the Corollary that the binary (±1 valued) sequence set corresponding to V is a unipolar WBE sequence set of M = 64 sequences of length 15. It follows further from (4.2) that, when this sequence set is used in a S-CDMA system, the interuser interference experienced by user i has variance

$$\sigma^2(i) = 15 (64 - 15) = 735 \tag{4.6}$$

for $1 \leq i \leq M$.

5 WBE-Preserving Transformation and Combination of Sequence Sets

We first consider operations on a sequence (multi-)set that preserve the WBE property.

Proposition 3: If the binary (±1 valued) sequences $x^{(1)}, x^{(2)}, \ldots, x^{(M)}$ of length L form a WBE sequence set or multi-set, then the sequence set or multi-set obtained by performing any of the following operations on the former set or multiset is also a WBE sequence set or multi-set:

(i) Replacing $x^{(i)}$ by $-x^{(i)}$ for any i;

(ii) Replacing $x^{(i)} = [x_1^{(i)}, x_2^{(i)}, \ldots, x_L^{(i)}]$ by its left cyclic shift
$T(x^{(i)}) = [x_2^{(i)}, \ldots, x_L^{(i)}, x_1^{(i)}]$ for $1 \le i \le M$;

(iii) Replacing $x^{(i)} = [x_1^{(i)}, x_2^{(i)}, \ldots, x_L^{(i)}]$ by its left *negacyclic shift*
$N(x) = [x_2^{(i)}, \ldots, x_L^{(i)}, -x_1^{(i)}]$ for $1 \le i \le M$;

(iv) Deleting the k-th component $x_k^{(i)}$ of $x^{(i)}$ for $1 \le i \le M$ and for any k; and

(v) Replacing $x^{(i)}$, which corresponds to $b^{(i)}$ in $GF(2)^L$, by the binary (±1 valued) sequence corresponding to $b^{(i)} + u$ for $1 \le i \le M$ and any u in $GF(2)^L$.

Proof: Operation (i) changes only the sign of $\langle x^{(j)}, x^{(i)} \rangle$ for all $j \ne i$ and hence does not alter the sum in (3.8). Operation (ii) causes the columns of the $M \times L$ array whose rows are $x^{(1)}, x^{(2)}, \ldots, x^{(M)}$ to be cyclically shifted, but this does not alter the orthogonality of the columns nor their equi-energy. Operation (iii) additionally changes the sign of one column in this array, but again this does not alter the orthogonality of the columns nor their equi-energy. Operation (iv) deletes one column of this array but again this does not alter the orthogonality nor the equi-energy of the remaining columns. Finally, operation (v) changes only the signs of those columns of the $M \times L$ array whose rows are $x^{(1)}, x^{(2)}, \ldots, x^{(M)}$ corresponding to positions in which u contains a 1, which clearly does not alter the orthogonality of the columns nor their equi-energy. □

Example 3: Adding any vector u in $GF(2)^L$ to the codewords of the linear code V of length $L = 2^m - 1$ considered in Example 1 gives the coset

$$u + V = \{u, u + v, u + T(v), \ldots, u + T^{L-1}(v)\},$$

which by Proposition 3(v) also corresponds to a WBE sequence set S of $M = 2^m$ sequences. We note that if m is not divisible by 4 and if the sequence u is the $(2^e + 1)$-st decimation of the m-sequence v, then u is also an m-sequence and the corresponding binary (±1 valued) sequence set, when augmented with the sequence corresponding to v is a so-called *Gold sequence set* where the name honors the originator of these sequence sets [3]. Note however, that, for any choice of u, the sequence set S is WBE. Note also that the full Gold sequence set cannot be WBE because it contains an odd number of sequences.

The next proposition gives a powerful way of combining smaller WBE sequence sets to produce larger ones.

Proposition 4: If $x^{(1)}, x^{(2)}, \ldots, x^{(M)}$ and $x'^{(1)}, x'^{(2)}, \ldots, x'^{(M')}$ are both WBE sequence (multi-)sets containing M and M', respectively, binary (± 1 valued) sequences of the same length L, then their "union" $x^{(1)}, x^{(2)}, \ldots, x^{(M)}, x'^{(1)}, x'^{(2)}, \ldots, x'^{(M')}$ is a WBE sequence (multi-)set with M + M' sequences.

Proof: Let $y^{(k)}$ and $y'^{(k)}$ denote the k-th columns of the $M \times L$ and the $M' \times L$ arrays whose rows are $x^{(1)}, x^{(2)}, \ldots, x^{(M)}$ and $x'^{(1)}, x'^{(2)}, \ldots, x'^{(M')}$, respectively. Then the k-th column of the $(M + M') \times L$ array for the "union" sequence set is $Y^{(k)} = (y^{(k)}, y'^{(k)})$. Thus,

$$\langle Y^{(l)}, Y^{(k)} \rangle = \langle y^{(l)}, y^{(k)} \rangle + \langle y'^{(l)}, y'^{(k)} \rangle = 0$$

for $k \neq l$ so that the "union" of these two sequence sets in indeed WBE. $\square$

Propositions 3 and 4 are very useful when constructing and/or analyzing sequence sets for various types of non-synchronous CDMA systems. We will illustrate this applicability with an example for so-called *quasi-synchronous CDMA*, which is defined in the same manner as S-CDMA in Section 2 except that there can now be a relative time misalignment of at most one chip between the symbols of any two users. Again we take the worst-case interuser interference experienced by any user over all admissible misalignments as the quantity to be minimized. A rather tedious argument that we will not repeat here shows that, for any user i, the worst interuser interference experienced by that user will occur when his symbol edge is either first or last among those of all M users and those other users are each either in full symbol synchronization with the specified user or exactly one chip misaligned with that user [4]. Assuming that L is large so that we can ignore whether the single chip from an adjacent expanded data symbol that overlaps the expanded data symbol of a specified user corresponds to a data bit with the same or with the opposite sign as that for the adjacent expanded data symbol whose L - 1 chips overlap the same expanded data symbol of the specified user, it follows that the worst interuser interference experienced by user i will have a variance

$$\sigma^2_{wc}(i) = \max_{\Delta \in \{-1,+1\}} \sum_{\substack{j=1 \\ j \neq i}}^{M} \max_{\theta \in \{0, \Delta\}} |\langle T^\theta(x^{(j)}), x^{(i)} \rangle|^2 \tag{5.1}$$

where again T is the (left) cyclic shift operator. The sequence set should thus be chosen to minimize

$$\sigma^2_{wc} = \max_i \sigma^2_{wc}(i). \tag{5.2}$$

Example 4: Let S be the WBE sequence set with $M = L + 1 = 2^m$ sequences corresponding to $u + V$ in Example 3, where u is an arbitrary vector in $GF(2)^L$.

We consider first the sum in (5.1) when the lag/lead parameter Δ is +1. Let T(S) corresponding to T(V) be the set of left cyclic shifts of the sequences in S. By

Proposition 4, the union sequence set $U = S \cup T(S)$ containing 2M sequences is also WBE. For the WBE sequence set U, (3.7) gives

$$2 \sum_{i=1}^{M} \sum_{j=1}^{M} |\langle x^{(j)}, x^{(i)} \rangle|^2 + 2 \sum_{i=1}^{M} \sum_{j=1}^{M} |\langle T(x^{(j)}), x^{(i)} \rangle|^2 = (2M)^2 L, \qquad (5.3)$$

where we have used the fact that $\langle T(x^{(j)}), T(x^{(i)}) \rangle = \langle x^{(j)}, x^{(i)} \rangle$. Now the first double sum on the left in (5.3) is just $M^2 L$, as follows from the fact that S is a WBE sequence set of M sequences. Moreover,

$$\sum_{j=1}^{M} |\langle T(x^{(j)}), x^{(i)} \rangle|^2$$

is independent of i, as follows from (4.3) and the fact that the linear code V is closed under cyclic shifting so that the Hamming distances from any vector in $u + V$ to all the vectors in $T(u + V) = T(u) + T(V) = T(u) + V$ does not depend on the particular choice of the former vector. It thus follows from (5.3) that

$$\sum_{j=1}^{M} |\langle T(x^{(j)}), x^{(i)} \rangle|^2 = M L \qquad (5.4)$$

for $1 \le i \le M$. But, because $\langle x^{(j)}, x^{(i)} \rangle = -1$ for all $i \ne j$ according to (4.5) and because $\langle x^{(j)}, x^{(i)} \rangle$ must have odd parity since L is odd, it follows that

$$\max_{\theta \in \{0, +1\}} |\langle T^{\theta}(x^{(j)}), x^{(i)} \rangle|^2 = |\langle T(x^{(j)}), x^{(i)} \rangle|^2 \qquad (5.5)$$

holds for all $j \ne i$. An entirely similar argument for $\Delta = -1$, which again exploits the fact that V is closed under cyclic shifting, shows that

$$\sum_{j=1}^{M} |\langle T^{-1}(x^{(j)}), x^{(i)} \rangle|^2 = M L \qquad (5.6)$$

and that

$$\max_{\theta \in \{0, -1\}} |\langle T^{\theta}(x^{(j)}), x^{(i)} \rangle|^2 = |\langle T^{-1}(x^{(j)}), x^{(i)} \rangle|^2 \qquad (5.7)$$

also holds for all $j \ne i$. Because of (5.1), (5.4)-(5.7) and the fact that $|\langle T^{-1}(x^{(i)}), x^{(i)} \rangle| = |\langle T(x^{(i)}), x^{(i)} \rangle|$, it follows that

$$\sigma^2_{wc}(i) = M L - |\langle T(x^{(i)}), x^{(i)} \rangle|^2 \qquad (5.8)$$

for $1 \le i \le M$. We see from (5.8) that the worst user will be that user i for which the "autocorrelation" magnitude $|\langle T(x^{(i)}), x^{(i)} \rangle|$ is *smallest*. But, for every i, the fact that L is odd implies

$$1 \leq |\langle T(x^{(i)}), x^{(i)}\rangle| \leq L \tag{5.9}$$

with equality on the left when $x^{(i)}$ corresponds to an m-sequence, as it would, for instance, for that user whose sequence corresponds to the additive vector u when u is an m-sequence (as it is in the Gold sequence set). We conclude then from (5.2), (5.8), (5.9) and the fact that $M = L + 1$ that

$$L \leq \sigma^2_{wc} \leq L^2 + L - 1 \tag{5.10}$$

with equality on the right if u is an m-sequence (and hence if S is the Gold sequence set). It may come as a small surprise to some readers that the Gold sequence set gives the poorest worst-case performance among the sequence sets corresponding to different choices of the vector u. What is more significant is that (5.10) applies for any $L = 2^m - 1$ with $m \geq 2$; there is no requirement that m not be divisible by 4 as is required for the Gold sequence set to exist.

6 Remarks

We have given rather abundant evidence to show the importance of sequence (multi-)sets that achieve equality in Welch's bound and we have shown that such sequence sets are surprisingly easy to construct. There seems no reason, in most CDMA systems, to settle for a sequence (multi-)set that does not achieve equality in Welch's bound.

It may come as a major surprise to some readers that the sequence set corresponding to *any* binary linear code V whose dual code has minimum distance at least 3 corresponds to a WBE sequence set and hence is optimum for use in S-CDMA systems. There is no requirement that the code V have any other special distance properties, as would be required for instance if one attempted to make the "crosscorrelation" magnitude $|\langle x^{(j)}, x^{(i)}\rangle|$ small for all $j \neq i$. Such an attempt requires, by (4.3), that the Hamming distance $d(b^{(j)}, d^{(i)})$ between the corresponding codewords in V be made as near to $L/2$ as possible for all $j \neq i$, or equivalently that the Hamming weights of the non-0 codewords be made as near as possible to $L/2$. The reader may object that if these crosscorrelations are not as nearly uniformly small in magnitude as possible, then the validity of invoking the central-limit theorem to justify the assumption that the interuser interference $\xi_k^{(i)}$ in (2.4) is Gaussian becomes suspect. We would counter by pointing out that, for a given variance, a Gaussian random variable has the maximum possible entropy [5, Section 20.5]. Thus, the channel created for user i by the S-CDMA system and in which $\xi_k^{(i)}$ is an additive noise term actually has its *minimum capacity* when $\xi_k^{(i)}$ is Gaussian. For a given variance of the interuser interference, its non-Gaussianness (if properly exploited) is a virtue, not a vice!

Acknowledgment

The research reported here was supported by the European Space Research and Technology Centre (ESTEC) in Noordwijk, The Netherlands, under ESTEC Contract No. 8696/89/NL/US. The interest and encouragement of Messrs. R. Viola and R. DeGaudenzi of ESTEC are gratefully acknowledged.

References

[1] L. R. Welch, "Lower Bounds on the Maximum Cross Correlation of Signals," *IEEE Trans. on Information Theory*, vol. IT-20, pp. 397-399, May 1974.

[2] W. W. Peterson and E. J. Weldon, Jr., *Error-Correcting Codes* (2nd Ed.). Cambridge, Mass.: M.I.T. Press, 1972.

[3] R. Gold, "Optimal Binary Sequences for Spread-Spectrum Multiplexing," *IEEE Trans. on Information Theory*, vol. IT-13, pp. 619-621, October 1967.

[4] J. L. Massey and T. Mittelholzer, Final Report ESTEC Contract No. 8696/89/NL/US: Technical Assistance for the CDMA Communication System Analysis, Signal and Information Processing Laboratory, Swiss Federal Institute of Technology, Zürich, Switzerland, 19 March 1991.

[5] C. E. Shannon, "A Mathematical Theory of Communication," *Bell System Technical Journal*, vol. 27, pp. 379-423 and pp. 623-656, July and October, 1948.

Average-case interactive communication

Alon Orlitsky

AT&T Bell Laboratories

Abstract

X and Y are random variables. Person P_X knows X, Person P_Y knows Y, and both know the joint probability distribution of the pair (X, Y). Using a predetermined protocol, they communicate over a binary, error-free, channel in order for P_Y to learn X. P_X may or may not learn Y. How many information bits must be transmitted (by both persons) on the average?

At least $H(X|Y)$ bits must be transmitted and $H(X) + 1$ bits always suffice[1]. If the support set of (X, Y) is a cartesian product of two sets, then $H(X)$ bits must be transmitted. If the random pair (X, Y) is uniformly distributed over its support set, then $H(X|Y) + 3\log{(H(X|Y) + 1)} + 15.5$ bits suffice. Furthermore, this number of bits is achieved when P_X and P_Y exchange four messages (sequences of binary bits).

The last two results show that when the *arithmetic* average number of bits is considered: (1) there is no asymptotic advantage to P_X knowing Y in advance; (2) four messages are asymptotically optimum. By contrast, for the *worst-case* number of bits: (1) communication can be significantly reduced if P_X knows Y in advance; (2) it is not known whether a constant number of messages is asymptotically optimum.

1 Introduction

In the following subsections we describe the problem, its background, and the results derived.

---

[1] $H(X)$ is the entropy of X and $H(X|Y)$ is the conditional entropy of X given Y. Entropies are binary.

1.1   The problem

Consider two *communicators*: an *informant* $P_\mathcal{X}$ having a random variable X and a *recipient* $P_\mathcal{Y}$ having a, possibly dependent, random variable Y. Both communicators want the recipient, $P_\mathcal{Y}$, to learn X with no probability of error, whereas the informant, $P_\mathcal{X}$, may or may not learn Y. To that end they communicate over an error-free channel. How many information bits must be transmitted on the average?

This problem is a variation on a scenario considered by El Gamal and Orlitsky [1] where both communicators, not just $P_\mathcal{Y}$, want to learn the other's random variable. As elaborated on below, some of our results either follow from or extend results therein.

We assume that the communicators alternate in transmitting *messages*: finite sequences of bits. Messages are determined by an agreed-upon, deterministic, protocol. A formal definition of protocols for the current model is given in [2]. Essentially, a *protocol for* (X, Y) (i.e., a protocol for transmitting X to a person who knows Y) guarantees that the following properties hold. (1) Separate transmissions: each message is based on the random variable known to its transmitter and on previous messages. (2) Implicit termination: when one communicator transmits a message, the other knows when it ends, and when the last message ends, both communicators know that communication has ended. (3) Correct decision: when communication ends, the recipient, $P_\mathcal{Y}$, knows X.

For every *input* — a possible value assignment for X and Y — the protocol determines a finite sequence of transmitted messages. The protocol is *m-message* if, for all inputs, the number of messages transmitted is at most m. The *average complexity* of the protocol is the expected number of bits it requires both communicators to transmit (expectation is taken over all inputs). $\bar{C}_m(X|Y)$, the *m-message average complexity* of (X, Y), is the minimum average complexity of an m-message protocol for (X, Y). It is the minimum average number of bits transmitted by both communicators using a protocol that never exchanges more than m messages. Since empty messages are allowed, $\bar{C}_m(X|Y)$ is a decreasing function of m bounded below by 0. We can therefore define $\bar{C}_\infty(X|Y)$, the *unbounded-message complexity of* (X, Y), to be the limit of $\bar{C}_m(X|Y)$ as $m \to \infty$. It is the *minimum* number of bits that must be transmitted on the average for $P_\mathcal{Y}$ to know X, even if no restrictions are placed on the number of messages exchanged. In summary,

$$\bar{C}_1(X|Y) \geq \bar{C}_2(X|Y) \geq \bar{C}_3(X|Y) \geq \cdots \geq \bar{C}_\infty(X|Y) .$$

A precise definition of complexities is given in Subsection 1.4. Here, we demonstrate them with the following example.

Example 1   A league has t teams named $1, \ldots, t$. Every week two random teams play each other. The outcome of the match is announced over the radio. All announcements have the same format: "The match between team I and team J was won by team K" where $1 \leq I < J \leq t$ and K is either I or J. The distribution of games and winners is uniform:

$$\Pr(k, (i,j)) = \begin{cases} \frac{1}{n(n-1)} & \text{if } 1 \leq i < j \leq t \text{ and } k \in \{i, j\}, \\ 0 & \text{otherwise.} \end{cases}$$

One day, while $P_\mathcal{Y}$ listens to a match announcement, $P_\mathcal{X}$ grabs the radio from him. Consequently, $P_\mathcal{Y}$ hears the first part of the announcement: "The match between team I and team J" and $P_\mathcal{X}$ hears the second part: "was won by team K."

$P_\mathcal{X}$ and $P_\mathcal{Y}$ agree that $P_\mathcal{Y}$ should know the winner. They are looking for a protocol that, on the average, requires few transmitted bits.

If no interaction is allowed, $P_\mathcal{X}$ has to send a single message enabling $P_\mathcal{Y}$ to uniquely determine the winner. This message is based solely on the winner (for that is all $P_\mathcal{X}$ knows). Let $\sigma(i)$ be the message sent by $P_\mathcal{X}$ when he hears "was won by team i." If the messages $\sigma(i)$ and $\sigma(j)$ are the same for $i \neq j$, then in the event of a match between teams i and j, $P_\mathcal{Y}$ cannot tell who the winner is. Also, if $\sigma(i)$ is a prefix of $\sigma(j)$ for $i \neq j$ then in the event of a match between i and j $P_\mathcal{Y}$ does not know when the message ends. Therefore, the messages $\sigma(1), \ldots, \sigma(t)$ must all be different and none can be a proper prefix of another. Hence the average-case one-way complexity satisfies

$$H(K) \leq \bar{C}_1(K|I, J) \leq H(K) + 1 \ .$$

where the upper bound derives from the one-way protocol where $P_\mathcal{X}$ transmits the Huffman code representation of K. In our case, K is uniformly distributed over $\{1, \ldots, t\}$, hence

$$\log t \leq \bar{C}_1(K|I, J) \leq \log t + 1 \ .$$

A two-message protocol that requires only $\lceil \log \log t \rceil + 1$ bits in the worst case was described in [2]. $P_\mathcal{Y}$ considers the binary representations $I_1, \ldots, I_{\lceil \log t \rceil}$ and $J_1, \ldots, J_{\lceil \log t \rceil}$ of I and J. Since $I \neq J$, there must be a first bit location L where the binary representations differ: $I_L \neq J_L$. Using $\lceil \log \log t \rceil$ bits, $P_\mathcal{Y}$ transmits L. $P_\mathcal{X}$ responds by transmitting K_L –

the L'th bit in the binary representation of the winning team K. The total number of bits exchanged is $\lceil \log \log t \rceil + 1$. It was also shown that no other protocol requires fewer bits in the worst case.

M. Costa [3] used the following protocol to reduce the average number of bits transmitted by P_Y. Sequentially over ℓ, P_Y considers the ℓth bit of I and J. He transmits 0 if $I_\ell = J_\ell$ and transmits 1 if $I_\ell \neq J_\ell$, stopping after the first transmitted 1. Assume, for simplicity, that t is a power of two. For $\ell = 1, \ldots, \log t$, the probability that the first bit location where I and J differ is ℓ is:

$$\Pr(L = \ell) = \frac{t}{t-1} \cdot \frac{1}{2^\ell} .$$

Therefore, the expected number of bits transmitted by P_Y is

$$\frac{t}{t-1} \sum_{l=1}^{\log t} \frac{\ell}{2^\ell} = \frac{t}{t-1} \cdot \frac{2t - \log t - 2}{t} = 2 - \frac{\log t}{t-1}.$$

Including the bit transmitted by P_X, we obtain

$$\bar{C}_2(K|I,J) \leq 3 - \frac{\log t}{t-1} . \qquad \qquad \square$$

Several other examples are described in [1]. They are concerned with communicators P_X and P_Y who want to learn each other's random variable, but can be easily modified to our case where only P_Y wants to learn X. In all these examples, however, a single message is optimum: it achieves the minimum number of bits. In the last example, a single message requires $\lceil \log t \rceil$ bits on the average while two messages require at most three bits.

1.2   Results

In Section 2, we prove that for all (X, Y) pairs,

$$H(X|Y) \leq \bar{C}_\infty(X|Y) \leq \ldots \leq \bar{C}_1(X|Y) \leq H(X) + 1 \qquad (1)$$

where $H(X)$ is the binary entropy of X and $H(X|Y)$ is the conditional binary entropy of X given Y. These bounds are not tight as $H(X|Y)$ can be much smaller than $H(X)$. However, Sections 3 and 4 show that they are the tightest expressible in terms of Shannon entropies.

The *support set* of (X, Y) is

$$S_{X,Y} \stackrel{\text{def}}{=} \{(x, y) : p(x, y) > 0\} ,$$

82

the set of all possible inputs. The support set is a *cartesian product* if $S_{X,Y} = A \times B$ for some sets A and B. In Section 3 we show that if $S_{X,Y}$ is a cartesian product then the upper bound in (1) is tight:

$$H(X) \leq \bar{C}_\infty(X|Y) \leq \ldots \leq \bar{C}_1(X|Y) \leq H(X) + 1 . \tag{2}$$

The pair (X, Y) is *uniformly distributed* (or *uniform*) if all possible values of (X, Y) are equally likely:

$$p(x, y) = \begin{cases} \frac{1}{|S_{X,Y}|} & \text{if } (x, y) \in S_{X,Y}, \\ 0 & \text{if } (x, y) \notin S_{X,Y}. \end{cases}$$

In Section 4 we show that whenever (X, Y) is uniformly distributed, the lower bound of Inequality (1) can almost be achieved:

$$H(X|Y) \leq \bar{C}_\infty(X|Y) \leq \ldots \leq \bar{C}_4(X|Y) \leq H(X|Y) + 3 \log \left(H(X|Y) + 1\right) + 15.5. \tag{3}$$

When (X, Y) is uniformly distributed, the average number of bits transmitted is simply the number of bits transmitted for an input (x, y), *arithmetically* averaged over all inputs in $S_{X,Y}$. The last inequality shows that whenever this arithmetic average number of bits is considered:

1. No asymptotic advantage is gained when P_X knows Y in advance: roughly the same number of bits is required whether or not P_X knows Y before communication begins.

2. While Example 1 showed that one message may require arbitrarily more bits than the minimum necessary, four messages require only negligibly more bits than the minimum.

We note that, with more complicated proofs, the constants 3 and 13.5 in Inequality (3) can be reduced. The following example illustrates Inequalities (2) and (3) in two simple cases.

Example 2 (Modified from [1] to fit our model.) Consider two random pairs.

1. (X_1, Y_1) is distributed over $\{1, \ldots, n\} \times \{1, \ldots, n\}$ according to

$$p_1(x, y) \stackrel{\text{def}}{=} \begin{cases} \frac{1}{n} & \text{if } x = y, \\ 0 & \text{if } x \neq y. \end{cases}$$

2. Let $\epsilon > 0$ be arbitrarily small. (X_2, Y_2) is distributed over $\{1, \ldots, n\} \times \{1, \ldots, n\}$ according to

$$p_2(x, y) \stackrel{\text{def}}{=} \begin{cases} \frac{1-\epsilon}{n} & \text{if } x = y, \\ \frac{\epsilon}{n(n-1)} & \text{if } x \neq y. \end{cases}$$

	1	2	3	.	n
1	$\frac{1}{n}$	0	0	.	0
2	0	$\frac{1}{n}$	0	.	0
3	0	0	$\frac{1}{n}$	.	0
.	.	.	.	.	.
n	0	0	0	.	$\frac{1}{n}$

p_1

	1	2	3	.	n
1	$\frac{1-\epsilon}{n}$	δ	δ	.	δ
2	δ	$\frac{1-\epsilon}{n}$	δ	.	δ
3	δ	δ	$\frac{1-\epsilon}{n}$	.	δ
.	.	.	.	.	.
n	δ	δ	δ	.	$\frac{1-\epsilon}{n}$

p_2

Figure 1: Similar random pairs with different complexities; $\delta \overset{\text{def}}{=} \frac{\epsilon}{n(n-1)}$.

The probability distributions underlying the two random pairs are illustrated in Figure 1. Being so close to each other, the random pairs have similar entropies:

$$H(X_2) \approx H(X_1) = \log n \ ,$$

and

$$H(X_2|Y_2) \approx H(X_1|Y_1) = 0 \ .$$

Their m-message complexities, however, are quite different. The pair (X_1, Y_1) is trivially uniformly distributed. Hence Inequality (3) implies[2] that the average complexities are close to the conditional entropy $H(X_1|Y_1) = 0$. Indeed, if P_y knows Y_1, he also knows X_1, hence

$$\bar{C}_\infty(X_1|Y_1) = \ldots = \bar{C}_1(X_1|Y_1) = 0 \ .$$

On the other hand, S_{X_2,Y_2} is a cartesian product, hence Inequality (2) implies that all average complexities are about $H(X_2)$:

$$\log n \leq \bar{C}_\infty(X_2|Y_2) \leq \ldots \leq \bar{C}_1(X_2|Y_2) \leq \log n + 1 \ . \qquad \square$$

1.3   Related results

El Gamal and Orlitsky [1] consider the average number of transmitted bits required when *both* communicators want to know the other's random variable. It is shown that $H(X|Y) +$

---

[2]This is an extremely degenerate example of a uniform pair, hence does not capture the essence of the upper bound.

$H(Y|X)$ transmitted bits are always needed and that $H(X,Y) + 2$ bits always suffice. If $S_{X,Y}$ is a cartesian product then $H(X,Y)$ bits are needed, while if (X,Y) is uniform over its support set, then $H(X|Y) + H(Y|X) + 3.1 \log \log |S_{X,Y}| + c$ bits always suffice.

Both bounds expressed in Inequality (2) and the lower bound in Inequality (3) follow from an easy modification of the corresponding results in [1]. But the upper bound of Inequality (3) does not. The new upper bound shows that the exchange of X and Y can be performed in two stages with no asymptotic loss of efficiency. First, P_X conveys X to P_y, then P_y conveys Y to P_X. More importantly, the new bound reduces the number of transmitted bits even after breaking the problem into these two stages. The extraneous term in [1] is proportional to $\log \log |S_{X,Y}|$ bits which can be arbitrarily higher than $H(X|Y) + H(Y|X)$, the minimum number necessary (e.g., $\log \log n$ bits versus none for the pair (X_1, Y_1) in Example 2). In the new upper bound the extraneous term is proportional to $\log \min\{H(X|Y), H(Y|X)\}$ bits which is negligible in comparison with $H(X|Y) + H(Y|X)$.

Additionally, the results presented here, apply even when only one of the random variables X and Y needs to be conveyed. This allows for an application of our results to the single-event analog of a problem by Slepian and Wolf [4] that will be considered in the full version of this manuscript.

Another aspect of the problem is the number $\hat{C}_m(X|Y)$ of bits that must be transmitted in the *worst-case* for P_y to learn X when only m messages are allowed. As shown in [2, 5]:

1. A single message may require exponentially more bits than the minimum number needed: for all (X,Y) pairs, $\hat{C}_1(X|Y) \le 2^{\hat{C}_\infty(X|Y)-1}$ with equality for some pairs.

2. With just two messages, the number of bits required is at most four times the minimum: for all (X,Y) pairs, $\hat{C}_2(X|Y) \le 4\hat{C}_\infty(X|Y) + 3$.

3. Two messages are not always optimum: for some (X,Y) pairs, $\hat{C}_2(X|Y) \ge 2\hat{C}_\infty(X|Y)$.

4. For some (X,Y) pairs, the number of bits required when P_X knows Y in advance is logarithmically smaller than that $\hat{C}_\infty(X|Y)$. In the worst-case analog of Example 1, for instance, $\lceil \log \log t \rceil + 1$ bits are required when (as we assume) P_X does not know the losing team in advance, whereas only one bit is needed if he does.

These results sharply contrast average- and worst-case complexities in at least two respects:

1. For average-case complexity, there is almost no difference between the number of bits required when P_X knows Y in advance and when he does not. For worst-case complexity there can be a logarithmic reduction in the number of bits needed if P_X knows Y before communication begins.

2. For average-case complexity, four messages are asymptotically optimum. For worst-case complexity, it is not known whether a constant number of messages are asymptotically optimum. Namely, whether there is an m such that for all (X, Y) pairs,

$$\hat{C}_m(X|Y) \leq \hat{C}_\infty(X|Y) + o(\hat{C}_\infty(X|Y)) .$$

A large discrepancy between the number of bits required with m- and $(m + 1)$ messages occurs in worst-case communication complexity. There, P_X knows X while P_Y knows Y and wants to learn the value of a predetermined boolean function $f(X, Y)$. A succession of papers [6, 7, 8] has shown that for every number m of messages, there is a boolean function f whose m-message worst-case complexity is exponentially higher than its $(m + 1)$-message worst-case complexity. Furthermore, m-message complexity is never more than exponentially higher than $(m + 1)$-message complexity. By contrast, in average-case interactive communication there may be an unbounded discrepancy between $\bar{C}_1(X|Y)$ and $\bar{C}_2(X|Y)$ (e.g., $\log t$ versus three bits in the league problem). Yet four messages are asymptotically optimum hence m messages, for any m, are at most negligibly better than four. Although the corresponding result is not known to hold for worst-case interactive communication, it can still be distinguished from communication complexity: as stated above, two messages require at most four times the minimum communication.

1.4 Complexity measures

The support set of a random pair (X, Y) with underlying probability distribution $p(x, y)$ was defined earlier as the set

$$S_{X,Y} = \{(x, y) : p(x, y) > 0\}$$

of ordered pairs occurring with positive probability. An *input* is an element of $S_{X,Y}$ viewed as value assignments to X and Y. For every input (x, y), the protocol used, ϕ, determines a finite sequence $\sigma_1(x,y), \ldots, \sigma_{m_\phi(x,y)}(x,y)$ of transmitted messages.[3] Since $P_\mathcal{X}$ and $P_\mathcal{Y}$ alternate in transmitting these messages, all even numbered messages in this sequence are transmitted by one communicator and all odd numbered messages are transmitted by the other. $m_\phi(x,y)$ is the number of messages exchanged for (x,y). The protocol ϕ is *m-message* if $m_\phi(x,y) \leq m$ for all $(x,y) \in S_{X,Y}$. The *length*, $|\sigma|$, of a message σ is the number of bits it contains. The *transmission length* of the input (x,y) is

$$l_\phi(x,y) \overset{\text{def}}{=} \sum_{i=1}^{m_\phi(x,y)} |\sigma_i(x,y)| \, ,$$

the number of bits transmitted by both communicators when $X = x$, $Y = y$, and the protocol used is ϕ.

The *average complexity of* ϕ is the expectation, over all inputs, of the number of bits transmitted during the execution of ϕ:

$$\bar{L}_\phi \overset{\text{def}}{=} \sum_{(x,y) \in S_{X,Y}} p(x,y) \cdot l_\phi(x,y) \, .$$

The *m-message average complexity of* (X,Y) is

$$\bar{C}_m(X|Y) \overset{\text{def}}{=} \inf\{\bar{L}_\phi : \phi \text{ is an } m\text{-message protocol for } (X,Y)\} \, ,$$

the number of bits that the two communicators must transmit on the average if restricted to protocols that require at most m messages. Since empty messages are allowed, $\bar{C}_m(X|Y)$ is a decreasing function of m bounded below by 0. We can therefore define $\bar{C}_\infty(X|Y)$, the *unbounded-message average complexity* of (X,Y), to be the limit of $\bar{C}_m(X|Y)$ as $m \to \infty$. It is the *minimum* number of bits that must be transmitted on the average for $P_\mathcal{Y}$ to know X. A protocol for (X,Y) whose average complexity is $\bar{C}_\infty(X|Y)$ is an *optimal* protocol for (X,Y).

2 General pairs

We prove that for all (X,Y) pairs,

$$H(X|Y) \leq \bar{C}_\infty(X|Y) \leq \bar{C}_1(X|Y) \leq H(X) + 1 \, . \tag{4}$$

---

[3]For notational simplicity, the protocol ϕ is implicit in $\sigma_i(x,y)$. Also, we do not assume that the same number of messages is transmitted for all inputs or that the same communicator transmits the first message for all inputs.

These bounds are not tight as $H(X|Y)$ can be much smaller than $H(X)$. However, Sections 3 and 4 show that for some (X, Y) pairs the upper bound is tight, for others the lower bound can be achieved. In that sense, the bounds of Inequality (4) are the tightest expressible in terms of Shannon entropies.

The upper bound is achieved by a Huffman code based on the marginal probabilities of X. The lower bound is intuitive too. Even if P_X knows Y, the expected number of bits he must transmit when $Y = y$ is at least $H(X|Y = y)$. Therefore, the expectation, over all inputs, of the number of transmitted bits, is at least $H(X|Y)$. This, however, is only a heuristic argument. When P_X and P_Y communicate, they alternate in transmitting messages, therefore the sequence of bits transmitted by P_X is parsed into messages. It is conceivable that the "commas" separating the messages can be used to mimic a ternary alphabet, thus decrease the number of transmitted bits (we count only "0"s and "1"s, not commas). To turn the above argument into a proof, we need to show that P_X cannot use the parsing to further encode his information. This is done in the next lemma. Lemma 2 incorporates this result in the argument above.

Let ϕ be a protocol for (X, Y). In Subsection 1.4 we defined $\sigma_1(x, y), \ldots, \sigma_{m_\phi(x,y)}(x, y)$ to be the sequence of messages transmitted by P_X and P_Y for the input (x, y). Let $\sigma_\phi^X(x, y)$ denote the concatenation of all messages transmitted by P_X for that input.

Lemma 1   Let ϕ be a protocol for (X, Y). If $(x, y), (x', y) \in S_{X,Y}$ and $x \neq x'$, then neither $\sigma_\phi^X(x, y)$ nor $\sigma_\phi^X(x', y)$ is a prefix of the other (in particular, they are not equal).

Proof: We show that if one of $\sigma_\phi^X(x, y)$ and $\sigma_\phi^X(x', y)$ is a prefix of the other then

1) $m_\phi(x, y) = m_\phi(x', y)$,

2) for $j = 1, \ldots, m_\phi(x, y)$, $\sigma_j(x, y) = \sigma_j(x', y)$.

Therefore, the correct-decision property implies that $x = x'$.

To prove (2) we show by induction on $i \leq m_\phi(x, y)$ that $\sigma_j(x, y) = \sigma_j(x', y)$ for $j = 1, \ldots, i$. The induction hypothesis holds for $i = 0$. Assume it holds for $i - 1$. If $\sigma_i(x, y)$ is transmitted by P_Y then, by the separate-transmissions property, he also transmits $\sigma_i(x', y)$, and $\sigma_i(x, y) = \sigma_i(x', y)$. If $\sigma_i(x, y)$ is transmitted by P_X then, since one of $\sigma_\phi^X(x, y)$ and $\sigma_\phi^X(x', y)$ is a prefix of the other and the previous messages are identical, one of $\sigma_i(x, y)$ and $\sigma_i(x', y)$ is a prefix of the other. By the implicit-termination property, $\sigma_i(x, y) = \sigma_i(x', y)$.

Now, (1) follows from the implicit-termination property as we have just shown that $\sigma_j(x, y) = \sigma_j(x', y)$ for $j = 1, \ldots, m_\phi(x, y)$.   □

Let (X, Y) be a random pair. The *support set of Y* is

$$S_Y = \{y : \text{for some } x, \, (x,y) \in S_{X,Y}\} \,,$$

the set of possible values of Y. P_y's *ambiguity set* when his random variable attains the value $y \in S_Y$ is:

$$S_{X|Y}(y) = \{x : (x,y) \in S_{X,Y}\} \,,$$

the set of possible X values when $Y = y$. Lemma 1 implies that for every $y \in S_Y$, the multiset[4] $\{\sigma_\phi^X(x,y) : x \in S_{X|Y}(y)\}$ is prefix free. Namely, it contains $|S_{X|Y}(y)|$ distinct elements, none of which is a prefix of another.

Lemma 2 For all (X, Y) pairs,

$$H(X|Y) \leq \bar{C}_\infty(X|Y) \leq \bar{C}_1(X|Y) \leq H(X) + 1 \,.$$

Proof: We mentioned that the upper bound is achieved by an Huffman code. We prove the lower bound. Let ϕ be a protocol for (X, Y). The preceding discussion and the data-compression theorem imply that for all $y \in S_Y$,

$$\sum_{x \in S_{X|Y}(y)} p(x|y) \cdot l_\phi^X(x,y) \geq H(X|Y = y)$$

where $l_\phi^X(x,y) \stackrel{\text{def}}{=} |\sigma_\phi^X(x,y)|$ denotes the number of bits transmitted by P_X for the input (x,y). The average number of bits transmitted by P_X under ϕ is:

$$
\begin{aligned}
\sum_{(x,y) \in S_{X,Y}} p(x,y) \cdot l_\phi^X(x,y) &= \sum_{y \in S_Y} p(y) \sum_{x \in S_{X|Y}(y)} p(x|y) \cdot l_\phi^X(x,y) \\
&\geq \sum_{y \in S_Y} p(y) H(X|Y = y) \\
&= H(X|Y) \,. \qquad \qquad \square
\end{aligned}
$$

Note that we proved a stronger version of the lower bound, than claimed in the lemma. We showed that P_X alone must transmit an average of at least $H(X|Y)$ bits, regardless of P_y's transmissions.

---

[4] A *multiset* allows for multiplicity of elements, e.g., $\{0, 1, 1\}$.

3 Pairs with Cartesian-product support sets

A support set $S_{X,Y}$ is a *Cartesian product* if for some sets $A \subseteq S_X$ and $B \subseteq S_Y$,

$$S_{X,Y} = A \times B \ .$$

Figure 2 illustrates one such set. We show that for all (X,Y) pairs with Cartesian-product support sets, the upper bound of Inequality (1) is tight:

$$H(X) \le \bar{C}_\infty(X|Y) \le \bar{C}_1(X|Y) \le H(X) + 1 \ .$$

Cartesian-product support sets are important mainly because for many (X,Y) pairs all

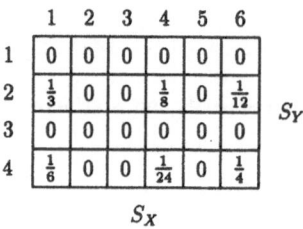

Figure 2: A probability distribution with a Cartesian-product support set.

inputs $(x,y) \in S_{X,Y}$ have positive probability, i.e., $S_{X,Y} = S_X \times S_Y$. In all these cases, at least $H(X)$ bits must be transmitted on the average. The same as the number of bits required when P_y has no information at all, and when no interaction is allowed. In other words, when $S_{X,Y}$ is a Cartesian product, neither P_y's knowledge of Y, nor interaction, help reduce communication.

Lemma 3 Let ϕ be a protocol for (X,Y). If $(x',y),(x,y),(x,y') \in S_{X,Y}$ and $x \ne x'$, then neither $\sigma_\phi(x',y)$ nor $\sigma_\phi(x,y')$ is a prefix of the other (in particular, they are not equal).
Proof: The proof is analogous to that of Lemma 1, hence omitted. $\square$

Just as S_Y denotes the set of possible values of Y and $S_{X|Y}(y)$ is P_y's ambiguity set when he knows the value $y \in S_Y$, we define

$$S_X \overset{\text{def}}{=} \{x : \text{ for some } y, (x,y) \in S_{X,Y}\}$$

to be the set of possible values of X and let P_X's *ambiguity set* when $X = x$ be

$$S_{Y|X}(x) \stackrel{\text{def}}{=} \{y : (x,y) \in S_{X,Y}\} .$$

Corollary 1   If the support set of (X, Y) is a Cartesian product then

$$\bar{C}_\infty(X|Y) \geq H(X) .$$

Proof: Let ϕ be a protocol for (X, Y). Recall that $l_\phi(x,y)$ denotes the number of bits transmitted under ϕ for the input (x,y) and that $\bar{L}_\phi$ is the number of transmitted bits, averaged over all inputs. We show that $\bar{L}_\phi \geq H(X)$.

Since $S_{X,Y}$ is a Cartesian product, $S_{X,Y} = A \times B$ for some sets $A \subseteq S_X$ and $B \subseteq S_Y$. For every $x \in A$, let

$$l_\phi(x) \stackrel{\text{def}}{=} \min\{l_\phi(x,y) : y \in S_{Y|X}(x)\}$$

be the minimum number of bits transmitted by both P_X and P_Y when $X = x$. Since $S_{X,Y}$ is a Cartesian product, Lemma 3 implies that for all pairs $(x_1, y_1), (x_2, y_2) \in S_{X,Y}$ such that $x_1 \neq x_2$, neither $\sigma_\phi(x_1, y_1)$ nor $\sigma_\phi(x_2, y_2)$ is a prefix of the other. Hence, the $l_\phi(x)$'s are the lengths of $|A|$ strings, none of which is a prefix of the other. By the data-compression theorem:

$$\sum_{x \in A} p(x) l_\phi(x) \geq H(X) .$$

Therefore,

$$
\begin{aligned}
\bar{L}_\phi &= \sum_{(x,y) \in A \times B} p(x,y) l_\phi(x,y) \\
&= \sum_{x \in A} p(x) \sum_{y \in S_{Y|X}(x)} p(y|x) l_\phi(x,y) \\
&\geq \sum_{x \in A} p(x) l_\phi(x) \\
&\geq H(X) .
\end{aligned}
$$
$\square$

Of course, we cannot prove, as we did in the last section, that this lower bound holds even for the average number of bits transmitted by P_X alone. There is always a protocol in which P_X transmits at most $H(X|Y) + 1$ bits on the average.

4   Uniformly-distributed pairs

In the last section we proved that if the support set of (X, Y) is a Cartesian product, the upper bound of Inequality (1) is tight: $H(X) \leq \bar{C}_\infty(X|Y) \leq \bar{C}_1(X|Y) \leq H(X) + 1$. We now show that if (X, Y) is uniformly distributed, the lower bound of Inequality (1) can be almost achieved:

$$H(X|Y) \leq \bar{C}_\infty(X|Y) \leq \bar{C}_4(X|Y) \leq H(X|Y) + 3 \log (H(X|Y) + 1) + 15.5 .$$

As we mentioned in the introduction, this shows that

1. P_X can communicate X to P_Y using roughly the number of bits needed when P_X knows Y in advance.

2. Four messages are asymptotically optimal for average-case complexity. The number $\bar{C}_1(X|Y)$ of bits needed when only one message is allowed may be arbitrarily larger then the minimum number necessary, $\bar{C}_\infty(X|Y)$, yet with four message only negligibly more than $\bar{C}_\infty(X|Y)$ bits are needed. We remark that the corresponding result is not known to hold for worst-case complexity.

We need a few definitions. Let Z be a random variable distributed over a support set S_Z according to a probability distribution p. A *(binary, prefix-free) encoding* of Z is a mapping ϕ from S_Z to $\{0,1\}^*$ such that for all $z_1, z_2 \in S_Z$, the *codeword*, $\phi(z_1)$, of z_1 is not a prefix of $\phi(z_2)$ – the codeword of z_2. For $z \in S_X$, let $\ell_\phi(z)$ be the length of $\phi(z)$. The *expected length* of ϕ is

$$\bar{\ell}_\phi(Z) \overset{\text{def}}{=} \sum_{z \in S_Z} p(z)\ell_\phi(z) .$$

The *minimum encoding length* of Z is

$$\mathcal{L}(Z) \overset{\text{def}}{=} \min\{\bar{\ell}_\phi(Z) : \phi \text{ is an encoding of } Z\} .$$

It is well known that

$$H(Z) \leq \mathcal{L}(Z) \leq H(Z) + 1 . \tag{5}$$

Recall that the support set, $S_{X,Y}$, of a random pair (X, Y) is the set of all possible inputs. (X, Y) is uniformly distributed if all elements in $S_{X,Y}$ are equally likely. The support set of X is the set S_X of all possible values of X, S_Y was similarly defined. P_Y's ambiguity set when his random variable attains the value $y \in S_Y$ is the set $S_{X|Y}(y)$ of possible X values in that case. Denote the collection of P_Y's ambiguity sets by

$$S_{X|Y} \overset{\text{def}}{=} \{S_{X|Y}(y) : y \in S_Y\} .$$

A collection of functions, each defined over S_X, *perfectly hashes* $S_{X|Y}$ if for every $y \in S_Y$ there is a function in the collection that is one-to-one over (or *hashes*) $S_{X|Y}(y)$.

Let b be an integer and let $\mathcal{F}$ be a collection of functions from S_X to $\{1,\dots,b\}$ that perfectly hashes $S_{X|Y}$. Assume also that the mapping hash(y) assigns, for each $y \in S_Y$, a function in $\mathcal{F}$ that hashes $S_{X|Y}(y)$. Then the random variable hash(Y) denotes a function in $\mathcal{F}$ that hashes P_y's ambiguity set $S_{X|Y}(Y)$. We show that

$$\bar{C}_2(X|Y) \leq \mathcal{L}(\text{hash}(Y)) + \lceil \log b \rceil . \tag{6}$$

P_X and P_y agree in advance on the collection $\mathcal{F}$, on the mapping hash(y), and on an encoding of hash(Y) whose expected length is $\mathcal{L}(\text{hash}(Y))$. When P_X is given X and P_y is given Y, they execute the following two-message protocol.

P_y transmits the encoding of hash(Y). Now P_X knows X and $f_{\text{hash}(Y)}$. He computes and transmits the binary representation of $f_{\text{hash}(Y)}(X)$. Since $f_{\text{hash}(Y)}$ is one-to-one over $S_{X|Y}(Y)$, P_y can recover X. The number of bits transmitted by P_X is always at most $\lceil \log b \rceil$. The expected number of bits transmitted by P_y is $\mathcal{L}(\text{hash}(Y))$.

To upper bound $\bar{C}_2(X|Y)$, we therefore demonstrate a collection $\mathcal{F}$ that perfectly hashes $S_{X|Y}$, and a mapping hash(y). By construction, each function in $\mathcal{F}$ has a small range while Lemmas 4 and 5 show that $\mathcal{L}(\text{hash}(Y))$ is low. Theorem 1 combines this result with Inequality (6).

One way to guarantee that $\mathcal{L}(\text{hash}(Y))$ is low, is to ensure that hash(Y) assumes certain values with high probability. The next lemma shows that for any collection of subsets, there is a function that hashes a relatively large fraction of these subsets.

Lemma 4 Let $\mathcal{S}$ be a finite collection of finite sets and let Pr be a probability distribution over $\mathcal{S}$, namely, $\Pr(S) \geq 0$ for every $S \in \mathcal{S}$ and $\sum_{S \in \mathcal{S}} \Pr(S) = 1$.

Define $\hat{\mu} \stackrel{\text{def}}{=} \max\{|S| : S \in \mathcal{S}\}$ to be the size of the largest set in $\mathcal{S}$. Then, for every $b \geq \hat{\mu}$ there is a function $f : \bigcup_{S \in \mathcal{S}} S \to \{1,\dots,b\}$ that perfectly hashes a subcollection of $\mathcal{S}$ whose probability is at least $\frac{b^{\underline{\hat{\mu}}}}{b^{\hat{\mu}}}$, that is,

$$\sum_{S \in \mathcal{S}} \Pr(S) 1_{f \text{ hashes } s} \geq \frac{b^{\underline{\hat{\mu}}}}{b^{\hat{\mu}}}$$

where $b^{\underline{\hat{\mu}}} \stackrel{\text{def}}{=} b \cdot (b-1) \cdot \dots \cdot (b - \hat{\mu} + 1)$ is the $\hat{\mu}$th falling power of b and

$$1_{f \text{ hashes } s} \stackrel{\text{def}}{=} \begin{cases} 1 & \text{if } f \text{ hashes } S \\ 0 & \text{otherwise.} \end{cases}$$

is an indicator function.

Proof: Let $\mathcal{F}$ be the set of all functions from $\bigcup_{S \in \mathcal{S}} S$ to $\{1, \ldots, b\}$. Generate a random function F in $\mathcal{F}$ by assigning to every element in $\bigcup_{S \in \mathcal{S}} S$ an element of $\{1, \ldots, b\}$ uniformly and independently.

For every given set $S \in \mathcal{S}$, the probability that F hashes S is

$$\sum_{f \in \mathcal{F}} \Pr(f) 1_{f \text{ hashes } s} = \frac{b^{\underline{|S|}}}{b^{|S|}} \geq \frac{b^{\underline{\hat{\mu}}}}{b^{\hat{\mu}}} .$$

Therefore,

$$
\begin{aligned}
\sum_{f \in \mathcal{F}} \Pr(f) \sum_{S \in \mathcal{S}} \Pr(S) 1_{f \text{ hashes } s} &= \sum_{S \in \mathcal{S}} \Pr(S) \sum_{f \in \mathcal{F}} \Pr(f) 1_{f \text{ hashes } s} \\
&\geq \sum_{S \in \mathcal{S}} \Pr(S) \frac{b^{\underline{\hat{\mu}}}}{b^{\hat{\mu}}} \\
&= \frac{b^{\underline{\hat{\mu}}}}{b^{\hat{\mu}}} .
\end{aligned}
$$

Hence the required inequality holds for at least one function $f \in \mathcal{F}$. $\qquad\square$

Using this result, we construct a sequence $f_1, , \ldots, f_k$ of functions that perfectly hashes $S_{X|Y}$. The probability that f_i is the first function to hash $S_{X|Y}(Y)$ rapidly decreases with i.

P_y's *ambiguity* when his random variable attains the value $y \in S_Y$ is

$$\mu_{X|Y}(y) \stackrel{\text{def}}{=} |S_{X|Y}(y)| ,$$

the number of possible X values when $Y = y$. The *maximum ambiguity* of P_y is

$$\hat{\mu}_{X|Y} \stackrel{\text{def}}{=} \sup\{\mu_{X|Y}(y) : y \in S_Y\},$$

the maximum number of X values possible with any given Y value. P_x's *ambiguity*, $\mu_{Y|X}(x)$, when $X = x$ and his *maximum ambiguity*, $\hat{\mu}_{Y|X}$, are defined symmetrically. In the sequel we frequently use the following abbreviation:

$$\beta \stackrel{\text{def}}{=} \frac{b^{\hat{\mu}_{X|Y}}}{b^{\underline{\hat{\mu}}_{X|Y}}} . \tag{7}$$

Let $\mathcal{F}$ be the set of functions from S_X to $\{1, \ldots, b\}$. For each function $f \in \mathcal{F}$ let

$$\alpha_1(f) \stackrel{\text{def}}{=} \Pr\Big(f \text{ hashes } S_{X|Y}(Y)\Big)$$

be the probability that f hashes Py's ambiguity set. Let f_1 be a function in $\mathcal{F}$ that maximizes α_1. The last lemma implies that $\alpha_1(f_1) \geq \beta$. Recursively, if $f_1, \ldots, f_{i-1}$ do not hash all ambiguity sets, let

$$\alpha_i(f) \overset{\text{def}}{=} \Pr\Big(f \text{ hashes } S_{X|Y}(Y) \mid \text{none of } f_1, \ldots, f_{i-1} \text{ hashes } S_{X|Y}(Y)\Big)$$

be the probability that Py's ambiguity set is hashed by the function $f \in \mathcal{F}$, given that it is not hashed by any of $f_1, \ldots, f_{i-1}$. Let f_i be a function in $\mathcal{F}$ that maximizes α_i. Again, the lemma guarantees that $\alpha_i(f_i) \geq \beta$.

Every time a function f_i is determined, it hashes at least one additional ambiguity set. Since $S_{X,Y}$ is finite, there are finitely many ambiguity sets, hence the process stops after some finite number, say k, of steps.

Once the sequence $f_1, \ldots, f_k$ has been defined, let $\text{hash}(y)$ be the first function in $f_1, \ldots, f_k$ that hashes $S_{X|Y}(y)$. Lemma 5 uses a sequence of three claims to show that $\mathcal{L}(\text{hash}(Y))$ is low.

We write $a_1, \ldots, a_k$ to denote the k element sequence whose ith element is a_i; we use the notation $a_1, a_2, , \ldots,$ if the sequence can be either finite or infinite. A sequence $q_1, q_2, \ldots$ *majorizes* a sequence $p_1, p_2, \ldots$ if for all $j \geq 1$

$$\sum_{i=1}^{j} q_i \geq \sum_{i=1}^{j} p_i$$

where 0's are appended to the tails of finite sequences.

For $i \in \{1, \ldots, k\}$ let

$$q_i \overset{\text{def}}{=} \Pr(\text{hash}(Y) = f_i)) \tag{8}$$

be the probability that f_i is the first function that hashes Py's ambiguity set. And for $i \in \mathcal{Z}^+$, let

$$p_i = \beta(1 - \beta)^{i-1} \tag{9}$$

be the probability that a geometric random variable with probability β of success attains the value i.

The first claim in the proof of Lemma 5 implies that $q_1, \ldots, q_k$ majorizes $p_1, p_2, \ldots$. The second claim shows that if $q_1, \ldots, q_k$ majorizes $p_1, p_2, \ldots$, then $\mathcal{L}(q_1, \ldots, q_k) \leq \mathcal{L}(p_1, p_2, \ldots)$. (The minimum encoding length of a distribution is defined as that of a random variable). The third claim combines Inequality (5) with the entropy of the geometric distribution to show that

Lemma 5 For all (X, Y) pairs

$$\mathcal{L}(\text{hash}(Y)) \leq \log \frac{1}{\beta} + \log e + 1$$

where β was defined in Equation (7).

Proof: We first show that $q_1, \ldots, q_k$, defined in Equation (8), majorizes $p_1, p_2, \ldots$, defined in Equation (9).

Claim 1 Let $p_1, p_2, \ldots$ and $q_1, q_2, \ldots$ be probability distributions such that

$$\frac{q_i}{1 - q_1 - \ldots - q_{i-1}} \geq \frac{p_i}{1 - p_1 - \ldots - p_{i-1}}$$

for all $i \geq 1$. Then $q_1, q_2, \ldots$ majorizes $p_1, p_2, \ldots$.

Proof: An induction on i shows that

$$p_{i+1} + p_{i+2} + \ldots = \prod_{j=1}^{i} \left(1 - \frac{p_j}{1 - p_1 - \ldots - p_{j-1}} \right)$$

and the corresponding equality holds for $q_{i+1} + q_{i+2} + \ldots$, hence, for all $i \geq 1$,

$$q_{i+1} + q_{i+2} + \ldots \leq p_{i+1} + p_{i+2} + \ldots \qquad \square$$

The discussion preceding this lemma showed

$$\frac{q_i}{1 - q_1 - \ldots - q_{i-1}} \geq \beta = \frac{p_i}{1 - p_1 - \ldots - p_{i-1}} \ ,$$

hence $q_1, \ldots, q_k$ majorizes $p_1, p_2, \ldots$. We use this fact to upper bound the minimum encoding length of $\text{hash}(Y)$.

Claim 2 Let $p_1, p_2, \ldots$ be a nonincreasing probability distribution ($p_i \geq p_{i+1} \geq 0$ and $\sum p_i = 1$). If the probability distribution $q_1, \ldots, q_k$ majorizes $p_1, p_2, \ldots$ then

$$\mathcal{L}(q_1, \ldots, q_k) \leq \mathcal{L}(p_1, p_2, \ldots) \ .$$

Proof: By majorization, $\sum_{i=1}^{k} p_i \leq 1$, hence p_i is positive for all $i \in \{1, \ldots, k\}$. For i in this range, let ℓ_i be the length of the codeword corresponding to p_i in

96

an encoding achieving $\mathcal{L}(p_1, p_2, \ldots)$. Since $p_i \leq p_{i-1}$, we must have $\ell_i \geq \ell_{i-1}$. Therefore,

$$
\begin{aligned}
\mathcal{L}(q_1, \ldots, q_k) & \leq \sum_{i=1}^{k} q_i \ell_i \\
& = \sum_{i=1}^{k} (\sum_{j=i}^{k} q_j)(\ell_i - \ell_{i-1}) \\
& \leq \sum_{i=1}^{k} (\sum_{j=i}^{k} p_j)(\ell_i - \ell_{i-1}) \\
& = \sum_{i=1}^{k} p_i \ell_i \\
& = \mathcal{L}(p_1, p_2, \ldots) .
\end{aligned}
$$
□

The two claims show that the minimum encoding length of hash(Y) is upper bounded by that of a geometric random variable with probability β of success. The next claim, given for completeness, estimates this length.

Claim 3 As defined in Equation (9), let $p_1, p_2, \ldots$ be the probability distribution of a geometric random variable with probability β of success. Then

$$
\mathcal{L}(p_1, p_2, \ldots) \leq \log \frac{1}{\beta} + \log e + 1 .
$$

Proof: We combine Inequality (5) and the following standard calculation.

$$
\begin{aligned}
H(p_1, p_2, \ldots) & = -\sum_{i=1}^{\infty} \beta(1 - \beta)^{i-1} \log(\beta(1 - \beta)^{i-1}) \\
& = -\beta \log \beta \sum_{i=1}^{\infty} (1 - \beta)^{i-1} - \beta \log(1 - \beta) \sum_{i=1}^{\infty} (i - 1)(1 - \beta)^{i-1} \\
& = \log \frac{1}{\beta} + \frac{1 - \beta}{\beta} \log \frac{1}{(1 - \beta)} \\
& \leq \log \frac{1}{\beta} + \log e
\end{aligned}
$$

where we used the (natural logarithm) inequality:

$$
\frac{1 - \beta}{\beta} \ln \frac{1}{1 - \beta} = \frac{1 - \beta}{\beta} \ln \left(1 + \frac{\beta}{1 - \beta}\right) \leq \frac{1 - \beta}{\beta} \cdot \frac{\beta}{1 - \beta} = 1 .
$$
□

The lemma follows.
□

We can now derive our first theorem:

Theorem 1   For all (X, Y) pairs,

$$\bar{C}_2(X|Y) \leq 2 \log \hat{\mu}_{X|Y} + 5.5 \ .$$

Proof: Combining Inequality (6) and Lemma 5, and abbreviating $\hat{\mu}_{X|Y}$ as $\hat{\mu}$, we obtain

$$
\begin{aligned}
\bar{C}_2(X|Y) &\leq \log \left(\frac{b^{\hat{\mu}}}{b^{\hat{\mu}}} \right) + \lceil \log b \rceil + \log e + 1 \\
&\leq \log \frac{b^{\hat{\mu}}}{(b - \hat{\mu})^{\hat{\mu}}} + \log b + \log e + 2 \\
&= \hat{\mu} \log \frac{b}{b - \hat{\mu}} + \log b + \log e + 2 \ .
\end{aligned}
$$

Choosing $b = \hat{\mu}^2 + \hat{\mu}$, we derive:

$$
\begin{aligned}
\bar{C}_2(X|Y) &\leq \hat{\mu} \log \left(1 + \frac{1}{\hat{\mu}} \right) + \log(\hat{\mu}^2 + \hat{\mu}) + \log e + 2 \\
&\leq 2 \log \hat{\mu} + 2 \log e + \log \left(1 + \frac{1}{\hat{\mu}} \right) + 2 \\
&\leq 2 \log \hat{\mu} + 5.5 \ . \qquad\qquad \square
\end{aligned}
$$

The theorem can be used to prove that for all (X, Y) pairs

$$\bar{C}_2(X|Y) \leq 2H(X|Y) + 2 \log \left(H(X|Y) + 1 \right) + 7.5 \ . \tag{10}$$

This result is of independent interest. One message may require arbitrarily more bits than the minimum necessary (e.g. $\bar{C}_1(X|Y) = \log t$ while $\bar{C}_2(X|Y) \leq 3$ in the League problem of Example 1). Yet Inequality (10) shows that two messages require at most twice the minimum number of bits. We proceed to show that by adding two more messages, communication can be reduced roughly by half, thereby asymptotically achieving the minimum possible.

Let k be an integer; a function f is *k-smooth* over a subset S of its domain if for every r in its range,

$$|S \bigcap f^{-1}(r)| \leq k.$$

Namely, if f does not assign the same value to more than k elements of S. The function f is *k-smooth* over a collection of subsets if it is k-smooth over each subset.

Lemma 6   Let $\hat{\mu}$ be an integer ≥ 3, let $\mathcal{S}$ be a finite collection of finite sets, each of size $\leq \hat{\mu}$, and let Pr be a probability distribution over $\mathcal{S}$. Then, there is a function $f : \bigcup\limits_{S \in \mathcal{S}} S \rightarrow$

$\{1,\ldots,\hat{\mu}\}$ that is $\frac{4\log\hat{\mu}}{\log\log\hat{\mu}}$-smooth over a subcollection of of $\mathcal{S}$ whose probability is at least $\frac{1}{2}$. Namely, if $S \in \mathcal{S}$ is selected at random according to Pr, then

$$\mathrm{Pr}\left(f \text{ is } \frac{4\log\hat{\mu}}{\log\log\hat{\mu}}\text{-smooth over } S\right) \geq \frac{1}{2} \ .$$

Proof: Generate a random function F from $\bigcup_{S\in\mathcal{S}} S$ to $\{1,\ldots,\hat{\mu}\}$ by assigning to every element in $\bigcup_{S\in\mathcal{S}} S$ an element of $\{1,\ldots,\hat{\mu}\}$ uniformly and independently. For every set $S \in \mathcal{S}$, every $i \in \{1,\ldots,\hat{\mu}\}$, and every integer k,

$$\mathrm{Pr}\left(|S \cap F^{-1}(i)| \geq k\right) \leq \binom{\hat{\mu}}{k}\frac{1}{\hat{\mu}^k} \leq \frac{1}{k!} \ .$$

Note that Pr now represents the joint (product) probability of S and F. Using the union bound,

$$\mathrm{Pr}(F \text{ is } \textit{not } k\text{-smooth over } S) \leq \frac{\hat{\mu}}{k!} \leq \hat{\mu}\frac{e^k}{k^k} \ .$$

For $k \geq \frac{4\log\hat{\mu}}{\log\log\hat{\mu}}$, this probability is at most $\frac{1}{2}$. The lemma now follows from an argument similar to the one used to prove Lemma 4. □

As in the discussion following Lemma 4, we can now assign to every y a function $\mathrm{smooth}(y)$ that is $\frac{4\log\hat{\mu}_{X|Y}}{\log\log\hat{\mu}_{X|Y}}$-smooth over $S_{X|Y}(y)$. The arguments used in Lemma 5 then show that the probability distribution underlying $\mathrm{smooth}(Y)$ majorizes the geometric distribution with probability $\frac{1}{2}$ of success, hence:

Lemma 7 For all (X,Y) pairs with $\hat{\mu}_{X|Y} \geq 3$,

$$\mathcal{L}(\mathrm{smooth}(Y)) \leq 2. \qquad \square$$

Theorem 2 For all (X,Y) pairs with $\hat{\mu}_{X|Y} \geq 3$

$$\bar{C}_4(X|Y) \leq \log\hat{\mu}_{X|Y} + 2\log\log\hat{\mu}_{X|Y} - 2\log\log\log\hat{\mu}_{X|Y} + 11.5 \ .$$

Proof: P_Y transmits the encoding of the function $f = \mathrm{smooth}(Y)$ using at most 2 bits on the average. Then P_X transmits $\lceil\log\hat{\mu}_{X|Y}\rceil$ bits describing $f(X)$. Now P_X and P_Y concentrate on a random pair whose maximum ambiguity is at most $\lceil\frac{4\log\hat{\mu}_{X|Y}}{\log\log\hat{\mu}_{X|Y}}\rceil$. Theorem 1 says that X can be conveyed to P_Y using an expected number of bits of at most

$$2\log\left\lceil\frac{4\log\hat{\mu}_{X|Y}}{\log\log\hat{\mu}_{X|Y}}\right\rceil + 5.5 \leq 2\log\log\hat{\mu}_{X|Y} - 2\log\log\log\hat{\mu}_{X|Y} + 8.5 \ . \qquad \square$$

99

For the first time, we now use the uniformity of (X, Y). It implies that

$$\log \mu_{X|Y}(y) = H(X|y)$$

for all $y \in S_Y$. Hence, if P_y's ambiguity, $\mu_{X|Y}(y)$, is the same for all $y \in S_Y$, then

$$\log \hat{\mu}_{X|Y} = H(X|Y)$$

and the theorem becomes:

$$\bar{C}_4(X|Y) \leq H(X|Y) + 2 \log H(X|Y) - 2 \log \log H(X|Y) + 10.5 .$$

This bound applies when $\hat{\mu}_{X|Y} \geq 3$. For $\hat{\mu}_{X|Y} \leq 2$, Theorem 1 implies

$$\bar{C}_4(X|Y) \leq \bar{C}_2(X|Y) \leq 2 \log \hat{\mu}_{X|Y} + 5.5 \leq 7.5.$$

If $\mu_{X|Y}(y)$ varies with y, then $\log \hat{\mu}_{X|Y}$ can be much larger then $H(X|Y)$. In that case we prefix the above protocol with a stage that identifies $\mu_{X|Y}(y)$ to within a factor of two.

Theorem 3 For all *uniform* (X, Y) pairs

$$\bar{C}_4(X|Y) \leq H(X|Y) + 3 \log (H(X|Y) + 1) + 15.5 .$$

Proof: First we describe a prefix-free encoding of the non-negative integers where the codeword of 0 is 2-bits long and the codeword of any other i is at $2\lceil \log(i+1)\rceil$-bits long. Consider the $\lceil \log(i + 1)\rceil$-bit binary representation of i. Encode every 0 (or 1) in this representation, except the last, as 00 (or 01). Encode the last 0 (or 1) as 10 (or 11). For example, 0, 1, and 2, are encoded as 10, 11, and 0110, respectively.

This construction can be used to derive a prefix-free encoding of the integers where the codewords of 0 and 1 are 3-bits long and any other i is encoded into $\lceil \log(i+1)\rceil + 2\lceil \log \log(i+1)\rceil$ bits. Given i, we first encode the integer $\lceil \log(i + 1)\rceil - 1$ as above. Then we provide the $\lceil \log(i+1)\rceil$ bits in the binary representation of i. More sophisticated constructions exist (e.g., [9]), but they will only marginally improve our results.

The protocol proceeds as follows. First, P_y uses the encoding above to convey $\lceil log \mu_{X|Y}(y)\rceil$. Then the communicators use the protocol of Theorem 2 to convey X to P_y as if the maximum ambiguity was $2^{\lceil log \mu_{X|Y}(y)\rceil}$. The total number of bits transmitted is at most

$$\bar{L}_\phi \leq \sum_{(x,y) \in S_{X,Y}} p(x, y) \Big(\log \mu_{X|Y}(y) + 3 \log \big(1 + \log \mu_{X|Y}(y)\big) + 15.5 \Big)$$

100

$$= \sum_{y \in S_Y} p(y)\Big(\log \mu_{X|Y}(y) + 3\log\big(1 + \log \mu_{X|Y}(y)\big) + 15.5\Big)$$

$$= \sum_{y \in S_Y} p(y)(H(X|y) + 3\log(H(X|y)+1) + 15.5)$$

$$\leq \sum_{y \in S_Y} p(y)H(X|y) + 3\log \sum_{y \in S_Y} p(y)(H(X|y)+1) + 15.5$$

$$= H(X|Y) + 3\log(H(X|Y)+1) + 15.5$$

where the second inequality uses the convexity of the logarithm. Note that the first two phases of the protocol consist of bits transmitted by P_y and are combined into one message.

□

We relate the number of bits required by a four-message protocol to the optimal number.

Corollary 2   For all *uniform* (X, Y) pairs

$$\bar{C}_4(X|Y) \leq \bar{C}_\infty(X|Y) + 3\log\big(\bar{C}_\infty(X|Y) + 1\big) + 15.5 \ . \qquad\qquad □$$

Remarks:

1. The results proven in this section apply when the pair (X, Y) is uniform only for each given value of Y, that is,

$$p(x|y) = \begin{cases} \frac{1}{|S_{X|Y}(y)|} & \text{if } x \in S_{X|Y}(y), \\ 0 & \text{if } x \notin S_{X|Y}(y). \end{cases}$$

2. Except for the last theorem and corollary, all other results in this section apply to all random pairs, not only uniform ones.

5   Open Problems

We mention two open problems.

1. Our bounds on $\bar{C}_\infty(X|Y)$ are tight in only two cases: when (X, Y) is uniform, and when its support set is a Cartesian product. For other random pairs, these bounds may be far asunder.

   Can tight bounds on $\bar{C}_\infty(X|Y)$ be derived for arbitrary random pairs? Can bounds be derived if some $\epsilon > 0$ probability of error is allowed?

101

2. In Equation (10), we have shown that if (X, Y) is uniform then

$$\bar{C}_2(X|Y) \leq 2H(X|Y) + 2\log\left(H(X|Y) + 1\right) + 7.5 \ .$$

Can the factor of two be improved, or is there a family of (X, Y) pairs with increasing $H(X|Y)$ such that for each pair in the family,

$$\bar{C}_2(X|Y) \geq 2H(X|Y) - o(H(X|Y) \ ?$$

Note that for *worst-case* complexities there are random pairs for which two messages require twice the minimum number of bits (cf., [5]).

6   Acknowledgements

I thank Abbas El Gamal for many ideas emanating from our joint work on [1] that have been put to use here, and Max Costa for sharing his league-problem protocol. I also acknowledge helpful discussions with Larry Shepp and Aaron Wyner.

References

[1] A. El Gamal and A. Orlitsky. Interactive data compression. In *Proc. of the 25th Annual Symposium on Foundations of Computer Science*, pages 100–108, 1984.

[2] A. Orlitsky. Worst-case interactive communication I: Two messages are almost optimal. *IEEE Transactions on Information Theory*, 36(5):1111–1126, September 1990.

[3] M. Costa. Private Communication, 1989.

[4] D. Slepian and J. Wolf. Noiseless coding of correlated information sources. *IEEE Transactions on Information Theory*, 19:471–480, 1973.

[5] A. Orlitsky. Worst-case interactive communication II: Two messages are not optimal. To appear, IEEE Transactions on Information Theory, July 1991.

[6] C.H. Papadimitriou and M. Sipser. Communication complexity. In *Proc. of the 14th Annual ACM Symposium on Theory of Computing*, pages 196–200, 1982.

[7] P. Duris, Z. Galil, and G. Schnitger. Lower bounds on communication complexity. In *Proc. of the 16th Annual ACM Symposium on Theory of Computing*, pages 81–91, 1984.

[8] N. Nisan and A. Wigderson. Rounds in communication complexity revisited. In *Proc. of the 23rd Annual ACM Symposium on Theory of Computing*, 1991.

[9] P. Elias. Universal codeword sets and representations of the integers. *IEEE Transactions on Information Theory*, 21:194–203, 1975.

Adaptive Lossless Data Compression over a Noisy Channel

James A. Storer
Computer Science Dept.
Brandeis University
Waltham, MA 02254

John H. Reif
Computer Science Dept.
Duke University
Durham, NC 27707

Abstract: With *dynamic communication* a sender and receiver work in a "lock-step" cooperation to maintain identical copies of a *dictionary D* (which is constantly changing). A key application of dynamic communication is *adaptive data compression*. A potential drawback of dynamic communication is *error propagation* (that causes the sender and receiver dictionaries to diverge and possibly corrupt all data to follow). Protocols that require the receiver to request re-transmission from the sender when an error is detected can be impractical for many applications where such two way communication is not possible or self-defeating (e.g., with data compression, re-transmission is tantamount to losing the data that could have been transmitted in the mean time). We present a new model of error *resilient* communication where even though errors may not be detected, there are strong guarantees that their effects will not propagate.

1. Introduction

Dynamic communication is when data that is sent by a sender to a receiver over a communication channel is encoded / decoded with respect to dynamically changing data. An encoder and decoder cooperate to maintain identical copies of a *dictionary D* (which is constantly changing). The encoder reads characters from the input stream that form an entry of D, transmits the index of this entry, and updates D (with some method that depends only on the current contents of D and the current match). Similarly, the decoder repeatedly receives an index, retrieves the corresponding entry of D to write to the output stream, and then performs the same algorithm as the encoder to update its copy of D.

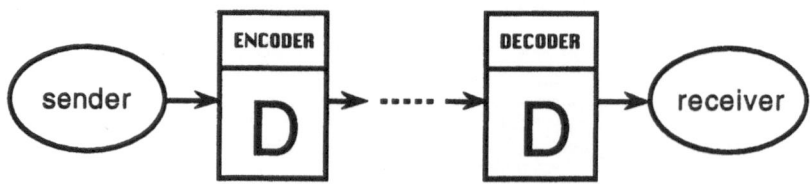

Dynamic Communication

Applications of dynamic communication include:

- *learning theory*

- *adaptive data compression*

- *robotics*
  (For example, reporting of data by an autonomous remote robot that is mapping unexplored terrain and transmitting coordinates as displacements from previous locations.)

Adaptive data compression can in some sense be viewed as a canonical example of dynamic communication and is the application that originally motivated this work; see the book of Storer [1988] for an introduction to the subject and references to the literature.

An apparent drawback of dynamic communication is error propagation; that is, a single error on the communication channel (e.g., add, delete, or change of an index sent on the channel) could cause the dictionaries of the encoder and decoder to differ, which in the worst case could corrupt all data to follow. With *one-way communication*, where the decoder cannot send messages to the encoder, the problem becomes even more critical. Not only are applications that require one-way communication becoming more common (e.g., space communications), but with many existing communication systems where the full bandwidth of the channel is consumed, even if a two-way channel is available to let the encoder know that an error has occurred, re-transmission of data can often be tantamount to losing data (i.e., losing new data that could have been transmitted during the time used to re-transmit the old data). Re-transmission can also have error propagation problems of its own.

In some sense, dynamic one-way communication can be viewed as "raising the stakes" for the effects of errors on the channel; even if the chance of an error is made very small via a standard error detection / correction protocol, if an error does occur on a single data item, it can have the catastrophic effect of corrupting an unbounded number of additional data items. We propose a new model for "error resiliency", where no attempt is made to detect or correct errors but strong guarantees are provided that errors don't propagate. This technique can be combined with standard error detection / correction protocols to yield a system with a low rate of errors that do not propagate.

The next section formally defines the dynamic communication model and the following section presents a basic model for error resilient communication. Section 4 then presents a scheme for protecting against error propagation from k errors for any constant k, under any distribution of errors. In Section 5, we employ Chernoff bounds to "expand" the k-error technique to give propagation protection with very high probability against a fixed error rate. Section 6 notes that although the constants required for the proof in Section 5 are reasonable, it is likely that there is room for improvement in practice.

2. One-Way Dynamic Communication

Throughout this paper when discussing one-way dynamic communication with respect to a dictionary D, we use the notation:

$|D|$ = the current number of entries in D

$|D_{max}|$ = the maximum number of entries that D may contain

Generic encoding and decoding algorithms for one-way dynamic communication are shown on the following page. For simplicity we assume that indices of D are encoded using exactly $\lceil log_2|D_{max}|\rceil$ bits, and we refer to such codes as *pointers*. However, during the period when D is not yet full, the generic algorithms are perfectly well defined if only $\lceil log_2|D|\rceil$ bits are sent for each index.

The *match heuristic* reads a string from the input stream that is in the dictionary (it must always be possible to read such a string since all strings of length 1 are always in the dictionary). Examples of match heuristics that might be used for data compression are:

- *Greedy*: Read the longest match possible.

- *Lookahead*: Employ a lookahead buffer to check if taking a shorter match now will pay off with better compression later.

- *Delimeter*: Use special delimeter characters (which may be supplied in advance or adaptively learned) to guide the "parsing" of the input.

After the match has been encoded by the encoder or decoded by the decoder, "learning" consists of modifying the dictionary by adding one or more new strings with an update heuristic and, if the dictionary is full, deleting some strings to make room. Examples of update heuristics that might be used for data compression are:

- *First Character* (FC): Add the last match concatenated with the first character of the current match. Note that in practice, the *next character* heuristic (NC), which adds the current match concatenated with the next character of the input, performs similarly to FC but complicates Step 2b of the decoder.

- *Identity* (ID): Add the last match concatenated with the current match.

- *All Prefixes* (AP): Add the set of strings consisting of the last match concatenated with each of the prefixes of the current match.

Examples of deletion heuristics that might be used for data compression are:

- *Least Recently Used* (LRU): Delete the string that has been matched least recently.

- *Swap when Full* (SWAP): Keep two dictionaries. When the *primary* dictionary becomes full, start learning new entries in the *auxiliary* dictionary but continue compressing data with the primary dictionary. From this point on, each time the auxiliary dictionary becomes full, the roles of the primary and auxiliary dictionaries are reversed, and the secondary dictionary is reset to be empty. The SWAP heuristic can be viewed as a discrete version of LRU.

(1) Initialize the *local dictionary* D to have one entry
for each character of the input alphabet.

(2) repeat forever

    (a) {Get the current match string s:}

        Use a *match heuristic* MH to read s from the input.

        Transmit $\lceil log_2|D| \rceil$ bits for the index of s.

    (b) {Update D:}

        Add each of the strings specified by an *update heuristic* UH to D
(if D is full, then first employ a *deletion heuristic* DH to make space).

Generic Encoding Algorithm

(1) Initialize the local dictionary D by performing Step 1 of the encoding algorithm.

(2) repeat forever

    (a) {Get the current match string s:}

        Receive $\lceil log_2|D| \rceil$ bits for the index of s.

        Retrieve s from D and output the characters of s.

    (b) {Update D:}

        Perform Step 2b of the encoding algorithm.

Generic Decoding Algorithm

The example heuristics discussed above can be used to model a broad class of data compression methods that employ *textual substitution* (e.g, Lempel and Ziv [1976], Ziv and Lempel [1977, 1978], Storer and Szymanski [1978], Welch [1984], Miller and Wegman [1985], Storer [1988]). Reif and Storer [1990] describe a custom VLSI design for textual substitution that is capable of processing input streams of over 300 million bits per second.

The exact choice of the update heuristic is not important for the work described here, so long as the following axioms are satisfied:

Succinctness: Except for the entries that contain the characters of the input alphabet, each entry of D can be represented as a pair of pointers to two other entries.

Robustness: For any fixed constants $0 < h \leq 1$ and $1 \leq k$, it suffices to update with probability h and only after the pair has been seen k times. Furthermore, there is a constant $1 \leq \alpha$, called the *learning constant*, such that D reaches size $h|D_{max}|$ after at most $\alpha k|D_{max}|$ pointers are transmitted.

Sufficieny: The swap deletion heuristic suffices and there exits a constant $\epsilon > 0$, called the *dictionary constant*, such that $\epsilon|D^{max}|$ entries may be left unfilled.

In the second two axioms we have intentionally left out a definition of what it means to "suffice", since this is application dependent. For adaptive data compression it means that the compression ratio is not significantly affected for large input strings (no worse that a factor of $1 + \delta$ for arbitrary small δ). The FC, NC, ID, and AP heuristics all satisfy the three axioms; note that the learning constant for all of these heuristics is ≤ 2, and we shall henceforth assume that $\alpha \leq 2$ to simplify the presentation. Note that in practice, a dictionary constant $\epsilon \leq .1$ will have a negligible effect on compression since only a small fraction of a bit per pointer is "wasted".

The techniques to be presented can be used in systems where once the dictionary is full, it remains the same until the system is restarted. The motivation being the specification of the swap heuristic in the sufficiency axiom is that it is realistic to assume that extremely secure swap signals (e.g., several hundred bits to encode a single swap bit) can be sent from the encoder to decoder once in a while, at an insignificant amortized cost, to insure that the decoder swaps at the same time as the encoder. We shall not address this issue further and henceforth limit our attention to the prevention of error propagation during the period when the dictionary is filling up.

3. A Model For Error Resilient Communication

We assume that pointers sent from the encoder to the decoder are subject to the following errors:

add: An extra pointer, chosen at random from the space of all pointers, is inserted into the communication stream.

delete: A pointer is deleted from the communication stream.

change: A new pointer, chosen at random from the space of all pointers, replaces a pointer.

We consider the following classes of error distributions:

uniform: Errors occur randomly.

arbitrary: Errors may be arbitrarily correlated.

Definition: A communication channel is *stable* if no further errors implies no further corrupted output from the decoder. ○

The goal is to provide guarantees that the system will always be stable (that is, no bytes except those corrupted are lost).

4. Protection Against K Errors

In this section we present an encoding scheme which, for any fixed integer constant $k \geq 0$, guarantees that k or fewer errors on the channel will not cause *any* error propagation. The two key ideas behind this construction are:

- Universal hashing is used to compute where a new entry is to be placed in the dictionary; this has the effect of eliminating dependence between addresses that is normally present in dynamic communication, so that if a given index is not used right away, it will have no effect on what indices are used in the future.

- Counts are maintained for all pointer pairs seen thus far and a pair is used only if it "warms up" to be a clear winner over pairs that hash to the same address.

Definition: Suppose that counts (initially 0) are maintained in the encoder for all pointer pairs sent thus far and in the decoder for all pointer pairs received thus far. For a sequence S of pointers with errors that is sent by the encoder, $LAG(S)$ is the maximum amount that any count in the decoder is incorrectly increased due to an error on the channel plus the maximum amount any count is decreased due to an error on the channel. For any integer $k >= 0$, the *warming value for k*, $w(k)$, is the smallest real number such that $LAG(S) \leq w(k)$ over all sequences of pointers with at most k errors. ○

Lag Theorem: For any $k \geq 1$, $w(k) \leq 3k$.

Proof: Let

$S = C_0 O_1 C_1 \ldots O_m C_m$ be the sequence of pointers *sent* by the encoder.

$R = C_0 I_1 C_1 \ldots I_m C_m$ be the sequence of pointers *received* by the decoder.

where:

$$|\prod_{i=1}^{m} I_m| \leq k$$

$$|C_i| \geq 1, \; 0 < i < m$$

$$|I_i| \geq 1, \; 1 \leq i \leq m$$

That is, the C's are blocks of correctly sent blocks of pointers and the O's are blocks of tokens that are incorrectly received as the I blocks, where *all* pointers are incorrect. Note that a legal such partitioning of S and R may not be unique but at least one exists and any one suffices for our construction.

Now consider a particular block O_i that is received as I_i. Let

x = the number of pointers that are changed
y = the number of pointers that are deleted
z = the number of pointers that are added

and write:

$$O_i = q_1 \ldots q_{x+y}$$
$$I_i = r_1 \ldots r_{x+z}$$

Note that any way of choosing x, y, and z that corresponds to a way to convert O_i to I_i suffices for our construction. Let p be the last pointer of C_{i-1} and s be the first pointer of C_i (it may be that p and/or s is empty; the construction to follow is easily modified for these cases). The only counts that might not be incremented are due to the loss of

$$pq_1, \; q_1 q_2 \; \ldots \; q_{x+y-1} q_{x+y}, \; q_{x+y} s$$

which in the worst case can increase the lag by $x + y + 1$. The only counts that might be incorrectly incremented are due to the addition of

$$pr_1, \; r_1 r_2 \; \ldots \; r_{x+z-1} r_{x+z}, \; r_{x+z} s$$

which in the worst case can increase the lag by $x + z + 1$. So the total change in lage for the transformation of O_i to I_i is at most $2x + y + z + 2$. In fact, for the case $y = z = 0$ (there are only change errors), this upper bound can be reduced by observing that if

$$pq_1 \; = \; q_1 q_2 \; = \; \ldots \; = \; q_{x+y-1} q_{x+y} \; = \; q_{x+y} s$$

then it follows that $p = q_1 \ldots = q_{x+y} = s$, and so it can't be that $pr_1 = r_1 r_2$ (or that $pr_1 = r_1 s$) or r_1 would not be an incorrect pointer; so the lag can be at most $2x + 1$. Hence, we have:

$$LAG(O_i) \leq \left\{ \begin{array}{l} 2x + 1 \; if \; (y + z) = 0 \\ 2x + y + z + 2 \; if \; (y + z) > 0 \end{array} \right\}$$

Let $w = x + y + z$. For the case $(y + z) = 0$, assuming $k > 0$ (otherwise, $LAG(O_i) = 0$), $2x + 1 \leq 2w + 1 \leq 3w$. For the case $(y + z) > 0$:

$$\begin{aligned} 2x + y + z + 2 &= 2(w - (y + z)) + (y + z) + 2 \\ &= 2w - (y + z) + 2 \\ &\leq 2w + 1 \\ &\leq 3w \end{aligned}$$

Hence the warming value for each block of w errors is $3w$, and the theorem follows. $\bigcirc$

Corollary: If there are just add or just delete errors, then $w(k) \leq k$.

Proof: Follows from a closer inspection of the proof of the Lag Theorem. $\bigcirc$

Given the Lag Theorem, the following page shows the Generic Encoding and Decoding Algorithms modified to employ the *k-error protocol* to insure perfect protection against any k errors; that is, to insure that the channel is always stable. Note that, as discussed earlier, since we are assuming that a secure version of the SWAP heuristic is employed if required, the deletion heuristic is not addressed in this protocol.

For correctness of the protocol, observe that for each pointer p in a sequence S of pointers sent by the encoder, the pointer pair qr that p represents is a clear "winner" among all pointer-pairs that hash to p; that is, the count of qr is more than $LAG(S)$ greater than the count of any other pair uv that hashes to p. Hence, when the decoder receives an uncorrupted pointer p, the pointer pair qr with the largest count that hashes to p must be the correct pair for p because k errors cannot cause some other pair uv that hashes to p to have a count equal or greater than that of qr.

As for compression performance, the learning process is slowed down as as k gets larger. In addition, it is possible for entries to "race"; that is, the counts of both entries become very large but the difference between them never becomes large enough for one to win. We conjecture that in practice, performance will not be significantly affected for small values of k and reasonably size dictionaries (e.g., 2^{16} or more entries). However, we shall not address bounds on the compression performance of the k-error protocol in this section since in the next section we shall consider techniques that generalize the protocol and avoid many of these problems.

111

Encoding Algorithm with K-Error Protocol:

(1) Initialize the *local dictionary* D to have one entry for each character of the input alphabet and to have an empty hash table that is capable of storing pointer pairs.

(2) **repeat forever**

    (a) {Get the current match string s:}

        Read the longest string s for which there exists pointer pair qr such that:

        1. $s = STRING(qr)$

        2. $count(qr) > count(uv) + w(k)$ for all uv such that $h(uv) = h(qr)$

        Transmit $\lceil log_2 |D| \rceil$ bits for the index of $h(qr)$.

    (b) {Update D:}

        if this is the first time qr has been used

            then $count(qr) := 0$

            else $count(qr) := count(qr) + 1$

Decoding Algorithm with K-Error Protocol:

(1) Initialize the local dictionary D by performing Step 1 of the encoding algorithm.

(2) **repeat forever**

    (a) {Get the current match string s:}

        Receive $\lceil log_2 |D| \rceil$ bits for the next pointer p.

        Output $STRING(qr)$ for the pointer pair qr such that:

        1. $h(qr) = p$

        2. $count(qr)$ is largest among all pairs that hash to p

        (Output nothing if no such pointer exists.)

    (b) {Update D:}

        Perform Step 2b of the encoding algorithm.

Notes:

- h denotes a universal hash function
- $STRING(qr)$ denotes the string corresponding to the pointer pair qr

5. High Probability Protection Against an Error Rate

In this section we employ Chernoff bounds to examine more carefully the proof of the lag theorem of the last section. The idea is to show that the k-error protocol actually gives, with high probability, strong protection against a fixed error rate during the period when the dictionary is changing and is vulnerable to error propagation.

Recall the Chernoff bound on the number of heads in a sequence of independent coin tossings (see Hagerup and Rub [1989]):

Let:

$p = $ the probability of coming up heads $(1 - p$ for tails)

$n = $ number of coin tossings

$X = $ number of heads

$u = $ mean value of $X = np$

$e(n) = (\frac{n}{n-1})^n = 2.7182\ldots$

$e = $ the natural logarithm base $= \lim_{n->\infty} e(n)$

(Note that $e < e(n)$ for all $n > 1$.)

Then for any real $h \geq 0$:

$$Prob(X \geq (1+h)u) \ \leq \ \left(\frac{e^h}{(1+h)^{(1+h)}}\right)^u \ = \ \frac{1}{e}\left(\frac{e}{(1+h)}\right)^{(1+h)u}$$

In particular, for $z \geq eu$ we shall employ the bound in its following form:

$$Prob(X \geq z) \ \leq \ \frac{1}{e}\left(\frac{z}{eu}\right)^{-z}$$

Damping Theorem: If a sufficiently large dictionary is employed, then if the k-error protocol of the last section is used on a channel with a uniform error probability of $1/r$ (on the average, 1 error for every r pointers), then the probability that the system looses stability (i.e., the probability that any errors propagate) is:

$$< \ \frac{1}{\left(\frac{r}{8e}\right)^{\left(\frac{w(k)}{2}+1\right)}}$$

In particular, for $r > 10^{12}$ (a reasonable assumption in practice for a clean channel with a low-overhead error-correction mechanism), choosing $w(k) > (\frac{9}{4}k - 2)$ [1] yields a probability that is:

$$< \frac{1}{r^k}$$

For example, if $k = 4$, then by using $w(k) = 7$, an error rate of $1/r^{12}$ is effectively "damped" to $r = 1/10^{48}$.

Proof Sketch: We follow a "brute-force" approach to essentially eliminate the need to address hash conflicts in our analysis of the protocol; a small constant c extra bits are added to each pointer so that the $|D_{max}|$ dictionary entries are hashed into a space of $|D_{max}|2^c$ indices. This approach is at the cost of either a reduction in the amount of compression or the necessity of using a large dictionary (so that c can be amoritized against a large pointer size). In the next section, we note that this brute-force approach is reasonable but that addressing hash conflicts in the analysis may yield better constants in practice.

Consider a particular index i that is used for the first time in a sequence of n pairs that are hashed into the dictionary. The probability that all other pairs do *not* hash to index i is:

$$\geq \left(1 - \frac{1}{n2^c}\right)^{n-1}$$

$$> \left(1 - \frac{1}{n2^c}\right)^{n}$$

$$= \left(\left(1 - \frac{1}{n2^c}\right)^{n2^c}\right)^{2^{-c}}$$

$$\geq \left(\frac{1}{e(n)}\right)^{2^{-c}}$$

$$\geq \frac{1 - 2^{-c}}{\left(\frac{e(n)}{e}\right)^{2^{-c}}}$$

Note that the last inequality follows because for all $0 < x < 1$, $1/e^x > 1 - x$.

Hence, the probability that two pairs hash to index i is:

$$\leq 1 - \frac{1 - 2^{-c}}{\left(\frac{e(n)}{e}\right)^{2^{-c}}}$$

$$< 2^{-c} + \left(\left(\frac{e(n)}{e}\right)^{2^{-c}} - 1\right)$$

$$= 2^{-c} + errorterm$$

Thus the expected number of hash conflicts is less than

---

[1] This is an approximate value; any $w(k) > 2.2509k - 2$ will do.

$$n(2^{-c} + errorterm)$$

and by the Chernoff bound it follows that the probability that there are more than ϵn hash conflicts is:

$$< \left(\frac{\epsilon}{e\,(2^{-c} + errorterm)} \right)^{-\epsilon n}$$

For any constant $d > 1$, this expression can be made less than $d^{-\epsilon n}$ if c is made sufficiently large (by choosing n sufficiently large).

Given that the probability of more than ϵn hash conflicts can be made arbitrarily low, our approach is employ the very "conservative" strategy of "throwing out" all indices that correspond to a hash conflict (even though many or all of these entries may not cause any problems for the decoder); we compute a bound on the two quantities:

X_{ij} = the number of times that the count of pair i, j is incorrectly increased by the decoder due to errors on the channel before there are $\epsilon|D_{max}|$ valid entries in the dictionary of the encoder.

Y_{ij} = the number of times that the count of a valid pair i, j is incorrectly *not* increased by the decoder due to errors on the channel before its count is $w(k)+1$ in the dictionary of the encoder.

Let:

$E_\epsilon(D)$ = The expected number of pointers transmitted by the encoder before there are $\epsilon|D_{max}|$ valid entries in the dictionary of the encoder.

From the proof of the LAG Theorem, we know that each error can cause at most 2 incorrect increases in counts and at most 2 incorrect decreases. So the expected value of X_{ij}, which we denote by $u(X_{ij})$, is given by: [2]

$$u(X_{ij}) = \frac{2E_\epsilon(D)}{r|D_{max}|}$$

Since $E_\epsilon(D) \leq \alpha w(k)|D_{max}|$, we have by the robustness axiom:

$$u(X_{ij}) \leq \frac{2\alpha w(k)}{r}$$

Similar to X_{ij}, we can compute:

$$u(Y_{ij}) \leq \frac{2w(k)}{r}$$

Hence, the probability that X_{ij} or Y_{ij} is $\geq (\frac{w(k)}{2} + 1)$ is:

---

[2] Note that although it is tempting to use $|D_{max}|^2$ in this expression, this can't be done without further assumptions.

$$\leq \frac{1}{e} \left(\frac{\frac{w(k)}{2}+1}{\frac{2e\alpha w(k)}{r}} \right)^{\frac{w(k)}{2}+1}$$

$$\leq \left(\frac{r}{8e} \right)^{\frac{w(k)}{2}+1}$$

From the above bound, the theorem follows, since a LAG of more than $w(k)$ cannot occur if no values of X_{ij} or Y_{ij} are more than $w(k)/2$.

As for the statement for the theorem under the assumption that $r > 10^{12}$, if we write $w(k) = 2xk - 2$, we can solve for the minimum value of x by simplifying the expression above as follows:

$$= r^k \left(\frac{r^{w(k)/2-k-1}}{(8e)^{w(k)/2+1}} \right)$$

$$= r^k \left(\frac{r^{x-1}}{(8e)^x} \right)^k$$

By letting $y = \frac{12}{\log_{10}(8e)}$, then the numerator and denominator of the expression above are equal when $x = y/(y-1)$, and so $x \geq 2.2509 \sim 9/4$, and a bound of $\frac{1}{r^k}$ follows for $r > 10^{12}$ and $w(k) > (\frac{9}{4}k - 2)$. $\bigcirc$

6. Practical Considerations

We conclude this paper by examining just how effective the Damping Theorem of the last section might be in practice. The weakness of the theoem is that it adds c extra bits to each pointer, which can be made to have an insignificant effect on compression by using a very large dictionary (and hence a very large pointer size) but will most likely be significant for dictionary sizes that are used in practice. For example, if we take $|D_{max}| = 2^{16}$ (a common and practical choice for data compression), and $\epsilon = .1$ (again, a reasonable value in practice), then we must choose $c \geq 5$; taking $c = 5$ and computing the *errorterm* used in the proof of the Damping Theorem, we see that the probability that there are more than ϵn hash conflicts is $< 2^{-1,500}$; far more than is needed by the remainder of the proof. However, the cost for this security against many hash conflicts is an extra 5 bits for every 16-bit pointer. If we consider a typical example for lossless compression with the ID heuristic using a dictionary of size 2^{16}, we might expect the compressed data to be 30 percent of the size of the original data, but now with the extra 5 bits per pointer, the compressed size will be $\frac{21}{16}30 = 39$ percent of the original size. Although a cost of 10 percent in compression ratio such as this might be reasonable in practice, it would be nice to avoid it.

We conjecture that this overhead can be reduced by a tighter analysis in the proof of the damping theorem that does not throw out all indices with hash conflicts, but rather, only throws out those that delay the dictionary filling process due to "racing" and "thrashing" of counts. We leave such analysis as a subject for future research.

7. References

T. Hagerup and C. Rub [1989]. "A Guided Tour of Chernoff Bounds", *Information Processing Letters* 33, 305-308.

A. Lempel and J. Ziv [1976]. "On the Complexity of Finite Sequences", *IEEE Transactions on Information Theory* 22:1, 75-81.

V. S. Miller and M. N. Wegman [1985]. "Variations on a Theme by Lempel and Ziv", *Combinatorial Algorithms on Words*, Springer-Verlag (A. Apostolico and Z. Galil, editors), 131-140.

J. Reif and J. A. Storer [1990]. "A Parallel Architecture for High Speed Data Compression", *Proceedings Third Symposium on the Frontiers of Massively Parallel Computation*, College Park, MD.

J. A. Storer [1988]. *Data Compression: Methods and Theory*, Computer Science Press, Rockville, MD.

J. A. Storer and T. G. Szymanski [1978]. "The Macro Model for Data Compression", *Proceedings Tenth Annual ACM Symposium on the Theory of Computing*, San Diego, CA, 30-39.

T. A. Welch [1984]. "A Technique for High-Performance Data Compression", *IEEE Computer* 17:6, 8-19.

J. Ziv and A. Lempel [1977]. "A Universal Algorithm for Sequential Data Compression", *IEEE Transactions on Information Theory* 23:3, 337-343.

J. Ziv and A. Lempel [1978]. "Compression of Individual Sequences Via Variable-Rate Coding", *IEEE Transactions on Information Theory* 24:5, 530-536.

Computer Science

Parallel string matching algorithms[*]

Dany Breslauer[†]
Columbia University

Zvi Galil
Columbia University and
Tel-Aviv University

Abstract

The string matching problem is one of the most studied problems in computer science. While it is very easily stated and many of the simple algorithms perform very well in practice, numerous works have been published on the subject and research is still very active. In this paper we survey recent results on parallel algorithms for the string matching problem.

1 Introduction

You are given a copy of Encyclopedia Britanica and a word and requested to find all the occurrences of the word. This is an instance of the string matching problem. More formally, the input to the string matching problem consists of two strings $TEXT[1..n]$ and $PATTERN[1..m]$; the output should list all occurrences of the pattern string in the text string. The symbols in the strings are chosen from some set which is called an *alphabet*. The choice of the alphabet sometimes allows us to solve the problem more efficiently as we will see later.

A naive algorithm for solving the string matching problem can proceed as follows: Consider the first $n - m + 1$ positions of the text string. Occurrences of the pattern can start only at these positions. The algorithm checks each of these positions for an occurrence of the pattern. Since it can take up to m comparisons to verify that there is actually an occurrence, the time complexity of this algorithm is $O(nm)$. Note that the only operations involving the input strings in this algorithm are comparisons of two symbols.

The only assumption we made about the alphabet in the algorithm described above is that alphabet symbols can be compared and the comparison results in an equal or unequal answer. This assumption, often referred to as the *general alphabet* assumption, is the weakest assumption we will have on the alphabet, and, as we have seen, is sufficient to solve the string matching problem. However, although the definition of the string matching problem does not

---

[*]Work partially supported by NSF Grant CCR-90-14605.
[†]Partially supported by an IBM Graduate Fellowship.

121

require the alphabet to be ordered, an arbitrary order is exploited in several algorithms [3, 4, 21, 22] which make use of some combinatorial properties of strings over an ordered alphabet [40]. This assumption is reasonable since the alphabet symbols are encoded numerically, which introduces a natural order. Other algorithms use a more restricted model where the alphabet symbols are small integers. Those algorithms usually take advantage of the fact that symbols can be used as indices of an array [2, 6, 42, 48] or that many symbols can be packed together in one register [26]. This case is usually called *fixed alphabet*.

Many sequential algorithms exist for the string matching problem and are widely used in practice. The better known are those of Knuth, Morris and Pratt [35] and Boyer and Moore [12]. These algorithms achieve $O(n+m)$ time which is the best possible in the worse case and the latter algorithm performs even better on average. Another well known algorithm which was discovered by Aho and Corasik [2] searches for multiple patterns over a fixed alphabet. Many variations on these algorithms exist and an excellent survey paper by Aho [1] covers most of the techniques used.

All these algorithms use an $O(m)$ auxiliary space. At a certain time it was known that a logarithmic space solution was possible [28], and the problem was conjectured to have a time-space trade off [10]. This conjecture was later disproved when a linear-time constant-space algorithm was discovered [29] (see also [21]). It was shown that even a 6-head two-way finite automaton can perform string matching in linear time. It is still an open problem whether a k-head one-way finite automaton can perform string matching. The only known cases are for $k = 1, 2, 3$ [30, 38, 39] where the answer is negative.

Recently, few papers have been published on the exact complexity of the string matching problem. Namely, the exact number of comparisons necessary in the case of a general alphabet. Surprisingly, the upper bound of about $2n$ comparisons, the best known before [5, 35], was improved to $\frac{4}{3}n$ comparisons by Colussi, Galil and Giancarlo [16]. In a recent work Cole [17] proved that the number of comparisons performed by the original Boyer-Moore algorithm is about $3n$.

In this paper we will focus on parallel algorithms for the string matching problem. Many other related problems have been investigated and are out of the scope of this paper [1, 27]. For an introduction to parallel algorithms see surveys by Karp and Ramachandran [33] and Eppstein and Galil [24].

In parallel computation, one has to be more careful about the definition of the problem. We assume the the input strings are stored in memory and the required output is a Boolean array $MATCH[1..n]$ which will have a 'true' value at each position where the pattern occurs and 'false' where there is no occurrence.

All algorithms considered in this paper are for the parallel random access machine (PRAM) computation model. This model consists of some processors with access to a shared memory. There are several versions of this model which differ in their simultaneous access to a memory location. The weakest is the exclusive-read exclusive-write (EREW-PRAM) model where at each step simultaneous read operation and write operations at the same memory location are not allowed. A more powerful model is the concurrent-read exclusive-write (CREW-PRAM) model where only simultaneous read operations are allowed. The most powerful model is the concurrent-read concurrent-write (CRCW-PRAM) model where read and write operations can be simultaneously executed.

In the case of the CRCW-PRAM model, there are several ways of how write conflicts are resolved. The weakest model, called the *common* CRCW-PRAM assumes that when several processors attempt to write to a certain memory location simultaneously, they all write the same value. A stronger model called the *arbitrary* CRCW-PRAM assumes an arbitrary value will be written. An even stronger model, the *priority* CRCW-PRAM assumes each processor has a priority and the highest priority processor succeeds to write. Most of the CRCW-PRAM algorithms described in this paper can be implemented in the common model. In fact these algorithms can be implemented even if we assume that the same constant value is always used in case of concurrent writes. However, to simplify the presentation we will sometimes use the more powerful priority model.

For the algorithms discussed in this paper we assume that the length of the text string is $n = 2m$, where m is the length of the pattern string. This is possible since the text string can be broken into overlapping segments of length $2m$ and each segment can be searched in parallel.

Lower bounds for some basic parallel computational problems can be applied to string matching. A lower bound of $\Omega(\frac{\log n}{\log \log n})$ for computing the parity of n input bits on a CRCW-PRAM with any polynomial number of processors [7] implies that one cannot count the number of occurrences faster than $O(\frac{\log n}{\log \log n})$. Another lower bound of $\Omega(\log n)$ for computing a Boolean AND of n input bits on any CREW-PRAM [20] implies an $\Omega(\log n)$ lower bound for string matching in this parallel computation model.

These lower bounds make the possibility of sublogarithmic parallel algorithms for any problem very unlikely. However several problems are known to have such algorithms [8, 9, 11, 36, 44, 45] including string matching. In fact, a very simple algorithm can solve the string matching problem in constant time using nm processors on a CRCW-PRAM: similarly to the naive sequential algorithm, consider each possible start of an occurrence. Assign m processors to each such position to verify the occurrence. Verifying an occurrence is simple; perform all m comparisons in parallel and any mismatch changes a value of the $MATCH$ array to indicate that an occurrence is impossible.

The following theorem will be used throughout the paper.

Theorem 1.1 (Brent [13]): Any PRAM algorithm of time t that consists of x elementary operations can be implemented on p processors in $\lceil x/p \rceil + t$ time.

Using this theorem for example, we can slow down the constant-time algorithm describe above to run in $O(s)$ time on $\frac{nm}{s}$ processors.

In the design of a parallel algorithm, one is also concerned about the total number of operations performed, which is the time-processor product. The best one can wish for is the number of operations performed by the fastest sequential algorithm. A parallel algorithm is called *optimal* if that bound is achieved. Therefore, in the case of the string matching problem, an algorithm is optimal if the time-processor product is linear in the length of the input strings.

An optimal parallel algorithm discovered by Galil [26] solves the problem in $O(\log m)$ time using $\frac{n}{\log m}$ processors. This algorithm works for fixed alphabet and was later improved by Vishkin [46] for general alphabet. Optimal algorithms by Karp and Rabin [32] and other algorithms based on Karp, Miller and Rosenberg's [31] method [23, 34] also work in $O(\log m)$ time for fixed alphabet. Breslauer and Galil [14] obtained an optimal $O(\log \log m)$ time algorithm for general alphabet. Vishkin [47] developed an optimal $O(\log^* m)^1$ time algorithm. Unlike the case of the other algorithms this time bound does not account for the preprocessing of the pattern. The preprocessing in Vishkin's algorithm takes $O(\frac{\log^2 m}{\log \log m})$. Vishkin's super fast algorithm raised the question whether an optimal constant-time algorithm is possible. This question was partially settled in a recent paper by Breslauer and Galil [15] showing an $\Omega(\log \log m)$ lower bound for parallel string matching over a general alphabet. The lower bound proves that a slower preprocessing is crucial for Vishkin's algorithm.

This paper is organized as follows. In Section 2 we describe the logarithmic time algorithms. Section 3 is devoted to Breslauer and Galil's $O(\log \log m)$ time algorithm. Section 4 covers the matching lower bound. Section 5 outlines the ideas in Vishkin's $O(\log^* m)$ algorithm. In some cases we will describe a parallel algorithm that achieves the claimed time bound using n processors. The optimal version, using $O(\frac{n}{t})$ processors, can be derived using standard techniques. Many questions are still open and some are listed in the last section of this paper.

2 Logarithmic time algorithms

The simplest parallel string matching algorithm is probably the randomized algorithm of Karp and Rabin [32]. The parallel version of their algorithm assumes the alphabet is binary and translates the input symbols into a 2×2 non-singular matrices. The following representation is used, which assures a unique representation for any string as a product of the matrices representing it.

$$f(0) = \begin{pmatrix} 1 & 0 \\ 1 & 1 \end{pmatrix} \qquad f(1) = \begin{pmatrix} 1 & 1 \\ 0 & 1 \end{pmatrix}$$

$$f(s_1 s_2 \cdots s_i) = f(s_1) f(s_2) \cdots f(s_i)$$

Most of the work in the algorithm is performed using a well known method for parallel prefix computation summarized in the following theorem:

Theorem 2.1 (Folklore, see also [37]): Suppose a sequence of n elements $x_1, x_2, \cdots, x_n$ are drawn from a set with an associative operation $*$, computable in constant time. Let $p_i = x_1 * x_2 * \cdots x_i$, usually called a prefix sum. Then an EREW-PRAM can compute all $p_i \quad i = 1 \cdots n$, in $O(\log n)$ time using $\frac{n}{\log n}$ processors.

---

[1] The function $\log^* m$ is defined as the smallest k such that $\log^{(k)} m \leq 2$, where $\log^{(1)} m = \log m$ and $\log^{(i+1)} m = \log \log^{(i)} m$.

Karp and Rabin's algorithm multiplies the matrices representing the pattern to get a single matrix which is called the *fingerprint* of the pattern. By Theorem 2.1 this can be done by a $\frac{m}{\log m}$-processor EREW-PRAM in $O(\log m)$ time. The text string is also converted to the same representation and matches can be reported based only on comparison of two matrices; the fingerprint of the pattern and the fingerprint of each text position. To compute the fingerprint of a text position j, which is the product of the matrix representation of the substring starting at position j and consisting of the next m symbols, first compute all prefix products for the matrix representation of the text and call them P_i. Then compute the inverse of each P_i; the inverse exists since each P_i is a product of invertible matrices. The fingerprint for a position j, $2 \leq j \leq n - m + 1$ is given by $P_{j-1}^{-1} P_{j+m-1}$; the finger print of the first position is P_m. By Theorem 2.1 the prefix products also take optimal $O(\log m)$ time on an EREW-PRAM. Since the remaining work can be done in constant optimal time, the algorithm works in optimal $O(\log m)$ total time.

However, there is a problem with the algorithm described above. The entries of those matrices may grow too large to be represented in a single register; so the numbers are truncated modulo some random prime p. All computations are done in the field $\mathcal{Z}_p$ which assures that the matrices are still invertible.

This truncated representation does not assure uniqueness, but Karp and Rabin show that the probability of their algorithm erroneously reporting a nonexisting occurrence is very small if p is chosen from a range which is large enough. This algorithm is in fact the only parallel algorithm which works in optimal logarithmic time on an EREW-PRAM; all the algorithms we describe later need a CRCW-PRAM.

The method used by Karp, Miller and Rosenberg [31] for sequential string matching can be adopted also for parallel string matching. Although the original algorithm worked in $O(n \log n)$ time, Kedem, Landau and Palem [34] were able to obtain an optimal $O(\log m)$ time parallel algorithm using a similar method. Chrochemore and Rytter [23] independently suggested a parallel implementation in $O(\log m)$ time using n processors. Another parallel algorithm which uses a similar method is the suffix tree construction algorithm of Apostolico et al. [6] which can also be used to solve the string matching problem. All these parallel algorithms need an arbitrary CRCW-PRAM and are for fixed alphabets; they also need a large memory space. It seems that this method cannot be used to obtain faster than $O(\log n)$ string matching algorithms, however it is applicable to other problems [23].

We describe a logarithmic time implementation of the Karp, Miller and Rosenberg [31] method for an n-processor arbitrary CRCW-PRAM. Consider the input as one string of length $l = n + m$ which is a text of length n concatenated with a pattern of length m. Two indices of the input string are called k-equivalent if the substring of length k starting at those indices are equal; this is in fact an equivalence relation on the set of indices of the input string. The algorithm assigns unique names to each index in the same equivalence class. The goal is to find all indices which are in the same m-equivalence class of the index where the pattern starts.

We denote by $n(i,j)$ the unique name assigned to the substring of length j starting at position i of the input string; assume $n(i,j)$ is defined only for $i + j \leq l + 1$ and the names are integers in the range $1 \cdots l$. Suppose $n(i, r)$ and $n(i, s)$ are known for all positions i of the input string. One can easily combine these names to obtain $n(i, r + s)$ for all positions i in constant time using l processors as follows: Assume a two dimensional array of size $l \times l$ is available; assign a processor to each position of the input string. Note that the string of length $r + s$ starting at position i is actually the string of length r starting at position i concatenated with the string of length s starting at position $i + r$. Each processor will try to write the position number it is assigned to in the entry at row $n(i, r)$ and column $n(i + r, s)$ of the matrix. If more then one processors attempts to write the same entry, assume an arbitrary one succeeds. Now $n(i, r + s)$ is assigned the value written in row $n(i, r)$ and column $n(i + r, s)$ of the matrix. That is, $n(i, r + s)$ is an index of the input string, not necessarily i, which is $(r + s)$-equivalent to i.

The algorithm start with $n(i, 1)$ which is the symbol at position i of the string, assuming the alphabet is the set of integers between 1 and l. It proceeds with $O(\log m)$ steps computing $n(i, 2)$, $n(i, 4)$, $\cdots n(i, 2^j)$ for $j \leq \lfloor \log_2 m \rfloor$, by merging names of two 2^j-equivalence classes into a names of 2^{j+1}-equivalence classes. In another $O(\log m)$ steps it computes $n(i, m)$ by merging a subset of the names of power-of-two equivalence classes computed before, and reports all indices which are in the same m-equivalence class of the starting index of the pattern.

This algorithm requires $O(m^2)$ space which can be reduced to $O(m^{1+\epsilon})$ for a time tradeoff as described in the suffix tree construction algorithm of Apostolico et al. [6].

In the rest of this section we will describe the algorithm of Vishkin [46], on which the faster algorithms, described later, are based.

As we have seen before, if we have nm processor CRCW-PRAM, we can solve the string matching problem in constant time using the following method:

- First, mark all possible occurrences of the pattern as 'true'.

- To each such possible beginning of the pattern, assign m processors. Each processor compares one symbol of the pattern with the corresponding symbol of the text. If a mismatch is encountered, it marks the appropriate beginning as 'false'.

Assuming that we can eliminate all but l of the possible occurrences we can use the method described above to get a constant time parallel algorithm with lm processors. Both Galil [26] and Vishkin [46] use this approach. The only problem is that one can have many occurrences of the pattern in the text, even much more than the $\frac{n}{m}$ needed for optimality in the discussion above.

To overcome this problem, we introduce the notion of the period used in these two papers. A string u is called a *period* of a string w if w is a prefix of u^k for some positive integer k or equivalently if w is a prefix of uw. We call the shortest period of a string w *the period* of w. For example, the period of the string *abacabacaba* is *abac*. The string *abacabac* is also a period, so is the string *abacabacab*.

Lemma 2.2 (Lyndon and Schutzenberger [41]): If w has two periods of length p and q and $|w| \geq p + q$, then w has a period of length $\gcd(p, q)$.

If a pattern w occurs in positions i and j of some string and $0 < j - i < |w|$ then the occurrences must overlap. This implies that w has a period of length $j - i$. Therefore, we cannot have occurrences of w at positions j and i if $0 < j - i < |u|$ and u is the period of the pattern. Clearly there are no more then $\frac{n}{|u|}$ occurrences of the pattern in a string of length n.

If the pattern is longer than twice its period length then instead of matching the whole pattern w we look only for occurrences of u^2, its period repeated twice. (Note that u^2 has the same period length as w by Lemma 2.2.) This case where the pattern is longer than twice its period is called the periodic case.

Assuming we could eliminate many of the occurrencesof u^2 and have only $n/|u|$ possible occurrences left, we can use the constant-time algorithm described above to verify these occurrences using only $2n$ processors. Then, by counting the number of consecutive matches of u^2, we can match the whole pattern.

Vishkin [46] shows how to count the consecutive matches in optimal $O(\log m)$ time on an EREW-PRAM using ideas which are similar to prefix computation. Breslauer and Galil [14] show how it can be done in constant optimal time on a CRCW-PRAM (and thus in optimal $O(\log m)$ time on an EREW-PRAM). Assume without loss of generality that the text is of length $n \leq \frac{3}{2}m$ and the pattern is $u^k v$ where u is its period length. Call an occurrence of u^2 at position i an *initial occurrence* if there is no occurrence of u^2 at position $i - |u|$ and a *final occurrence* if there is no occurrence at position $i + |u|$. There is at most one initial occurrence which can start an actual occurrence of the pattern: the rightmost initial occurrence in the first $\frac{m}{2}$ positions. Any initial occurrence in a position greater than $\frac{m}{2}$ cannot start an occurrence of the pattern since the text is not long enough. Any initial occurrence on the left cannot start an occurrence of the pattern since u, the period length of the pattern, is not repeated enough times. The corresponding final occurrence is the leftmost final occurrence to the right of the initial occurrence. By substructing the positions of the initial occurrence from the final occurrences and verifying an occurrence of v starting after the final occurrence, one can tell how many times the period is repeated and what are the actual occurrences of the pattern.

For the rest of the description we assume without loss of generality that the pattern is shorter than twice its period length, what is called the non periodic case.

Suppose u is the period of the pattern w. If we compare two copies of w shifted with respect to each other by i positions for $0 < i < |u|$, there must be at least one mismatch. Vishkin [46] takes one of these mismatches and calls it a *witness* to the fact that i is not a period length. More formally, let k be the index of one such mismatch, then

$$PATTERN[k] \neq PATTERN[k - i].$$

We call this k a witness, and define

$$WITNESS[i + 1] = k.$$

Using this witness information Vishkin suggests a method which he calls a *duel* to eliminate at least one of two close possible occurrences. Suppose i and j are possible occurrences and $0 < j - i < |u|$. Then, $r = WITNESS[j - i + 1]$ is defined. Since $PATTERN[r] \neq PATTERN[r + i - j]$, at most one of them is equal to $TEXT[i + r - 1]$ (see figure 2.1), and at least one of the possible occurrences can be ruled out (As in a real duel sometimes both can be ruled out.).

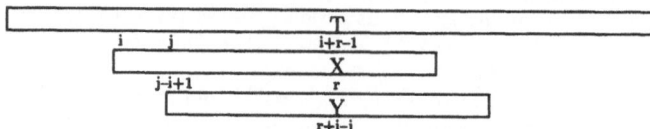

Figure 2.1. $X \neq Y$ and therefore we cannot have $T = X$ and $T = Y$.

Vishkin's algorithm [46] consists of two phases. The first is the pattern analysis phase in which the witness information is computed to help later with a text analysis phase which finds the actual occurrences.

We start with a description of the text analysis phase. Let $\mathcal{P} = |u|$ be the period length of the pattern. The pattern analysis phase described later computes witnesses only for the first half of the pattern. If the pattern has a period which is longer than half its length, we define $\mathcal{P} = \lceil \frac{m}{2} \rceil$.

The text analysis phase works in stages. There are $\lfloor \log \mathcal{P} \rfloor$ stages. At stage i the text string is partitioned into consecutive blocks of length 2^i. Each such block has only one possible start of an occurrence. We start at stage 0 where the blocks are of size one, and each position of the string is a possible occurrence.

At stage i, consider a block of size 2^{i+1} which consists of two blocks of size 2^i. It has at most two possible occurrences of the pattern, one in each block of size 2^i. A duel is performed between these two possible occurrences, leaving at most one in the 2^{i+1} block.

At the end of $\lfloor \log \mathcal{P} \rfloor$ stages, we are left with at most $\frac{2n}{|u|}$ possible occurrences of u which can be verified in constant-time using n processors. Note that the total number of operations performed is $O(n)$ and the time is $O(\log m)$. By Theorem 1.1 an optimal implementation is possible. Moreover, it is even possible to implement this phase on a CREW-PRAM within the same time bound. It is the pattern analysis phase which requires a CRCW-PRAM.

The pattern analysis phase is similar to the text analysis phase. It takes $\lfloor \log m \rfloor$ stages. The description below outlines a logarithmic time implementation using m processors.

The $WITNESS$ array which we used in the text processing stage is computed incrementally. Knowing that some witnesses are already computed in previous stages, one can easily compute more witnesses. Let i and j be two indices in the pattern such that $i < j \leq \lceil m/2 \rceil$. If $s = WITNESS[j - i + 1]$ is already computed then we can find at least one of $WITNESS[i]$ or $WITNESS[j]$ using a duel on the pattern as follows:

- If $s + i - 1 \leq m$ then $s + i - 1$ is also a witness either for i or for j.

- If $s + i - 1 > m$ then either s is a witness for j or $s - j + i$ is a witness for i (see figure 2.2).

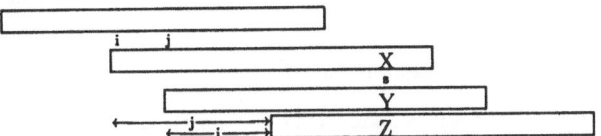

Figure 2.2. $X \neq Y$ and therefore we cannot have $Z = X$ and $Z = Y$.

The pattern analysis proceeds as follows: At stage i the pattern is partitioned into consecutive blocks of size 2^i. Each block has at most one yet-to-be-computed witness. The first block never has $WITNESS[1]$ computed. Consider the first block of size 2^{i+1}. It has at most one other yet-to-be-computed witness, say $WITNESS[k]$. We first try to compute this witness by comparing two copies of the pattern shifted with respect to each other by $k - 1$ positions. This can be easily done in constant time on an arbitrary CRCW-PRAM with m processors. If a witness is not found, then k is the period length of the pattern and the pattern analysis terminates. If a witness was found, a duel is performed in each block of size 2^{i+1} between the two yet-to-be-computed witnesses in each such block. It results in each block of size 2^{i+1} having at most one yet-to-be-computed witness. After $O(\lfloor \log m \rfloor)$ stages the witness information is computed for the first half of the pattern and the algorithm can proceed with the text analysis.

The optimal implementation of the pattern analysis is very similar to Galil's [26] original algorithm. Each iteration of the pattern analysis described above has actually two steps: the first step tries to verify a period length using a naive algorithm which compares long strings; if fails, a witness is found and it is used in a step which is identical to the actual string analysis phase.

Suppose the naive algorithm would be applied at stage i just to verify a period length of a prefix of the pattern of length 2^{i+1} instead of the whole pattern. If a mismatch is found it can be used as a witness as described before. If no mismatch has been found, continue to a *periodic* stage $i + 1$ which will try to verify the same period length of a prefix of double length. At some point either a mismatch is found or the period length is verified for the whole string and the pattern analysis is terminated. If a mismatch was found, it follows from Lemma 2.1 that the first mismatch can be used as a witness value for all uncomputed witnesses in the first block; and the algorithm can catch up to stage $i + 1$ (with the current value of i) by performing duels.

Galil's [26] original algorithm had only one stage which consisted of similar two steps; application of a naive algorithm to verify a period length of a prefix of the pattern of increasing length and elimination of close possible occurrences which would imply a short period length. The main difference is that Galil's algorithm had to compare long strings also in the steps Vishkin's algorithm uses the witness information. So n operations are performed at each round making the algorithm non-optimal. Galil [26] suggests an improvement for a finite alphabet which packs $O(\log m)$ symbols in a single integer and thus uses less processors to perform the comparisons, making an optimal implementation possible in $O(\log m)$ time.

3 An $O(\log \log m)$ time algorithm

The $O(\log \log m)$ time algorithm of Breslauer and Galil [14] is similar to Vishkin's algorithm from the previous section. The method is based also on an algorithm for finding the maximum suggested by Valiant [45] for a comparison model and implemented by Shiloach and Vishkin [44] on a CRCW-PRAM.

As before, we have two stages. The first stage, the pattern analysis, computes the witness information which is used in the text analysis to find the actual occurrences. Let $\mathcal{P} = |u|$ be the period length of the pattern. As before if the pattern has a period length longer than half its length, we define $\mathcal{P} = \lceil \frac{m}{2} \rceil$.

Partition the text into blocks of size $\mathcal{P}$ and consider each one separately. In each block consider each position as a possible occurrence. Assuming we had $\mathcal{P}^2$ processors for each such block a duel can be made between all pairs of possible occurrences resulting with at most one occurrence in each block. Since we have only n processors, partition the blocks into groups of size $\sqrt{\mathcal{P}}$ and repeat recursively. The recursion bottoms out with one processor per block of size 1. When done we are left with one possible occurrence at most in each group of size $\sqrt{\mathcal{P}}$, thus $\sqrt{\mathcal{P}}$ possible occurrences all together. Then in constant time make all duels as described above. We are left with a single possible occurrence (or none) in each block of size $\mathcal{P}$ and proceed with counting the consecutive occurrences of the period described in section 2.

To make the text analysis run in optimal $O(\log \log m)$ time we start with an $O(\log \log \mathcal{P})$ time sequential algorithm which runs in parallel in all subblocks of length $\log \log \mathcal{P}$ leaving only $\frac{\mathcal{P}}{\log \log \mathcal{P}}$ possible occurrences in each block by performing duels. Then proceed with the procedure above starting with the reduced number of possible occurrences.

The pattern analysis can be done also in optimal $O(\log \log m)$ time. We describe here only an m processor algorithm. It works in stages and it takes at most $\log \log m$ stages. Let $k_i = m^{1-2^{-i}}$, $k_0 = 1$. At the end of stage i, we have at most one yet-to-be-computed witness in each block of size k_i. The only yet-to-be-computed index in the first block is 1.

1. At the beginning of stage i we have at most k_i/k_{i-1} yet-to-be-computed witnesses in the first k_i–block. Try to compute them using the naive algorithm on $PATTERN(1 \cdots 2k_i)$ only. This takes constant time using $2k_i \frac{k_i}{k_{i-1}} = 2m$ processors.

2. If we succeed in producing witnesses for all the indices in the first block (all but the first for which there is no witness), compute witnesses in each following block of the same size using the optimal duel algorithm described above for the text processing. This takes $O(\log \log m)$ time only for the first stage. In the following stages, we will

130

have at most $\sqrt{m}$ indices for which we have no witness, and duels can be done in $O(1)$ time.

3. If we fail to produce a witness for some $2 \leq j \leq k_i$, it follows that $PATTERN(1 \cdots 2k_i)$ is periodic with period length p, where $p = j - 1$ and j is the smallest index of an yet-to-be-computed witness. By Lemma 2.1 all yet-to-be-computed indices within the first block are of the form $kp + 1$. Check periodicity with period length p to the end of the pattern. If p turns out to be the length of the period of the pattern, the pattern analysis is done and we can proceed with the text analysis. Otherwise, the smallest witness found is good also for all the indices of the form $kp + 1$ which are in the first k_i–block, and we can proceed with the duels as in 2.

If p processors are available and $m \leq p \leq m^2$, this algorithm can be modified to work in $O(\log \log_{\frac{2p}{m}} m)$ time. If the number of processors is smaller than $\frac{m}{\log \log m}$ the algorithm can be slowed down to work in $\frac{m}{p}$ time. When the number of processors is larger than n^2 the naive algorithm solves the problem in constant time. All these bounds can be summerized in one expression: $O(\lceil \frac{m}{p} \rceil + \log \log_{\lceil 1+p/m \rceil} 2p)$.

4 A lower bound

In this section we describe the lower bound of Breslauer and Galil [15] for a model which is similar to Valiant's parallel comparison tree model [45]. We assume the only access the algorithm has to the input strings is by comparisons which check whether two symbols are equal or not. The algorithm is allowed m comparisons in each round, after which it can proceed to the next round or terminate with the answer. We give a lower bound on the minimum number of rounds necessary in the worst case.

Consider a CRCW-PRAM that solves the string matching problem over a general alphabet. In this case the PRAM can perform comparisons, but not computation, with its input symbols. Thus, its execution can be partitioned into comparison rounds followed by computation rounds. Therefore, a lower bound for the number of rounds in the parallel comparison model immediately translates into a lower bound for the time of the CRCW-PRAM. If the pattern is given in advance and any preprocessing is free, then this lower bound does not hold, as Vishkin's $O(\log^* m)$ algorithm shows. The lower bound also does not hold for CRCW-PRAM over a fixed alphabet strings. Similarly, finding the maximum in the parallel decision tree model has exactly the same lower bound [45], but for small integers the maximum can be found in constant time on a CRCW-PRAM [25].

We start by proving a lower bound for a related problem of finding the period length of a string. Given a string $S[1..m]$ we prove that $\Omega(\log \log m)$ rounds are necessary for determining whether it has a period length smaller than $\frac{m}{2}$. Later we show how this lower bound translates into a lower bound for string matching.

We show a strategy for an adversary to answer $\frac{1}{4} \log \log m$ rounds of comparisons after which it still has the choice of fixing the input string S in two ways: in one the resulting

131

string has a period of length smaller than $\frac{m}{2}$ and in the other it does not have any such period. This implies that any algorithm which terminates in less rounds can be fooled.

We say that an integer k is a possible period length if we can fix S consistently with answers to previous comparisons in such a way that k is a period length of S. For such k to be a period length we need each residue class modulo k to be fixed to the same symbol, thus if $l \equiv j \mod k$ then $S[l] = S[j]$.

At the beginning of round i the adversary will maintain an integer k_i which is a possible period length. The adversary answers the comparisons of round i in such a way that some k_{i+1} is a possible period length and few symbols of S are fixed. Let $K_i = m^{1-4^{-(i-1)}}$. The adversary will maintain the following invariants which hold at the beginning of round number i:

1. k_i satisfies $\frac{1}{2}K_i \leq k_i \leq K_i$.

2. If $S[l]$ was fixed then for every $j \equiv l \mod k_i$ $S[j]$ was fixed to the same symbol.

3. If a comparison was answered as equal then both symbols compared were fixed to the same value.

4. If a comparison was answered as unequal, then

    a. it was between different residues modulo k_i;

    b. if the symbols were fixed then they were fixed to different values.

5. The number of fixed symbols f_i satisfies $f_i \leq K_i$.

Note that invariants 3 and 4 imply consistency of the answers given so far. Invariants 2, 3 and 4 imply that k_i is a possible period length: if we fix all symbols in each unfixed residue class modulo k_i to a new symbol, a different symbol for different residue classes, we obtain a string consistent with the answers given so far that has a period length k_i.

We start at round number 1 with $k_1 = K_1 = 1$. It is easy to see that the invariants hold initially. We show how to answer the comparisons of round i and how to choose k_{i+1} so that the invariants still hold. All multiples of k_i in the range $\frac{1}{2}K_{i+1} \ldots K_{i+1}$ are candidates for the new k_{i+1}. A comparison $S[l] = S[j]$ must be answered as equal if $l \equiv j \mod k_{i+1}$. We say that k_{i+1} forces this comparison.

Theorem 4.1 (see [43]): For large enough n, the number of primes between 1 and n denoted by $\pi(n)$ satisfies, $\frac{n}{\ln n} \leq \pi(n) \leq \frac{5}{4}\frac{n}{\ln n}$.

Corollary: The number of primes between $\frac{1}{2}n$ and n is greater than $\frac{1}{4}\frac{n}{\log n}$.

Lemma 4.2: If $p, q \geq \sqrt{\frac{m}{k_i}}$ and are relatively prime, then a comparison $S[l] = S[k]$ is forced by at most one of pk_i and qk_i.

Proof: Assume $l \equiv k \mod pk_i$, $l \equiv k \mod qk_i$ for $1 \leq l, k \leq m$. Then also $l \equiv k \mod pqk_i$. But $pqk_i \geq m$ and $1 \leq l, k \leq m$ which implies $l = k$, a contradiction. $\square$

132

Lemma 4.3: The number of candidates for k_{i+1} which are prime multiples of k_i and satisfy $\frac{1}{2}K_{i+1} \le k_{i+1} \le K_{i+1}$ is greater than $\frac{K_{i+1}}{4K_i \log m}$. Each such candidate satisfies the condition of Lemma 4.2.

Proof: These candidates are of the form pk_i for prime p. The number of such prime values of p can be estimated using the corollary to Lemma 4.1. It is at least

$$\frac{1}{4} \frac{K_{i+1}}{k_i \log \frac{K_{i+1}}{k_i}} \ge \frac{K_{i+1}}{4K_i \log m}.$$

Each one of these candidates also satisfies the condition of Lemma 4.2 since $k_i \le K_i$, $pk_i \ge \frac{K_{i+1}}{2}$ and

$$p^2 \ge \frac{1}{k_i} \frac{K_{i+1}^2}{4K_i} = \frac{1}{k_i} \frac{m^{2-2\cdot4^{-i}}}{4m^{1-4^{-(i-1)}}} = \frac{m}{k_i} \frac{1}{4} m^{2\cdot4^{-i}} \ge \frac{m}{k_i}. \quad \square$$

Lemma 4.4: There exists a candidate for k_{i+1} in the range $\frac{1}{2}K_{i+1} \ldots K_{i+1}$ that forces at most $\frac{4mK_i \log m}{K_{i+1}}$ comparisons.

Proof: By Lemma 4.3 there are at least $\frac{K_{i+1}}{4K_i \log m}$ such candidates which are prime multiples of k_i and satisfy the condition of Lemma 4.2. By Lemma 4.2 each of the m comparisons is forced by at most one of them. So the total number of comparisons forced by all these candidates is at most m. Thus, there is a candidate that forces at most $\frac{4mK_i \log m}{K_{i+1}}$ comparisons. $\quad \square$

Lemma 4.5: For m large enough and $i \le \frac{1}{4} \log \log m$, $1 + m^{2\cdot4^{-i}} 16 \log m \le m^{3\cdot4^{-i}}$.

Proof: For m large enough,

$$\log \log (1 + 16 \log m) < \frac{1}{2} \log \log m = \left(1 - \frac{2}{4}\right) \log \log m$$

$$\log (1 + 16 \log m) < 4^{-\frac{1}{4} \log \log m} \log m$$

$$1 + 16 \log m < m^{4^{-\frac{1}{4} \log \log m}} \le m^{4^{-i}},$$

from which the lemma follows. $\quad \square$

Lemma 4.6: Assume the invariants hold at the beginning of round i and the adversary chooses k_{i+1} to be a candidate which forces at most $\frac{4mK_i \log m}{K_{i+1}}$ comparisons. Then the adversary can answer the comparisons in round i so that the invariants also hold at the beginning of round $i + 1$.

Proof: By Lemma 4.4 such k_{i+1} exists. For each comparison that is forced by k_{i+1} and is of the form $S[l] = S[j]$ where $l \equiv j \mod k_{i+1}$ the adversary fixes the residue class modulo k_{i+1} to the same new symbol (a different symbol for different residue classes). The adversary answers comparisons between fixed symbols based on the value they are fixed to. All other comparisons involve two positions in different residue classes modulo k_{i+1} (and at least one unfixed symbol) and are always answered as unequal.

Since k_{i+1} is a multiple of k_i, the residue classes modulo k_i split; each class splits into $\frac{k_{i+1}}{k_i}$ residue classes modulo k_{i+1}. Note that if two indices are in different residue classes modulo k_i, then they are also in different residue classes modulo k_{i+1}; if two indices are in the same residue class modulo k_{i+1}, then they are also in the same residue class modulo k_i.

We show that the invariants still hold.

1. The candidate we chose for k_{i+1} was in the required range.

2. Residue classes which were fixed before split into several residue classes, all are fixed. Any symbol fixed at this round causes its entire residue class modulo k_{i+1} to be fixed to the same symbol.

3. Equal answers of previous rounds are not affected since the symbols involved were fixed to the same value by the invariants held before. Equal answers of this round are either between symbols which were fixed before to the same value or are within the same residue class modulo k_{i+1} and the entire residue class is fixed to the same value.

4. a. Unequal answers of previous rounds are between different residue classes modulo k_{i+1} since residue classes modulo k_i split. Unequal answers of this round are between different residue classes because comparisons within the same residue class modulo k_{i+1} are always answered as equal.

   b. Unequal answers which involve symbols which were fixed before this round are consistent because fixed values dictate the answers to the comparisons. Unequal answers which involve symbols that are fixed at the end of this round and at least one was fixed at this round are consistent since a new symbol is used for each residue class fixed.

5. We prove inductively that $f_{i+1} \leq K_{i+1}$. We fix at most $\frac{4mK_i \log m}{K_{i+1}}$ residue classes modulo k_{i+1}. There are k_{i+1} such classes and each class has at most $\lceil \frac{m}{k_{i+1}} \rceil \leq \frac{2m}{k_{i+1}}$ elements. By Lemma 4.5 and simple algebra the number of fixed elements satisfies:

$$
\begin{aligned}
f_{i+1} &\leq f_i + \frac{2m}{k_{i+1}} \frac{4mK_i \log m}{K_{i+1}} \\
&\leq K_i \left[1 + \left(\frac{m}{K_{i+1}} \right)^2 16 \log m \right] \\
&\leq m^{1-4^{-(i-1)}} (1 + m^{2 \cdot 4^{-i}} 16 \log m) \\
&\leq m^{1-4^{-i}} = K_{i+1}. \quad \square
\end{aligned}
$$

Theorem 4.7: Any comparison-based parallel algorithm for finding the period length of a string $S[1..m]$ using m comparisons in each round requires $\frac{1}{4} \log \log m$ rounds.

Proof: Fix an algorithm which finds the period of S and let the adversary described above answer the comparisons. After $i = \frac{1}{4}\log\log m$ rounds $f_{i+1}, k_{i+1} \leq m^{1-4^{-\frac{1}{4}\log\log m}} = \frac{m}{\sqrt{\log m}} \leq \frac{m}{2}$. The adversary can still fix S to have a period length k_{i+1} by fixing each remaining residue class modulo k_{i+1} to the same symbol, different symbol for each class. Alternatively, the adversary can fix all unfixed symbols to different symbols. Note that this choice is consistent with all the the comparisons answered so far by invariants 3 and 4, and the string does not have any period length smaller than $\frac{m}{2}$. Consequently, any algorithm which terminates in less than $\frac{1}{4}\log\log m$ rounds can be fooled. $\square$

Theorem 4.8: The lower bound holds also for any comparison-based string matching algorithm when $n = O(m)$.

Proof: Fix a string matching algorithm. We present to the algorithm a pattern $P[1..m]$ which is $S[1..m]$ and a text $T[1..2m-1]$ which is $S[2..2m]$, where S is a string of length $2m$ generated by the adversary in the way described above (We use the same adversary that we used in the previous proof; the adversary sees all comparisons as comparisons between symbols in S.). After $\frac{1}{4}\log\log 2m$ rounds the adversary still has the choice of fixing S to have a period length smaller than m, in which case we will have an occurrence of P in T, or to fix all unfixed symbols to completely different characters, what implies that there would be no such occurrence. Thus, the lower bound holds also for any such string matching algorithm. $\square$

This lower bound actually holds even if the algorithm is allowed to perform order comparison which can result in a *less than*, *equal* or *greater than* answers as shown in Breslauer and Galil's paper [15]. When the number of comparisons in each round is p and $n = O(m)$, the bound becomes $\Omega(\lceil\frac{m}{p}\rceil + \log\log_{\lceil 1+p/m\rceil} 2p)$, matching the upper bound.

5   A faster algorithm

The fast string matching algorithm of Vishkin [47] has two stages. The pattern analysis stage is slow and takes optimal $O(\frac{\log^2 m}{\log\log m})$ time while the text analysis is very fast and works in optimal $O(\log^* m)$ time. An alternative randomized implementation of the pattern analysis that works in optimal $O(\log m)$ time with very high probability will not be covered in this paper.

As we have seen before, we can assume without loss of generality that the pattern is shorter than twice its period length. Thus witnesses can be computed for all indices which are smaller than $\frac{m}{2}$.

Definition: A *deterministic sample* $DS = [ds(1), ds(2), \cdots, ds(l)]$ is a set of positions of the pattern string such that if the pattern is aligned at position i of the text and the symbols at positions $ds(1) \cdots ds(l)$ of the pattern are verified, that is $PATTERN[ds(j)] = TEXT[i + ds(j) - 1]$ for $1 \leq j \leq l$, then i is the only possible occurrence of the pattern in an interval of length $\frac{m}{2}$ around i.

Deterministic samples are useful since one can always find a small one.

Lemma 5.1: For any pattern of length m, a deterministic sample of size $\log m - 1$ exists.

Proof: We show how to find a deterministic sample of length $\log m - 1$. If this sample is verified for position i of the text then i is the only possible occurrence in an interval of length $\frac{m}{2}$ around i.

Consider $\frac{m}{2}$ copies of the pattern placed under each other, each shifted ahead by one position with respect to the previous one. Thus copy number k is aligned at position k of copy number one. Call the symbols of all copies aligned over position number i of the first copy *column* i (see figure 5.1). Since we assume that the pattern is shorter than twice its period length and there are $\frac{m}{2}$ copies, for any two copies there is a witness to their mismatch.

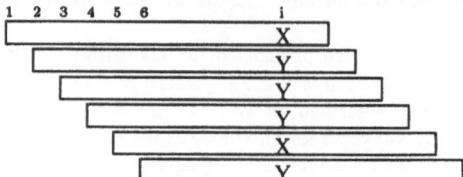

Figure 5.1. Aligned copies of the pattern and a column i.

Take the first and last copies and a witness to their mismatch. The column of the mismatch has at least two different symbols and thus one of the symbols in that column, in either the first or the last copy, appears in the column in at most half of the copies. Keep only the copies which have the same symbol in that column to get a set of at most half the number of original copies, which all have the same symbol at the witness column. This procedure can be repeated at most $\log m - 1$ times until there is a single copy left, say copy number k. Note that all columns chosen hit copy number k. The deterministic sample is the indices in copy number k of the columns considered. There are at most $\log m - 1$ such columns. If this sample is verified for position i of a text string no other occurrence at positions $i - k + 1$ to $i - k + \frac{m}{2}$ is possible. $\square$

One can find such a deterministic sample in parallel by the constructive proof of Lemma 5.1. Assume the witness information is produced by either Vishkin's [46] or Breslaur and Galil's algorithm [14] (It does not matter which algorithm since the time bound is dominated by the following steps.). There are $O(\log m)$ steps in the construction, each step counts how many symbols are equal to the witness symbol in the first and last copies, and can be implemented using Theorem 2.1 in optimal $O(\log m)$ time, or even faster by an algorithm of Cole and Vishkin [19] for prefix sums of small integers which works in optimal $O(\frac{\log m}{\log \log m})$ time.

Since the sums are taken at each round only for copies which are left, the total amount of operations performed at each round is half the number of operations of the previous round and sums up over all rounds to be linear. By Brent's Theorem the total time required for the pattern analysis is $O(\frac{\log^2 m}{\log \log m})$ with optimal number of processors.

One can use the deterministic sample to find all occurrences of the pattern in a text string in constant time and $O(n \log m)$ processors: for each position verify the deterministic sample for that position resulting in a few possible occurrences which can be verified in constant time using a linear number of processors.

We now show how to use the data structure constructed in the pattern analysis phase to search for all occurrences of the pattern starting in any position of a block of size $m/2$ of the text. We describe only an $O(\log^* m)$ version using m processors. An optimal implementation can be obtained using standard techniques.

Assume that the output of the pattern analysis phase is a sequence of compact arrays $A_0, A_1, \cdots, A_l$ where $A_0 = \{-k+1, \cdots, \frac{m}{2} - k\}$ is the set of all copies of the pattern considered at the start of the construction of the deterministic sample (relative to k, the copy that survived) and $A_i \subseteq A_0$ is the set of all copies remaining at the end of step i. These compact arrays can be generated in the same bounds of the pattern analysis described above and are used to efficiently assign processors to their tasks.

Initially all positions in the block are candidates for a potential occurrence and after each stage only part of the candidates will survive.

The algorithm starts with verifying $ds(1)$ for each candidate. This takes constant time using m processors. Call the candidates for which there is a match a matching candidate. Let l and r be the index of the leftmost and rightmost matching candidate respectively. Consider A_1 as a template around l and around r of all possible occurrences which have the same symbol in the column under $ds(1)$ relative to l or r. Note that since all other positions, even matching candidates (for which $ds(1)$ was just verified) cannot be real occurrences since they will disagree with the verified $ds(1)$ for position l or r. (The two templates cover the $\frac{m}{2}$ text positions under consideration, because there are no occurrences before l or after r and $l + \frac{m}{2} - k \geq r - k + 1$.) Thus, the candidates that survive for the next stage are those among the matching candidates aligned with a position in A_1 relative to l or r.

We can continue in this manner. In stage i there will be a set of candidates. The leftmost is l and the rightmost r and the set of candidates is aligned with the subset of A_i relative to l or r for which $Ds(1), \ldots, Ds(i)$ have been verified. (We described stage 0.) However this will take $\log m$ stages. Note that in the second stage we have at most $\frac{m}{4}$ candidates. So we can achieve double progress with the same processors: In stage 1 we can verify $ds(2)$ and $ds(3)$.

At the start of a general stage, assume the leftmost and rightmost candidates are l and r and the candidates are positions aligned with elements of A_s (relative to l or r) for which $Ds(1), \ldots, Ds(s)$ have been verified. Since $|A_s| \leq \frac{m}{2^s+1}$, the m processors now verify $Ds(s+1), \ldots, ds(s+2^s)$ for all the candidates. For the purpose of efficient processor assignment they will be assigned to verify 2^s positions for *all* the elements in A_s. Those assigned to a non candidate simply do nothing. As above, we define matching candidates as the candidates for which all positions were verified as matches, l and r as the leftmost and rightmost matching candidates and the candidates surviving for the next stage are the matching candidates that are aligned with A_{s+2^s} (relative to l or r). Since the new value of s is larger than 2^s, we have at most $O(\log^* m)$ stages, each of which takes constant time.

The number of processor is m.

This exponential acceleration phenomenon was called the *accelerating cascade* design principle by Cole and Vishkin [18] where by carefully choosing the parameters they were able to get an optimal $O(\log^* m)$ time parallel algorithms for another problem. For the complete description of the algorithm see Vishkin's paper [47].

6 Open questions

- String matching over a fixed alphabet. The lower bound of Section 4 assumes the input strings are drawn from a general alphabet and the only access to them is by comparisons. The lower and upper bounds for the string matching problem over a general alphabet are identical to those for comparison based maximum finding algorithm obtained by Valiant [45]. A constant time algorithm can find the maximum of integers in a restricted range [25] which suggests the possibility of a faster string matching algorithm.

- Faster randomized algorithm. The similarity to the maximum finding algorithm and the existence of a constant expected time randomized algorithm for that problem suggests the possibility of a faster randomized string matching algorithm.

- String matching with long text strings. If the text string is much longer than the pattern, the lower bound of Section 4 does not apply. Indeed, on a comparison model where all computation is free one can do the preprocessing for Vishkin's fast algorithm in constant time using m^2 processors. If $n = m^2$ the n processors are available to preprocess the short pattern. However, it is not known if the preprocessing can be performed on a CRCW-PRAM, or how is can be done faster with less then m^2 processors on a comparison model.

- String matching with preprocessing. What are the exact bounds if preprocessing is free like in Vishkin's fast algorithms. A constant time optimal algorithm is still possible.

- String matching on CREW and EREW-PRAM. The fastest optimal CREW-PRAM deterministic algorithm is obtained by slowing down the CRCW-PRAM algorithm to $O(\log m \log \log m)$ time. What is the exact bound on these models.

7 Acknowledgements

We would like to thank Terry Boult and Thanasis Tsantilas for valuable comments and suggestions for this paper.

References

[1] Aho, A. (1990), Algorithms for finding patterns in strings, *Handbook of theoretical computer science*, 257-300.

[2] Aho, A. and Corasik, M. J. (1975), Efficient string matching: an aid to bibliographic search, *Comm. ACM 18:6*, 333-340.

[3] Apostolico, A. (1989), Optimal parallel detection of squares in strings; Part I: Testing square freedom, *CSD-TR-932, purdue*.

[4] Apostolico, A. (1990), Optimal parallel detection of squares in strings; Part II: Detecting all squares, *CSD-TR-1012, purdue*.

[5] Apostolico, A. and Giancarlo, R. (1986), The Boyer-Moore-Galil string searching strategies revisited, *SIAM J. Comput. 15:1*, 98-105.

[6] Apostolico, A., Iliopoulos, C., Landau, G. M., Schieber, B. and Vishkin, U. (1988), Parallel construction of a suffix tree with applications, *Algorithmica 3*, 347-365.

[7] Beame, P., and Hastad, J. (1987), Optimal Bound for Decision Problems on the CRCW PRAM, *Proc. 19th ACM Symp. on Theory of Computing*, 83-93.

[8] Berkman, O., Breslauer, D., Galil, Z. Schieber, B., and Vishkin, U. (1989), Highly parallelizeable problems, *Proc. 21st ACM Symp. on Theory of Computing*, 309-319.

[9] Berkman, O., Schieber, B., and Vishkin, U. (1988), Some doubly logarithmic optimal parallel algorithms based on finding nearest smallers, *manuscript*.

[10] Borodin, A. B., Fischer, M. J., Kirkpatrick, D. G., Lynch, N. A. and Tompa, M. (1979), A time-space tradeoff for sorting on non-oblivious machines, *Proc. 20th IEEE Symp. on Foundations of Computer Science*, 294-301.

[11] Borodin, A., and Hopcroft, J. E. (1985), Routing, merging, and sorting on parallel models of comparison, *J. of Comp. and System Sci. 30*, 130-145.

[12] Boyer, R. S., and Moore, J. S. (1977), A fast string searching algorithm, *Comm. ACM 20*, 762-772.

[13] Brent, R. P. (1974), The parallel evaluation of general arithmetic expressions, *J. ACM 21*, 201-206.

[14] Breslauer, D. and Galil, Z. (1990), An optimal $O(\log \log n)$ parallel string matching algorithm, *SIAM J. Comput. 19:6*, 1051-1058.

[15] Breslauer, D. and Galil, Z. (1991), A lower bound for parallel string matching, *Proc. 23nd ACM Symp. on Theory of Computation*, to appear.

[16] Colussi, L., Galil, Z. and Giancarlo, R. (1990), On the exact complexity of string matching, *Proc. 31st IEEE Symp. on Foundations of Computer Science*, 135-143.

[17] Cole, R. (1991), Tight bounds on the complexity of the Boyer-Moore string matching algorithm, *Proc. 2nd annual ACM-SIAM symp. on discrete algorithms*, 224-233.

[18] Cole, R. and Vishkin, U. (1986), Deterministic coin tossing and accelerating cascades: micro and macro techniques for designing parallel algorithms, *Proc. 18th ACM Symp. on Theory of Computing*, 206-219.

[19] Cole, R. and Vishkin, U. (1989), Faster optimal prefix sums and list ranking, *Inform. and Comput. 81*, 334-352.

[20] Cook, S. A., Dwork, C. and Reischuk, R. (1986), Upper and lower time bounds for parallel random access machines without simultaneous writes, *SIAM J. Comput. 15:1*, 87-97.

[21] Crochemore, M. (1989), String-Matching and Periods, In *Bulletin of EATCS*.

[22] Crochemore, M. and Perrin, D. (1989), Two way pattern matching, *JACM*, to appear.

[23] Crochemore, M. and Rytter, W. (1990), Usefulness of the Karp-Miller-Rosenberg algorithm in parallel computations on strings and arrays, manuscript.

[24] Eppstein, D. and Galil, Z. (1988), Parallel algorithmic techniques for combinatorial computation, *In Ann. Rev. Comput. Sci. 3*, 233-283.

[25] Fich, F. E., Ragde, R. L., and Wigderson, A. (1984), Relations between concurrent-write models of parallel computation, *Proc. 3rd ACM Symp. on Principles of Distributed Computing*, 179-189.

[26] Galil, Z. (1985), Optimal parallel algorithms for string matching, *Information and Control 67*, 144-157.

[27] Galil, Z. and Giancarlo, R. (1988), Data structures and algorithms for approximate string matching, *Journal of Complexity 4*, 33-72.

[28] Galil, Z. and Seiferas, J. (1980), Saving space in fast string-matching, *SIAM J. Comput. 2*, 417-438.

[29] Galil, Z. and Seiferas, J. (1983), Time-space-optimal string matching, *J. Comput. Syst. Sci. 26*, 280-294.

[30] Geréb-Graus, M. and Li, M. (1990), Three one-way heads cannot do string matching, manuscript.

[31] Karp, R. M., Miller, R. E. and Rosenberg, A. L. (1972), Rapid identification of repeated patterns in strings, trees and arrays, *Proceedings of the 4th ACM Symposium on Theory of Computation*, 125-136.

[32] Karp, R. M. and Rabin, M. O. (1987), Efficient randomized pattern matching algorithms, *IBM J. Res. Develop. 31:2*, 249-260.

[33] Karp, R. M. and Ramachandran, V. (1990), A survey of parallel algorithms for shared-memory machines, *Handbook of theoretical computer science*.

[34] Kedem, Z., Landau, G. and Palem, K. (1988), Optimal parallel suffix-prefix matching algorithm and applications. *manuscript.*

[35] Knuth, D. E., Morris, J. H. and Pratt, V. R. (1977), Fast pattern matching in strings, *SIAM J. Comput. 6*, 322-350.

[36] Kruskal, C. P. (1983), Searching, merging, and sorting in parallel computation, *IEEE trans. on computers 32*, 942-946.

[37] Lander, R. E. and Fischer, M. J. (1980), Parallel Prefix Computation, *J. ACM 27:4*, 831-838.

[38] Li, M. (1984), Lower bounds on string-matching, *TR 84-636 Department of Computer Science, Cornell University.*

[39] Li, M. and Yesha, Y. (1986), String-matching cannot be done by a two-head one-way deterministic finite automaton, *Information Processing Letters 22*, 231-235.

[40] Lothaire, M. (1983), Combinatorics on Words, *Encyclopedia of mathematics and its applications, Vol. 17*, Addison Wesley.

[41] Lyndon, R. C. and Schutzenberger, M. P. (1962), The equation $a^M = b^N c^P$ in a free group, *Michigan Math. J. 9*, 289-298.

[42] McCreight, E. M. (1976), A space-economical suffix tree construction algorithm, *Journal of ACM, 33:3*, 262-272.

[43] Rosser, J. B. and Schoenfeld, L. (1962), Approximate formulas for some functions of prime numbers, *Illinois Journal of Mathematics*, 6:64-94.

[44] Shiloach, Y. and Vishkin, U. (1981), Finding the maximum, merging and sorting in a parallel computation model, *J. Algorithms 2*, 88-102.

[45] Valiant, L. G. (1975), Parallelism in comparison models, *SIAM J. Comput. 4*, 348-355.

[46] Vishkin, U. (1985), Optimal parallel pattern matching in strings, *Information and Control 67*, 91-113.

[47] Vishkin, U. (1990), Deterministic sampling - A new technique for fast pattern matching, *SIAM J. Comput. 20:1*, 22-40.

[48] Weiner, P. (1973), Linear pattern matching algorithms, *Proc. 14th IEEE symp. on switching and automata theory*, 1-11.

Some applications of Rabin's fingerprinting method

Andrei Z. Broder*

Abstract

Rabin's fingerprinting scheme is based on arithmetic modulo an irreducible polynomial with coefficients in Z_2. This paper presents an implementation and several applications of this scheme that take considerable advantage of its algebraic properties.

1 Introduction

Fingerprints are short tags for larger objects. They have the property that if two fingerprints are different then the corresponding objects are certainly different and there is only a small probability that two different objects have the same fingerprint. (The latter event is called a collision.)

More precisely, a fingerprinting scheme is a certain collection of functions $\mathcal{F} = \{f : \Omega \to \{0,1\}^k\}$, where Ω is the set of all possible objects of interest and k is the length of the fingerprint, such that, for any choice of a fixed set $S \subset \Omega$ of n distinct objects, if f is chosen uniformly at random in $\mathcal{F}$, then with high probability

$$|f(S)| = |S|.$$

In other words, if an adversary chooses a set $S \subset \Omega$ of n distinct objects, and we choose $f \in \mathcal{F}$ uniformly at random then

$$f(A) \neq f(B) \Longrightarrow A \neq B \tag{1}$$

$$\Pr(f(A) = f(B) \mid A \neq B) = \text{very small} \tag{2}$$

(The requirements and the model used here, are similar to those for "universal hashing" [1]; however the emphasis and the relation between n and k are different: for hashing we are interested in bounding the number of collisions, and typically n is a small fraction of 2^k; for fingerprinting we want to avoid collisions altogether, and we must take $n \ll 2^k$.)

---

*DEC Systems Research Center, 130 Lytton Ave, Palo Alto, CA.

2 Rabin's fingerprinting scheme

The following fingerprinting scheme for strings is due to Michael Rabin [4].

Let $A = (a_1, a_2, \ldots, a_m)$ be a binary string. We assume that $a_1 = 1$. (In certain application this assumption might be false, hence the implementation must actually prefix every string to be fingerprinted by a 1. We ignore this technicality for the rest of the paper.) We associate to the string A a polynomial $A(t)$ of degree $m - 1$ with coefficients in $\mathbf{Z}_2$,

$$A(t) = a_1 t^{m-1} + a_2 t^{m-2} + \cdots + a_m. \tag{3}$$

Let $P(t)$ be an irreducible polynomial of degree k, over $\mathbf{Z}_2$. (Such a polynomial can be easily found. See [3].) Having fixed P, we define the fingerprint of A to be the polynomial

$$f(A) = A(t) \bmod P(t). \tag{4}$$

Assume that an adversary chooses a set S of n distinct binary strings. After the adversary chooses S, we choose uniformly at random an irreducible polynomial P of degree k. We claim that for a proper choice of k the probability (over the random choices of P) that there exists a pair of distinct strings in S that have the same fingerprint, is extremely small.

Indeed, consider the product

$$Q = \prod_{\{A,B\}} (A(t) - B(t)), \tag{5}$$

taken over all the unordered pairs $A, B \in S$, with $A \neq B$. This product is a polynomial with coefficients in $\mathbf{Z}_2$. Its degree can be bound by

$$
\begin{aligned}
\deg Q &\leq \sum_{\{A,B\}} \max \{ \deg A(t), \deg B(t) \} \\
&\leq \sum_{\{A,B\}} (\deg A(t) + \deg B(t)) \leq n \sum_{A \in S} |A|.
\end{aligned}
\tag{6}
$$

In particular if all the strings in S have length less than m, then the degree of Q is less than $n^2 m$.

If there are two distinct strings, $A, B \in S$, such that $f(A) = f(B)$, it means that P divides $A(t) - B(t)$ and hence P divides Q. The adversary, by his choices of strings, has fixed a particular Q. This Q can not have more than $\deg(Q)/k$ irreducible factors of degree k. However the total number of irreducible polynomials of degree k with coefficients in $\mathbf{Z}_2$ is greater than $(2^k - 2^{k/2})/k$. Hence the probability that a randomly chosen irreducible polynomial of degree k divides Q, which is the same as the probability that two distinct strings have the same fingerprint, is less than

$$\frac{\deg Q}{k} \cdot \frac{k}{2^k - 2^{k/2}} \approx \frac{\deg Q}{2^k} \leq \frac{nm^2}{2^k}. \tag{7}$$

144

For instance, if the adversary chooses any collection of 2^{15} binary strings of total length less than 2^{25} bits, and we choose P uniformly at random among the irreducible polynomials of degree 64 over $\mathbf{Z}_2$, the probability that we made a bad choice of P is less than 2^{-23}.

3   Properties of Rabin's scheme

Besides satisfying the basic conditions (1) and (2), Rabin's scheme has additional properties that are useful in some applications and facilitate its implementation. (In this section all arithmetic operations are in $\mathbf{Z}_2$.)

1. At the hardware level the representation of the string A and the polynomial $A(t)$ with coefficients over $\mathbf{Z}_2$ is identical. The basic operations with polynomials have simple implementations: addition is equivalent with bit-wise exclusive or, and multiplication by t is equivalent with shift left one bit.

2. Fingerprinting is distributive over addition (in $\mathbf{Z}_2$):

$$f(A + B) = f(A) + f(B). \tag{8}$$

3. Fingerprints can be computed in linear time. Consider the bit string A = $[b_1, \ldots, b_l]$; If

$$\begin{aligned} f([b_1, &\ldots, b_l]) \\ &= (b_1 t^{l-1} + b_2 t^{l-2} + \ldots + b_l) \bmod P(t) \\ &= r_1 t^{k-1} + r_2 t^{k-2} + \ldots + r_k \end{aligned}$$

then

$$\begin{aligned} f([b_1, &\ldots, b_{l+1}]) \\ &= (f(b_1, \ldots, b_l)t + b_{l+1}) \bmod P(t) \\ &= r_2 t^{k-1} + r_3 t^{k-2} + \ldots + r_k t + b_{l+1} \\ &\quad + (r_1 t^k) \bmod P(t) \end{aligned}$$

Observe that

$$t^k \bmod P(t) = t^k - P(t) = P(t) - t^k,$$

so $t^k \bmod P$ is equivalent to P with the leading coefficient removed.

At the hardware level, the fingerprint of A can be kept in a shift register. Computing the fingerprint of A extended by b_l consists of one shift left operation with b_l as input bit and r_1 as output bit, and then, conditioned upon $r_1 = 1$, a bit-wise exclusive or operation, the second operand being P with the leading coefficient removed. (The next section discusses software implementation for a typical 32 bit word computer.)

4. More generally, the fingerprint of the concatenation of two strings can be computed via the equality

$$f(\text{concat}(A, B)) = f(\text{concat}(f(A), B)). \tag{9}$$

5. If we are given $f(A)$ and $f(B)$, and the length l of B then

$$f(\text{concat}(A, B)) = A(t) * t^l + B(t) \bmod P(t)$$
$$= f(f(A) * f(t^l)) + f(B) \tag{10}$$

(In practical applications it is useful to precompute $f(t^l)$ for a suitable range of values.)

4 Implementation issues

A straightforward implementation of the linear algorithm described above in software for a 32 bit per word computer requires two word bit shifts and two word xors (exclusive ors) *per bit*, or 16 shifts and 16 xors per byte. This is too slow for many applications.

However it is easy to process more than one bit at a time, using precomputed tables. A good trade-off is to process 32 bits at a time divided into four bytes. It is convenient to take k (that is, the degree of P, and thus the length of the fingerprint) to be a multiple of 32, in order to have fingerprints that consist of an integral number of words. Below we take $k = 64$.

The notations $W_{b1}, \ldots, W_{b4}$ refer to the four bytes of the 32-bit word W, and the notations $W_1, \ldots, W_{32}$ refer to the 32 bits of of the same word.

The input is the string $A[1], \ldots, A[s]$ where each $A[i]$ is a 32-bit word.

The final algorithm is

$$W[1] \leftarrow 0; W[2] \leftarrow 0;$$
for $s = 1, \ldots, m$ do

$\quad h, i, j, l \leftarrow W[1]_{b1}, W[1]_{b2}, W[1]_{b3}, W[1]_{b4};$

$\quad W[1] \leftarrow W[2]$ xor $TA[h, 1]$ xor $TB[i, 1]$ xor $TC[j, 1]$ xor $TD[l, 1];$

$\quad W[2] \leftarrow A[s]$ xor $TA[h, 2]$ xor $TB[i, 2]$ xor $TC[j, 2]$ xor $TD[l, 2];$

$\quad$ od

The invariant maintained by this algorithm is that

$$W[1]_1 t^{63} + W[1]_2 t^{62} + \cdots + W[1]_{32} t^{32}$$
$$+ W[2]_1 t^{31} + \cdots + W[2]_{32}$$
$$= f((A[1], A[2], \ldots, A[s])) \bmod P(t).$$

The tables TA, TB, TC, and TD are defined by

$$TA[i,1]_1 t^{63} + TA[i,1]_2 t^{62} + \cdots + TA[i,1]_{32} t^{32}$$
$$+ TA[i,2]_1 t^{31} + \cdots + TA[i,2]_{32}$$
$$= i_1 t^{95} + i_2 t^{94} + \cdots + i_8 t^{88} \bmod P(t);$$

$$TB[i,1]_1 t^{63} + TB[i,1]_2 t^{62} + \cdots + TB[i,1]_{32} t^{32}$$
$$+ TB[i,2]_1 t^{31} + \cdots + TB[i,2]_{32}$$
$$= i_1 t^{87} + i_2 t^{86} + \cdots + i_8 t^{80} \bmod P(t);$$

$$TC[i,1]_1 t^{63} + TC[i,1]_2 t^{62} + \cdots + TC[i,1]_{32} t^{32}$$
$$+ TC[i,2]_1 t^{31} + \cdots + TC[i,2]_{32}$$
$$= i_1 t^{79} + i_2 t^{78} + \cdots + i_8 t^{72} \bmod P(t);$$

$$TD[i,1]_1 t^{63} + TD[i,1]_2 t^{62} + \cdots + TD[i,1]_{32} t^{32}$$
$$+ TD[i,2]_1 t^{31} + \cdots + TD[i,2]_{32}$$
$$= i_1 t^{71} + i_2 t^{70} + \cdots + i_8 t^{64} \bmod P(t),$$

where $0 \leq i < 256$ and i_j denotes the j-th bit in the binary representation of i.

This algorithm requires 2 xors and 1 table indexing per byte. For most applications this is acceptable, particularly if the inner loop is coded in assembly language. For instance in our implementation the time required to fingerprint the contents of a file is under 3% of the time required to read the file in memory.

Finding a random irreducible polynomial $P(t)$ is simple (see [4]) and computing the associated tables is straightforward. However if all we want is to fingerprint relatively short strings, the overhead is prohibitive. Nevertheless in typical applications it is not necessary to pick a new random $P(t)$ every time the fingerprinting package of procedures is used. Instead, the polynomial $P(t)$ can be "wired-in," thus saving the computation of the tables $TA, \ldots, TD$. The idea is to assume that the virtual adversary has committed to all the strings that will fingerprinted by the package in the future in a certain context, and use equation (7) to bound the probability of collision. For instance, we can imagine that the adversary has committed to all variable names used in all Modula-2 programs in the next twenty years at DEC-SRC. We want to avoid fingerprint collisions within the same module. Assuming that the typical module has 2^9 variable names of 2^6 bits average length, the probability of collision within one module is less than $2^{25}/2^k$. If we assume that the number of modified modules is 2^{10} per day and that we use 64 bits fingerprints, the probability of even one collision in 20 years is less than 2^{-16}, which is negligible compared to the likelihood that the compiler produces erroneous results from other causes.

However, in order to maintain the fiction that the adversary has committed to all the future strings, it is crucial to ensure that no string depends on our choice for

$P(t)$. In particular, we must maintain the rule that no fingerprinting of strings that contain fingerprints is allowed.

If fingerprints are computed piecemeal via equations (9) and (10) then it makes sense to make the length of the underlying string a part of the fingerprint, and have the fingerprint package keep track of it. Since

$$x^{2^k - 1} \equiv 1 \bmod P(t),$$

it suffices to record the length modulo $2^k - 1$ even if larger lengths might appear. (However notice that in this case the bound (7) becomes useless.)

5 Applications

Below there are several fingerprinting usages that take advantage of the algebraic properties of Rabin's scheme. For more applications see [2] and [5].

5.1 Fingerprinting dags

This application of fingerprints was motivated by *Vesta*, a software development environment in construction at DEC Systems Research Center.

Vesta needs to make identifying tags for derived objects, that is, files mechanically derived from other files (e.g. via compilation or linking). Essentially, the relations between all the derived objects (ever) and all the sources can be put in the form of a gigantic dag (directed acyclic graph) with labeled edges. The vertices that have no predecessors are sources. To any vertex we can associate an edge-labeled tree: The tree of a vertex is formed in the obvious manner from the trees of its direct predecessors.

Vesta requires the vertices of the dag to have identical tags if and only if their associated trees are isomorphic. (In other words, two derived objects have the same tag if and only if they were derived in the same way from the same sources.)

5.1.1 Simplistic solution

It is straightforward to encode these trees as strings; the string for a parent is the concatenation of the strings corresponding to its children separated by labels and delimiters. We can use as a tag the fingerprint of the string. Fortunately it is not necessary to keep the strings themselves around. The tag of the parent can be computed from the tags of the children, provided that the length of the strings involved is known.

More precisely let A represent a string and α its associated polynomial. Let "$\|$" denote concatenation. Then ignoring delimiters

$$A(\text{parent}) = A(\text{child1}) \parallel A(\text{child2})$$

148

or

$$\alpha(\text{parent}) = \alpha(\text{child1}) * t^{\text{length}(A(\text{child2}))} + \alpha(\text{child2}),$$

and hence

$$f(\text{parent}) = (f(\text{child1}) * t^{\text{length}(A(\text{child2}))} + f(\text{child2})) \bmod P(t).$$

The problem with this approach is that the length of the strings involved might be extremely large and hence the formula (7) does not give any meaningful bound on the probability of collision. (The length of the describing string may grow exponentially in the size of the dag.)

5.1.2 General solution

Assume that an adversary chooses a dag G on n nodes. He also associates source strings to the nodes with no predecessors. After the adversary chooses G, we choose uniformly at random an irreducible polynomial $P(t)$ of degree k and a random permutation T of the set with 2^k elements. (For the *Vesta* application described above, a reasonable choice is $k = 96$.)

Let f be the Rabin's fingerprinting function. The tag of a node is computed as follows:

$$Tag(\text{parent}) = T(f(\text{Label1} \parallel Tag(\text{child1}) \parallel \text{Label2} \parallel Tag(\text{child2}) \parallel ...)).$$

For source nodes:

$$Tag(\text{source}) = T(f(\text{text}))$$

Observe that each node has a fingerprint and a tag.

Because T is chosen uniformly at random we can imagine that we construct T iteratively as follows: Start by assigning each sink (i.e. each *Vesta* source) in the dag a distinct random tag. Process the nodes of the dag in reverse topological order. If the result of f at a certain node v (i.e. either $f(\text{text})$ or $f(\text{label1} \parallel ...)$) is a new value x, we let $T(x)$ be a new random tag; otherwise T(x) is already determined.

Alternatively we can think that all these random tag are chosen simultaneously. But with high probability for this assignment of tags all the node fingerprints are different: we are fingerprinting n strings of (reasonable) average length m so the formula (7) applies and with high probability there are no collisions.

This would be a perfect solution except that generating a truly random permutation on 2^k elements requires time exponential in k. There are many ways of doing pseudo-random permutations: In our implementation we use first a non-linear mapping (each byte of the fingerprint is mapped via a random permutation of the set $\{0, ..., 255\}$) followed by an arithmetic mapping (the fingerprint is viewed as a vector of numbers in $\mathbf{Z}_{2^{32}}$ and multiplied by a non-singular matrix).

5.2 Fingerprinting subsets

Consider a programming application where there is a need to fingerprint n subsets $A_1, \ldots, A_n$, of a given ground set $\Omega = \{\omega_1, \omega_2, \ldots, \omega_r\}$. (An example of such an application is given below.) Associate to each ω_i a distinct irreducible polynomial $R_i(t)$ of degree m and define

$$R(A) = \prod_{\omega_i \in A} R_i.$$

Let the fingerprint of A be defined by

$$f(A) = R(A) \bmod P(t) = \prod_{\omega_i \in A} R_i \bmod P(t), \tag{11}$$

where as before $P(t)$ is a random irreducible polynomial of degree k.

With this setup we obtain that

$$\Pr(\exists i, j \text{ such that } A_i \neq A_j \text{ and } f(A_i) = f(A_j)) \approx \frac{n^2 r m}{2^k}.$$

It is also possible to take the R_i to be just random polynomials of degree k. The increase in the probability of collision is only $1/2^k$ per pair, so the above formula becomes

$$\Pr(\exists i, j \text{ such that } A_i \neq A_j \text{ and } f(A_i) = f(A_j)) \approx \frac{n^2 r k}{2^k}.$$

Note that if the fingerprints have to be computed from scratch, then the bit complexity of the approach above, $O(rk^2)$, is worse than simply representing each set as a sorted list and computing the fingerprint of the list in the obvious manner, in $O(rk \log r)$ bit operations.

Yet another possibility is to define the fingerprint of a set via

$$f(A) = \sum_{\omega_i \in A} t^i \bmod P(t). \tag{12}$$

This is the most efficient approach when the ground set is explicitly available and its elements can be easily numbered.

However there are situations when the sets encountered are derived from previously fingerprinted sets and when the ground set is not fully known at once. We can then take advantage of the following relations

- If $C = A \cup B$ for $A \cap B = \emptyset$, then

$$f(C) = (f(A) * f(B)) \bmod P.$$

- If $C = A \setminus B$ for $B \subset A$, then

$$f(C) = (f(A) * f(B)^{-1}) \bmod P,$$

where the inversion is in the field of residues modulo P.

As an example consider fingerprinting the spanning trees of a graph $G(V, E)$. Each spanning tree can be viewed as a set of $|V| - 1$ edges. If we add an edge e to a tree T it forms a cycle with the tree edges; the cycle can be broken by removing a suitable edge f thus obtaining a new spanning tree, T'. (If the edges e and f are chosen at random among eligible edges, the process becomes a Markov Chain on the set of spanning trees. We were interested in fingerprinting these trees in order to check certain statistics.)

Using the definition (11) we have

$$f(T') = (f(T) * R_e * R_f^{-1}) \bmod P(t),$$

where R_e and R_f are irreducible (or random) polynomials associated to e and f respectively. Note that R_f^{-1} (or even all pairs $R_e * R_f^{-1}$) can be precomputed.

For definition (12) we have

$$f(T') = (f(T) + t^{N(e)} - t^{N(f)}) \bmod P(t),$$

where $N(\cdot)$ is just a numbering of the edges. In this case this approach is more efficient, in particular if we precompute $t^i \bmod P(t)$ for $i = 1, \ldots, r$.

5.3 Fingerprinting binary trees

We can fingerprint binary trees using one of the two methods described in section 5.1. For small trees, Jim O'Toole has found another approach (unpublished) that avoids using delimiters and keeping track of the lengths of the strings involved.

Consider two strings A and B. Pad the shorter string with 0's to the left, to ensure equal length. Define the *intercalation* of A and B to be the string obtained by alternating the bits of A and B, that is, if $A = (a_1, \ldots, a_m)$ and $B = (b_1, \ldots, b_m)$ then

$$S(A, B) = (a_1, b_1, a_2, b_2, \ldots, a_m, b_m). \tag{13}$$

Let C be the string associated to the parent of two nodes with associated strings A and B. The encoding rule is

$$C = S(A, B), \tag{14}$$

and the fingerprint of the tree rooted at the parent

$$f(C) = C(t) \bmod P(t). \tag{15}$$

In order to make this encoding unambiguous we enlarge the input tree to a full binary tree (that is, each node has degree either 2 or 0) by adding dummy leaves. We associate the string 11 to the true leaves and 10 to the dummy leaves. It is easy to check that with these initial conditions, equation (14) defines an unambiguous encoding of binary trees.

151

Note that (14) implies that

$$C(t) = A(t^2) * t + B(t).$$

But in $\mathbf{Z}_2$ we have $A(t^2) = (A(t))^2$, and therefore

$$C(t) \bmod P(t) = \big((A(t) \bmod P(t))^2 * t + B(t) \bmod P(t) \big) \bmod P(t)$$
$$= S(f(A), f(B)) \bmod P(t)$$

In other words the fingerprint of C is obtained by fingerprinting the intercalation of the fingerprints of A and B.

The disadvantage of this approach is that the length of the encoding might be exponential in the size of the tree, while the length of the encoding presented in section 5.1.1 is only linear in the size of the tree.

Acknowledgement

I am indebted to Jim O'Toole for allowing me to present here his results on fingerprinting binary trees and for several stimulating discussions.

References

[1] L. Carter and M. Wegman, Universal Classes of Hash Functions. *JCSS*, 18:143–154, 1979.

[2] R. M. Karp and M. O. Rabin, Efficient randomized pattern matching algorithms. Center for Research in Computing Technology, Harvard University, Report TR-31-81, 1981.

[3] M. O. Rabin, Probabilistic algorithms in finite fields. *SIAM J. of Computing*, 9:273–280, 1980.

[4] M. O. Rabin, Fingerprinting by random polynomials. Center for Research in Computing Technology, Harvard University, Report TR-15-81, 1981.

[5] M. O. Rabin, Discovering repetitions in strings. In A. Apostolico and Z. Galil, (eds.), "Combinatorial Algorithms on Words," Springer-Verlag, 279–288, 1985.

Periodic Prefixes in Texts

Maxime Crochemore[1] & Wojciech Rytter[2]

(Preliminary version)

We give some new results on the combinatorics of periodic prefixes in texts. This is related to the time-space optimal string-searching algorithm of Galil and Seiferas (called GS algorithm [GS 83]). The algorithm contains a parameter k that says how many times some prefixes repeat at the beginning of the text. The complexity of GS algorithm essentially depends on k and originally the GS algorithm works only for $k \geq 4$. We improve on the value of k and we show that $k=3$ is the least integer for which GS algorithm works. This value of the parameter k also minimizes the time of the search phase of the string-searching algorithm. With parameter $k=2$ we consider a simpler version of the algorithm working in linear time and logarithmic space. This algorithm is based on the following fact : any word of length n starts by less than $\log_\phi n$ squares of primitive prefixes. Fibonacci words have a logarithmic number of square prefixes. Hence the combinatorics of square prefixes and cube prefixes is essential for string-matching with small memory. We give a time-space optimal sequential computation of the period of a word based on GS algorithm. The latter corrects the algorithm given in [GS 83] for the computation of periods. We also present an optimal parallel algorithm for the pattern preprocessing.

Introduction

The GS algorithm can be treated as a space efficient implementation of the algorithm of Knuth, Morris and Pratt [KMP 77]. KMP algorithm essentially works in stages, each composed of a left-to-right scan of the pattern against the text followed by a shift of the pattern to the right. The main operation in one stage consists in making a suitable safe shift of the pattern. We refer the reader to [KMP 77] and [Ah 90] for the definition of shift functions or failure functions related to the algorithm. More recently [Si 90] has improved on KMP, using a clever implementation of the automaton underlying KPM algorithm. In [Cr 91], is presented a time-space optimal string-searching algorithm that can also be considered as an efficient implementation of the KMP algorithm.

[1] LITP, Institut Blaise Pascal, France,
Université Paris 7, 2 place Jussieu, 75251 PARIS cedex 05.
Work by this author is supported by PRC "Mathématiques-Informatique" & NATO Grant CRG 900293.

[2] Institute of Informatics, Warsaw University, ul. Banacha 2, 00913 WARSAW 59, Poland.

153

Space efficient implementations of KMP are based on properties of periods of prefixes $p[1..i]$ of the pattern p. If we know that periods of these segments are always big with respect to i then it is easy to reduce space, memorizing a single constant of proportionality. If the period of a prefix is small then the pattern starts with a segment which repeats in the pattern several times, say k times. Such a segment is defined later as a *highly repeating prefix* (HRP). The main parameter of the HRP is the number k which essentially describes the continuation of the HRP as a period. If $k=2$ the HRP yields a square prefix; if $k=3$ it leads to a cube prefix. Hence the combinatorics of cubes and squares in strings plays an important role in the problem. In particular if the string is cube-free or square-free then the string-searching is much simpler (no pattern preprocessing is even needed).

We show that the number of HRP's of a word x is $O(\log|x|)$. The result is very easy to prove for $k \geq 3$. But the proof becomes rather intricate for $k=2$. In this latter case, the bound on the maximum number of HRP's is shown to be less than $\log_\Phi |x|$ where Φ is the golden ratio.

In algorithms that perform left-to-right scans the current situation is when a mismatch is met during the scan. The next step is then a shift of the pattern to the right. At this point the algorithm has discovered inside the text t an occurrence of a prefix y of the pattern p followed by a letter b. If y is the pattern itself then an occurrence of the pattern is found. If not, ya is a prefix of p for some letter $a \neq b$. The shift that comes next must if possible keep a prefix of the pattern match the text. This is realized by both KMP algorithm and Simon algorithm.

The following fact belons to the folkore of stringology:

Let $x=vz$ with v nonempty. Then, z is a prefix of x iff x is a prefix of some power v^e of v. The word z above, which is both a proper prefix and a suffix of x, is called a *border* of x. Borders and periods are in one-to-one correspondence. Finally, we denote by $per(x)$ the smallest period of x.

The base of many combinatorial properties of repetitions in words is given by the well-known periodicity lemma of Fine and Wilf (see [Lo 83]). Its weak version can formulated as follows:

Let p and q be two periods of a word x. Then, if $p+q \leq |x|$, $\gcd(p, q)$ is also a period of x.

KMP string-matching algorithm is based on a linear time computation of periods of all the prefixes of pattern p. And since all these periods (or associated borders) can be pairwise distinct, the space complexity of the algorithm is inherently $O(|p|)$.

The idea used in GS algorithm to save on space is to get rid of some periods of the pattern, namely its large periods. When the above prefix y of the pattern has a large period, the algorithm shifts the pattern to the right a number of places less than the shortest period of y. This kind of shift is not optimal but no occurrence of the pattern in the text is missed and this avoids memorizing large periods.

Large periods are defined relatively to an integer k that will be always considered as greater than 1 in the following. It is greater than 3 in [GS]. And we show that the same approach works for $k=3$.

Recall that a word v is said to be *primitive* if it is not a power of another word, that is, if $v=u^i$ implies both $i=1$ (and $u=v$). Note that the empty word is not primitive.

Highly repeating prefixes and string-searching

We introduce now the basic notions to deal with GS algorithm. Let v^k be a prefix of x with v a primitive word ($k>1$). The word v is called a *k-highly repeating prefix* of x, a *k-HRP* of x, or even a *HRP* of x when k is clear from the context. When v is a k-HRP of x, the prefix v^2 of x has smallest period $|v|$. Thus, we can consider the longest prefix z of x which has prefix period v. Then, *the scope of v* is the interval of integers $[L, R]$ defined by

$$L = |v^2| \text{ and } R = |z|.$$

Note that by definition any prefix of x whose length falls inside the scope of v has smallest period $|v|$. Some shorter prefixes may also have the same period, but we do not deal with them. Let us take $k=2$ and look at the Fibonacci word Fib_9 (recall that the sequence of Fibonacci words is defined by $\text{Fib}_1=b$, $\text{Fib}_2=a$, and $\text{Fib}_i=\text{Fib}_{i-1}\text{Fib}_{i-2}$ for $i>2$). In Figure 1 are presented the scopes of its 2-HRP's.

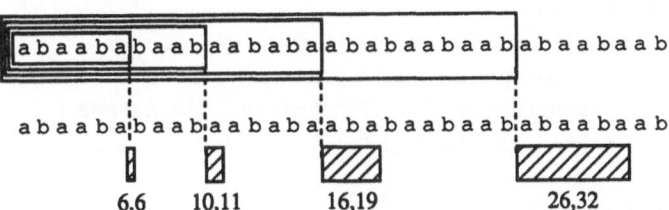

Figure 1. Repeating prefixes of the 9-th Fibonacci word and their scopes.

The following lemma shows that all scopes of highly repeating prefixes of a word x are pairwise disjoint (see Figure 1, for instance). And lemma 2 gives a lower bound on periods of prefixes whose length does not belong to any scope of a highly repeating prefix. These two lemmas provide the basic elements for a proof of Algorithm SIMPLE-TEXT-SEARCH shown in Figure 2.

155

Lemma 1

Let $[L_1, R_1]$ and $[L_2, R_2]$ be the respective scopes of two different highly repeating prefixes v_1 and v_2 of x. Assume that $|v_1|<|v_2|$. Then $R_1<L_2$.

Lemma 2

Let $(v_i / i=1,...,r)$ be the sequence of k-highly repeating prefixes of x in order of increasing lengths. Let $([L_i, R_i] / i=1,...,r)$ be the corresponding sequence of scopes. Then any non-empty prefix u of x satisfies:

$per(u) = L_i/2$ if $|u|$ is in $[L_i,R_i]$ for some i,

$per(u) > |u|/k$ if not.

In Figure 2 is shown the version SIMPLE-TEXT-SEARCH of KMP algorithm. In the algorithm we compute the length of shifts using the rule stated in Lemma 2 on the list of scopes of k-highly repeating prefixes. We assume that these intervals have been computed previously.

Algorithm SIMPLE-TEXT-SEARCH
/* Search text t for pattern p. */
/* An $O(r)$-space version of KMP algorithm, where r is the number of k-highly */
/* repeating prefixes of p. $([L_i, R_i] / i=1,...,r)$ is their sequence of scopes.   */
    $pos:=0; j:=0;$
    while $pos \leq n-m$ **do**
     {   **while** $j<m$ **and** $p[j+1]=t[pos+j+1]$ **do** $j:=j+1$;
          if $j=m$ **then** report the match at position pos;
          if j belongs to some.$[L_i, R_i]$ **then** { $pos:=pos+L_i/2; j:=j-L_i/2;$ }
          else { $pos:=pos+\lfloor j/k\rfloor+1; j:=0;$ }
     }
    return false;
end.

Figure 2. Searching phase algorithm.

Inside algorithm SIMPLE-TEXT-SEARCH, the test "j belongs to some $[L_i, R_i]$" can be implemented in a straightforward way. It needs $O(1)$ space and does not affect the asymptotic time complexity of the algorithm. The following is proved in [GS 83].

Assume that the pattern p has r k-highly repeating prefixes and that their list of scopes and their list of lengths are already computed. Then the algorithm SIMPLE-TEXT-SEARCH solves the string-searching problem on p and t using $O(r)$ space and performing at most $k.|t|$ symbol comparisons. The total complexity of the algorithm is also linear in $|t|$.

We call PREPROCESS the preprocessing phase of the entire algorithm. Algorithm PREPROCESS computes the scopes of all HRP's of the pattern p. It works in the same way as algorithm SIMPLE-TEXT-SEARCH, id est, as if the pattern is being searched for inside itself. Its time complexity is linear in the size of p.

Algorithms SIMPLE-TEXT-SEARCH and PREPROCESS require $O(r)$ extra space to work. This space is used to store the scopes of the HRP's of the pattern p. A simple application of periodicity lemma shows that for $k \geq 3$ a non-empty word x has no more than $\log_{k-1}|x|$ k-highly repeating prefixes. The following fact can be proved : if u and v are two HRP's of x and $|u|<|v|$, then the stronger inequality holds $(k-1).|u|<|v|$. However, the case k=2 is more difficult as we will see later.

Denote by HRP1(x) and HRP2(x) respectively when defined, the shortest and the second shortest k-highly repeating prefixes of a given word x. The basic tools are now properties and interplay between HRP1's and HRP2's starting at some specific positions inside the pattern p.
We can use algorithm PREPROCESS to compute HRP1's and HRP2's. This proves the following : HRP1(p) and HRP2(p) can be computed in $O(1)$ space and time proportional to the length of HRP2(p). The trick is simply to stop the execution of the algorithm PREPROCESS the first time it discovers the second HRP.

Pattern decomposition

What happens if the pattern p has at most one k-HRP? We say that such a pattern is k-simple. In this case obviously we can make string-searching in linear time using only $O(1)$ memory by applying algorithms PREPROCESS and SIMPLE-TEXT-SEARCH. The most memory consuming was the list of scopes, but now this list is reduced to only one pair (or even no pair) of integers specifying only one scope. Unfortunately not all patterns are simple. For instance, if we take $k=2$ then the Fibonacci words are not simple. However if we take $k=3$ they become simple. On the contrary, the word $((a^e b)^e c)^e$ is not simple for any $k \leq e$.
When $k \geq 3$ words satisfy a remarkable combinatorial property originally discovered by Galil and Seiferas for $k \geq 4$ (see [GS 83]):
        each pattern p can be decomposed into uv where u is "short" and v is a simple word.

With such a decomposition of the pattern p the search phase, called TEXT-SEARCH, can be realized conceptually in two phases as follows. First find all occurrences of v, which is efficient due to the simplicity of v. The algorithm SIMPLE-TEXT-SEARCH can be used for this purpose. Next, check whether the occurrences of v are preceded by u which is efficient due to the "shortness" of u. We define here what we mean by the shortness of u.
We say that the decomposition $p = uv$ is k-perfect iff v is k-simple and $|u| \leq 2.per(v)$.

157

It is proved in [GS 83] that the algorithm TEXT-SEARCH computes the positions of occurrences of pattern p inside text t. It uses constant extra memory space. It is linear in time and makes at most $(k+2).|t|$ symbol comparisons.

The next two lemmas are used in the proof of the decomposition theorem.

Lemma 3

Assume that $z=HRP1(x)$ is defined and let $x=zz'$.
a) If $HRP2(x)$ is defined, $|HRP2(x)| > 2.|HRP1(x)|$.
b) If $HRP1(z')$ is defined. Then $HRP1(x)$ is a prefix of $HRP1(z')$.

The structure of the decomposition algorithm is based on a sequence $V(x)$ of HRP1's. The elements of the sequence $V(x)=(v_1, v_2,...)$ are called *working factors* and are defined as follows. The first element v_1 is $HRP1(x)$. Let $x=v_1x'$, then v_2 is $HRP1(x')$, and so on until there is no HRP1. In particular the sequence is empty if x does not start with a k-th power. Lemma 3 shows that the sizes of v_i's are in increasing order. The next lemma says more on properties of their size: some v_i eventually reaches the size of $HRP2(x)$. Note that the length of $HRP2(x)$ is greater than twice the length of v_1 (Lemma 3).

Lemma 4 (Key lemma)

Let $V(x) = (v_1, v_2,...)$ be the sequence of working factors of x and assume that $HRP2(x)$ exists. Let i be the greatest integer such that $|v_1...v_i| < |HRP2(x)|$. Then, if v_{i+1} exists, $|v_{i+1}| \geq |HRP2(x)|$.

Lemma 4 yields the notion of *special positions* in x. The first special position of x is defined as the length of the word $v_1...v_i$ of Lemma 4. If $x=v_1...v_ilx'$, then the second special position is defined on x' in the same manner, and so on until no application of Lemma 4 is possible.

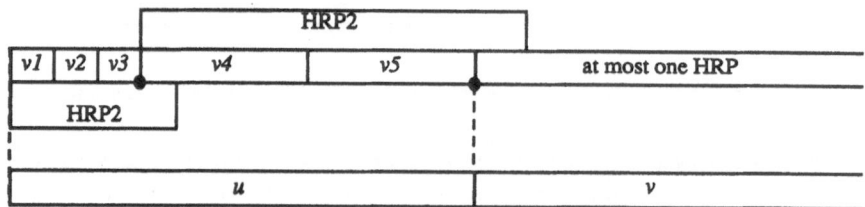

Figure 3. The sequence of working factors $v1, v2,...$ leads to special positions •, and to the 3-perfect decomposition uv of x.

The algorithm given by Galil and Seiferas [GS 83] to compute perfect decompositions can be described informally as follows:

- compute the sequence of working factors from left to right (see Figure 3);
- at each special position test whether there is an HRP2.

A more formal description of the algorithm is given in Figure 4.

Algorithm DECOMPOSE(x).

/* Computes a k-perfect decomposition of pattern x ($k\geq3$), HRP is related to k. */

$i:=0$; $hrp1:=|HRP1(x)|$; $hrp2:=|HRP2(x)|$;

while $hrp1$ and $hrp2$ exist **do**

{ $i:=i+hrp1$; $hrp1:=|HRP1(x_{i+1}...x_n)|$;

  if $hrp1 \geq hrp2$ **then** $hrp2:=|HRP2(x_{i+1}...x_n)|$;

}

return the decomposition $(x_1...x_i, x_{i+1}...x_n)$ of x;

end.

Figure 4. Computing perfect decompositions.

The correctness and complexity of algorithm DECOMPOSE immediately follows from the decomposition theorem.

Theorem 5 (Decomposition theorem for $k=3$)

(i) For $k\geq3$, each pattern p has a k-perfect decomposition computable in $O(1)$ space and $O(|p|)$ time.

(ii) Let j be the last special position of x. Let $u=x[1..j]$ and $v=x[j+1..n]$. Then, the decomposition uv of x is a 3-perfect decomposition.

Proof

By definition of the sequence of special positions of x, v is certainly 3-simple because it has at most one HRP. It remains to prove that $|u|<2.per(v)$. The key point is that each next HRP2 is at least twice bigger than the preceding one. This follows from Lemmas 3 and 4.

The cost of computation of working factors is proportional to the total length of consecutive HRP1's, which is linear; the cost of computation of consecutive HRP2's is proportional to their total length, which is also linear. Hence the algorithm DECOMPOSE takes linear time. ♦

Square prefixes

We show here that $k=3$ is the smallest bound for the decomposition theorem. This means that GS algorithm does not work in constant space for $k=2$. It however works in logarithmic space as we will see later.

The consequence of the following proposition is that, if uv is a factorization of x with v having at most one 2-HRP, u cannot be always be kept short.

159

Theorem 6

The decomposition theorem does not work for $k=2$: for any constant $c>0$ there exist infinitely many words x with the following property: if uv is factorization of x such that v has at most one 2-HRP then $|u|/per(v)>c$. (indeed, an even stronger inequality holds: $|u|/|v| >c$).

Proof.

Let $w = aabaabab$. Note that the smallest period of this word, and that of w^n ($n>0$), is $|w|=8$. Consider a factorization of the word w^3 of the form $uu'w^2$ (with $u, u' \in A^*$ and $w = uu'$). The word $v = u'w^2$ obviously starts with the square $(u'u)^2$. But it also starts with a strictly shorter square. For instance, if $u = aa$ and $u' = baabab$, v starts with the square $(baaba)^2$.

The factorization ($aabaababa, baababaabaabab$) of w^3 is such that the right part has at most one 2-HRP. The word $v = baababaabaabab$ is even the longest suffix of w^3 with this property. Thus, to meet the result of the proposition, one simply has to choose $x = w^n$ for a large enough integer n. ♦

When $u=HRP1(x)$, u^2 is a prefix of x and u is a primitive word. We call u^2 a *square prefix*. The number of square prefixes of a word can be logarithmic as it happens for Fibonacci words (see Figure 1). This follows easily from the recurrences defining Fibonacci words and from the fact that two consecutive Fibonacci words "almost" commute (up to the exchange of the last two symbols). We show that the maximal number of square prefixes of a word is also logarithmic. Moreover we characterize those words that reach the bound. As a consequence this shows that algorithms SIMPLE-TEXT-SEARCH and PREPROCESS of Section 2 work in logarithmic space for $k=2$. The result is based on the following lemma. Its proof comes from several applications of the periodicity lemma.

Lemma 7

Let $u1^2, u2^2, u3^2$ be three prefixes of x such that $u1, u2, u3$ are primitive and $|u3|<|u2|<|u1|$. Then $|u2|+|u3|\leq|u1|$.

The inequality in Lemma 15 cannot be improved. For instance, the Fibonacci word of Figure 1 has four square prefixes of respective lengths 6, 10, 16, 26. Moreover, it is well-known that f_i^2 is a prefix the Fibonacci word f_j, for all i in the interval $(4,...j-2)$.

Example

The word $ababaabababaab$ begins with squares of lengths 4, 10 and 14 corresponding to the 2-HRP's ab, $ababa$ and $ababaab$. Computation proves that no shorter word can have three 2-HRP's like it. ♦

160

Fact

The word *aabaabaaabaabaabaaab* begins with squares of lengths 2, 6, 14 and 20 corresponding to the 2-HRP's *a*, *aab*, *aabaaba* and *aabaabaaab*. Computation proves that no shorter word can have four 2-HRP's like it. Moreover, it is the unique word of length 20 on the alphabet $\{a, b\}$ (up to the exchange of letters *a* and *b*) having this property. ♦

Theorem 8

(i) The number of square prefixes of a non-empty word x is less than $\log_\Phi |x|$ (where Φ stands for the golden ratio ($\Phi = (1+\sqrt{5})/2 = 1.618...$).

(ii) Let $(u_i \,/\, i \geq 0)$ the sequence of words on the alphabet $\{a, b\}$ defined by

$$u_0 = a, \; u_1 = aab, \; u_2 = aabaaba \text{ and } u_i = u_{i-1}u_{i-2} \text{ for } i \geq 3.$$

Then, for each i, u_i^2 has the maximum number of square prefixes among the words of the same length. Moreover, for $i \geq 3$, u_i^2 is the unique word on the alphabet $\{a, b\}$ having this property (up to a permutation of letters *a* and *b*).

Proof.

(i) We first prove that, for $i>1$, if x has i 2-HRP's then $|x|>2.F_{i+1}$ (where F_i is the i-th Fibonacci number, $F_i=|f_i|$).

For $i=1$, $|x| \geq 2 = 2.F_2$. For $i=2$, $|x| \geq 6 > 2.F_3 = 4$. By the fact stated above, for $i=3$, $|x| \geq 14 > 2.F_4$. The rest of the proof is by induction on i. Let $i \geq 4$. Let $u1$, $u2$ and $u3$ be the three longest 2-HRP's of x (in order of decreasing lengths). The word $u3^2$ has exactly i-2 2-HRP's. By induction its length is then greater than $2.F_i$. For the same reason $|u2^2|>2.F_{i+1}$. By lemma 7, we get $|u1| \geq |u3|+|u2|$ which implies $2.|u1|>2.F_{i-1}+2.F_i=2.F_{i+1}$. Thus $|x| \geq 2.F_{i+1}$, as expected.

The Fibonacci number F_{i+1} is greater than (or equal to) Φ^{i-1}. Thus $|x|>2.\Phi^{i-1}$ which yields $i<\log_\Phi |x|$ inequality that also holds for $i=1$.

(ii) Consequence of the above fact and of Lemma 7. ♦

It is known from [KMP 77] that the number of certain periods of a word is bounded by $\log_\Phi |x|$. These periods of x are all greater than (or equal to) $per(x)$ the smallest one. From this point of view, the result of Theorem 8 is a dual result that gives the same bound on the number of small starting periods of x that are all not greater than $per(x)$.

Optimal computation of periods

The evaluation of the periods of a word in strongly related to the computation of all overhanging occurrences of a pattern in a text. Indeed periods of a word correspond to overhanging occurrences of the word over itself. An overhanging occurrence of the pattern p in the text t is specified by a suffix z of t that is also a proper prefix of p. If t is written yz, the position of the overhanging occurrence of p is $|y|$. And if $t=p$, the position $|y|$ is also a period of p.

161

String-searching algorithms which use left-to-right scanning of the pattern against the text naturally compute overhanging occurrences of the pattern. They can be used to compute all the periods of a word. It is the case for the algorithm of Knuth, Morris and Pratt [KMP 77] and its variations, and also for the time-space optimal algorithm of Crochemore [Cr 91]. We show that the algorithm of Galil and Seiferas is also adapted for this purpose, provided the parameter k is chosen large enough ($k \geq 7$). This thus leads to a new time-space optimal algorithm for computing periods in words.

The computation of periods through GS algorithm (see [GS 83]) was based on the following assertion: "$per(x)$ is the position of the second occurrence of x inside xx". Unfortunately this assertion does not always hold. For instance, $x=ababa$ has period 2 and the second occurrence of x inside xx is at position 5. The assertion precisely holds when x is not primitive.

Let uv be a k-perfect decomposition of the pattern p. If u is empty, the algorithm SIMPLE-TEXT-SEARCH naturally extends to the computation of overhanging occurrences. It works in linear time and constant space. The situation is different when u is non empty because v can have many overhanging occurrences in t and thus u could be scanned too many times by the algorithm. To avoid this situation we impose a condition to the decomposition of the pattern and this yields the choice $k=7$ for the parameter.

Assume u is not empty and $k \geq 3$. The pattern p can then be written $p = uv = u'w^kv'$ with u' a proper prefix of u, $|u| < |w|.(k-1)/(k-2)$ and $k \leq k'$. The word w is the last k-HRP2 computed by algorithm DECOMPOSE in Figure 4.

Condition (*): $|u| + 2.|w| \leq |p|/2$.

Lemma 9

Let uv be the k-perfect decomposition of p computed by algorithm DECOMPOSE. If u is non-empty and $k \geq 7$, condition (*) is satisfied.

To compute overhanging occurrences of p in t we may assume without loss of generality that $|t|=|p|$. The core of the algorithm is a procedure that computes the overhanging occurrences of p starting in the left half part of t.

Lemma 10

On entry p and t, algorithm Left_Occ works in constant space and time $O(n)$ where $n = |t| = |p|$.

To compute all overhanging occurrences of p in t, we first apply the algorithm Left_Occ and then repeat the same process with both the right half part of t and the left half part of p.

162

Function Left_Occ (p, t);

/* It is assumed that $|t|=|p|$ */

/* Computes the set $\{|y| \mid t = yz,\ z$ prefix of p and $|y| \leq |t|/2\}$ */

/* of positions of p in the left half part of t */

    $P := \emptyset$;

    compute 7-perfect decomposition (u, v) of p;

    find all overhanging occurrences of v starting in the left half part of t

        with Algorithm SIMPLE-TEXT-SEARCH;

    for each position i of an occurrence of v in t **do**

        if u is a suffix of $t[1]...t[i]$ **then** add i to P;

    return P;

end.

Figure 5. Computing left overhanging occurrences.

Function Ov_Occ (p, t);

/* computes all the overhanging occurrences of p in t; we assume $|t| = |p|$ */

    $P := $ Left_Occ(p, t);

    while 7-perfect decomposition of p is different from (ε, v) **do**

    $\{\ n := |t|/2;\ t := $ right half part of t, $t[n+1..|t|]$; $\ p := $ left half part of p of length $|t|-n$;

    add Left_Occ$(p, t)+n$ to P; $\}$

    return(P);

end.

Figure 6. Computing overhanging occurrences.

Theorem 11

Algorithm Ov_Occ find all overhanging occurrences of p in t in constant space and time $O(n)$ where $n = |p| = |t|$.

Proof

The correctness is left for the reader. Constant space is enough to memorize left or right parts of input strings. The time complexity is bounded by

$$c.n + c.n/2 + ... + c.n/2^i$$

where c is the constant of proportionality for Algorithm Left_Occ. This quantity if less than $2.c.n$ which proves the result. ◆

Corollary 12

Algorithm Ov_Occ finds all periods of a word of length n in constant space and time $O(n)$.

Proof

Apply Ov_Occ to the pair (p, p) and note that the position of an overhanging occurrence of p over itself is a period of p (see Figure 16). ◆

Parallel perfect decomposition

We consider now the parallel computation of the k-perfect decomposition of pattern p. For that we choose $k=4$. Assume that $v_1=HRP1(x)$ exists. Consider the maximum integer e such that v_1^e is prefix of x ($e\geq4$ because $k=4$). Define $LongHRP1(x)$ to be $w_1=v_1^{e-3}$. Let x' be such that $x=w_1x'$. By definition of v_1 and e, v_1 is not the $HRP1(x')$. And since $HRP1(x')$, if it exists, cannot be shorter than v_1, it is longer than v_1. This leads to an iterative definition of the parallel working sequence of the word $x : P(x)=(w_1,w_2,...)$.

We will compute the elements of this sequence or more precisely the sequence of positions of LongHRP1's in the sequence.

The crucial point in the parallel algorithm is the following fact due to hypothesis $k=4$ (and the periodicity lemma):

Fact

Let $P(x)=(w_1, w_2,...)$ be the parallel working sequence of x. For $i>0$, $2.|w_i|<|w_{i+1}|$ if w_{i+1} is defined. The length of $P(x)$ is $O(\log|x|)$.

We use the components of the string-searching algorithm of Vishkin to compute the basic functions used in the preprocessing phase of GS algorithm.

Lemma 13

Assume $HRP1(x)$, $HRP2(x)$ and $LongHRP1(x)$ exists. Let $n1$, $n2$ and $n3$ be their respective sizes. Let $n=|x|$. Then $HRP1(x)$, $HRP2(x)$ and $LongHRP1(x)$ can be computed by an optimal parallel algorithm in $O(\log n)$ time on a CRCW PRAM. They use respectively $n1/\log(n)$, $n2/\log(n)$ and $n3/\log(n)$ processors.

If $HRP1(x)$ or $HRP2(x)$ does not exist, then the number of processors is $O(n/\log(n))$.

If we discover the first prefix period v_1 then we check whether it is an $HRP1(x)$ and compute its continuation C. This can be done using the procedure PeriodCont from [GR 88, page 228] and a kind of binary search. If we know the continuation C of the prefix period v_1 and its size then computing $LongHRP1(x)$ is a simple matter.

The computation of $HRP2(x)$ can be done essentially in the same way, the procedure MakeSparsePeriodic from [GR 88] can be used. The table of witnesses can be computed for the prefix C and the algorithm looks for the next supposed prefix period. We refer the reader to [GR 88, pages 225-230] for technical details.

The time complexity and the number of processors are of the same order as that of the whole string-searching algorithm of Vishkin, restricted to the part of the pattern which is inspected. Hence the total number of operations is proportional to the size of computed objects, if they are nonempty. If they are empty then the number of all operations is just proportional to the total size n of the input pattern. This completes the proof. ◆

Using parallel versions of functions HRP1, HRP2 and LongHRP1 the parallel decomposition algorithm is a simple version of the sequential pattern preprocessing. Its correctness follows from the correctness of its sequential counterpart.

Algorithm Parallel 4-perfect decomposition;
    $i:=0$; $hrp1:=|HRP1(x)|$; $longhrp1:=|LongHRP1(x)|$; $hrp2:=|HRP2(x)|$;
    while $longhrp1$ and $hrp2$ exist **do**
    { $i:=i+longhrp1$; $hrp1:=|HRP1(x_{i+1}...x_n)|$; $longhrp1:=|LongHRP1(x_{i+1}...x_n)|$;
      if $hrp1 \geq hrp2$ **then** $hrp2:=|HRP2(x_{i+1}...x_n)|$ }
    return the decomposition $(x_1...x_i, x_{i+1}...x_n)$;
end.

Figure 7. Parallel perfect decomposition.

Theorem 14

The 4-perfect decomposition of a word x of length n can be found in $\log^2 n$ time with $n/\log^2 n$ processors of a CRCW PRAM or in $\log^3 n$ time with $n/\log^3 n$ processors of a CREW PRAM.

References

[Ah 90]   A.V. AHO, Algorithms for finding patterns in strings, in: (J. VAN LEEUWEN, editor, *Handbook of Theoretical Computer Science*, vol A, *Algorithms and complexity*, Elsevier, Amsterdam, 1990) 255-300.

[CP 89]   M. CROCHEMORE, D. PERRIN, Two-way string-matching, *J. ACM* (1989). to appear.

[Cr 91]   M. CROCHEMORE, String-Matching on Ordered Alphabets, *Theoret. Comput. Sci.* (1991). to appear.

[GS 80]   Z. GALIL, J. SEIFERAS, Saving space in fast string matching, *SIAM J.Comput.* 9 (1980) 417-438.

[GS 83]   Z. GALIL, J. SEIFERAS, Time-space optimal string matching, *J. Comput. Syst. Sci.* 26 (1983) 280-294.

[GR 88]   A. GIBBONS, W. RYTTER, *Efficient parallel algorithms*, Cambridge University Press, Cambridge, 1988.

[KMP 77] D.E. KNUTH, J.H. MORRIS Jr, V.R. PRATT, Fast pattern matching in strings, *SIAM J.Comput.* 6 (1977) 323-350.

[Lo 83]   M. LOTHAIRE, *Combinatorics on words*, Addison-Wesley, Reading, Mass., 1983.

[Si 89]   I. SIMON, personal communication, 1989.

Reconstructing sequences from shotgun data

Paul Cull
pc@cs.orst.edu

Jim Holloway
holloway@cs.orst.edu

Department of Computer Science
Oregon State University
Corvallis, OR 97331

Abstract

One method of sequencing DNA produces a linear list of bases, {A, C, G, T}, for many short overlapping fragments of the original DNA. To find the sequence of the original piece of DNA, the many fragments must be reassembled. While this problem of reassembly is similar to the NP–complete shortest common superstring problem [GMS80], we believe that biologists are actually trying to solve a simpler problem. Biologists assume that short overlaps between substrings are insignificant. Further, they assume that there is a unique string from which substrings could have been produced. We consider a reconstruction problem with these restrictions. We devise algorithms for this problem both when the overlaps must be exact and when there may be errors in the overlaps. Our algorithms are based on Rabin–Karp string matching, and on suffix arrays. We investigate the running times of our algorithms, and show that in expected case they have running times proportional to the length of the reconstructed sequence. We give the timings of some test runs and note that the suffix array algorithms seem to be faster.

1 Introduction

Biological and physical limitations require that DNA be sequenced in fragments. Because of these limitations there are two approaches used to obtain the appropriate sized fragments of DNA to sequence.

One class of methods for sequencing DNA is loosely termed ordered sequencing. These methods use an oligonucleotide primer to initiate the DNA sequencing at a known point and the sequencing reaction proceeds from this point. The leading several hundred bases of the DNA strand are sequenced and then removed using exonucleases exposing the next segment of the DNA to be sequenced. The process of sequencing several hundred bases and removing them is continued until the entire DNA sequence has been determined. We will not consider ordered sequencing in this paper.

166

Another class of methods for sequencing DNA is loosely referred to as shotgun sequencing. Many identical copies of the DNA to be sequenced are cleaved by one or more restriction endonucleases. A restriction endonuclease will cleave the DNA sequence at each occurrence of a specific six to eight base subsequence (sonication can also be used to create unordered fragments). This results in a multiset of DNA fragments that are not ordered. DNA fragments are essentially selected at random from this multiset and sequenced. A sequence that is believed to represent the original DNA sequence is assembled by finding overlaps between the DNA fragments that have been sequenced.

In this paper we will present several algorithms that could be used to assemble the sequenced DNA fragments into a contiguous sequence. In section 2 we will review the work that has been done on the shortest common superstring problem. Section 3 will define the exact shotgun sequencing problem, a problem closely related to the shortest common superstring problem and give two algorithms to solve it. In section 4 the inexact shotgun sequencing problem will be considered when we allow a small number of errors to occur in the string matching.

We will use calligraphic letters such as $\mathcal{S}$ to represent multisets of strings. Capital letters late in the alphabet will be used to indicate strings, while lower case letters will be used for characters of a string such as $S = s_1 s_2 \ldots s_m$. The strings are composed of characters from the alphabet Σ. We will use the following notation,

- n be the number of strings in the multiset $\mathcal{S} = \{S_1, S_2, \ldots, S_n\}$

- $N = \sum_{i=1}^{n} |S_i|$

- L_{min} = length of shortest common superstring

- L = length of superstring found by heuristic

- C = compression = $N - L_{min}$

- $C_H = N - L$ = compression found by heuristic.

2   Shortest Common Superstring Problem

In 1980 Gallant, Maier & Storer [GMS80] showed that the shortest common superstring problem is NP–complete. They first defined superstring,

> a *superstring* of a set of strings $\mathcal{S} = \{S_1, S_2, \ldots, S_n\}$ is a string S containing each S_i, $1 \leq i \leq n$, as a substring,

and then defined the shortest common superstring problem,

> Given a set of strings $\mathcal{S}$ and a positive integer l, does $\mathcal{S}$ have a superstring of length l?

With these definitions they are able to show that the shortest common superstring problem is NP–complete even when the alphabet is restricted to {0,1}. The NP–completeness result suggests that there is no polynomial time algorithm for this problem. Therefore, one should probably attack this problem in one of several ways:

1. Add assumptions about the substrings so that the problem with the extra assumptions is no longer NP–complete.

2. Show that the hard instances of the problem are rare, so that the problem can usually be solved quickly.

3. Instead of finding the superstring with the minimum length, find a superstring with a length that can be shown to be close to the minimum length.

Although we mention three approaches to deal with an NP–complete problem, the papers discussed all deal with the third approach, approximation. Later we will assume that there is a unique construction of the superstring with a minimum acceptable overlap. This is an example of approach one, adding assumptions so that the problem is no longer NP–complete.

In a paper by Peltola, Söderlund, Tarhio & Ukkonen [PSTU83] a heuristic for a generalized minimal length superstring problem is given. The problem is generalized by allowing errors in the string matching. They define superstring,

a superstring of a set of strings $S = \{S_1, S_2, \ldots, S_n\}$ is a string S containing each S_i, $1 \le i \le n$, as an **approximate** substring.

An approximate substring S_i of S with an error ratio δ is defined,

a substring of S that can be transformed into S_i with at most $\delta|S_i|$ delete, insert, replace, and transpose operations.

The minimum number of delete, insert, and replace operations needed to transform one string into another is frequently called the minimum edit distance. They then define the generalized minimal length superstring problem,

Given a set of strings S, a positive integer l, and an error ratio δ, $0 < \delta < 1$, does S have a common superstring of length at most l?

No performance guarantees are given for the heuristic. The running time of the heuristic is $O(\delta N^2)$.

Papers by Tarhio & Ukkonen [TU88] and Turner [Tur89] develop approximation algorithms that they conjecture will find an approximate shortest common superstring that is, at worst, twice the length of the actual shortest common superstring. Tarhio & Ukkonen [TU88] analyzed their algorithm in terms of the compression of the strings in S instead of the length of the shortest common superstring of S. The main result of the paper is that $C_H \ge \frac{1}{2}C$. The running time of the algorithm presented by Tarhio & Ukkonen is $O(nN)$.

Turner [Tur89] presents an algorithm to approximate the shortest common superstring with the same performance guarantees as the algorithm of Tarhio &

Ukkonen [TU88]. Turner uses suffix arrays to reduce the number of suffix/prefix comparisons that need to be done. The running time of the algorithm is $O(N \log N)$ or $O(N \log n)$, depending on whether direct indexing over the alphabet Σ is allowed.

Later, in a paper by Ukkonen [Ukk90], an $O(N)$ or $O(N \min(\log n, \log |\Sigma|))$ algorithm, depending on whether direct indexing over the alphabet Σ is allowed, to solve the approximate shortest common superstring problem is presented. This reduction in time is achieved by a clever use of the Aho–Corasick [AC75] string matching automaton. Again, this algorithm achieves the same compression ratio as the algorithms of Tarhio & Ukkonen and Turner.

The first approximation algorithm that approximated the shortest common superstring of S instead of the maximal compression of the strings in S was given by Li [Li90]. Li was able to give an algorithm to compute an approximate shortest common superstring of length $O(L_{min} \log L_{min})$. The algorithm is similar to the greedy algorithms given above by Tarhio & Ukkonen, Turner, and Ukkonen. It differs when the strings with the maximum overlap are joined, not only are the strings that were joined removed from the set of strings, but all substrings of the resulting joined string are removed from the set of strings. This results in the size of the set of strings decreasing fast enough to show the $O(L_{min} \log L_{min})$ bound on the length of the approximate shortest common superstring. Although the running time of the algorithm is not considered, it is clearly polynomial in the size of the input.

Blum et. al.[BJL+91] recently showed that the greedy algorithm of Tarhio & Ukkonen [TU88] and Turner [Tur89] does find superstrings that are, at worst, a multiplicative factor of four longer than the minimum length superstring. Blum et. al. also give a modification of the algorithm to get the multiplicative factor down to three.

3   Exact Shotgun Sequencing Problem

Finding the minimum length superstring of a set of strings seems to require us to look at many of the possible alignments of the strings in the set. When molecular biologists try to solve the similar problem of aligning their sequence fragments into a contiguous sequence, they assume that matches of a length greater than some constant are "significant". In this section will use this assumption to construct an algorithm to build a contiguous sequence. This assumption will allow us to find alignments that are "good enough" and not require us to search the entire space of alignments. We will see that, in these algorithms, the run time is directly related to the compression.

If we let k be the minimum length match that we consider significant, one naive algorithm to solve this problem would compare the $\frac{n(n-1)}{2}$ pairs of strings to see if there is a prefix/suffix match of length k. If the $n-1$ prefix/suffix matches have not been found, k is incremented and the process is repeated.

In this section we will present two new algorithms to solve the exact shotgun sequencing problem. The first uses the ideas of Rabin–Karp [KR87] string matching. The second algorithm sorts the strings before building a consensus string. We will use the term "consensus string" to refer to the string that is constructed by aligning the sequence fragments. The ideas of Knuth, Morris & Pratt [KMP77], and Boyer & Moore [BM77] were considered, but it is not clear to us how to construct a small finite automata that would match strings with errors.

3.1 Assumptions

We will make the following assumptions before we define the exact shotgun sequencing problem.

1. An integer k can be supplied that defines the minimum acceptable overlap between two strings.

2. There is a unique alignment of the strings in S such that all suffix/prefix overlaps are of length k or greater.

3. All suffix/prefix overlaps are exact matches.

We feel that, if k is chosen carefully, these assumptions are reasonable. For large k, it will be very unlikely that any overlap other than overlaps that arise from the shotgun procedure will appear. If these assumptions are too strong, a simple modification to the algorithms described here (essentially increasing the value of k) will allow the algorithms to produce a series of alignments, each successive alignment being more compressed than the previous.

3.2 The problem definition

We are given a multiset of strings, $S = \{S_1, S_2, \ldots, S_n\}$, and an integer k. We make the following assumptions about S and k

1. S_i is not a substring of S_j for $1 \leq i, j \leq n, i \neq j$.

2. An ordering, H, of the strings in S exists such that

$$\forall_{1 \leq i < n} i \ \exists_{j \geq k} j \ \mathsf{suffix}(S_{H_i}, j) = \mathsf{prefix}(S_{H_{i+1}}, j).$$

The problem is, given the multiset S and the integer k, find the ordering H.

3.3 Rabin–Karp type algorithm

Given two strings S and T, $i = |S|, j = |T|, i < j$, the Rabin–Karp algorithm for string searching computes a hash value for the shorter string, S. A hash value for each length i substring of T is computed and compared with the hash value for S. By cleverly choosing the hashing function, the hash value for the substring of T ending at position h can be computed from the hash value for the substring of T ending at position $h - 1$ and the character t_h. The hash function Rabin and Karp [KR87] chose was $\mathsf{hash}(m) = m \bmod p$ where p is a large prime and m is an integer representation of the string $T_{h-i+1} \ldots T_h$. With this in mind, it is easy to compute the hash value of the length i substring of T ending at position h

$$\text{hash}(h) = ((\text{hash}(h-1) - \text{index}(t_{h-i}) \cdot \sigma^{i-1}) \cdot \sigma + \text{index}(t_h)) \bmod p$$

Associated with each $c \in \Sigma$ is a unique integer l, $0 \leq l < \sigma$. The function $\text{index}(c)$ will return the integer associated with c.

The probability of two strings drawn randomly from Σ^i having the same hash value is shown by Gonnet & Baeza–Yates [GBY90] to be

$$\frac{1}{p} + O\left(\frac{1}{\sigma^i}\right).$$

With the appropriate choices of p and k, the frequency of collisions will be small.

The Rabin–Karp algorithm has two properties that make it particularly well suited to the exact shotgun sequencing problem.

- The hash value for the prefix and suffix of a string can be computed incrementally. Given the hash value for the length i prefix of a string, the hash value for the length $i + 1$ prefix can be computed with a constant number of operations.

- The hash value of a substring can be incrementally computed equally well from the right or the left end of the substring.

Figure 1 gives an algorithm based on the ideas of Rabin and Karp for the exact shotgun sequencing problem. Since the strings we are comparing are always prefixes (suffixes), we do not need to subtract the value of the leading (trailing) character of the string. The jth forward hash value of the string S (fhv[S]) is the hash value of the string $s_1 s_2 \ldots s_j$. The function we will use to compute the jth fhv[S] from the $(j-1)^{st}$ fhv[S] is

$$\text{fhv}[S] = (\text{fhv}[S] \cdot \sigma + \text{index}(s_j)) \bmod p.$$

Let $i = |S| - j + 1$. The jth backward hash value of the string S (bhv[S]) is the hash value of the string $s_i s_{i+1} s_{|S|}$. The function we will use to compute the jth bhv[S] from the $(j-1)^{th}$ bhv[S] is

$$\text{bhv}[S] = ((\text{index}(s_i) \cdot \sigma^{j-1}) + \text{bhv}[S]) \bmod p.$$

The initial length k forward and backward hash values are computed for each $S \in \mathcal{S}$. Each fhv[S] and bhv[S] is added to a binary tree. Whenever the hash value being added to the tree matches a hash value already in the tree, the associated strings are compared. If the prefix of one matches the suffix of the other, the strings are removed from $\mathcal{S}$ and joined, the resulting string is added back to $\mathcal{S}$. When all of the possible prefixes and suffixes of length k have been joined, the value of k is incremented, the search tree is cleared and the hash values for the new value of k are computed.

If we assume that no collisions occur, the number of operations used by the Rabin–Karp type algorithm for the exact shotgun sequencing problem is

$$O(C_H \cdot \log n).$$

```
consensus_RK (S)

dm ← 1
k ← minimum suffix–prefix length
compute initial values of fhv and bhv
while |S| > 1
    hashtree ← nil
    for each S ∈ S
        j ← |S| - k + 1
        fhv[S] ← (fhv[S] * σ + index (s_k)) mod p
        bhv[S] ← (bhv[S] + (index (s_j)·dm)) mod p
        if (fhv[S] ∉ hashtree) add (fhv[S], hashtree)
        else join (S, matched string from hash tree)
        if (bhv[S] ∉ hashtree) add (bhv[S], hashtree)
        else join (S, matched string from hash tree)
    dm ← (dm·σ) mod p
    k ← k + 1
```

Figure 1: Rabin–Karp type algorithm for the exact shotgun sequencing problem

The lines of the inner for loop will be executed once for each position of compression in the final consensus sequence. The "if (fhv[S] ∉ hashtree) add (fhv[s], hashtree)" and "if (bhv[S] ∉ hashtree) add (bhv[s], hashtree)" lines take $O(\log n)$ time to execute and each of the other lines of the inner loop will take constant time, therefore the running time of the algorithm in Figure 1 will be $O(C_H \cdot \log n)$.

The hash tree is set up as a binary tree. Associated with each node is a hash value and a linked list of all strings that have the hash value associated with the node. When a string has the same hash value as a node, a linear search is performed on the strings associated with the node and if a match is found, it is returned. If no match is found, the string is added to the linked list. In the extraordinary case where each string hashes to the same value, the algorithm will perform just as the naive algorithm does [CH91].

The algorithm in Figure 1 has been implemented in C on a Sun 3/260. Figure 2 shows the results of running the program on strings of length 507. The length of the overlaps between strings was between 180 and 200 characters. The minimal acceptable overlap, k, was set to 20. The number of strings was varied between 25 and 1000 and the amount of CPU time used to compute the consensus string is plotted. Figure 3 show the time divided by the log of the number of strings as the number of strings was varied.

3.4 Algorithm based on sorting

In the naive algorithm a great deal of time is spent searching for a string in S with a particular prefix. If the list of prefixes were sorted, the time needed to search S for a string with a particular suffix could be significantly reduced. In this section we will

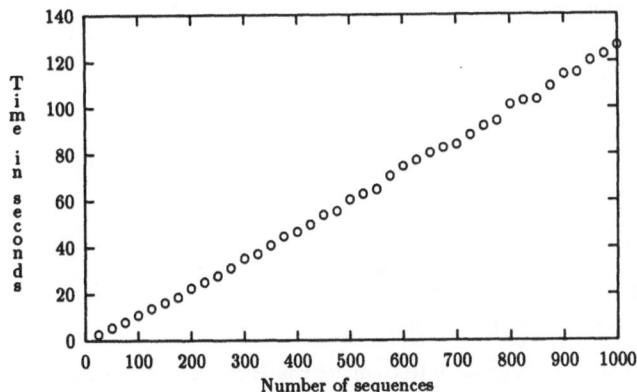

Figure 2: Running time for RK algorithm.

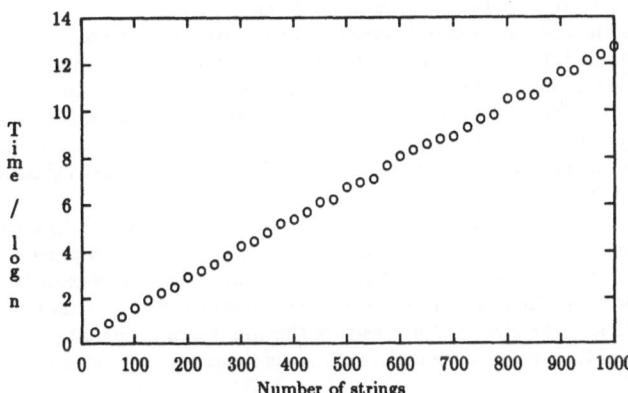

Figure 3: Running time / log (n) for RK algorithm.

```
consensus_trie (S)

trie ← sort_seqs (S, 0)
k ← minimum match length acceptable
while |S| > 1
    for each S ∈ S
        R ← search (S, trie, k)
        if R ≠ Λ
            T ← join (R, S)
            add T to S
            remove R and S from S
    k ← k + 1
print one remaining entry in S
```

Figure 4: Algorithm to compute consensus sequence using a trie

use a trie to speed the search.

By sorting the strings using a bucket sort and keeping track of the positions of the buckets within the sorted list, we can find a prefix of length i using $O(i)$ operations. During the standard bucket sort of the strings we will build a trie [Knu73] where a node at depth i represents a bucket containing all strings in S with a particular prefix of length i. The children of a depth i node represent the strings in S with prefixes of length $i + 1$ where the strings associated with the parent node and the strings associated with the child node agree in positions 1 through i.

The function sort_seqs in Figure 5 will recursively sort the sequences and build the trie. When $|S| > 1$, the strings in S will be sorted by the position indicated by the variable "column". The strings are then divided into at most σ groups, one group for each distinct character appearing at position "column" in some $S \in S$. A node in the trie is created for each of the non–empty groups of strings. Each of these nodes is a child of the node representing S. The function sort_seqs is then called recursively for each of the groups of strings.

The function sort_by_position will order the strings passed to it by the characters in the position passed to it. The function sort_by_position returns a node of the trie that points to the beginning of σ buckets along with the size of each bucket. The function sort_by_position takes time proportional to the number of strings being sorted. The time to sort the strings and construct the trie is proportional to N, the sum of the lengths of the strings, since each character of the strings in S is compared at most once.

The algorithm used to search the trie is given in Figure 6. It is essentially just a σ-ary tree search for the length j suffix of the string S. It does assume that you can index by the characters in Σ in constant time. If it is not possible to index by the characters in Σ then an $O(\log |\Sigma|)$ search would have to be used to index the buckets and trie pointers. The worst case time required to determine if a prefix of length l exists in the $|S|$ strings is $O(l)$. This can be seen by noting that in the search procedure, at most one comparison is done at each character position and only

```
sort_seqs (S, column)
```
/* column is the position in the strings that the strings are to be ordered by */

```
    root ← nil
    if |S| > 1
        root ← sort_by_position (S, column)
        for each i, such that 0 ≤ i < σ
            S_i ← root.bucket_pos_i
            root.child_i ← sort_seqs (S_i, column + 1)
    return root
```

Figure 5: Algorithm to build trie and sort strings

```
search (S, trie, j)
```

/* Traverse trie */

```
    while (trie ≠ Λ) ∧ (j > 0)
        prev ← trie
        c ← s_{|S|-j}
        trie ← trie.next_c
        j ← j - 1
    j ← j + 1
    T ← prev.bucket_pos_c
    c_s ← s_{|S|-j}
    h ← 0
    c_t ← t_{h-j}
```

/* Compare strings */

```
    while (c_s ≠ c_t) ∧ (j > 1)
        j ← j - 1
        h ← h + 1
        c_s ← s_{|S|-j}
        c_t ← t_h
    if (c_s = c_t) ∧ (j = 1) return T
    else return nil
```

Figure 6: Algorithm to search a trie for a string

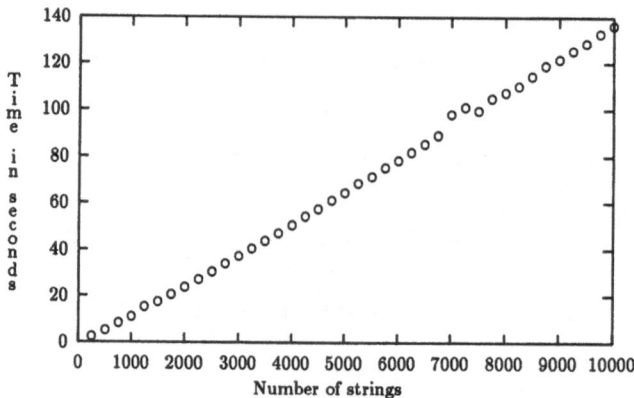

Figure 7: Running time for sort based algorithm.

characters in the prefix are compared.

The algorithm to find the consensus sequence using the sorted list of strings and the trie is given in Figure 4. For a given value of k, the suffix of length k of each string is searched for in the sorted list of prefixes. If a match is found the strings are removed from S, joined, and the result of the join is added back to S. When all of the the suffixes of length k have been searched for, k is incremented and the process is repeated.

Each iteration of the inner loop will result in one character of compression. In the worst case, the time to build the consensus string is

$$O(\sigma n C_H).$$

This situation arises when the the branching factor on the trie is nearly one. When the branching factor is closer to σ, the expected time to build the consensus string is

$$O(C_H \log n).$$

The algorithm in Figures 4, 5 and 6 has been implemented in C on a Sun 3/260. Figure 7 shows the results of running the program on strings of length 507. The length of the overlaps between strings was between 180 and 200 characters. The minimum acceptable overlap, k, was set to 20. The number of strings was varied between 250 and 10000 and the amount of CPU time used to compute the consensus string is plotted. Figure 8 shows the time divided by the log of the number of strings as the number of strings was varied.

4 Inexact Shotgun Sequencing Problem

The process of sequencing DNA and RNA is not perfect and mistakes are occasionally made in processing the gels, reading the gels and entering the data into a database. In

176

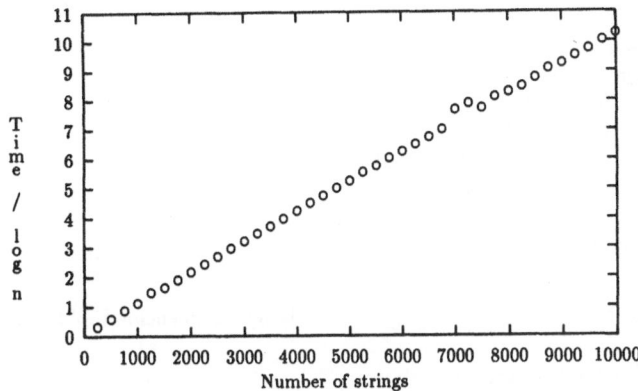

Figure 8: Running time / log (n) for sort based algorithm.

this section, we will not assume that all suffix/prefix overlaps match exactly, but that there are at most $\log(v)$ positions where the suffix and prefix do not match where v is the length of the overlap between the two strings. We will briefly discuss an algorithm to solve the log inexact shotgun sequencing problem based on the Rabin–Karp string matching algorithm. We will present in more detail another algorithm based loosely on the sorting algorithm presented in section 3.4.

We will say that the two length m string S and T log inexact match or $S \approx T$ if

$$\sum_{i=0}^{m-1} (s_i \oplus t_i) \leq \log m$$

$$\text{where } s_i \oplus t_i = \begin{cases} 0 & \text{if } s_i = t_i \\ 1 & \text{if } s_i \neq t_i \end{cases}.$$

4.1 Assumptions

We will make the following assumptions and then define the log inexact shotgun sequencing problem.

- An integer k can be supplied that defines the minimum acceptable overlap between two strings.

- There is a unique alignment of the strings in S such that all suffix/prefix overlaps are of length k or greater.

- All suffix/prefix overlaps are log inexact matches.

177

```
naive (S, k)
    while |S| > 1
        for each S_i ∈ S
            for each S_j ∈ {S − S_i}
                if verify (S_i, S_j, k)
                    add (join (S_i, S_j), S)
                    remove (S_i, S)
                    remove (S_j, S)
        k ← k + 1
```

Figure 9: Naive algorithm for log inexact shotgun sequencing problem

4.2 Problem definition

We are given a multiset of strings, $S = \{S_1, S_2, \ldots, S_n\}$, and an integer k. We make the following assumptions about S and k

1. S_i is not a substring of S_j for $1 \leq i,\ j \leq n,\ i \neq j$.

2. An ordering, H, of the strings in S exists such that

$$\forall_{1 \leq i < n} i\ \exists_{j \geq k} j\ \text{suffix}(S_{H_i}, j) \approx \text{prefix}(S_{H_{i+1}}, j).$$

The problem is, given the multiset S and the integer k, find the ordering H.

4.3 Naive log inexact algorithm

The naive algorithm presented in Figure 9 will solve the log inexact shotgun sequencing problem. The algorithm simply tries all possible length k prefix/suffix log inexact matches, merges the log inexact matches that are found and increments k. This process is iterated until the only string remaining in S is the single consensus string.

The naive algorithm will use $O(nC_H N)$ time in the worst case and $O(nC_H \log k)$ time in the expected case. A more detailed discussion of this naive algorithm can be found in [CH91].

The naive algorithm was implemented in C and run on a Sun 3/260. Figure 10 shows the square root of the running time as the number of strings is varied from 4 to 120. Each string was 100 characters long and had an overlap length of 50 with its neighbor in the consensus sequence. The minimum overlap accepted, k, was 24 characters.

4.4 Rabin Karp log inexact algorithm

We can use the ideas developed in section 3.3 and extend the definition of string equality to build an algorithm that will solve the log inexact shotgun sequencing

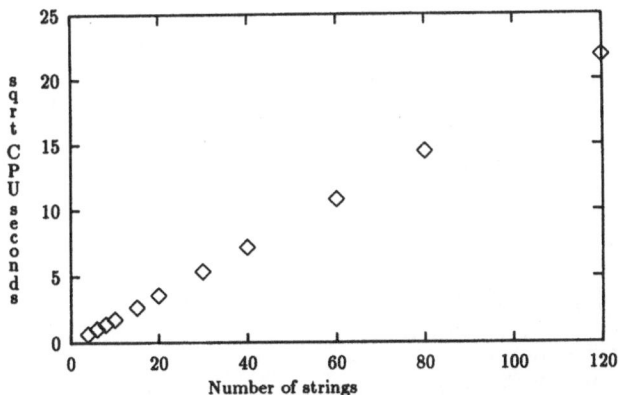

Figure 10: Square root running time of naive algorithm

problem. Let the function F have the domain of all $S \in \Sigma^*$ and the range of non-negative integers. $F(S)$ will be the base σ integer representation of the string S. Given two strings S and T, from Σ^q we let

$$
\begin{aligned}
F(S) &= \text{index}(s_1) \cdot \sigma^{q-1} + \text{index}(s_2) \cdot \sigma^{q-2} + \cdots + \text{index}(s_q) \\
F(T) &= \text{index}(t_1) \cdot \sigma^{q-1} + \text{index}(t_2) \cdot \sigma^{q-2} + \cdots + \text{index}(t_q).
\end{aligned}
$$

When there is exactly one position i such that $s_i \neq t_i$, $F(S)$ and $F(T)$ will differ by $d\sigma^{q-i}$, $0 \leq d < \sigma$. If $s_i \neq t_i$ and $s_j = t_j$ for all j, $0 \leq j < i$, $i < j \leq q$, then

$$
F(S) = F(T) + (\text{index}(s_i) - \text{index}(t_i))\sigma^{q-i}
$$

If the two strings S and T differ in positions i_1 and i_2 then $F(S)$ and $F(T)$ will differ by some additive combination of $d_1\sigma^{q-i_1}$ and $d_2\sigma^{q-i_2}$, $1 \leq d_1, d_2 \leq \sigma$, and so on for larger numbers of differences.

The RK log inexact algorithm presented in Figure 11 will find an ordering of the strings in S that solves the log inexact shotgun sequencing problem. This algorithm will use $O(C_H|S|(\log \log^2 p) + |S| \log p)$ expected time to compute the log inexact consensus string when the length of the overlap between strings is at most $\log p$. The worst case time is $O(C_H|S| \log p)$.

This algorithm and a parallel version of the algorithm are considered in much greater depth in [CH91].

4.5 Suffix array based log inexact algorithm

Two strings, S_i and S_j of length m, with at most $\log m$ positions that do not match must have a common substring S', $|S'| \geq \frac{m}{\log m}$. In this section we will develop an algorithm based on this simple observation to solve the log inexact shotgun sequencing problem.

match $(S, k, \text{errorlist})$

```
for each S ∈ 𝒮
      fhv[S] ← index(s₁) · σ^(k-1) + index(s₂) · σ^(k-2) + ··· + index(s_k)   (mod p)
      bhv[S] ← index(s_{|S|-k+1}) · σ^(k-1) + index(s_{|S|-k+2}) · σ^(k-2) + ··· + index(s_{|S|})   (mod p)
dm ← 1
while (|𝒮| > 1) ∧ (k < c · log p)
      k ← k + 1
      for each S ∈ 𝒮
            fhv[S] ← fhv[S] * σ + index(s_k)   (mod p)
            for each T ∈ (𝒮 - S)
                  bhv[T] ← bhv[T] + index(t_{|T|-k}) * dm   (mod p)
                  diff ← |fhv[S] - bhv[T]|
                  if diff ∈ errorlist
                        if verify (S, T, k)
                              add (join (S, T), 𝒮)
                              remove (S, 𝒮)
                              remove (T, 𝒮)
      dm ← dm * σ   (mod p)
```

Figure 11: RK based algorithm for log inexact match problem

log-inexact-suffix-array $(𝒮, k)$

```
S_tot ← S₁ ∘ γ ∘ S₂ ∘ γ ∘ ··· ∘ γ ∘ S_n
build a suffix-array for S_tot
for each string S_i ∈ 𝒮
      position[i] ← search-suffix-array (S_i, S_tot, k, suffix_array)
```

Figure 12: Algorithm to solve the log inexact match problem

4.5.1 Algorithm

Figures 12 and 13 give an algorithm for the log inexact shotgun sequencing problem. Given the set of strings $𝒮 = \{S_1, S_2, \ldots, S_n\}$ and an integer k that specifies the minimum length overlap, the algorithm will compute the log inexact consensus sequence. The string

$$S_{tot} = S_1 \circ \gamma \circ S_2 \circ \gamma \cdots \gamma \circ S_n,$$

where $\gamma \notin \Sigma$, is composed and a suffix array is created for S_{tot} (See [MM90]). Each string $S_i \in 𝒮$ is positioned in S_{tot} so that a prefix of S_i log inexact matches a prefix of some suffix of S_{tot}. The log inexact match must be at least k characters in length and must not include the character γ. Knowing the position of each S_i in S_{tot} will allow us to easily construct a log inexact consensus sequence.

Positioning each $S_i \in S_{tot}$ is done using the algorithm in Figure 13. The string search-suffix-array $(S_i, S_{tot}, k,$ suffix_array)

```
ptr ← 0
blocksize ← ⌊k/log k⌋
while ptr < |S_i|
    P = search (S_i [ptr] ... S_i [ptr + blocksize - 1], suffix_array)
    for each P_j ∈ P
        if verify (S_i, S_tot [P_j - ptr])
            return P_j - ptr
    ptr ← ptr + blocksize
return nil
```

Figure 13: Algorithm to find log inexact matches using a suffix array

S_i is partitioned into $\frac{|S_i|\log k}{k}$ substrings, each of length $\frac{k}{\log k}$. While a log inexact match has not been found for S_i, the suffix array is searched for the substrings of S_i. For each exact match between a substring of S_i and a substring of S_{tot}, the the location of the exact match is used to verify that a log inexact match exist between a prefix of S_i and a suffix of some S_j in S_{tot}.

4.5.2 Worst case running time

Theorem 1 *The worst case running time of the algorithm in Figures 12 and 13 is* $O\left(\frac{N^3 \log k}{k}\right)$.

Proof. The worst case time to build the suffix array is $O(N \log N)$ [MM90]. The function verify can easily be computed in time linear in the size of the strings using a simple character by character comparison. For each $S_i \in S$ there will be at most $N \frac{|S_i|\log k}{k}$ calls to verify since there will be at most N substrings returned from the function search and at most $\frac{|S_i|\log k}{k}$ calls to search will be made. The total time spent in the function verify will be at most

$$N|S_1|\frac{|S_1|\log k}{k} + N|S_2|\frac{|S_2|\log k}{k} + \ldots + N|S_n|\frac{|S_n|\log k}{k}$$

$$= \frac{\log k}{k} \sum_{i=1}^{n} N|S_i|^2$$

$$= O\left(\frac{N^3 \log k}{k}\right)$$

Searching the suffix array (calling the function search) for S_i takes $O(|S_i| + \log |S_{tot}|)$ [MM90] time. For each S_i there will be at most $\frac{|S_i|\log k}{k}$ calls to search so the total time spent in search will be at most

$$\sum_{i=1}^{n}\left[\left(\frac{|S_i|}{\log |S_i|} + \log |S_{tot}|\right)\frac{|S_i|\log k}{k}\right] = \frac{\log k}{k}\left(\sum_{i=1}^{n}\frac{|S_i|^2}{\log |S_i|} + \sum_{i=1}^{n}|S_i|\log N\right)$$

$$\leq O(\frac{N^3 \log k}{k})$$

4.5.3 Expected running time

The expected running time is significantly better than the worst case running time. Before we give the expected running time we need to prove the following lemma that will be used to show that we can expect to look at a constant number of the $P_j's$ in Figure 13.

Lemma 1 *Given m urns and $m-1$ balls, each ball placed in a randomly selected urn, the expected number of empty urns after each ball has been placed in an urn is me^{-1} as m goes to infinity.*

Proof. Let P_i be the probability that urn U_i is empty.

$$P_i = \left(1 - \frac{1}{m}\right)^{m-1}$$

The expected number of empty urns is

$$
\begin{aligned}
\sum_{i=1}^{m} P_i &= m\left(1 - \frac{1}{m}\right)^{m-1} \\
&= \frac{m}{1 - \frac{1}{m}}\left(1 - \frac{1}{m}\right)^{m}
\end{aligned}
$$

$$\lim_{m \to \infty}\left(\frac{m}{1 - \frac{1}{m}}\left(1 - \frac{1}{m}\right)^{m}\right) = me^{-1} \quad \blacksquare$$

Theorem 2 *The expected running time of the algorithm in Figures 12 and 13 is*

$$O\left(\frac{nN \log k}{|\Sigma|^{\frac{k}{\log k}}} + N\right).$$

Proof. S_i is segmented into $\log k$ pieces of length $\frac{k}{\log k}$. We are likely to find a segment of S_i that exactly matches some substring of S_{tot} looking at a constant number of segments. This can be seen by lemma 1, let $m = \log k$, treat each segment of S_i as an urn, and each of the $\log k$ errors in the log inexact match as a ball. From lemma 1 we expect $\frac{\log k}{e}$ segments to contain none of the $\log k$ errors. Therefore, since the expected ratio of segments with no errors to total segments is a constant, we expect to look at a constant number of the segments to find a match with no errors.

We expect that **search** will return

$$\frac{N}{|\Sigma|^{\frac{k}{\log k}}}$$

substrings of S_{tot} since we are assuming that all length $\frac{k}{\log k}$ strings are equally likely to be substrings of S_{tot}. Each call to **search** will take $O\left(\frac{k}{\log k}\right)$ time. We expect each unsuccessful call to **verify** to take $\log k$ time since at each position of the potential match there is a $\frac{|\Sigma|-1}{|\Sigma|}$ chance that the characters do not match and we only need to find $\log k$ positions where the characters do not match. The total time spent in successful calls to the function verify will be less than $O(N)$. The time to build the suffix array is expected to be $O(N)$[MM90]. So, the expected time to solve the log inexact match problem using the algorithm in Figures 12 and 13 is

$$O\left(\frac{nN\log k}{|\Sigma|^{\frac{k}{\log k}}} + N\right) \quad \blacksquare$$

The seeming improvement as $|\Sigma| \to \infty$ is an artifact of the assumption that direct addressing via Σ is possible.

4.5.4 Implementation and Discussion

The algorithm discussed in section 4.5 has been implemented in C and run on a Sun 3/260. Figure 14 shows the running time of the algorithm presented in figures 12 and 13 as the number of strings is varied from 100 to 2500 while the length of the strings is 100, the minimum acceptable overlap is 24, and the overlap between adjacent strings is 50. Figure 15 shows the running time of the algorithm as the size of the strings is varied from 100 to 2000 characters while the number of strings is 100, the minimal acceptable overlap is 24, and the actual overlap is 50. Figure 16 shows the running time of the algorithm as the size of the minimum acceptable overlap is varied from 6 to 48. The number of strings is 200, the size of the strings is 200, and the actual overlap is 150.

The sets of strings are generated by initially generating a string that is the length of the strings in S. Each successive string is generated by using a suffix of the previous string as the prefix of the current string. The length of the suffix that is used as a prefix of the current string is the length of the overlap between strings. When generating data for the algorithms to solve the log inexact shotgun sequencing problem the suffix from the previous string was modified before being used as the prefix of the current string. Once all of the strings had been generated, the order of the strings is randomized.

Although the suffix array based log inexact shotgun sequencing algorithm was initially designed to construct a consensus sequence from sequence fragments, it can be used to align similar sequences. As an example of this, we have used the algorithm to align the following three sequences from GenBank [BB88].

1. *Saccharomyces cerevisiae* TATA-box factor (TFIID) gene, 5' flank. This 1157 base pair DNA sequence was published by Schmidt et. al. [SKPB89] and has GenBank accession number M26403.

2. *S. cerevisiae* TATA-binding protein (TFIID) gene, complete cds. This 2439 base pair DNA sequence was published by Hahn et. al. [HBSG89] and has GenBank accession number M27135.

3. *S. cerevisiae* transcription initiation factor IID (TFIID). This 1140 base pair DNA sequence was published by Horikoshi et. al. [HWF+89] and has GenBank accession number X16860.

183

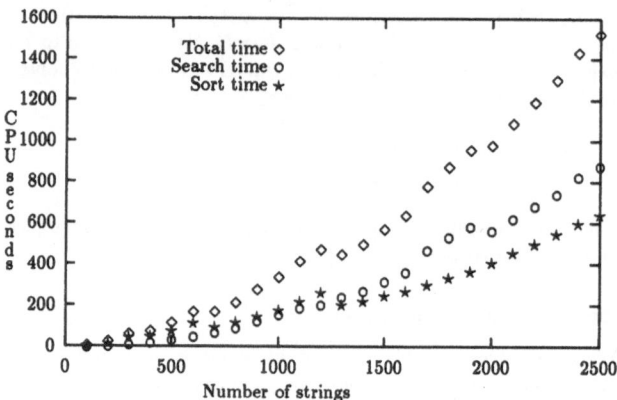

Figure 14: Running time of suffix array algorithm varying number of strings

The sequences M26403 and M27135 can be aligned with 6 differences, the sequences M26403 and X16860 can be aligned with 7 differences and the sequences M27135 and X16860 can be aligned with 1 difference. Running the suffix array based log inexact algorithm uses 2.62 CPU seconds to build the suffix array and 0.08 seconds to align the sequences with a minimum acceptable overlap (the value k) of 24. The naive algorithm uses over 37 CPU seconds with the minimum acceptable overlap set to 24. To help the naive algorithm we could remove the last 1000 base pairs of the sequence M27135 and set the minimum acceptable overlap to 900. With this help the naive algorithm still takes over 11 CPU seconds to compute the alignment.

4.5.5   Generalization

We have used $\log n$ errors in a match of length n simply because we used $\log n$ errors in the RK log inexact shotgun sequencing algorithm in section 4.4. Any function of n, $f(n)$, such that $f(n) \leq n$ could be used as the maximum number of errors allowed in a length n match. Following the same arguments that were used in sections 4.5.2 and 4.5.3 it can be shown that allowing $f(n)$ errors in matches of length n, the algorithm in Figures 12 and 13 will use

$$O\left(\frac{N^3 f(k)}{k}\right)$$

worst case time and

$$O\left(\frac{nNf(k)}{|\Sigma|^{\frac{k}{f(k)}}} + N\right)$$

expected time.

We allowed n^ϵ, $0 < \epsilon < 1$, errors for matches of length n. Figure 17 shows the log of the running time of the algorithm in Figure 12 and 13 when allowing n^ϵ errors

184

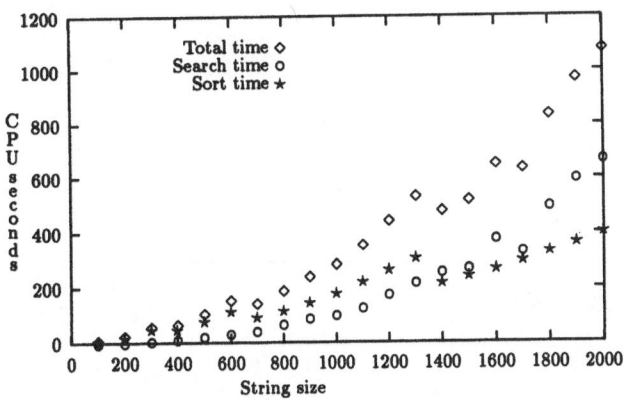

Figure 15: Running time of suffix array algorithm varying size of strings

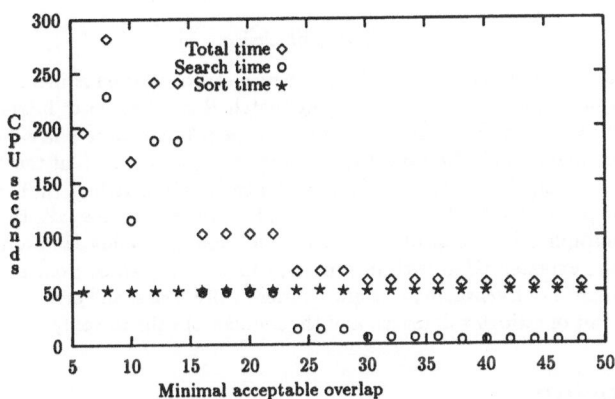

Figure 16: Running time of suffix array algorithm varying size of minimum overlap

185

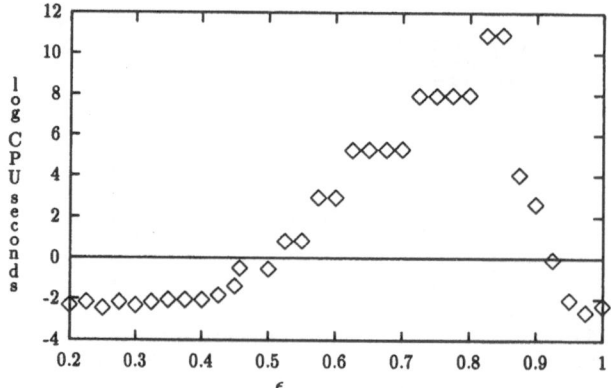

Figure 17: Log of running time of suffix array algorithm allowing n^ϵ errors

in length n matches while varying ϵ. The data shown in Figure 17 is for 10 sequences, each 410 bases long with overlaps of 180 bases and a minimum acceptable overlap of 40 bases.

Figure 17 has a knee at about $\epsilon = 0.45$. As the number of errors allowed increases, the likelihood that some substring of S_{tot} will be falsely matched by the string

$$R = S_i[\text{ptr}] \ldots S_i[\text{ptr} + \text{blocksize} - 1]$$

(as used in Figure 13) increases. When the number of errors allowed is increased by one, the number of substrings in S_{tot} that falsely match R is expected to increase by a factor of σ. There is some value μ of ϵ where we expect there to be one substring in S_{tot} that falsely matches R. We expect the running time to be constant for $0 < \epsilon\mu$ since we expect one call, the correct call, to verify for each pair of strings that match. As ϵ grows above μ we expect false calls to verify. The number of false calls to verify will grow by a multiplicative factor of σ for each additional error allowed in a match. This is seen in the exponential growth in running time as ϵ increases from $\epsilon = 0.45$ to $\epsilon = 0.85$ in Figure 17. Eventually, as ϵ grows, there will be so many errors allowed that nearly any pair of strings will match and the number of calls to verify will fall.

5   Conclusion

We have defined the shotgun sequencing problem and presented two algorithms to solve it. We let

- n be the number of strings in the multiset $\mathcal{S} = \{S_1, S_2, \ldots, S_n\}$
- $N = \sum_{i=1}^{n} |S_i|$

186

- L = length of superstring found by heuristic

- C_H be the compression, $C_H = N - L$

- k be the minimum acceptable overlap length between strings

The first algorithm using ideas of Rabin–Karp exact string matching, is expected to solve the problem in $O(C_H \log |\mathcal{S}|)$ time. The second algorithm is also expected to run in $O(C_H \log |\mathcal{S}|)$ time, but in practice runs faster than the Rabin–Karp type algorithm by a factor of about ten.

We also define a similar problem, the log inexact shotgun sequencing problem, and present an algorithm to solve this problem. The algorithm uses suffix arrays as developed by [MM90] and is expected to run in $O\left(\frac{nN \log k}{|\Sigma|^{\frac{k}{\log k}}} + N\right)$ time. This algorithm is generalized to allow not only $\log k$ errors but any reasonable function of k errors. Finally we give a small, unexpected, example use of this algorithm to align three nearly identical DNA sequences of the TFIID gene in yeast.

These algorithms allow a small number of bases to be transformed but do not allow bases to be deleted. If, for example, a base is missed while reading a GC rich region (an occasional problem) the algorithms presented in this paper will not work since part of the match will be offset by one or more positions. One area of future research will be designing algorithms for this problem that allow a small number of deletions as well as transformations.

References

[AC75]   A. V. Aho and M. J. Corasick. Efficient string matching: an aid to bibliographic search. *Communications of the ACM*, 18:333–340, 1975.

[BB88]   H. S. Bilofsky and C. Burks. The genbank genetic sequence data bank. *Nucleic Acids Research*, 16:1861–1863, 1988.

[BJL+91]   A. Blum, T. Jiang, M. Li, J. Tromp, and M. Yannakakis. Linear approximation of shortest superstrings. In *Proceedings of the ACM Symposium on Theory of Computing*, pages 328–336, Baltimore, MD, 1991. ACM press.

[BM77]   R. S. Boyer and J. S. Moore. A fast string searching algorithm. *Communications of the ACM*, 20:762–772, 1977.

[CH91]   P. Cull and J. L. Holloway. Algorithms for constructing a consensus sequence. Technical Report TR-91-20-1, Oregon State University, Department of Computer Science, 1991.

[GBY90]   G. H. Gonnet and R. A. Baeza-Yates. An analysis of the Karp–Rabin string matching algorithm. *Information Processing Letters*, 34:271–274, 1990.

[GMS80]   J. Gallant, D. Maier, and J. Storer. On finding minimal length super-strings. *Journal of Computer and System Science*, 20:50–58, 1980.

[HBSG89] S. Hahn, S. Buratowski, P. A. Sharp, and L. Guarente. Isolation of the gene encoding the yeast TATA binding protein TFIID: A gene identical to the SPT15 suppressor of ty element insertions. *Cell*, 58:1173–1181, 1989.

[HWF+89] M. Horikoshi, C. K. Wang, H. Fujii, J. A. Cromlish, P. A. Weil, and R. G. Roeder. Cloning and structure of a yeast gene encoding a general transcription initiation factor TFIID that binds to the TATA box. *Nature*, 341:299–303, 1989.

[KMP77]   D. E. Knuth, J. H. Morris, and V. R. Pratt. Fast pattern matching in strings. *SIAM Journal of Computing*, 6:323–350, 1977.

[Knu73]   D. E. Knuth. *The art of computer programming: searching and sorting*, volume 3. Addison–Wesley, 1973.

[KR87]   R. M. Karp and M. O. Rabin. Efficient randomized pattern–matching algorithms. *IBM Journal of Research and Development*, 32:249–260, 1987.

[Li90]   M. Li. Towards a DNA sequencing theory. In *IEEE Symposium on the Foundations of Computer Science*, pages 125–134, 1990.

[MM90]   U. Manber and G. Myers. Suffix arrays: A new method for on–line string searches. In *Proceedings of the First Annual ACM–SIAM Symposium on Discrete Algorithms*, pages 319–327. SIAM, 1990.

[PSTU83] H. Peltola, H. Soderlund, J. Tarhio, and E. Ukkonen. Algorithms for some string matching problems arising in molecular genetics. In *Information Processing 83*, pages 53–64, 1983.

[SKPB89] M. C. Schmidt, C. Kao, R. Pei, and A. J. Berk. Yeast TATA-box tran-scription factor gene. *Proceedings of the National Academy of Science*, 86:7785–7789, 1989.

[TU88]   J. Tarhio and E. Ukkonen. A greedy approximation algorithm for con-structing shortest common superstrings. *Theoretical Computer Science*, 57:131–145, 1988.

[Tur89]   J. Turner. Approximation algorithms for the shortest common superstring problem. *Information and Computation*, 83:1–20, 1989.

[Ukk90]   E. Ukkonen. A linear–time algorithm for finding approximate shortest common superstrings. *Algorithmica*, 5:313–323, 1990.

A Systematic Design and Explanation of the Atrubin Multiplier

Shimon Even and Ami Litman

Computer Science Dept., Technion, Haifa, Israel 32000
and
Bellcore, 445 South St., Morristown, NJ 07960-1910

ABSTRACT

The Atrubin systolic array, for multiplying two serially supplied integers in real-time, was invented in 1962, but to date, no simple explanation of its operation, or proof of its validity, has been published.

We present a methodical design of the array which yields a simple proof of its validity. First, we use a broadcast facility, and then we show how it can be removed by retiming which avoids the introduction of either slow-down or duplication.

A similar retiming technique can be used to remove instant-accumulation. These retiming methods are applicable to arrays of any dimension.

1. Introduction

In 1962, Allan J. Atrubin invented a synchronous system for real-time multiplication of integers. (It was published in 1965, [A].) The host (user) feeds the system two binary encoded multiplicands, x and y, serially, least significant bits first, and the system outputs the product $x \cdot y$, in binary, serially, least significant bit first. Clearly, the time it takes to multiply two n-bit multiplicands is $2 \cdot n$.

Informally, a (finite, or infinite) synchronous system, serving a host, is called *systolic*, if it has the following characteristics. The system consists of *segments*, connected to each other and to the host by communication *lines*. Each segment consists of a modest amount of hardware, which realizes a Moore finite state automaton; i.e. its current output signals, which appear at its output ports, depend only on its present state, and its next state depends on the present state and the present input signals, which appear presently at the input ports. Without loss of generality, we may assume that each output port of a segment is the output port of a (clocked) delay flip-flop. The lines go from one output port to one input port; there is no fan-in or fan-out in these connections.

Systolic systems are free of long paths on which a rippling effect can occur between clock ticks. In fact, if a clock rate is not too high for each single segment, then it is not too high for the whole system, no matter how large the system is.

189

The Atrubin system is systolic. Furthermore, its segments are all identical, and are arranged in a linear array. This simplifies the design and production of the system.

It is well known, [W], how to perform multiplication in $O(\log n)$ time, but the equipment required is of size $O(n^2)$. The asymptotically smaller circuit, [SS], of size $O(n \cdot \log n \cdot \log\log n)$, which performs multiplication in $O(\log n)$ time, is not a practical alternative. Thus, the Atrubin multiplier, which requires equipment of size $O(n)$, remains competitive.

The general layout of the multiplier is depicted in the following diagram. In order to multiply n-bit numbers, the array must consist of at least $\lceil \frac{n}{2} \rceil$ cells. All cells have the same structure. Each consists of a few hundred transistors, and realizes a finite Moore automaton.

Recently, Even showed how, with the addition of another systolic array, repeated modular multiplication can be computed almost as fast ([E]). The combination of the two arrays may be useful in cryptographic applications.

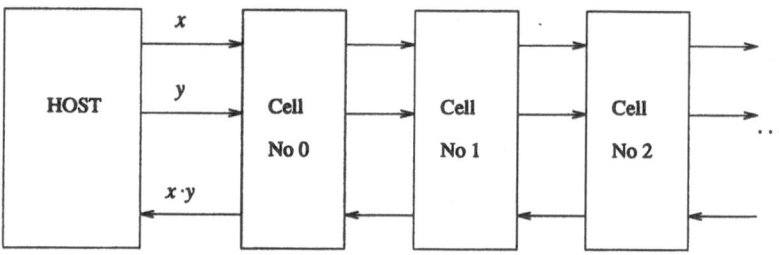

The Atrubin Multiplication Array

In spite of its reputation, the structure of the Atrubin array has remained a mystery. It is the purpose of this paper to explain this mystery away, and prove the validity of the multiplier. This is done by breaking the design into stages.

First, the operation and validity of a multi-input serial adder is discussed. Next, a simplified version of the multiplier is studied, in which a broadcast facility is used. Finally, by using retiming, the systolic version is achieved.

Finally, we show that our design methodology is helpful in building other useful systolic arrays. This is true for systems in which it is natural to use a broadcast facility in the intermediate design stage, as well as for systems in which it is natural to use instant-accumulation. The technique can be used, not just for linear arrays, but for arrays of any dimension. It is interesting to note that contrary to the general case, [EL], no duplication of hardware is necessary for the removal of broadcast or instant-accumulation from arrays.

190

2. Serial Adder

One of the components used in the design of the multiplier is a serial adder. It is a simple extension of a two-inputs serial full adder. Its structure is depicted in the diagram.

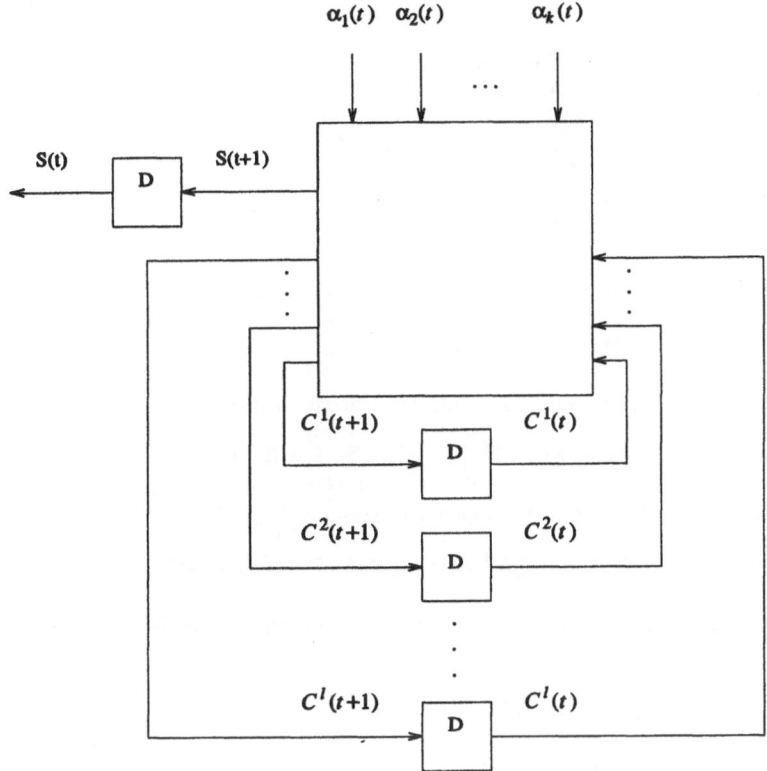

Serial adder

The main logic box computes the outputs $(S(t+1), C^1(t+1), C^2(t+1),..., C^l(t+1))$ from the inputs $(\alpha_1(t), \alpha_2(t),..., \alpha_k(t))$ to satisfy

$$S(t+1) + 2 \cdot [C^1(t+1) + \cdots + C^l(t+1)] = \alpha_1(t) + \cdots + \alpha_k(t) + C^1(t) + \cdots + C^l(t). \quad (1)$$

Note that the carries are represented in unary. A necessary condition is that $l \geq k-1$. The D boxes are clocked delay flip-flops (registers).

The initial conditions (which can be implemented via a global reset signal for the flip-flops) are:

$$S(0) \equiv C^1(0) \equiv C^2(0) \equiv \cdots \equiv C^l(0) \equiv 0. \tag{2}$$

Lemma 1:

Let the α inputs feed n-bit positive integers, in binary, least significant bits first, and let $m = \lceil \log_2 l \rceil$. The following equation holds:

$$\sum_{a=1}^{n+m} S(a) \cdot 2^{a-1} = \sum_{j=1}^{k} \sum_{b=0}^{n-1} \alpha_j(b) \cdot 2^b, \tag{3}$$

and for $t > n+m$, $S(t) = 0$.

Proof:

First, by induction on i, from 0 to $n-1$, the following holds:

$$\sum_{a=1}^{i+1} S(a) \cdot 2^{a-1} + 2^{i+1} \cdot \sum_{d=1}^{l} C^d(i+1) = \sum_{j=1}^{k} \sum_{b=0}^{i} \alpha_j(b) \cdot 2^b.$$

The basis follows from equations (1) and (2). The inductive hypothesis is

$$\sum_{a=1}^{i} S(a) \cdot 2^{a-1} + 2^i \cdot \sum_{d=1}^{l} C^d(i) = \sum_{j=1}^{k} \sum_{b=0}^{i-1} \alpha_j(b) \cdot 2^b.$$

Taking equation (1), for $t = i$, and multiplying all terms by 2^i, yields:

$$S(i+1) \cdot 2^i + 2^{i+1} \cdot \sum_{d=1}^{l} C^d(i+1) = \sum_{j=1}^{k} \alpha_j(i) \cdot 2^i + 2^i \cdot \sum_{d=1}^{l} C^d(i).$$

Adding the latter to the inductive hypothesis, completes the proof of the claim.

Thus, we have

$$\sum_{a=1}^{n} S(a) \cdot 2^{a-1} + 2^n \cdot \sum_{d=1}^{l} C^d(n) = \sum_{j=1}^{k} \sum_{b=0}^{n-1} \alpha_j(b) \cdot 2^b. \tag{4}$$

For $t \geq n$, (1) degenerates into

$$S(t+1) + 2 \cdot \sum_{d=1}^{l} C^d(t+1) = \sum_{d=1}^{l} C^d(t). \tag{5}$$

By induction on t, from n up, it follows (in a manner similar to the proof of the previous claim) that:

$$\sum_{a=1}^{t+1} S(a) \cdot 2^{a-1} + 2^{t+1} \cdot \sum_{d=1}^{l} C^d(t+1) = \sum_{j=1}^{k} \sum_{b=0}^{n-1} \alpha_j(b) \cdot 2^b.$$

And all that remains to be shown is that for $t = n+m-1$, the second term on the l.h.s. is zero. For $t \geq n$, (5) implies that

$$\sum_{d=1}^{l} C^d(t+1) \leq \frac{1}{2} \cdot \sum_{d=1}^{l} C^d(t).$$

Therefore,

$$\sum_{d=1}^{l} C^d(t+1) \le \left\lfloor \frac{1}{2^{t+1-n}} \cdot \sum_{d=1}^{l} C^d(n) \right\rfloor .$$

Note that $\sum_{d=1}^{l} C^d(n) \le l$. Thus, if $l < 2^{t+1-n}$ then the r.h.s. is zero. It suffices that $\lceil \log_2 l \rceil < t+1-n$, or $\lceil \log_2 l \rceil \le t-n$. Therefore, (3) follows, as well as the fact that for $t > n+m$, $S(t) = 0$.

■

3. A Nonsystolic Version of the Multiplier

Let us describe now a nonsystolic version of the multiplier. Its overall structure is depicted in the following diagram.

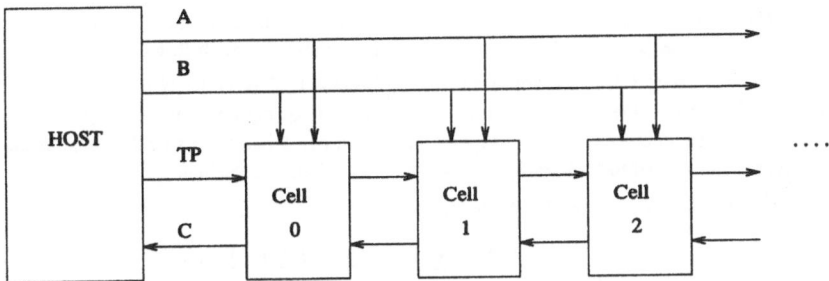

Nonsystolic multiplication array

The host has three binary output channels and one binary input channel. On channel A and B, the two nonnegative n-bit multiplicands are fed to the system, sequentially, least significant bits first, one bit on each tick of the (implied global) clock. The first bits are fed at time $t = 0$, and the last ones at time $t = n-1$. All subsequent bits are zeros. On channel TP, a timing pulse is fed to the system at time $t = -1$; i.e. the logical value is 1 at $t = -1$, and 0 at $t \ge 0$. This timing pulse is delayed one unit of time in each cell. Thus, the i-th cell gets it at time $t = i-1$. Cell number 0 of the system, feeds the product of the two multiplicands to the host, via channel C. The $2n$ bits of the product are delivered sequentially, least significant bit first, starting from time $t = 1$ and ending at time $t = 2n$.

The structure of all the cells is identical, and will be described shortly. The cell is a Moore finite automaton; i.e. its outputs go through clocked delay flip-flops. Thus, the outputs of each cell, at time t, depend only on its state, and not on its inputs at time t. The

193

only nonsystolic parts in this system are the broadcast channels, A and B. In the following section, it will be shown how the broadcast channels can be removed, to make the system completely systolic.

The structure of a typical cell is shown in the following diagram.

The signal which travels on the TP channel, has the logical value $t = i-1$ as it enters the i-th cell, and has the value $t = i$, as it leaves. This signal goes though a delay flip-flop which is not reset. All other flip-flops of the i-th cell are reset by the signal $t = i-1$; i.e. their output values at time $t = i$ is zero. Thus, for every $i \geq 0$,

$$S_i(i), \bar{A}_i(i), \bar{B}_i(i), C_i^1(i), C_i^2(i) \equiv 0. \tag{6}$$

The *sample and hold* (S&H) are clocked flip-flops which sample the l.h.s. input when the top input is equal to 1. Thus,

$$\bar{A}_i(t) = \begin{cases} 0 & \text{if } 0 \leq t \leq i \\ A(i) & \text{if } t > i \end{cases}.$$

And a similar statement holds for $\bar{B}_i(t)$.

The lower half of the diagram depicts a serial adder, just as in the previous section. It has 4 input lines and only 2 carries, but this is acceptable for the following reason.

Input number 1 has the value $A(t) \wedge B(t) \wedge (t=i)$. This value is 0 if $t \neq i$. Input number 2 has the value $\bar{A}_i(t) \wedge B(t)$, which is 0 if $t = i$, by equation (6). Therefore, input number 2 (and 3) are never equal to 1, if input number 1 has the value 1. Thus, effectively, the number of input lines is bounded by 3, and 2 carries suffice.

Our aim is to prove the following equation:

$$\sum_{t=1}^{2n} S_0(t) \cdot 2^{t-1} = [\sum_{t=0}^{n-1} A(t) \cdot 2^t] \cdot [\sum_{t=0}^{n-1} B(t) \cdot 2^t]. \tag{7}$$

Instead, we will first prove two lemmas. Equation (7) will follow from Equation (8), by substituting $i = 0$ and observing that the equality implies that all bits after the $2n$'th, must be 0.

Lemma 2:

For every $k \geq 0$ and for every $i \geq n$,

$$S_i(i+k), C_i^1(i+k), C_i^2(i+k) \equiv 0.$$

Proof:

By induction on k. For $k = 0$ the statement follows from equation (6). Consider the case of $k = 1$.

Since $i \geq n$, at time $t \geq i$, $A(t) \equiv B(t) \equiv 0$. Thus, inputs number 1,2 and 3 of the i-th cell are all zero. At time $t = i$, input number 4 is zero, owing to the gate $(\neg (t=i) \wedge S_{i+1}(t))$ producing it. Thus,

$$S_i(i+1), C_i^1(i+1), C_i^2(i+1) \equiv 0.$$

194

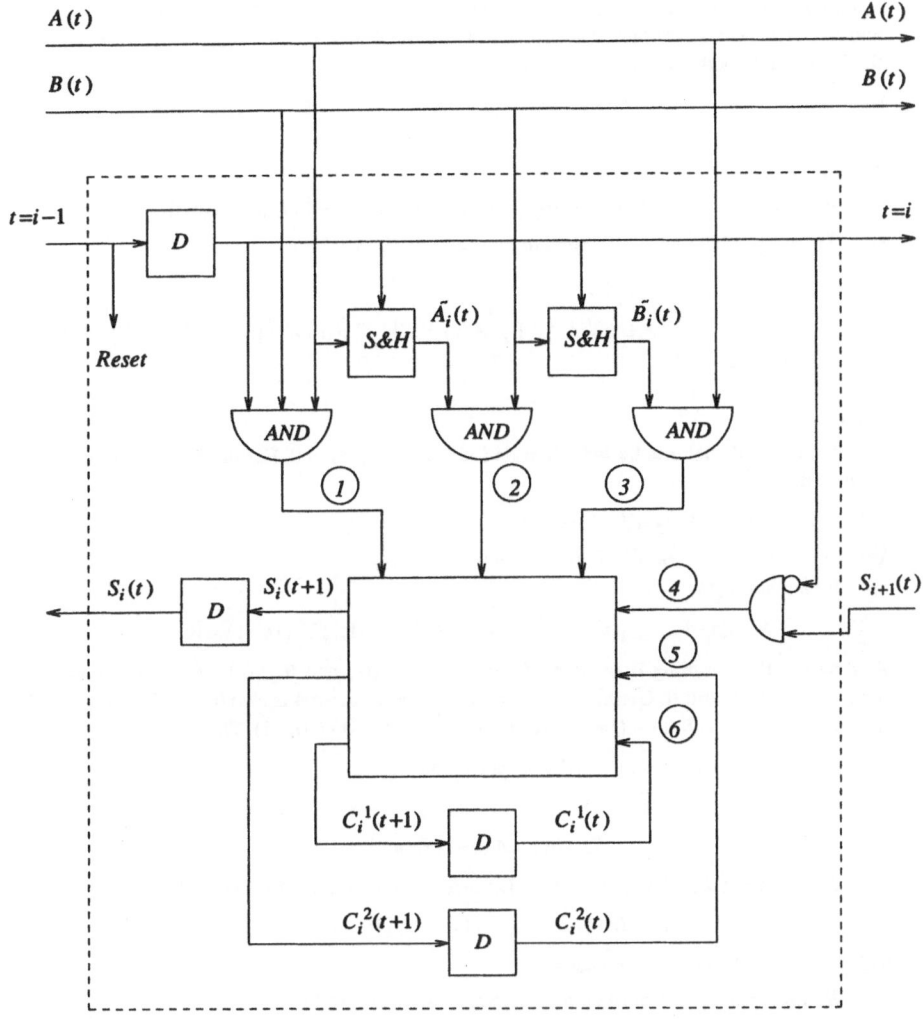

i-th cell of nonsystolic array

195

For the proof of the inductive step, for higher values of k, observe that inputs number 1,2 and 3 remain zero, as explained above. Input number 4 is zero, since $S_{i+1}(i+k-1) \equiv 0$, by the inductive hypothesis applied to $k-2$.

■

Lemma 3:

For every $n-1 \geq i \geq 0$, the sequence $\{S_i(t)\}$ satisfies the following conditions:

(a) There exists an integer N_i, such that for all $t > N_i$, $S_i(t) \equiv 0$.

(b)

$$\sum_{t=i+1}^{\infty} S_i(t) \cdot 2^{t-i-1} = [\sum_{t=i}^{n-1} A(t) \cdot 2^{t-i}] \cdot [\sum_{t=i}^{n-1} B(t) \cdot 2^{t-i}]. \tag{8}$$

Proof:

We prove the lemma by induction on i, from n-1 down to 0. For the basis it suffices to show that:

(1)   For $t = n$, $S_{n-1}(t) = A(n-1) \cdot B(n-1)$,

(2)   for $t > n$, $S_{n-1}(t) \equiv 0$.

By equation (6),

$$S_{n-1}(n-1), \bar{A}_{n-1}(n-1), \bar{B}_{n-1}(n-1), C_{n-1}^1(n-1), C_{n-1}^2(n-1) \equiv 0.$$

At $t = n-1$, the sampling is being performed, but $\bar{A}_{n-1}(t)$ and $\bar{B}_{n-1}(t)$ are still 0. Thus, inputs 2 and 3 are still 0. Owing to the gate producing it, input 4 is also 0. The only input that counts is $A(t) \wedge B(t) \wedge (t=i)$, which is equal to $A(n-1) \cdot B(n-1)$. Thus,

$$C_{n-1}^1(n) \equiv C_{n-1}^2(n) \equiv 0,$$

and

$$S_{n-1}(n) = A(n-1) \cdot B(n-1).$$

For $t \geq n$, inputs 1,2 and 3 are 0, and by Lemma 2, input number 4 is O too. Thus,

$$S_{n-1}(t+1) \equiv C_{n-1}^1(t+1) \equiv C_{n-1}^2(t+1) \equiv 0.$$

This completes the proof of the basis.

We turn to the proof of the inductive step. The inductive hypothesis is:

(a) There exists an integer N_{i+1}, such that for all $t > N_{i+1}$, $S_{i+1}(t) \equiv 0$.

(b)

$$\sum_{t=i+2}^{\infty} S_{i+1}(t) \cdot 2^{t-i-2} = [\sum_{t=i+1}^{n-1} A(t) \cdot 2^{t-i-1}] \cdot [\sum_{t=i+1}^{n-1} B(t) \cdot 2^{t-i-1}]. \tag{9}$$

Let us consider the situation in the i-th cell. For $t = i$, by equation (6),

$$S_i(t) \equiv C_i^1(t) \equiv C_i^2(t) \equiv 0.$$

196

At $t = i$, input number 1 is equal to $A(i) \cdot B(i)$. For $t > i$, input number 1 is equal to 0. Thus, the number fed sequentially to the adder, by input number 1, starting at time $t = i$, in binary, least significant bit first, is simply $A(i) \cdot B(i)$.

At $t = i$, $A(i)$ and $B(i)$ are sampled, and become the values of $\bar{A}_i$ and $\bar{B}_i$, respectively. However, inputs number 2 and 3 are still equal to 0. For $t > i$, they are $A(i) \cdot B(t)$ and $B(i) \cdot A(t)$, respectively. Thus, the numbers they feed, sequentially, starting at $t = i$, are

$$\sum_{t=i+1}^{n-1} A(i) \cdot B(t) \cdot 2^{t-i} \quad \text{and} \quad \sum_{t=i+1}^{n-1} B(i) \cdot A(t) \cdot 2^{t-i},$$

respectively.

Consider input number 4. Owing to the gate which produces it, it is 0 at $t = i$. By equation (6), it is 0 at $t = i+1$. By the inductive hypothesis, for large t's, $S_{i+1}(t) \equiv 0$. Thus, the (finite) number fed to the adder, from $t = i$, and on, is

$$\sum_{t=i+2}^{\infty} S_{i+1}(t) \cdot 2^{t-i}.$$

By Lemma 1, there exists an integer N_i, such that for $t > N_i$, $S_i(t) \equiv 0$, and

$$\sum_{t=i+1}^{\infty} S_i(t) \cdot 2^{t-i-1} = A(i) \cdot B(i) + \sum_{t=i+1}^{n-1} A(i) \cdot B(t) \cdot 2^{t-i} + \sum_{t=i+1}^{n-1} B(i) \cdot A(t) \cdot 2^{t-i} +$$

$$\sum_{t=i+2}^{\infty} S_{i+1}(t) \cdot 2^{t-i}. \tag{10}$$

It remains to show that equation (10) implies equation (8). By the inductive hypothesis,

$$\sum_{t=i+1}^{\infty} S_i(t) \cdot 2^{t-i-1} = A(i) \cdot B(i) + \sum_{t=i+1}^{n-1} A(i) \cdot B(t) \cdot 2^{t-i} + \sum_{t=i+1}^{n-1} B(i) \cdot A(t) \cdot 2^{t-i} +$$

$$2^2 \cdot [\sum_{t=i+1}^{n-1} A(t) \cdot 2^{t-i-1}] \cdot [\sum_{t=i+1}^{n-1} B(t) \cdot 2^{t-i-1}]$$

$$= A(i) \cdot B(i) + A(i) \cdot \sum_{t=i+1}^{n-1} B(t) \cdot 2^{t-i} + B(i) \cdot \sum_{t=i+1}^{n-1} A(t) \cdot 2^{t-i} +$$

$$[\sum_{t=i+1}^{n-1} A(t) \cdot 2^{t-i}] \cdot [\sum_{t=i+1}^{n-1} B(t) \cdot 2^{t-i}]$$

$$= [\sum_{t=i}^{n-1} A(t) \cdot 2^{t-i}] \cdot [\sum_{t=i}^{n-1} B(t) \cdot 2^{t-i}].$$

∎

4. From Nonsystolic to Systolic

In this section, we shall use a retiming technique. The idea is not new; it can be found in a primitive form in papers from the seventies (see, for example, [CHEP] and

[D]). A more explicit description of the technique, and its consequences, were presented in the work of Leiserson et. al. (see, for example, [LS] and [LRS]). We shall follow our own approach, as in [EL].

Consider a general one dimensional array, with broadcast, as depicted in the following diagram.

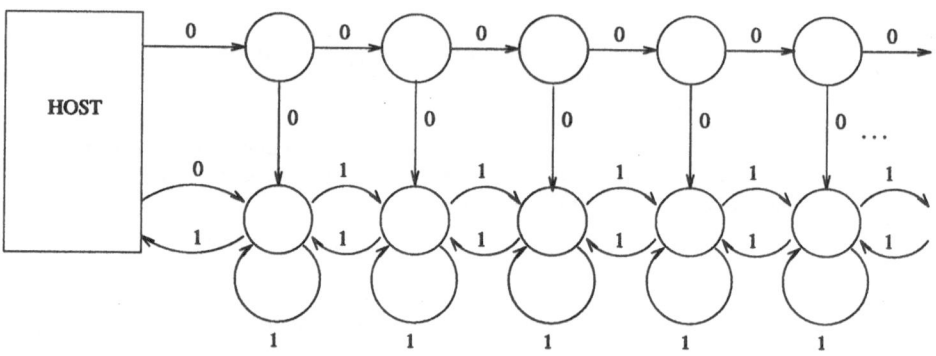

One dimensional array, with broadcast

The vertices (circles) represent memoryless combinational logic boxes. The edges (lines) represent information channels. The integer $d(e)$, next to an edge e, is the number of delay flip-flops (registers) on the corresponding line.

Observe that the nonsystolic multiplication array, of the previous section, fits into the general scheme described by this diagram. Each of the vertices in the upper level represents nothing but a simple junction, which transmits to both outputs (to the right, and down) the information received from its input (from the left); this transmission is instantaneous. Thus, the upper level represents the broadcast capability. The logic boxes represented by the vertices in the lower level are all identical. All output lines emanating from these boxes have exactly one delay flip-flop. Thus, the lower level of the system is completely systolic. The self-loops of the vertices in the lower level represent the memory of the individual cells.

Our purpose is to shift the delays (as tokens in a marked graph) in such a way that the array will divide into cells. All cells are to be identical. Each output line of a cell will have at least one delay on it. This transformation will yield a systolic array, although each cell may "contain" more than one stage of the original array. This shift of delays is best described by *retiming*: An assignment of nonnegative integers to the vertices (circles). The meaning of a retiming value $\lambda(v)$ to a vertex v is the number of ticks of the clock by which the operation of the corresponding logic is delayed. (In the marked graph model, this is the number of times vertex v is to be "fired" backwards.)

In the following diagram, the value of $\lambda(v)$ is shown in the circle representing v, and the corresponding delays of the edges are shown next to them. In the upper level the sequence of $\{\lambda(v)\}$ is 0,0,1,1,2,2,... and in the lower it is the same sequence with one zero omitted. The new number of delays, $d'(e)$, on an edge $u \xrightarrow{e} v$, is determined by the formula

$$d'(e) = d(e) + \lambda(v) - \lambda(u), \tag{13}$$

where $d(e)$ is the previous number of delays on the edge e.

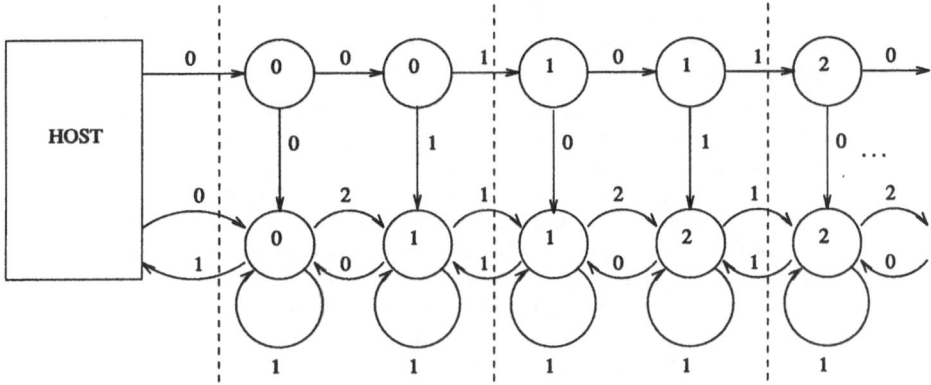

The array, after retiming

Observe that all the "new" cells, as between the dashed horizontal lines, are identical, and each of the edges emanating from them has a (one tick) delay on it. Thus, the new array is systolic. It is also easy to see that, as far as the host is concerned, nothing has changed; the host will observe the same input-output behavior. (The proof of this claim, for general systolic systems with broadcast, is fairly involved, and can be found in Section 6 of [EL], but is not required in the relatively simple situation at hand.) Thus, we have achieved our goal of building a systolic array for multiplication. This design is a slight improvement of Atrubin's design, since it does not use a global reset signal.

5. Comments on Similar Systems

The retiming used above is useful in other similar situations. For example, one can design a systolic queue (FIFO), or a stack (FILO), by following the same route we have taken above: First allow the broadcast facility, and design the array. Next, use the retiming of the previous section to make the whole system systolic.

A similar technique is useful in handling a situation which, in a sense, is the opposite of broadcast.

199

Instant-accumulation is a mechanism which instantaneously presents to the host, via a single edge, a value which depends on values produced for this purpose by all the segments. A common example is an AND function of binary values produced by all the segments. However, the exact computation performed by the instant-accumulation mechanism is immaterial; it may be any commutative and associative multi-variable function of the individual values produced by the segments.

We assume that all the inputs to the instant-accumulation mechanism pass through delays, as they come out of the segments. This is in concert with the assumption that every edge has at least one delay, except those emanating from the host.

Let us describe, briefly, a construction of a linear systolic array to recognize palindromes*. There are several known solutions (see, for example, [C], [S] and [LS]). Our solution differs mainly in the methodology of the construction.

First construct a linear array with instant-accumulation, as follows. Initially, each cell's output to the instant-accumulation mechanism, is "yes"; i.e. from the point of view of the cell, the input sequence read so far by the system is a palindrome. This remains the cell's output until the cell has seen two input symbols. The input symbols are fed, one at a time to the first cell. The first symbol to reach a cell is stored, and is not fed forward. (One can use a TP signal, which is delayed two ticks in each cell, to notify the cell that its "first" symbol is arriving.) Every consecutive symbol is fed forward to the next cell, via a one-tick delay. At each tick, following the storage of the cell's first symbol, the arriving symbol is compared to the stored one. If they are equal, a "yes" is output; if they are different, a "no" is output. These outputs are supplied to the instant-accumulation mechanism, which supplies a "yes" answer to the host, if and only if all the (delayed) output answers of the cells are "yes". The general layout of the array is depicted in the following diagram:

The next diagram shows, schematically, the number of delays on each edge, before retiming. (So far, the design methodology of the system for recognizing palindromes follows the steps in [LS].)

Now apply the following retiming: The vertices in the upper level are retimed by 0, -1, -1, -2, -2, ... , while the vertices in the lower level are retimed by -1, -1, -2, -2, -3 ... The resulting number of delays on the edges, as determined by Equation 13, are shown in the following diagram, as well as the division into new segments. Observe that the resulting system is systolic. Note that this retiming avoids slow-down ([LS]), as well as duplication ([EL]).

---

* A *palindrome* is a word which is identical to its reversal.

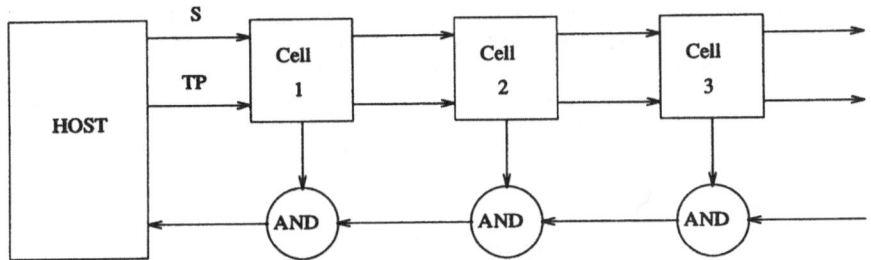

General layout, with instant-accumulation

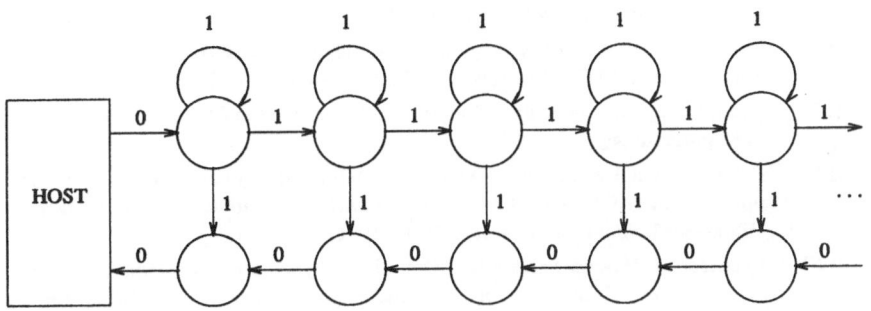

The delays, before retiming

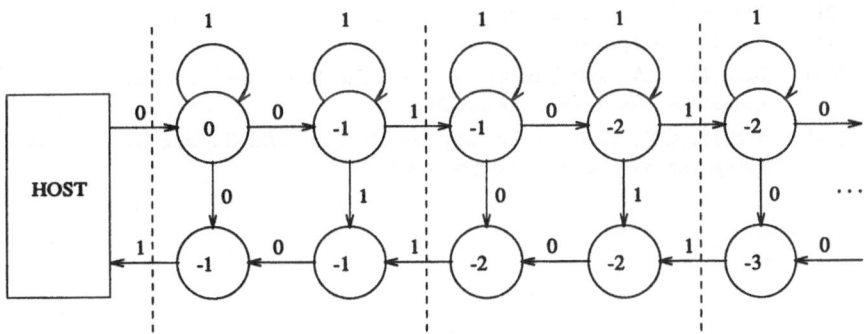

The delays, after retiming

201

The techniques described above, either for removing the broadcast facility, or removing the instant-accumulation facility (but not both in the same system), are applicable to arrays of any finite dimension. Each new segment of the k-dimensional systolic array, after the retiming, consists of 2^k segments of the initial array. Along each axis, the retiming is identical to the 1-dimensional case, and in general, the retiming of a vertex is the sum of the retiming of vertices at its projections on the axes.

References

[A]   Atrubin, A.J., "A One-Dimensional Real-Time Iterative Multiplier", *IEEE Trans. on Electronic Computers*, Vol. EC-14, No. 3, June 1965, pp. 394-399.

[C]   Cole, S.N., "Real-Time Computation by n-Dimensional Iterative Arrays of Finite-State Machines", *IEEE Trans. on Computers*, Vol. C-18, No. 4, 1969, pp. 349-365.

[CHEP] Commoner, F., Holt, A.W., Even, S., Pnueli, A., "Marked Directed Graphs", *J. of Computer and System Sciences*, Vol. 5, 1971, pp. 511-523.

[D]   Cohen, D., "Mathematical Approach to Computational Networks", Information Sciences Inst., ISI/RR-78-73, Nov. 1978. ARPA order No. 2223.

[E]   Even, S., "Systolic Modular Multiplication", presented in CRYPT'90. To appear in its proceedings.

[EL]   Even, S., and Litman, A., "On the Capabilities of Systolic Systems", to be presented in *3rd Annual ACM Symp. on Parallel Algorithms and Architectures*, Hilton Head, South Carolina, July 21-24, 1991.

[LS]   Leiserson, C.E., and Saxe, J.B., "Optimizing Synchronous Systems", *Twenty-Second Annual Symposium on Foundations of Computer Science*, IEEE, 1981, pp. 23-36. Also, *Journal of VLSI and Computer Systems*, Vol. 1, 1983, pp. 41-67.

[LRS] Leiserson, C.E., Rose, F.M., and Saxe, J.B., "Optimizing Synchronous Circuitry by Retiming", *Third Caltech Conference on Very Large Scale Integration*, ed. R. Bryant, Computer Science Press, 1983, pp. 87-116.

[S]   Sieferas, J.I., "Iterative Arrays with Direct Central Control", *Acta Informatica*, Vol. 8, 1977, pp. 177-192.

[SS]   Schonhage, A., and Strassen, V., "Schnelle Multiplikation grosser Zahlen", *Computing*, Vol. 7, 1971, pp. 281-292.

[W]   Wallace, C.S., "A Suggestion for a Fast Multiplier", *IEEE Trans. on Electronic Computers*, Vol. EC-13, Feb. 1964, pp. 14-17.

On the Shannon Capacity of Graph Formulae

Luisa Gargano*
Dipartimento di Informatica
Università di Salerno
84081 Baronissi (SA), Italy

János Körner †
IASI
CNR
00185 Roma, Italy

Ugo Vaccaro*
Dipartimento di Informatica
Università di Salerno
84081 Baronissi (SA), Italy

Abstract

In a recent paper we have solved several well–known combinatorial problems treating them as special cases of our generalization of Shannon's notion of graph capacity. We present a new simple formalism to deal with all such problems in a unified manner, considering graphs or families of graphs as special formulae the variables of which are pairs of vertices of their common vertex sets. In all of these problems, the maximum size of a set of n–length sequences from a fixed alphabet is to be determined under various restrictions on the element pairs present in the same coordinate of any two sequences from the set. For sufficiently homogeneous formulae capacity becomes computable.

1 Introduction

Extremal combinatorics and information theory often deal with the same problem from a different prospective. The two fields hardly ever touch. Still, in the sixties, Alfréd Rényi's inspiring work was showing how fruitful the connections between these two areas can be for both. Information-theoretic methods have been used ever since to deal with combinatorial problems more or less explicitly in the work of his students G. Katona, J. Komlós and others, e. g. J. Spencer; yet it is fair to say that the community of combinatorialists regards entropic arguments as an undesirable oddity eventually to be eliminated from proofs once "we get a better understanding" of the problem at hand. In this paper, we will have to use and actually compute entropies of graphs, (this notion was introduced in [9].) The calculations will be needed to determine the precise numerical values in our results; we will not exhibit them in too much detail. The interested reader will be referred to some earlier articles. Apart from the derivation of these numerical values, the paper will be self–contained.

*Work partially supported by the Italian Ministry of the University and Scientific Research, Project: Algoritmi, Modelli di Calcolo e Strutture Informative.
†On leave from the Mathematical Institute of HAS, 1364 Budapest, POB 127, Hungary.

Recently, in a series of papers ([7], [8], [11], or, for the earlier work: [4] and [13]) we have applied a novel information–theoretic construction method combined with traditional entropy-based bounding techniques to determine the precise exponential asymptotics in a large number of combinatorial problems of which the most outstanding is Rényi's question about the maximum number of pairwise qualitatively independent k–partitions of an n–set, a problem that had been open for 20 years. Other problems include various generalizations of Sperner's celebrated theorem on the largest family of subsets of a set under the restriction that no set in the family contain an other. All of these problems can be reformulated in such a way that the maximum size of a set of n–length sequences from a fixed alphabet is to be determined under various restrictions on the element pairs present in the same coordinate of any two sequences from the set. In [7], [8] and [11] one requires the presence of some (directed) edges of certain (directed) graphs between the coordinates. Such problems include Sperner's theorem and its various generalizations, the Shannon capacity of a graph, Rényi's question on qualitatively independent partitions (posed as a problem in his book [18]) and many more. Körner ([11]) has considered a further generalization of this set–up in which a different kind of condition appears. Not only does one require the presence of some edges of certain graphs between the sequences, but, simultaneously, one excludes the presence of other edges of other graphs in these coordinates. This latter kind of concepts, called inclusion–exclusion capacities in [11] are useful in dealing with the intersection number of certain product graphs and represent a natural generalization of previous concepts of capacity.

On the surface, it seems that these are all similar but different concepts. One of our main objectives in this paper is to introduce a new and unified formalism to deal with all of them. Further, we will solve some new problems not considered in the above papers.

In Shannon's original work capacity is a concept linked to the transmission or storage of information. These are the capacities of a noisy channel or of a finite automaton and are capacities in a very precise engineering sense. The capacity of a graph is indeed the zero–error capacity of a discrete memoryless channel. Mathematically, the channel can be described in terms of a graph as far as its zero–error data transmission potential goes. We have shown ([13],[8]) that a slight change in the definition of graph capacity leads to problems of equal mathematical interest but with no engineering interpretation. To this end we did nothing but formally generalize Shannon capacity to directed graphs. (We actually had to change edges to non–edges for such a generalization to become possible.)

From our point of view, (cf. [8]) all of the above problems can be looked at as variants of the same basic question. To explain what we have in mind let us first look at the example of Shannon's capacity.

Shannon capacity

Let G be a graph with vertex set $V(G) = V$ and edge set $E(G)$. For any $n \geq 1$ let us say that the sequences $\mathbf{x} \in V^n$, $\mathbf{x}' \in V^n$ are G-different if there is a coordinate $1 \leq i \leq n$ for which the vertices x_i and x_i' are connected by an edge of G. Let $N(G, n)$ denote the maximum number of pairwise G-different elements of V^n, and define

$$C(G) = \lim_{n \to \infty} \frac{1}{n} \log N(G, n).$$

Here and in the sequel all logarithms are to the base 2. (It is a well-known elementary fact that the above limit exists.)

Intuitively, $N(G, n)$ is the maximum number of n-length sequences of elements of V one can distinguish at the receiving end of a discrete memoryless channel through which these sequences are transmitted. Thus these sequences are *different* in a strong, precise information-theoretic sense; they are different for the receiver who can tell the one from the other despite the confusions caused by the operation of the channel. It is the concept of being different that changes from problem to problem as we develop our general approach. In the previous example it is defined via a channel. The following slight variation in the definition has proved very useful in extremal set theory.

Sperner capacity

Consider a directed graph G with vertex set $V(G) = V$ and edge set $E(G) \subset V^2$. Once again, let us say that the sequences $\mathbf{x} \in V^n$, $\mathbf{x}' \in V^n$ are G-different if there is a coordinate $1 \leq i \leq n$ for which the vertices x_i and x_i' are connected by an arc (i.e., directed edge) of G. Let $N(G, n)$ denote the maximum number of pairwise G-different elements of V^n, and define

$$\Sigma(G) = \lim_{n \to \infty} \frac{1}{n} \log N(G, n).$$

Once again, one immediately sees that the limit exists.

We have called $\Sigma(G)$ the Sperner capacity of the directed graph G, (cf. [8]. A similar concept had been introduced earlier by Körner and Simonyi ([13]).) It should be clear that Sperner capacity is a formal generalization of Shannon capacity. Indeed, if the graph G is symmetric (i.e., $(x, y) \in E(G) \iff (y, x) \in E(G)$) then the Sperner capacity of G coincides with the Shannon capacity of the corresponding undirected graph. Moreover, the Sperner capacity of a directed graphs depends on the orientation of its edges: It has been shown by Calderbank *et al.* [3] that for the two graphs T and T' shown below $\Sigma(T) = \log 3$ and $\Sigma(T') = 1$.

Given $\mathbf{x} \in V^n$, we shall denote by $P_{\mathbf{x}}$ the probability distribution on V given by

$$P_{\mathbf{x}}(a) = \frac{1}{n}|\{i \ : \ x_i = a, \ 1 \le i \le n\}|,$$

where $\mathbf{x} = x_1 \cdots x_n$. The distribution $P_{\mathbf{x}}$ is called the type of $\mathbf{x}$. Given a probability distribution P on V, denote by V_P^n the set of all sequences in V^n having type P (notice that such a set may be empty). Let $V^n(P, \epsilon)$ denote the set of those $\mathbf{x} \in V^n$ for which

$$|P_{\mathbf{x}} - P| = \max_{a \in V} |P_{\mathbf{x}}(a) - P(a)| \le \epsilon.$$

Hence V_P^n is a shorthand for $V^n(P, 0)$.

We denote by $d_G(x)$ the degree of the vertex $x \in V(G)$, by $d(G)$ the maximum of $d_G(x)$ taken over all vertices of G and by $\overline{G} = (V, \{(x, y) : x, y \in V, \mathbf{x} \ne y, \ (x, y) \notin E(G)\})$ the complement of G. The chromatic number $\chi(G)$ is the minimum number of induced edge–free subgraphs needed to cover the vertex set of G.

As an important technical tool in our treatment, we introduce the concept of Sperner capacity of a directed graph G within a fixed distribution P. Denote by $N(G, P, \epsilon, n)$ the maximum number of pairwise G–different elements of $V^n(P, \epsilon)$. The capacity of the graph G within the probability distribution P is defined as ([5], [8])

$$\Sigma(G, P) = \lim_{\epsilon \to \infty} \limsup_{n \to \infty} \frac{1}{n} \log N(G, P, \epsilon, n).$$

Clearly,

$$\Sigma(G) = \max_P \Sigma(G, P).$$

We have mentioned in [8] that Shannon's graph capacity can be looked at as some sort of an OR–capacity of a graph. In its definition we require the construction of a large number of sequences with the property that between any two sequences the first *or* the second *or* the third etc. edge of the graph occur in some coordinate. In the same paper we have introduced another concept of capacity of a graph where the same holds with *or* replaced by *and*. Actually, the latter was a special case of our new concept of capacity of graph families. Talking about the various concepts of capacity one quickly reaches the limit of confusion unless a common language is used that allows us to compare problems and results. It is therefore preferable to pass to the formal treatment. More importantly, the above indicates the possibility of using Boolean language to treat capacities. This will enable us to deal with exclusions in a formally identical way to the other conditions.

2 Edge Formulae

Given a vertex set V, an *edge formula* is an expression composed of directed edges $e \in V \times V$, parentheses, and the operators $\vee$ (logical OR), $\wedge$ (logical AND) and $\neg$ (negation). Given an edge e, an ordered pair of sequences $(\mathbf{x}, \mathbf{y}) \in V^n \times V^n$ is *compliant with* e if and only if there exists an index i, $1 \le i \le n$ such that $(x_i, y_i) = e$. Analogously, the pair $(\mathbf{x}, \mathbf{y})$ is compliant with $\neg e$ if and only if for every index i one has $(x_i, y_i) \ne e$. We will give a recursive definition for two sequences to be compliant with an arbitrary edge formula. If F_1 and F_2 are two edge formulae, the pair of sequences $(\mathbf{x}, \mathbf{y}) \in V^n \times V^n$ is compliant with $F_1 \vee F_2$ if and only if it is compliant with either F_1 or F_2. The definitions of being compliant with $F \wedge F_2$ and $\neg F_1$ are similar. We are interested in the *capacity* of edge formulae, that is, in determining the asymptotics of the maximum number of sequences which are pairwise compliant with the given formula, i.e., any ordered pair of which is compliant with the formula.

Definition 1 *Given a set V and an edge formula F we denote by $I(F,n)$ the maximum number of sequences in V^n which are pairwise compliant with F. The capacity of F is defined as*

$$C(F) = \limsup_{n \to \infty} \frac{1}{n} \log I(F, n).$$
(1)

It is not hard to see that actually the limit exists. The argument is the same as for Shannon capacity, cf. e. g. [2].

207

Definition 2 *Given a set V, a probability distribution P on V and an edge formula F, we denote by $I(F, P, \epsilon, n)$ the maximum number of sequences in $V^n(P, \epsilon)$ which are pairwise compliant with F. The capacity of F in the distribution P is defined as*

$$C(F, P) = \lim_{\epsilon \to \infty} \limsup_{n \to \infty} \frac{1}{n} \log I(F, P, \epsilon, n). \tag{2}$$

For brevity, we will write $I(F, P, n) = I(F, P, 0, n)$.

Lemma 1 *For each set V and edge formula F on V*

$$C(F) = \max_P C(F, P). \tag{3}$$

Proof. Omitted.

□

Notice that the Sperner capacity within the distribution P of a directed graph $G = (V, E)$ can be expressed, using the above notation as

$$C(G, P) = C\left(\bigvee_{e \in E} e, P\right). \tag{4}$$

This is what we mean by saying that the Sperner (Shannon) capacity of a single graph is an OR–capacity. We note that even though the determination of the capacity of one–edge graphs (within any distribution P) is a trivial task, the calculation of the Sperner capacity of most graphs is a tremendous unsolved problem ([15]). The following result, proved in [8], shows that computing the capacity of the AND–combination of edge formulae is "easy".

Theorem GKV *Given the edge formulae $F_1, \ldots, F_n$ not containing any negations and a probability distribution P on the common vertex set V*

$$C\left(\bigwedge_{i=1}^{n} F_i, P\right) = \min_{1 \leq i \leq n} C(F_i, P).$$

Corollary GKV *Given the edge formulae $F_1, \ldots, F_n$ not containing any negations*

$$C\left(\bigwedge_{i=1}^{n} F_i\right) = \max_P \min_{1 \leq i \leq n} C(F_i, P),$$

where the maximum is taken for all probability distributions on the vertex set V.

We have developed a language that will allow us to deal with many mathematical problems expressible in terms of finding the maximum number of sequences of some fixed length n from an arbitrary alphabet V under any restriction on the pairs of coordinates occurring in pairs of these sequences. We will illustrate the advantages of our language by solving some new problems. Our interest in these problems is motivated by an attempt to generalize the main result of [7] on qualitatively independent partitions to what one might be tempted to call cases of qualitative dependence.

3 Qualitatively dependent partitions

Rényi ([18]) called two partitions of a set into finitely many classes qualitatively independent (QI) if the number of the classes of their common refinement is maximum, i. e., it equals the product of their respective number of classes. In other words, two partitions of a set are QI if every class of the one intersects each class of the other. Intuitively, two partitions are QI if and only if they can be generated by two independent random variables. Since every element of V^n can be considered as the "characteristic vector" of a partition of the set $1, 2, \ldots, n$ into $|V|$ classes, one easily sees that the determination of the maximum number of pairwise QI partitions of an n–set into k classes (for some fixed k, as n goes to infinity) is a problem we can formulate equivalently in terms of Sperner capacities. Furthermore, the resulting capacity problem has a nice solution ([8]).

Recently, problems of this kind have got new attention in terms of managing the resource *"randomness"* in algorithms. To understand this aspect of our problem, it is enough to look at it in a dual manner. Then we can ask how many atoms a probability space must have in order to support N pairwise independent k–valued random variables? Clearly, this problem is equivalent to Rényi's original question. Pursueing Rényi's nice idea to ask combinatorial questions about probability spaces, it is natural to extend our scope from independence to dependence. In the combinatorial setting it is imperative to measure the *degree of dependence* of two partitions by the number of classes in their (roughest) common refinement, i.e., by the number of intersecting pairs among their respective classes. Then, the more pairs of classes of two partitions intersect, the closer the two partitions are to being QI. In this framework the first natural question to ask is how many more k–partitions can live on a probability space of n atoms if we allow them to be *weakly dependent*. We can say that two k–partitions are r–dependent if r pairs of their classes have non–empty intersections. Two partitions can be called r–weakly dependent if they are r'– dependent for some $r' \geq r$. Then two k–partitions are QI if they are k^2–weakly dependent. We are interested in the maximum number $W(k, r, n)$ of pairwise r–weakly dependent k–partitions of an n–set, asymptotically in n, for any fixed values of k and r.

The dual question is of equal interest. In other words, we might ask for the maximum number of pairwise *strongly dependent* k–partitions of a given set. Unlike for qualitative independence, the extremal case of the present problem is trivial for dependence is the strongest when the two partitions are identical. Two partitions can be called r–strongly dependent if they are s–dependent for some $s \le r$. We are then interested in the maximum number $S(k, r, n)$ of pairwise r–strongly dependent k–partitions of an n–set, asymptotically in n, for any fixed values of k and r.

These questions have an immediate translation into the language of capacities of graph formulas along the lines indicated above. We will deal with them later on. For the time being, we prefer to consider analogous but simpler questions involving undirected rather than directed graphs.

4 Weakly dependent sequences

In this section we will consider the undirected graph analogon of weakly dependent partitions. Let K_r be the complete undirected graph on the vertex set $[r] = \{1, 2, \ldots, r\}$. We investigate the asymptotics of the maximum number of sequences in $[r]^n$ such that between any pair of them there exist at least t different edges of K_r. We introduce the (usual) notation $\binom{V}{2}$ for the family of all the two–element subsets of the arbitrary set V.

Definition 3 *Given two sequences* $\mathbf{x}, \mathbf{y} \in [r]^n$, *we say that they are* t–*different if*

$$|\{e \in E(K_r) \ : \ e = \{x_i, y_i\} \text{ for some } i, \ 1 \le i \le n\}| = t.$$

Let $M_r(t, n)$ (resp. $M_r(t, P, n)$) be the maximum number of sequences in $[r]^n$ (resp. in $[r]_P^n$) such that any two of them are at least t–different, i.e., they are t'–different for some $t' \ge t$.

We shall investigate the quantity

$$M_r(t) = \limsup_{n \to \infty} \frac{1}{n} \log M_r(t, n).$$

Let us consider the family

$$\mathcal{E}_t = \left\{ E \subseteq \binom{[r]}{2} \ : \ |E| = \binom{r}{2} - (t - 1) \right\}.$$

The sequences $\mathbf{x}, \mathbf{y} \in [r]^n$ are at least t–different if and only if for each $E \in \mathcal{E}_t$ they are compliant with the edge formula $\bigvee_{e \in E} e$. Therefore, for each probability distribution P

$$M_r(t, P, n) = I \left(\bigwedge_{E \in \mathcal{E}_t} \bigvee_{e \in E} e, \quad P, \quad n \right).$$

Likewise, by (2)–(3) and Theorem GKV one gets

$$M_r(t) = \max_P C \left(\bigwedge_{E \in \mathcal{E}_t} \bigvee_{e \in E} e, \quad P \right) = \max_P \min_{E \in \mathcal{E}_t} C \left(\bigvee_{e \in E} e, \quad P \right). \tag{5}$$

Defining $\mathcal{G}_t = \{G \subseteq K_r : E(G) \in \mathcal{E}_t\}$ we can further write

$$M_r(t) = \max_P \min_{G \in \mathcal{G}_t} C(G, P). \tag{6}$$

Since the above expression involves all possible graphs with $\binom{r}{2} - (t-1)$ edges on the vertices in $[r]$, it is easy to see that the maximum in P is achieved by the uniform probability distribution $U_r = (1/r, \ldots, 1/r)$ on $[r]$. We thus have

$$M_r(t) = \min_{G \in \mathcal{G}_t} C(G, U_r). \tag{7}$$

The following result gives a lower bound on $C(G, U_r)$ that we shall use to determine $M_r(t)$ for several values of t. We first need the following definition. Given $\mathbf{a} = (a_1, \ldots, a_r)$, and $\mathbf{b} = (b_1, \ldots, b_r)$, with $a_1 \geq \cdots \geq a_r$, $b_1 \geq \cdots \geq b_r$, and $\sum_{i=1}^r a_i = \sum_{i=1}^r b_i$, we say that $\mathbf{a} \preceq \mathbf{b}$ if and only if $\sum_{i=1}^k a_i \leq \sum_{i=1}^k b_i$ for each $k = 1, \ldots, r$.

Lemma 2

$$\min_{G \in \mathcal{G}_t} C(G, U_r) \geq \log r - \frac{1}{r} \sum_{x \in [r]} \log(1 + d(x)),$$

where the vector $\mathbf{d} = \{d(x) : x \in [r]\}$ *yields the minimum with respect to the relation* $\preceq$ *among all the vectors* $\mathbf{d}'$ *such that* $d'(x)$ *is integer for each* $x \in [r]$, *and* $\sum_{x \in [r]} d'(x) = 2(t-1)$. *(This means that for a vector* $\mathbf{d}''$ *with the preceding properties* $\mathbf{d}'' \preceq \mathbf{d}$ *implies* $\mathbf{d}'' = \mathbf{d}$.)

Proof. Omitted. □

The following theorem describes the function $M_r(t)$ for many values of t and gives us a good idea of its shape.

Theorem 1

i) *If* $t \leq r/2 + 1$ *then* $M_r(t) = \log r - \frac{2}{r}(t-1)$;

ii) *If* $t = (cr/2) + 1$ *with* $r \equiv 0 \bmod (c+1)$ *then* $M_r(t) = \log \frac{r}{c+1}$.

iii) *If* $t = (r^2 - 1)/4 + 1$ *then* $M_r(t) = \log r - \frac{r+1}{2r} \log \frac{r+1}{2} - \frac{r-1}{2r} \log \frac{r-1}{2}$

Proof.

Case i). Let $t \leq r/2 + 1$. For each graph $G \in \mathcal{G}_t$ the complement has $|E(\overline{G})| = t - 1$ edges. The minimizing vector in Lemma 2 is the degree vector $\mathbf{d}$ having the first $r - 2(t-1)$ components equal to 0 and the last $2(t-1)$ equal to 1, thus giving

$$\min_{G \in \mathcal{G}_t} C(G, U_r) \geq \log r - \frac{2(t-1)}{r}. \tag{8}$$

On the other hand, consider the bipartite graph $F_t = ([r], \{(i, t-1+i) \ : \ i = 1, \ldots, t-1\})$. Clearly, F_t has $t-1$ edges, and thus for its complement we have $\overline{F}_t \in \mathcal{G}_t$. We claim that $C(\overline{F}_t, U_r)$ is easy to compute. Notice first that for every graph G and every probability distribution P on its vertex set the capacity $C(G, P)$ is upper bounded by the corresponding graph entropy $H(G, P)$, cf. [4] (or for a more detailed explanation [17]). An elementary calculation (cf. [9] or, for a more detailed explanation: [10]) shows that

$$H(\overline{F}_t, U_r) = \log r - \frac{2(t-1)}{r}.$$

Hence, we also have

$$C(\overline{F}_t, U_r) \leq \log r - \frac{2(t-1)}{r}.$$

Thus inequality (8) is indeed an equality and i) holds.

Case ii). Let $t = (cr/2) + 1$ with $r \equiv 0 \bmod (c+1)$. Since $|E(\overline{G})| = cr/2$ for each $G \in \mathcal{G}_t$, the minimum in Lemma 2 is reached by the vector in which every component has value c, thus giving

$$\min_{G \in \mathcal{G}_t} C(G, U_r) \geq \log r - \log(c+1).$$

Clearly, the minimizing vector $\mathbf{d}$ is the degree vector of a graph that is the vertex–disjoint union of cliques (complete subgraphs) of $c + 1$ vertices each. Bounding capacity by entropy as before and computing graph entropy of this very simple graph (cf. [9] for the actual computation) we get

$$C(G, U_r) \leq H(G, U_r) = \log r - \log(c+1),$$

proving the desired equality.

Case iii). Let $t = (r^2 - 1)/4 + 1$. For each $G \in \mathcal{G}_t$, one has $|E(\overline{G})| = (r^2 - 1)/4$. It is easy to see that the minimizing vector in Lemma 2 is the one having its first $(r+1)/2$ components equal to $(r-1)/2$ and the remaining $(r-1)/2$ coordinates equal to $(r-1)/2 - 1$. Thus

$$\min_{G \in \mathcal{G}} C(G, U_r) \geq \log r - \frac{1}{r}\left[\frac{r+1}{2} \log \frac{r+1}{2} + \frac{r-1}{2} \log \frac{r-1}{2}\right]. \tag{9}$$

On the other hand, one immediately sees that the minimizing vector is the degree vector of the complete bipartite graph J consisting of two disjoint independent sets of respective cardinality $(r+1)/2$ and $(r-1)/2$ with all the edges between the two present in the graph. Proceeding as before, we see that

$$C(J, U_r) \leq H(J, U_r) = \log r - \frac{1}{r}\left[\frac{r+1}{2} \log \frac{r+1}{2} + \frac{r-1}{2} \log \frac{r-1}{2}\right]$$

and thus (9) holds with equality.

$\square$

5 Formulae and their negations

In order to deal with what we should call strongly dependent sequences, we first state a general lemma on the connection of the capacity of a formula and that of its negation.

Lemma 3 *For each edge formula F and probability distribution P on a set V*

$$C(\neg F, P) + C(F, P) \leq H(P)$$

where $H(P) = -\sum_x P(x) \log P(x)$.

Proof. Omitted.

$\square$

We will not enter the question of tightness of the lemma. In the special case of graph capacity, the analogous inequality is known to be tight for every distribution if the graph is perfect. For more on these questions cf. [12], [6] and [17].

Let $m_r(t, n)$ be the maximum number of sequences in $[r]^n$ such that any two of them are at most t–different with respect to K_r, i.e., they are t'–different for some $1 \leq t' \leq t$.

We shall determine the quantity

$$m_r(t) = \limsup_{n \to \infty} \frac{1}{n} \log m_r(t, n).$$

Notice that two sequences in $[r]^n$ are at most t–different if they are compliant with the edge formula $\neg \left(\bigwedge_{E \in \mathcal{E}_{t+1}} \bigvee_{e \in E} e \right)$, so that

$$m_r(t) = C \left(\neg \left(\bigwedge_{E \in \mathcal{E}_{t+1}} \bigvee_{e \in E} e \right) \right). \tag{10}$$

Since all sequences in $[k]^n$ are pairwise at most $[k(k-1)/2]$–different it follows that $m_r(t) \geq \log k$ whenever $k(k-1)/2 \leq t < k(k+1)/2$. The main result of this section is the following theorem which shows that the above trivial lower bound is indeed tight.

Theorem 2 *For each r and $t \leq \binom{r}{2}$*

$$m_r(t) = \log k(t),$$

where $k(t) = \max \left\{ k \; : \; t \geq \binom{k}{2} \right\}$.

Proof. We have just seen that $\log k(t)$ is a valid lower bound on $m_r(t)$. The rest of this section is devoted to the proof of the corresponding upper bound $m_r(t) \leq \log k(t)$.

Given a probability distribution P on $[r]$, let $m_r(t, P, n)$ be the maximum number of sequences in $[r]_P^n$ such that any two of them are at most t–different with respect to K_r. Define

$$m_r(t, P) = \limsup_{n \to \infty} \frac{1}{n} \log m_r(t, P, n).$$

Clearly, $m_r(t) = \max_P m_r(t, P)$, with the maximum taken for all probability distributions over $[r]$. Notice that

$$m_r(t, P, n) = I \left(\neg \left(\bigwedge_{E \in \mathcal{E}_{t+1}} \bigvee_{e \in E} e \right), P, n \right),$$

where

$$I \left(\bigwedge_{E \in \mathcal{E}_{t+1}} \bigvee_{e \in E} e, P, n \right) = M_r(t+1, P, n).$$

Therefore, from Lemma 3 and formulae (5), and (4) we get

$$m_r(t, P) \leq \max_{G \in \mathcal{G}_{t+1}} [H(P) - C(G, P)]. \tag{11}$$

Let us notice now the elementary inequality (cf. [17])

$$H(P) - C(G, P) \leq H(\overline{G}, P).$$

214

Observing (cf. [9] or [10]) that the entropy of a graph is upper bounded by the logarithm of its chromatic number, the last inequality yields

$$H(P) - C(G, P) \leq \log[\chi(\overline{G})]. \tag{12}$$

Inequalities (11) and (12) imply

$$m_r(t) \leq \max_{G \in \mathcal{G}_{t+1}} \log \chi(\overline{G}) = \max_{J \in \mathcal{J}_{t+1}} \log \chi(J),$$

where $\mathcal{J}_{t+1} = \{\overline{G} \; : \; G \in \mathcal{G}_{t+1}\}$. Since every graph $J \in \mathcal{J}_{t+1}$ has t edges, we have $\chi(J) \leq k(t)$ for every $J \in \mathcal{J}_{t+1}$, (see [2], Theorem 4, p. 335) and we finally obtain

$$m_r(t) \leq \max_{J \in \mathcal{J}_{t+1}} \log \chi(J) \leq \log k(t).$$

6 Weakly dependent partitions

The case of partitions is analogous to that of sequences the only difference being that the role of the undirected graphs is now taken over by digraphs. In order to avoid repetitions the presentation will be very succint. We intend to return to these questions in a separate paper. Let us write

$$W(k, t) = \limsup_{n \to \infty} \frac{1}{n} \log W(k, t, n),$$

where $W(k, t, n)$ is the maximum number of pairwise t-weakly dependent k-partitions of an n-set. As in the case of sequences, we introduce the family B_t of directed graph defined as

$$B_t = \{G : V(G) = [k] \text{ and } |E(G)| = k(k - 1) - (t - 1)\}.$$

We have

$$W(k, t) = \min_{G \in B_{t-k}} \Sigma(G, U_k).$$

It is natural to believe that the partitions question reduces to the one on sequences. At first glance it seems tempting to make the following

Conjecture *For every t we have*

$$\min_{G \in B_t} \Sigma(G, U_k) = \min_{G \in \mathcal{G}_{\lceil t/2 \rceil}} C(G, U_k).$$

215

References

[1] R. Ahlswede, "Coloring hypergraphs: A new approach to multi-user source coding", *J. Comb. Inf. Syst. Sci.*, pt. I, **4**, (1979) pp. 76–115.

[2] C. Berge, *Graphs*, 2nd revised edition, North-Holland, Amsterdam (1985),

[3] A. R. Calderbank, R. L. Graham, L. A. Shepp, P. Frankl, and W.-C. W. Li, "The Cyclic Triangle Problem", manuscript, (1991).

[4] G. Cohen, J. Körner, G. Simonyi, " Zero–error capacities and very different sequences", in: *Sequences: combinatorics, compression, security and transmission*, R. M. Capocelli (Ed.), Springer–Verlag, (1990), pp. 144–155.

[5] I. Csiszár, J. Körner, *Information Theory. Coding Theorems for Discrete Memoryless Systems*, Academic Press, New York, (1982).

[6] I. Csiszár, J. Körner, L. Lovász, C. Marton, G. Simonyi "Entropy splitting for antiblocking corners and perfect graphs", *Combinatorica*, **10**, (1990), pp. 27–40.

[7] L. Gargano, J. Körner, U. Vaccaro, "Sperner capacities", *Graphs and Combinatorics*, to appear.

[8] L. Gargano, J. Körner, U. Vaccaro, "Capacities: from information theory to extremal set theory", submitted to *J. of AMS*.

[9] J. Körner, "Coding of an information source having ambiguous alphabet and the entropy of graphs". *Transactions of the 6th Prague Conference on Information Theory, etc.*, 1971, Academia, Prague, (1973), pp. 411–425

[10] J. Körner, "Fredman-Komlós bounds and information theory" , *SIAM J.on Algebraic and Discrete Meth.*, **7**, (1986), 560–570

[11] J. Körner, "Intersection number and capacities of graphs", submitted to *J. Comb. Theory*, Ser. **B**.

[12] J. Körner, K. Marton, "Graphs that split entropies", *SIAM J. Discrete Math.*, **1**, (1988), 71–79.

[13] J. Körner, G. Simonyi, "A Sperner–type theorem and qualitative independence", *J. Comb. Theory*, Ser. **A**, to appear

[14] L. Lovász, "On the ratio of optimal integral and fractional covers", *Discrete Mathematics*, **13**, (1975), pp. 383–390.

[15] L. Lovász, "On the Shannon capacity of a graph", *IEEE Trans. on Information Theory*, IT-**25**, (1979), pp. 1–7.

[16] A.W. Marshall, I. Olkin, *Inequalities: Theory of Majorization and Its Applications*, Academic Press, New York, (1979).

[17] K. Marton, *On the Shannon capacity of probabilistic graphs*, submitted to *J. Comb. Theory*, Series **A**.

[18] A. Rényi, *Foundations of Probability*, Wiley, New York, (1971)

[19] C. E. Shannon, "The zero-error capacity of a noisy channel", *IRE Trans. on Information Theory*, **2**, (1956), pp. 8–19.

[20] Z. Tuza, "Intersection properties and extremal problems for set systems", in: *Irregularities of Partitions* G. Halász and V. T. Sós (Eds.), Algorithms and Combinatorics **8**, Springer Verlag, (1989), pp. 141–151.

An Efficient Algorithm for the All Pairs Suffix-Prefix Problem

Dan Gusfield *   Gad M. Landau †   Baruch Schieber ‡

Abstract

For a pair of strings (S_1, S_2), define the suffix-prefix match of (S_1, S_2) to be the longest suffix of string S_1 that matches a prefix of string S_2. The following problem is considered in this paper. Given a collection of strings $S_1, S_2, \ldots, S_k$ of total length m, find the suffix-prefix match for each of the $k(k-1)$ ordered pairs of strings. We present an algorithm that solves the problem in $O(m + k^2)$ time, for any fixed alphabet. Since the size of the input is $O(m)$ and the size of the output is $O(k^2)$ this solution is optimal.

1. Introduction

For a pair of strings (S_1, S_2), define the suffix-prefix match of (S_1, S_2) to be the longest suffix of string S_1 that (exactly) matches a prefix of string S_2. We consider the **All Pairs Suffix-Prefix Problem** defined as follows. Given a collection of strings $S_1, S_2, \ldots, S_k$ of total length m, find the suffix-prefix match for each of the $k(k-1)$ ordered pairs of strings.

---

*Computer Science Division, University of California at Davis, Davis, CA 95616, Tel. (916) 752-7131, Email: gusfield@cs.ucdavis.edu. Partially supported by Dept. of Energy grant DE-FG03-90ER60999, and NSF grant CCR-8803704.

†Dept. of Computer Science, Polytechnic University, 333 Jay Street, Brooklyn, NY 11201, Tel. (718) 260-3154, Email: landau@pucs2.poly.edu. Partially supported by NSF grant CCR-8908286 and the New York State Science and Technology Foundation, Center for Advanced Technology in Telecommunications, Polytechnic University, Brooklyn, NY.

‡IBM - Research Division, T.J. Watson Research Center, P.O. Box 218, Yorktown, NY 10598, Tel. (914) 945-1169, Email: sbar@watson.ibm.com.

Using a variant of the well-known KMP string matching algorithm [KMP77] one can find the suffix-prefix match of a single pair (S_i, S_j) in $O(m_i + m_j)$ time, where m_i and m_j are the respective lengths of the strings. So overall, this approach leads to an $O(km)$ time solution. We present an algorithm that solves the problem in $O(m + k^2)$ time, for any fixed alphabet. Since the size of the input is $O(m)$ and the size of the output is $O(k^2)$ this solution is optimal.

The motivation for the all pairs suffix-prefix problem comes from two related sources, sequencing and mapping DNA [Les88]. Sequencing means getting a complete linear listing of the nucleotides in the string, and mapping means finding the linear order or exact positions of certain significant features in the string. Sequencing is essentially the extreme special case of mapping. In either case, long strings of DNA must first be cut into smaller strings since only relatively small strings can be sequenced as a single piece. Known ways of cutting up longer strings of DNA result in the pieces being randomly permuted. Hence, after obtaining and sequencing or mapping each of the smaller pieces one has the problem of determining the correct order of the small pieces. The most common approach to solving this problem is to first make many copies of the DNA, and then cut up each copy with a different cutter so that pieces obtained from one cutter overlap pieces obtained from another. Myriad choices of how to cut are available; some are chemical, some or enzymatic and some are physical (radiation or vibration). Then, each piece is sequenced or mapped, and in the case of mapping a string is created to indicate the order and type of the mapped features, where an alphabet has been created to represent the features.

Given the entire set of sequenced or mapped pieces, the problem of assembling the proper DNA string has been *modeled* as the *shortest common superstring problem:* find the shortest string which contains each of the other strings as a contiguous substring.

There are several approaches to solve this problem. In one approach [TU88, KM89, Tur89, Ukk90, BJLTY91, KM91] the problem is modeled as a maximum length Hamilton tour problem as follows. First, each string that is totally contained in another string is removed (finding these strings in linear time is easy - a simple byproduct of building the suffix tree of the given set of strings, as explained later). Let $\mathcal{S} = \{S_1, \ldots, S_k\}$ be the set of remaining strings. Define a complete weighted graph G_S with k vertices, where each vertex i in G_S corresponds to string S_i. Define the weight of each edge (i,j) to be the length of the suffix-prefix match of (S_i, S_j). Observe that the *maximum* length Hamilton tour in this graph gives the shortest superstring. Any approximations within factor $c < 1$ of the optimal Hamilton tour results in a superstring with *compression* size (i.e., number of symbols "saved") of at most most c times the optimal compression. Such approximations and other practical methods based on the Hamilton tour approach has given good results [KM89, KM91]. In a slightly different approach [BJLTY91] model the problem as a minimum length Hamilton tour problem. They consider the same graph G_S, but define the weight of each edge (i, j) to be the length of string S_i minus the length of the suffix-prefix match of (S_i, S_j). Observe that the *minimum* length Hamilton tour in this weighted graph gives the shortest superstring. Any approximations within factor $c > 1$ of the optimal Hamilton tour results in a superstring whose length is at most c times the optimal. Clearly, the all pair suffix-prefix problem is the first step in these methods.

Another approach for solving the problem is the greedy approach [TU88, Ukk90, BJLTY91]; substrings are built up by greedily joining the two strings (original or derived) with maximum suffix-prefix match. Although this approach does not require solving the all pairs suffix-prefix problem, the all-pairs result leads to an efficient implementation of this method, and generalizes the kinds of operations used in [TU88, Ukk90, BJLTY91].

The main data structure used in our algorithm is the *suffix tree* of a set of strings. To make the paper self contained we recollect the definition of suffix trees in Section 2, and describe our algorithm in Section 3.

2. Suffix trees

Let $C = c_1 c_2 \ldots c_n$ be a string of n characters, each taken from some fixed size alphabet Σ. Add to C a special symbol \$ (c_{n+1}) that occurs nowhere else in C. The suffix tree associated with C is the trie (digital search tree) with $n + 1$ leaves and at most n internal vertices such that: (1) each edge is labeled with a substring of C, (2) no two sibling edges have the same (nonempty) prefix, (3) each leaf is labeled with a distinct position of C and (4) the concatenation of the labels on the path from the root to leaf i describes the suffix of C starting at position i. (See Fig. 1 for an example.)

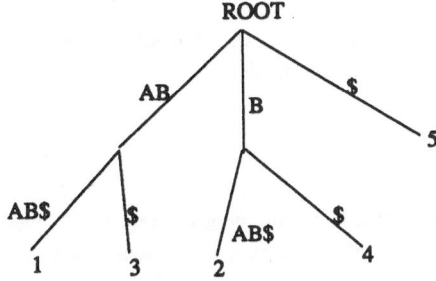

Figure 1. The Suffix Tree of the string ABAB$

Weiner [Wei73] and McCreight [McC76] presented linear time algorithms to compute a suffix tree, Apostolico et al. [AILSV88] compute a suffix tree in parallel.

Given a set of strings $S = \{S_1, S_2, \ldots, S_k\}$, define the *generalized suffix tree* of S to be the tree given by "superimposing" the suffix trees of all strings in S, identifying edges with the same labels. In other words, the *generalized suffix tree* of S is the trie consisting of suffixes of all strings in S. The generalized suffix tree can be computed in the following three steps.

STEP 1. Concatenate the strings in S into one long string $C = S_1\$_1 S_2\$_2 \cdots \$_{k-1}S_k$, where $\$_i$ $(1 \leq i \leq k-1)$ is a special symbol that occurs nowhere else in C.

STEP 2. Compute the suffix tree of C.

STEP 3. Extract the generalized suffix tree of S from the suffix tree of C by changing the labels of some edges in the tree as follows. Let (u, v) be an edge of the suffix tree whose label contains (at least) one of the special symbols $\$_i$. Consider the concatenation of the labels on the path from the root to v. Since the resulting string contains (at least) one special symbol, it occurs only once in C, and hence, vertex v must be a leaf. Change the label of (u, v) to be the prefix of its original label up to (and including) the first occurrence of a special symbol, and change the label of v accordingly. It is not difficult to see, that the part of the suffix tree of C that remains connected to the root is the generalized suffix tree of S.

Implementation Note. Computing the generalized suffix tree by Weiner's algorithm would

221

take $O(m \log k)$ time. However, since each of the special symbols $\$_i$ $(1 \leq i \leq k-1)$ appears only once in C one may add some minor changes to the algorithm in order to achieve linear time. We leave the details to the interested reader.

3. The algorithm

We start by constructing the generalized suffix tree (ST) for the given set of k strings. Consider the path from the root of ST to each internal vertex v. It defines a string $\text{Str}(v)$ whose length is $\text{Len}(v)$. (Note that $\text{Len}(v)$ gives the number of characters on the path to v, and not the number of edges on this path.) As we construct this tree we build up a list $L(v)$ for each internal vertex v. List $L(v)$ holds the index i if and only if the suffix of length $\text{Len}(v)$ of the string S_i is equal to $\text{Str}(v)$. That is, the vertex v is the vertex just before a leaf representing a suffix from string S_i. This step can be done in linear time for any fixed alphabet using the algorithms of [Wei73, McC76].

Consider a string S_j, and focus on the path from the root of ST to the leaf $l(j)$ representing the first suffix of string j, i.e., the entire string S_j. The key observation is the following: Let v be a vertex on this path. If $i \in L(v)$ then a suffix of string S_i, of length $\text{Len}(v)$, matches a prefix of string S_j of the same length. For each index i, we want to record the deepest vertex v on the path to $l(j)$ (i.e., furthest from the root) such that $i \in L(v)$. Clearly, $\text{Str}(v)$ is then the suffix-prefix match of (S_i, S_j). It is easy to see that by one traversal from the root to $l(j)$ we can find the deepest vertices for all $1 \leq i \leq k$ $(i \neq j)$. However, performing this computation for each string S_j separately will result in running time of $O(km)$. In order to achieve linear time we traverse the tree only once.

Traverse the tree ST in Depth First Search (DFS) order ([Tar83]), maintaining k stacks, one for each string. During the traversal when a vertex v is reached in a forward edge traversal, push v onto the ith stack, for each $i \in L(v)$. When a leaf $l(j)$ (representing the entire string S_j) is reached, scan the k stacks and record for each index i the current top of the ith stack. It is not difficult to see that the top of stack i contains the vertex v for which $\text{Str}(v)$ is the suffix-prefix match of (S_i, S_j). If the ith stack is empty, then there is no overlap between a suffix of string S_i and a prefix of string S_j. When the DFS backs up past a vertex v, we pop the top of any stack whose index is in $L(v)$.

Complexity: The total number of indices in all the lists $L(v)$ is $O(m)$. The number of edges in ST is also $O(m)$. Each push or pop of a stack is associated with a leaf of ST, and each leaf is associated with at most one pop and one push, hence traversing ST and updating the stacks takes $O(m)$ time. Recording of each of the $O(k^2)$ answers is done in $O(1)$ time per answer. Therefore, the total running time of the algorithm is $O(m + k^2)$ time.

Extensions. We note two extensions. Let $k' \leq k(k-1)$ be the number of ordered pairs of strings which have a non-zero length suffix-prefix match. By doubly linking the tops of each stack, all non-zero length suffix-prefix matches can be found in $O(m + k')$ time. Note that the order of the stacks in the linked list will vary, since a stack that goes from empty to non-empty must be linked at one of the ends of the list; hence we must also stack the name of the string associated with the stack.

At the other extreme, suppose we want to collect for every pair not just the longest suffix-prefix match, but all suffix-prefix matches. We modify the above solution so that when the tops of the stacks are scanned, we read out the entire contents of each scanned stack. This extension would be an initial step in producing a range of near-optimal superstrings, which is often of more use than just a single shortest superstring. If the output size is k^*, then the complexity for this problem is $O(m + k^*)$.

References

[AILSV88] A. Apostolico, C. Iliopoulos, G.M. Landau, B. Schieber, and U. Vishkin. Parallel construction of a suffix tree with applications. *Algorithmica*, 3:347–365, 1988.

[BJLTY91] A. Blum, T. Jiang, M. Li, J. Tromp, and M. Yanakakis. Linear approximation of shortest superstrings. In *Proc. of the 23rd ACM Symp. on Theory of Computing*, 328-336, 1991.

[KM89] J. Kececioglu and E. Myers. A procedural interface for a fragment assembly tool. Technical Report TR 89-5, University of Arizona, Computer Science Dept., April 1989.

[KM91] J. Kececioglu and E. Myers. A robust and automatic fragment assembly system, 1991. Manuscript.

[KMP77]   D.E. Knuth, J.H. Morris, and V.R. Pratt. Fast pattern matching in strings. *SIAM Journal on Computing*, 6:323–350, 1977.

[Les88]   A. Lesk, editor. *Computational Molecular Biology, Sources and Methods for Sequence Analysis*. Oxford University Press, Oxford, UK, 1988.

[McC76]   E.M. McCreight. A space-economical suffix tree construction algorithm. *Journal of the ACM*, 23:262–272, 1976.

[Tar83]   R.E. Tarjan. *Data Structures and Network Algorithms*. CBMS-NSF Regional Conference Series in Applied Math. SIAM, Philadelphia, PA, 1983.

[TU88]   J. Tarhio and E. Ukkonen. A greedy approximation algorithm for constructing shortest common superstrings. *Theoretical Computer Science*, 57:131–145, 1988.

[Tur89]   J. Turner. Approximation algorithms for the shortest common superstring problem. *Information and Computation*, 83(1):1–20, 1989.

[Ukk90]   E. Ukkonen. A linear time algorithm for finding approximate shortest common superstrings. *Algorithmica*, 5:313–323, 1990.

[Wei73]   P. Weiner. Linear pattern matching algorithm. In *Proc. 14th IEEE Symp. on Switching and Automata Theory*, pages 1–11, 1973.

Efficient Algorithms for Sequence Analysis[*]

David Eppstein [1]   Zvi Galil [2,3]

Raffaele Giancarlo [4,5]   Giuseppe F. Italiano [2,6]

[1] Department of Information and Computer Science, University of California, Irvine, CA 92717

[2] Department of Computer Science, Columbia University, New York, NY 10027

[3] Department of Computer Science, Tel-Aviv University, Tel-Aviv, Israel

[4] AT&T Bell Laboratories, Murray Hill, NJ 07974

[5] On leave from Dipartimento di Matematica, Università di Palermo, Palermo, Italy

[6] Dipartimento di Informatica e Sistemistica, Università di Roma "La Sapienza", Rome, Italy

Abstract: We consider new algorithms for the solution of many dynamic programming recurrences for sequence comparison and for RNA secondary structure prediction. The techniques upon which the algorithms are based effectively exploit the physical constraints of the problem to derive more efficient methods for sequence analysis.

1. INTRODUCTION

In this paper we consider algorithms for two problems in sequence analysis. The first problem is sequence alignment, and the second is the prediction of RNA structure. Although the two problems seem quite different from each other, their solutions share a common structure, which can be expressed as a system of dynamic programming recurrence equations. These equations also can be applied to other problems, including text formatting and data storage optimization.

We use a number of well motivated assumptions about the problems in order to provide efficient algorithms. The primary assumption is that of concavity or convexity. The recurrence relations for both sequence alignment and for RNA structure each include an energy cost, which in the sequence alignment problem is a function of the length of a gap in either input sequence, and in the RNA structure problem is a function of the length of a loop in the hypothetical structure. In practice this cost is taken to be the logarithm, square root, or some other simple function of the length. For our algorithms we make no such specific assumption, but we require that the function be either convex or concave. The second assumption is that of sparsity. In the sequence alignment problem we need only consider alignments involving some sparse set of exactly matching subsequences; analogously, in the RNA structure problem we need only consider structures involving some sparse set of possible base pairs. We show how the algorithms for both problems may be further sped up by taking advantage of this sparsity rather than simply working around it.

Our primary motivation in developing the sequence analysis algorithms we present is their application to molecular biology, although the same sequence analysis procedures also have important uses in other fields. The reasons for the particular interest in molecular biology are, first, that it

---

[*] Work partially supported by NSF Grant CCR-9014605.

is a growing and important area of scientific study, and second, the lengths of biological sequences are such that the need for efficient methods of sequence analysis is becoming acute.

The development and refinement of rapid techniques for DNA sequencing [39, 47] have contributed to a boom in nucleic acid research. A wealth of molecular data is presently stored in several data banks (reviewed by Hobish [25]), which are loosely connected with each other. The biggest of these banks are GenBank [7] and the EMBL data library [20]. As of 1986, GenBank contained 5,731 entries with a total of more than 5 million nucleotides. Most of these data banks double in size every 8-10 months, and there are currently no signs that this growth is slowing down. On the contrary, the introduction of automatic sequencing techniques is expected to accelerate the process [36]. The existence of these nucleotide and amino acid data banks allows scientists to compare sequences of molecules and find similarities between them. As the quantity of data grows, however, it is becoming increasingly difficult to compare a given sequence, usually a newly determined one, with the entire body of sequences within a particular data bank.

The current methods of RNA structure computation have similar limitations. In this case one needs only perform computations on single sequences of RNA, rather than performing computations on entire data banks at once. However the cost of present methods grows even more quickly with the size of the problem than does the cost of sequence comparison, and so at present we can only perform structure computations for relatively small sequences of RNA.

Present algorithms for biological sequence analysis tend to incorporate sophisticated knowledge of the domain of application, but their algorithmic content is limited. Typically they act by computing a set of simple recurrences using dynamic programming in a matrix of values. Our algorithms solve the same problems, using the same assumptions about the physical constraints of the domain. However, by using more sophisticated algorithmic techniques we can take better advantage of those constraints to derive more efficient methods for sequence analysis.

2. SEQUENCE ALIGNMENT

The first sequence analysis problem we study is that of sequence alignment. We only consider alignments between pairs of sequences, since the problem of multiple sequence alignment is NP-complete in general [18, 35] and therefore not likely to be solvable in polynomial time. We start by reviewing the main algorithms for sequence analysis. We refer the reader to references [51, 61, 62] for surveys on this topic and its significance to various fields. Next, we introduce the algorithms that we have obtained, briefly mentioning the technicalities involved in them. Our main focus here is on fast computation of sequence alignment rather than on establishing how meaningful a given alignment is. For this latter topic, so important for molecular biology, the reader is referred to [62].

Sequence alignment is an important tool in a wide variety of scientific applications [51, 64]. It seeks to compare two input sequences, either to compute a measure of similarity between them or to find some core sequence with which each input sequence shares characteristics. For instance, in molecular biology the sequences being compared are proteins or nucleotides; in geology, they represent the stratigraphic structure of core samples; in speech recognition, they are samples of digitized speech.

The first important use of such comparisons is to find a common subsequence or consensus sequence for the two input sequences. For instance, in molecular biology a common structure to a set of sequences can lead to an elucidation of the function of those sequences. A related use of sequence alignment is an application to computer file comparison, made by the widely used *diff* program [4]. The other use of sequence alignment is to compute a measure of similarity between the sequences. This can be used for instance in molecular biology, to compute the taxonomy of evolutionary descent of a set of species, or the genetic taxonomy of a set of proteins within a species [14]. It can also be used to group proteins by common structure, which also typically has a high correlation with common function.

2.1. Longest Common Subsequences and Edit Distance Computation

The simplest definition of an alignment is a matching of the symbols of one input sequence against the symbols of the other so that the total number of matched symbols is maximized. In other words, it is the longest common subsequence between the two sequences [5, 23, 26]. However, in practice this definition leaves something to be desired. One would like to take into account the process by which the two sequences were transformed one to the other, and match the symbols in the way that is most likely to correspond to this process. For instance, in molecular biology the sequences being matched are either proteins or nucleotides, which share a common genetic ancestor, and which differ from that ancestor by a sequence of mutations. The best alignment is therefore one that could have been formed by the most likely sequence of mutations.

More generally, one can define a small set of *edit operations* on the sequences, each with an associated cost. For molecular biology, these correspond to genetic mutations, for speech recognition they correspond to variations in speech production, and so forth. The alignment of the two sequences is the set of edit operations taking one sequence to the other, having the minimum cost. The measure of the distance between the sequences is the cost of the best alignment.

The first set of edit operations that was considered consisted of substitutions of one symbol for another (point mutations), deletion of a single symbol, and insertion of a single symbol. With these operations the problem can be solved in time $O(mn)$, where m and n are the lengths of the two input strings (we assume without loss of generality that $m < n$). This algorithm was discovered independently many times by researchers from a variety of fields [13, 21, 34, 41, 45, 46, 48, 56, 57] and is based upon dynamic programming, one of several used problem-solving technique in computer science and operations research. In particular, the following recurrence allows to find the best alignment between two given strings x and y:

$$D[i,j] = \min\{D[i-1,j-1] + sub(x_i,y_j), D[i-1,j] + del(y_j), D[i,j-1] + ins(x_i)\} \qquad (1)$$

where $D[i,j]$ is the best alignment between the prefix $x_1, x_2, \ldots, x_i$ and the prefix $y_1, y_2, \ldots y_j$, $del(y_j)$ is the cost of deleting the symbol y_j, $ins(x_i)$ is the cost of inserting symbol x_i, and $sub(x_i, y_j)$ is equal to 0 if $x_i = y_j$ and is equal to the cost of substituting symbol y_j for x_i. Further, one can find the subsequence of the first string having the best alignment with the second string in the same time.

If only the cost of the best sequence is desired, the $O(mn)$ dynamic programming algorithm for the sequence alignment problem need take only linear space: if we compute the values of the dynamic programming matrix in order by rows, we need only store the values in the single row above the one in which we are performing the computation. However if the sequence itself is

227

desired, it would seem that we need $O(mn)$ space, to store pointers for each cell in the matrix to the sequence leading to that cell. Hirschberg [22] showed that less storage was required, by giving an ingenious algorithm that computes the edit sequence as well as the alignment cost, in linear space, and remaining within the time bound of $O(mn)$.

Approaches other than dynamic programming are also possible. Masek and Paterson [38] have designed an algorithm that can be thought of as a finite automaton of $O(n)$ states, which finds the edit distance between the pattern and the strings in the database in $O(mn/\log n)$ time. Further speed-ups can be obtained using this automata-theoretic approach [32, 55]; however such improvements seem to require automata with exponentially many states, and an exponentially large preprocessing time to construct the automata; therefore they are only of theoretical interest, and are only applicable when m is much smaller than n.

2.2. Generalized Edit Operations and Gap Costs

Above we discussed sequence alignment with the operations of point substitution, single symbol insertion, and single symbol deletion. A number of further extensions to this set of operations have been considered. If one adds an operation that transposes two adjacent symbols, the problem becomes NP-complete [18, 58]. If one is allowed to take circular permutations of the two input sequences, before performing a sequence of substitutions, insertions, and deletions, the problem can be solved in time $O(nm \log n)$ [29]. We call this latter problem the *circular sequence alignment* problem.

An extension considered by Waterman [59] used dynamic programming to produce all the sequence alignments within a specified distance of the optimum. Waterman's algorithm allows one to determine "true" alignments, whenever these may disagree with the optimum (computed) solution because either the edit operations are not correctly weighted or some other constraints on the sequences were not taken into account.

Another important generalization includes subsequence-subsequence alignments as well as subsequence-sequence alignments [53, 54]. In this case, one is interested in aligning only pieces of the two input sequences, according to some maximization criteria; indeed the most common of such problems consists of finding maximally homologous subsequences. This is very important in molecular sequence analysis, where mutations involve changes in large blocks of a sequence. In this framework, even if it does not seem to be meaningful to look for a global alignment between two sequences, it is nevertheless important to find the pairs of maximally similar subsequences of the input sequences [54]. The mathematical formulation of this problem can be stated as follows. Given two input sequences x and y and a real valued function s expressing the similarity of each alignment, find the local maxima of s for x and y.

The solution proposed by Sellers [53] uses a technique called the intersection method. The basic idea is to solve a dynamic programming recurrence similar to (1) to compute entries in a dynamic programming matrix. Entry (i, j) of the matrix stores the score of the alignment between $x_1, x_2, \ldots, x_i$ and $y_1, y_2, \ldots, y_j$. Starting from the matrix, two graphs are constructed. The vertices of the graphs are entries in the matrix, while edges are defined according to the choice of the minimum in the recurrence. Paths in these graphs correspond to alignments of the two sequences. The first graph G_1 contains all the alignments with nonnegative score which are not intersected from above or the left by an alignment with greater score. Analogously, the second graph G_2 contains alignments with nonnegative score which are not intersected from below or the left by an alignment with greater score. Taking the intersection of these two graphs gives all the local optimal alignments plus some spurious alignments which must be identified and removed. This can be done by finding the strongly connected components of the intersection graph by using well known algorithms [3]. As a result, the total time required to find the local maxima for the alignment of the two strings is $O(mn)$. As a last remark, we notice that the intersection method is a general technique which can be used also with the other algorithms for sequence analysis described in this paper.

A further generalization of the set of operations has been considered, as follows. We call a consecutive set of deleted symbols in one sequence, or inserted symbols in the other sequence, a *gap*. With the operations and costs above, the cost of a gap is the sum of the costs of the individual insertions or deletions which compose it. However, in molecular biology for example, it is much more likely that a gap was generated by one mutation that deleted all the symbols in the gap, than that many individual mutations combined to create the gap. Similar motivations apply to other applications of sequence alignment. Experimental results by Fitch and Smith [15] indicate that the cost of a gap may depend on its endpoints (or location) and on its length. Therefore we would like to allow gap insertions or deletions to combine many individual symbol insertions or deletions, with the cost of a gap insertion or deletion being some function of the length of the gap. The cost $w(i, j)$ of a generic gap $x_i...x_j$ that satisfies such experimental findings must be of the form

$$w(i, j) = f^1(x_i) + f^2(x_j) + g(j - i) \tag{2}$$

where f^1 and f^2 give the cost of breaking the sequence at the endpoints of the gap and g gives a cost that is proportional to the gap length.

Moreover, the most likely choices for g are linear or convex functions of the gap lengths [15, 60]. With such a choice of g, the cost of a long gap will be less than or equal to the sums of the costs of any partition of the gap into smaller gaps, so that as desired the best alignment will treat each gap as a unit. This requirement on the function g is equivalent to saying that the function w satisfies the following inequality:

$$w(i, j') + w(i', j) \leq w(i, j) + w(i', j') \text{ for all } i < i' \leq j < j' \tag{3}$$

When a function w satisfies this inequality we say that it is convex; when it satisfies the inverse of (3) we say that it is concave; and when it satisfies (3) with equality we say that it is linear. As we have said, the cost (or weight) functions that are meaningful for molecular biology are either convex or linear; however it is also natural to consider other classes of gap cost function. We call the sequence alignment problem with gap insertions and deletions the *gap sequence alignment problem*, and similarly we name special cases of this problem by the class of cost functions considered, e.g. the *convex sequence alignment problem*, etc.

To solve the gap sequence alignment problem, the following dynamic programming equation was considered, where w' is a cost function analogous to w which satisfies (2).

$$D[i, j] = \min\{D[i - 1, j - 1] + sub(x_i, y_j), E[i, j], F[i, j]\} \tag{4}$$

where

$$E[i,j] = \min_{0 \le k \le j-1} \{D[i,k] + w(k,j)\} \tag{5}$$

$$F[i,j] = \min_{0 \le l \le i-1} \{D[l,j] + w'(l,i)\} \tag{6}$$

with initial conditions $D[i,0] = w'(0,i)$, $1 \le i \le m$ and $D[0,j] = w(0,j)$, $1 \le j \le n$.

The original sequence alignment problem treats the cost of a gap as the sum of the costs of the individual symbol insertions or deletions of which it is composed. Therefore if the gap cost function is some constant times the length of the gap, the gap sequence alignment problem can be solved in time $O(mn)$. This can be generalized to a slightly wider class of functions, the *linear* or *affine* gap cost functions. For these functions, the cost $g(x)$ of a gap of length x is $k_1 + k_2 x$ for some constants k_1 and k_2. A simple modification of the solution to the original sequence alignment problem also solves the linear sequence alignment problem in time $O(mn)$ [19].

For general functions, the gap sequence alignment problem can be solved in time $O(mn^2)$, by a simple dynamic programming algorithm [51]. The algorithm is similar to that for the original sequence alignment problem, but the computation of each entry in the dynamic programming matrix depends on all the previous entries in the same row or column, rather than simply on the adjacent entries in the matrix. This method was discovered by Waterman et al. [66], based on earlier work by Sellers [52]. But this time bound is an order of magnitude more than that for non-gap sequence alignment, and thus is useful only for much shorter sequences.

An early attempt to obtain an efficient algorithm for the problem of sequence alignment with convex gaps costs was made by Waterman [60]. As later shown by Miller and Myers [40], the algorithm in [60] still has an $O(mn^2)$ time behavior. However, the merit of [60] is that it proposes the use of convex gap cost functions for molecular biology. Later on, Galil and Giancarlo [16] considered the gap sequence alignment problem for both convex and concave cost functions. They reduced the problem to $O(n)$ subproblems, each of which can be expressed as a dynamic program generalizing the *least weight subsequence* problem, which had previously been applied to text formatting [24, 31] and optimal layout of B-trees [24]. Galil and Giancarlo solved this subproblem in time $O(n \log n)$, or linear time for many simple convex and concave functions such as $\log x$ and x^2. As a result they solved both the convex and concave sequence alignment problems in time $O(mn \log n)$, or $O(mn)$ for many simple functions. The algorithm by Galil and Giancarlo is very simple and thus likely to be useful also in practice. We point out that Miller and Myers [40] independently solved the same problem in similar time bounds.

Wilber [67] pointed out a resemblance between the least weight subsequence problem and a matrix searching technique that had been previously used to solve a number of problems in computational geometry [1]. He used this technique in an algorithm for solving the least weight subsequence problem in linear time. His algorithm also extends to the generalization of the least weight subsequence problem used as a subproblem by Galil and Giancarlo in their solution of the concave sequence alignment problem; however because Galil and Giancarlo use many interacting instances of the subproblem, Wilber's analysis breaks down and his algorithm can not be used for concave sequence alignment. Klawe and Kleitman [30] studied a similar problem for the convex case and they obtained an $O(n\alpha(n))$ algorithm for it, where $\alpha(n)$ denotes a very slowly growing function, namely the inverse of Ackermann's function. The method devised by Klawe and Kleitman yields an $O(mn\alpha(n))$ algorithm for the gap sequence alignment problem with convex costs. Eppstein [9] showed that it is possible to modify Wilber's algorithm so that it can be used to solve the gap sequence alignment with concave costs in $O(mn)$ time. He used this result to solve the gap sequence alignment problem for functions which are neither convex nor concave, but a mixture of both. More precisely, if the cost function can be split into s convex and concave pieces, then the gap sequence alignment can be solved in $O(mns\alpha(n/s))$ time. This time is never worse than the time of $O(mn^2)$ for the naive dynamic programming solution [51] known for the general case. When s is small, the bound will be much better than the bound obtained by the naive solution. All these algorithms are based upon matrix searching and so the constant factors in the time bounds are quite large. Therefore, the algorithms are mainly of theoretical interest.

2.3. Sparse Sequence Alignment

All of the alignment algorithms above take a time which is at least the product of the lengths of the two input sequences. This is not a serious problem when the sequences are relatively short, but the sequences used in molecular biology can be very long, and for such sequences these algorithms can take more computing time than what is available for their computation.

Wilbur and Lipman [68, 69] proposed a method for speeding up these computations, at the cost of a small loss of accuracy, by only considering matchings between certain subsequences of the two input sequences. Let $F = \{f_1, f_2, \ldots, f_b\}$ be a given set of *fragments*, each fragment being a string over an alphabet A of size s. Let the two input sequences be $x = x_1 x_2 \ldots x_m$ and $y = y_1 y_2 \ldots y_n$. We are interested in finding an optimal alignment of x and y using only fragments from F that occur in both strings. A fragment $f = (i, j, k)$ of length k *occurs* in x and y if $f = x_i x_{i+1} \ldots x_{i+k-1} = y_j y_{j+1} \ldots y_{j+k-1}$. Given F, x and y, we can find all occurrences of fragments from F in x and y in time $O(n + m + M)$, where M denotes the number of such occurrences, by using standard string matching techniques.

An occurrence of a fragment (i', j', k') is said to be *below* an occurrence of (i, j, k) if $i + k \leq i'$ and $j + k \leq j'$; i.e. the substrings in fragment (i', j', k') appear strictly after those of (i, j, k) in the input strings. Equivalently, we say that (i, j, k) is *above* (i', j', k'). The *length* of fragment (i, j, k) is the number k. The *diagonal* of a fragment (i, j, k) is the number $j - i$. An alignment of fragments is defined to be a sequence of fragments such that, if (i, j, k) and (i', j', k') are adjacent fragments in the sequence, either (i', j', k') is below (i, j, k) on a different diagonal (a *gap*), or the two fragments are on the same diagonal with $i' > i$ (a *mismatch*). The cost of an alignment is taken to be the sum of the costs of the gaps, minus the number of matched symbols in the fragments. The number of matched symbols may not necessarily be the sum of the fragment lengths, because two mismatched fragments may overlap. Nevertheless it is easily computed as the sum of fragment lengths minus the overlap lengths of mismatched fragment pairs. The cost of a gap is some function of the distance between diagonals $w(|(j - i) - (j' - i')|)$.

When the fragments are all of length 1, and are taken to be all pairs of matching symbols from the two strings, these definitions coincide with the usual definitions of sequence alignments. When the fragments are fewer, and with longer lengths, the fragment alignment will typically approximate fairly closely the usual sequence alignments, but the cost of computing such an alignment may be much less.

231

The method given by Wilbur and Lipman [69] for computing the least cost alignment of a set of fragments is as follows. Given two fragments, at most one will be able to appear after the other in any alignment, and this relation of possible dependence is transitive; therefore it is a partial order. Fragments are processed according to any topological sorting of this order. Some such orders are by rows (i), columns (j), or back diagonals $(i+j)$. For each fragment, the best alignment ending at that fragment is taken as the minimum, over each previous fragment, of the cost for the best alignment up to that previous fragment together with the gap or mismatch cost from that previous fragment. The mismatch cost is simply the length of the overlap between two mismatched fragments; if the fragment whose alignment is being computed is (i, j, k) and the previous fragment is $(i - l, j - l, k')$ then this length can be computed as $\max(0, k' - l)$. From this minimum cost we also subtract the length of the new fragment; thus the total cost includes a term linear in the total number of symbols aligned. Formally, we have

$$D(i, j, k) = -k + \min \left\{ \begin{array}{l} \min_{(i-l, j-l, k')} D[i - l, j - l, k'] - \max(0, k' - l) \\ \min_{(i', j', k') \text{ above } (i, j, k)} D[i', j', k'] + w(|(j - i) - (j' - i')|) \end{array} \right. \tag{7}$$

The naive dynamic programming algorithm for this computation, given by Wilbur and Lipman, takes time $O(M^2)$. If M is sufficiently small, this will be faster than many other sequence alignment techniques. We remark that the Wilbur-Lipman algorithm works for general cost functions w and does not take any advantage of the fact that w's used in practice satisfy inequality 3. We can show that by using this restriction on w we can compute recurrence 7 in time close to linear in M [11, 12]. This has the effect of making such computations even more practical for small M, and it also allows more exact computations to be made by allowing M to be larger.

The algorithms that we propose are quite different from the one given by Wilbur and Lipman and have the following common background. We consider recurrence 7 as a dynamic program on points in a two-dimensional matrix. Each fragment (i, j, k) gives rise to two points, (i, j) and $(i + k - 1, j + k - 1)$. We compute the best alignment for the fragment at point (i, j); however we do not add this alignment to the data structure of already computed fragments until we reach $(i+k-1, j+k-1)$. In this way, the computation for each fragment will only see other fragments that it is below. We compute separately the best mismatch for each fragment; this is always the previous fragment from the same diagonal, and so this computation can easily be performed in linear time. From now on we will ignore the distinction between the two kinds of points in the matrix, and the complication of the mismatch computation. Thus, we consider the following subproblem: Compute

$$E[i, j] = \min_{(i', j') \text{ above } (i, j)} D[i', j'] + w(|(j - i) - (j' - i')|), \tag{8}$$

where $D[i, j]$ is easily computable from $E[i, j]$.

We define the *range* of a point in which we have to compute recurrence 8 as the set of points below and to the right of it. Furthermore, we divide the range of a point into two portions, the *left influence* and the *right influence*. The left influence of (i, j) consists of those points in the range of (i, j) which are below and to the left of the forward diagonal $j - i$, and the right influence consists of the points above and to the right of the forward diagonal. Within each of the two influences, $w(|p - q|) = w(p - q)$ or $w(|p - q|) = w(q - p)$; i.e. the division of the range in two parts removes the complication of the absolute value from the cost function. Thus, we can now write recurrence 8 as:

$$E[i,j] = \min\{LI[i,j], RI[i,j]\}, \tag{9}$$

where

$$RI[i,j] = \min_{\substack{(i',i') \text{ above } (i,j) \\ j' - i' < j - i}} D(i',j') + w((j-i) - (j'-i')) \tag{10}$$

and

$$LI[i,j] = \min_{\substack{(i',i') \text{ above } (i,j) \\ j - i < j' - i'}} D(i',j') + w((j'-i') - (j-i)). \tag{11}$$

We observe that the order of computation of the points in the matrix must be the same for the two recurrences so that they can be put together into a single algorithm for the computation of recurrence 8.

Assuming that the cost function w is convex, we can compute recurrences 10 and 11 in time $O(n + m + M \log M \alpha(M))$. This time bound reduces to $O(n + m + M \log M)$ when w is concave. Our algorithm uses in a novel way an algorithmic technique devised by Bentley and Saxe [6], namely dynamic to static reduction. Matrix searching [1, 30] is also used. We remark that if the cost function is convex/concave and simple, matrix searching can be replaced by the algorithm of Galil and Giancarlo [16] to obtain an $O(n + m + M \log M)$ algorithm both for simple convex and concave cost functions. In the case that the cost function is not simple, we can still use the algorithm of Galil and Giancarlo instead of matrix searching. This results in a slow-down of our algorithm by a factor of $\log M$, but gives an algorithm which is likely to be more practical. The reader is referred to [12] for further details on the algorithm.

When the function w is linear, we can compute recurrences 10 and 11 in time $O(n + m + M \log \log \min(M, nm/M))$. This algorithm is based on the use of efficient data structures for the management of priority queues with integer keys [27]. As a by-product, we also obtain an improved implementation of the algorithm for the longest common subsequence devised by Apostolico and Guerra [5]. Our implementation runs in time $O(n \log s + d \log \log \min(d, nm/d))$. Here s is the minimum between m and the cardinality of the alphabet and d denotes the number of dominating matches defined in [23].

3. ALGORITHMS FOR COMPUTATION OF RNA SECONDARY STRUCTURE

In this section we are interested in algorithms for the computation of RNA secondary structure. More specifically, we will consider algorithms for loop dependent energy rules [70]. In order to make the presentation of our algorithms self contained, we briefly review the biological background common to all of them.

RNA molecules are among the primary constituents of living matter. RNA is used by cells to transport genetic information between the DNA repository in the nucleus of the cell and the

233

ribosomes which construct proteins from that information. It is also used within the process of protein construction, and may also have other important functions. An RNA molecule is a polymer of nucleic acids, each of which may be any of four possible choices: adenine, cytosine, guanine, and uracil. Thus an RNA molecule can be represented as a string over an alphabet of four symbols, corresponding to the four possible nucleic acid bases. In practice the alphabet may need to be somewhat larger, because of the sporadic appearance of certain other bases in the RNA sequence. This string or sequence information is known as the *primary structure* of the RNA. The primary structure of an RNA molecule can be determined by gene sequencing experiments. Throughout this section we denote an RNA molecule by the string $y = y_1 y_2, ..., y_n$ and we refer to its i-th base y_i.

In an actual RNA molecule, hydrogen bonding will cause further linkages to form between pairs of bases. Adenine typically pairs with uracil, and cytosine with guanine. Other pairings, in particular between guanine and uracil, may form, but they are much more rare. Each base in the RNA sequence will pair with at most one other base. Paired bases may come from positions of the RNA molecule that are far apart in the primary structure. The set of linkages between bases for a given RNA molecule is known as its *secondary structure*.

The *tertiary structure* of an RNA molecule consists of the relative physical locations in space of each of its constituent atoms, and thus also the overall shape of the molecule. The tertiary structure is determined by energetic (static) considerations involving the bonds between atoms and the angles between bonds, as well as kinematic (dynamic) considerations involving the thermal motion of atoms. Thus the tertiary structure may change over time; however for a given RNA molecule there will typically be a single structure that closely approximates the tertiary structure throughout its changes. The tertiary structure determines how the molecule will react with other molecules in its environment, and how in turn other molecules will react with it. Thus the tertiary structure controls enzymatic activity of RNA molecules as well as the splicing operations that take place between the time RNA is copied from the parent DNA molecule and the time that it is used as a blueprint for the construction of proteins.

Because of the importance of tertiary structure, and its close relation to molecular function, molecular biologists would like to be able to determine the tertiary structure of a given RNA molecule. Tertiary structures can be determined experimentally, but this requires complex crystalization and X-ray crystallography experiments, which are much more difficult than simply determining the sequence information of an RNA molecule. Further, the only known computational techniques for determining tertiary structure from primary structure involve simulations of molecular dynamics, which require enormous amounts of computing power and therefore can only be applied to very short sequences [44].

Because of the difficulty in computing tertiary structures, some biologists have resorted to the simpler computation of secondary structure, which also gives some information about the physical shape of the RNA molecule. Secondary structure computations also have their own applications: by comparing the secondary structures of two molecules with similar function one can determine how the function depends on the structure. In turn, a known or conjectured similarity in the secondary structures of two sequences can lead to more accurate computation of the structures themselves, of possible alignments between the sequences, and also of alignments between the structures of the sequences [49].

234

3.1. Secondary Structure Assumptions and the Structure Tree

A perfectly accurate computation of RNA structure would have to include as well a computation of tertiary structure, because the secondary structure is determined by the tertiary structure. As we have said this seems to be a hard problem. Instead, a number of assumptions have been made about the nature of the structure. An energy is assigned to each possible configuration allowed by the assumptions, and the predicted secondary structure is the one having the minimum energy.

The possible base pairs in the structure are usually taken to be simply those allowed by the possible hydrogen bonds among the four RNA bases; that is, a base pair is a pair of positions (i, j) where the bases at the positions are adenine and uracil, cytosine and guanine, or possibly guanine and uracil. We write the bases in order by their positions in the RNA sequence; i.e. if (i, j) is a possible base pair, then $i < j$. Each pair has a binding energy determined by the bases making up the pair.

Define the *loop* of a base pair (i, j) to be the set of bases in the sequence between i and j. The primary assumption of RNA secondary structure computation is that no two loops cross. In other words, if (i, j) and (i', j') are base pairs formed in the secondary structure, and some base k is contained in both loops, then either i' and j' are also contained in loop (i, j), or alternately i and j are both contained in loop (i', j'). This assumption is not entirely correct for all RNA [37], but it works well for a great majority of the RNA molecules found in nature.

A base at position k is *exposed* in loop (i, j) if k is in the loop, and k is not in any loop (i', j') with i' and j' also in loop (i, j). Because of the non-crossing assumption, each base can be exposed in at most one loop. We say that (i', j') is a *subloop* of (i, j) if both i' and j' are exposed in (i, j); if either i' or j' is exposed then by the non-crossing assumption both must be.

Therefore the set of base pairs in a secondary structure, together with the subloop relation, forms a forest of trees. Each root of the tree is a loop that is not a subloop of any other loop, and each interior node of the tree is a loop that has some other subloop within it. We further define a *hairpin* to be a loop with no subloops, that is, a leaf in the loop forest, and we define a *single loop* or *interior loop* to be a loop with exactly one subloop. Any other loop is called a *multiple loop*. A base pair (i, j) such that the two adjacent bases $(i + 1, j - 1)$ are also paired is called a *stacked pair*. A single loop such that one base of the subloop is adjacent to a base of the outer loop is called a *bulge*.

As we have said, each base pair in an RNA secondary structure has a binding energy which is a function of the bases in the pair. We also include in the total energy of the secondary structure a *loop cost*, which is usually assumed to be a function of the length of the loop. This length is simply the number of exposed bases in the loop. The loop cost may also depend on the type of the loop; in particular it may differ for hairpins, stacked pairs, bulges, single loops, and multiple loops. The loop costs in use today for hairpins, bulges and single loops are logarithms [70]. Therefore they are convex functions according to the definition given in the previous section. Moreover, they are simple convex functions. With these definitions one can easily compute the total energy of a structure, as the sum of the base pair binding energies and loop costs. The optimal RNA secondary structure is then that structure minimizing the total energy.

3.2. Computation of Secondary Structure

With the definitions above, the optimum secondary structure can be computed by a three-dimensional dynamic program with matrix entries for each triple (i, j, k), where i and j are positions in the RNA sequence (not necessarily forming a base pair) and k is the number of exposed bases in a possible loop containing i and j. This computation takes time $O(n^4)$. The algorithm was discovered by Waterman and Smith [65] and it is the first polynomial time algorithm obtained for this problem. Clearly, this time bound is so large that the computation of RNA structure using this algorithm is feasible only for very short sequences. Furthermore, the space bound of $O(n^3)$ also makes this algorithm impractical. Therefore, one needs further assumptions about the possible structures, or about the energy functions determining the optimum structure, in order to perform secondary structure computation more efficiently.

A particularly simple assumption is that the energy cost of a loop is zero or a constant, so that one need only consider the energy contribution of the base pairs in the structure. Nussinov et al. [42] showed how to compute a structure maximizing the total number of base pairs, in time $O(n^3)$; this algorithm was later extended to allow arbitrary binding energies for base pairs, while keeping the same time bound [43]. A less restrictive assumption, although not realistic, is that the cost of a multiple loop is a linear function of its length, rather than being a convex function of the base pairs in the loop and the length of the loop. Kruskal et al. [50] used such an assumption to derive a set of dynamic programming equations that yield the minimum energy RNA secondary structure in time $O(n^3)$. The $O(n^3)$ time bound is mainly due to the computation of internal loops and the computation of multiple loops. Indeed, each such computation takes $O(n)$ for each entry (i, j) of the dynamic programming matrix. We point out that the algorithm by Kruskal et al. is a variation of an earlier algorithm obtained by Zuker and Stiegler [71].

Instead of restricting the possible loop cost functions, one could restrict the possible types of loops. In particular, an important special case of RNA secondary structure computation is the computation of the best structure with no multiple loops. Such structures can be useful for the same applications as the more general RNA structure computation. Single loop RNA structures could be used to construct a small number of pieces of a structure which could then be combined to find a structure having multiple loops; in this case one sacrifices optimality of the resulting multiple loop structure for efficiency of the structure computation.

The single loop secondary structure computation can again be expressed as a dynamic programming recurrence relation [50, 63]. Again this relation seems to require time $O(n^4)$, but the space requirement is reduced from $O(n^3)$ to $O(n^2)$. In fact the time for solving the recurrence can also be reduced, to $O(n^3)$, as was shown by Waterman and Smith [65]. In this paper, the authors also conjectured that the given algorithm runs in $O(n^2)$ time for convex (and concave) functions. Eppstein et al. [10] have shown how to compute single loop RNA secondary structure, for convex or concave energy costs, in time $O(n^2 \log^2 n)$. For many simple cost functions, such as logarithms and square roots, they show how to improve this time bound to $O(n^2 \log n \log \log n)$. The algorithm obtained by [10] is based on a new and fast method for the computation of internal loops for convex or concave energy costs. These results have recently been improved by Aggarwal and Park [2], who gave an $O(n^2 \log n)$ algorithm, and further by Larmore and Schieber [33], who gave an $O(n^2)$ algorithm for the concave case and an $O(n^2 \alpha(n))$ algorithm for the convex case.

However, all these algorithms use matrix searching techniques, which lead to a high constant factor in the time bound.

We remark that the algorithms in [2, 10, 33] can also be used as a subroutine in the algorithm devised by Kruskal et al. [50]. Namely, one can compute the interior loops by using the algorithm given in [10] or [2] or [33] rather than the naive algorithm given in [50] for the same problem. Although such a modification does not yield any asymptotic speed up in the algorithm by Kruskal et al., it achieves a practical speed up since it reduces the computation time for internal loops. Similar considerations apply to the algorithm devised by Zuker and Stiegler.

Eppstein [9] has extended Aggarwal and Park's algorithm for single loop RNA structure to handle the case that the energy cost of a loop is not a convex or a concave function of the length, but can be split into a small number s of convex and concave pieces. His algorithm takes time $O(n^2 s \log n \alpha(n/s))$, or $O(n^2 s \log n \log(n/s))$ if matrix searching techniques are avoided. When s is small, these times will be much better than the $O(n^3)$ time known for general functions [65].

3.3. Sparseness in Secondary Structure

The recurrence relations that have been defined for the computation of RNA structure are all indexed by pairs of positions in the RNA sequence (and possibly also by numbers of exposed bases). For many of these recurrences, the entries in the associated dynamic programming matrix include a term for the binding energy of the corresponding base pair. If the given pair of positions do not form a base pair, this term is undefined, and the value of the cell in the matrix must be taken to be $+\infty$ so that the minimum energies computed for the other cells of the matrix do not depend on that value, and so that in turn no computed secondary structure includes a forbidden base pair.

Further, for the energy functions that are typically used, the energy cost of a loop will be more than the energy benefit of a base pair, so base pairs will not have sufficiently negative energy to form unless they are stacked without gaps at a height of three or more. Thus we could ignore base pairs that can not be so stacked, or equivalently assume that their binding energy is again $+\infty$, without changing the optimum secondary structure. This observation is similar to that of sparse sequence alignment, in which we only include pairs of matching symbols when they are part of a longer substring match.

The effect of such constraints on the computation of the secondary structure for RNA is twofold. First, they contribute to make the output of the algorithms using them more realistic from the biological point of view [70]. Second, they combine to greatly reduce the number of possible pairs, which we denote by M, that must be considered to a value much less than the upper bound of n^2. For instance, if we required base pairs to form even higher stacks, M would be further reduced. The computation and minimization in this case is taken only over positions (i, j) which can combine to form a base pair.

The algorithms listed earlier account for the constraints just mentioned by giving a value of $+\infty$ at positions (i, j) that cannot form a base pair. Then this value can never supply the minimum energy in future computations, so (i, j) will never be used as a base pair in the computed RNA structure. Nevertheless, the time complexities of those algorithms depend on the total number of possible pairs (i, j), including those that are disallowed from pairing. Fortunately, the algorithms

237

can be modified to ignore altogether such pairs. The net effect of such a modification is to replace n^2 by M in the time bounds for the algorithms, where by M we denote the number of base pairs that are allowed to form. That is, the time bounds for the algorithms in [50, 71] and [65] (for the single loop structure) becomes $O(Mn)$ whereas the time bound of the algorithm in [65] (for the general case) becomes $O(Mn^2)$. Besides the constraints given by the problem itself, algorithms actually used in practice (as the one by Zuker and Stiegler) also incorporate heuristics in order to reduce even further the computational effort. One particular heuristic for the computation of interior loops is reported in [70]. Again, such heuristics boil down to reducing the number of entries (i, j) one has to consider in order to compute the secondary structure.

Based on the preceding discussion, we model the computation of a dynamic programming matrix D yielding an RNA secondary structure as follows. We are given one or more recurrences stating how to compute D and we are also given a set S of M entries (i, j) on which we have to compute D. We assume that all entries not included in S do not matter for the final result. We notice that S may be obtained by imposing the physical constraints of the problem and/or by imposing heuristic consideration on the entries of D. We next show how to take advantage of the sparsity of D in the computation of single loop RNA structure. Waterman and Smith [63] obtained the following dynamic programming equation:

$$D[i, j] = \min\{D[i-1, j-1] + b(i, j), H[i, j], V[i, j], E[i, j]\} \qquad (12)$$

where

$$V[i, j] = \min_{0 < k < i} D[k, j-1] + w'(k, i) \qquad (13)$$

$$H[i, j] = \min_{0 < l < j} D[i-1, l] + w'(l, j) \qquad (14)$$

$$E[i, j] = \min_{\substack{0 < k < i-1 \\ 0 < l < j-1}} D[k, l] + w(k + l, i + j). \qquad (15)$$

The function w corresponds to the energy cost of an internal loop between the two base pairs, and w' corresponds to the cost of a bulge. Both w and w' typically combine terms for the loop length and for the binding energy of bases i and j. Experimental results [70] show that both w and w' are convex function, i.e. they satisfy equation 3. The function $b(i, j)$ contains only the base pair binding energy term, and corresponds to the energy gain of a stacked pair. As we have said before, recurrence 12 must be computed on all entries (i, j) in a given set S. Clearly, V, H and E must be computed on the same set of points.

First note that the computation of $V[i, j]$ within a fixed column j does not depend on that of other columns, except indirectly via the values of $D[i, j]$. We may perform this computation using the algorithm of Galil and Giancarlo [16]. If the number of points i in column j such that $(i, j) \in S$ is denoted by p_j, then the time for computing all values of $V[i, j]$ for a fixed j will be $O((p_j + p_{j-1}) \log M)$. The total time for these computations in all columns will then be $O(M \log M)$. We could achieve even better bounds using the more complicated algorithms by Klawe and Kleitman [30] or Eppstein [9] but this would not affect our total time bounds.

The computation of $H[i, j]$ is similar. Therefore the remaining difficulty is the computation of $E[i, j]$, as defined by 15. For this problem, each point in S may be considered as having a *range of influence* consisting of the region of the dynamic programming matrix E (and D) below and to the right of it. Thus, the range of each point is a quarter-plane with vertical and horizontal boundaries. For any given point (i, j), there is a point (i', j') containing (i, j) in its range and such that (i', j') provides the minimum in recurrence 15 for (i, j). Obviously, (i, j) can be contained in the range of influence of many points. Thus, when several points have intersecting ranges, we must compute which of them supply the minima in the intersection. The methods we use to perform this computation include matrix searching [1] together with binary search and divide and conquer. The reader is referred to [12] for a complete account of how such techniques yield an $O(n + M \log M \log \min(M, n^2/M))$ time algorithm for the computation of recurrence 12. When the cost function is simple, in addition to being convex, the binary search can be eliminated, and the time bound reduces to $O(n + M \log M \log \log \min(M, n^2/M))$. Since the function w typically used for the free energy of an internal loop is a logarithm and thus a simple convex function, we can solve the RNA secondary structure problem in $O(n + M \log M \log \log \min(M, n^2/M))$ time. We remark that, as in the case of the algorithm by [10], our algorithm can be used to speed up the algorithms by Kruskal et al. [50] and Zuker and Stiegler [71].

For concave cost functions w and w', we can compute recurrence 12 in the same time bounds as for convex cost function by using the same algorithm. When the cost functions are linear, we have a different algorithm that computes recurrence 12 in time $O(n + M \log \log \min(M, n^2/M))$ [11]. The concave result is of some merit from the combinatorial point of view but gives no contribution to computational biology since concave cost functions are not meaningful for the computation of *RNA* secondary structure. The assumption of linearity is also not as realistic as that of convexity, but it has been used in practice because of the increased efficiency of the corresponding algorithms (see for instance [28]). As can be seen from the bounds above, our new algorithms are again somewhat more efficient in the linear case than in the convex case.

We end this section by mentioning that very recently Larmore and Schieber [33] showed how to reduce the $O(n + M \log M \log \min(M, n^2/M))$ bound to $O(n + M \log \min(M, n^2/M))$ for the concave case and to $O(n + M\alpha(\min(M, n)) \log \min(M, n^2/M))$ for the convex case.

4. CONCLUSIONS AND OPEN PROBLEMS

We have considered two problems which benefit from mathematical methods applied to molecular biology: sequence alignment and computation of single loop RNA secondary structure. The common unifying framework for these two problems is that they both can be solved by computing a set of dynamic programming equations and that these equations need not be computed for all points in their domain of definition. Moreover, the cost functions that are typically used satisfy convexity or concavity constraints. We have shown that it is possible to obtain asymptotically fast algorithms for both problem. Our algorithms are robust in the sense that they do not depend on any heuristic knowledge used to make the domain of the dynamic programming equations sparse. The crucial idea behind the new algorithms is to consider the computation of the given dynamic programming equations as a geometric problem that consists of identifying and maintaining a map of regions in the dynamic programming matrix. The algorithmic techniques that we use are a sophisticated upgrade of basic computer science tools such as divide-and-conquer and efficient data structures for answering queries.

239

Our results represent a contribution to the general problem of finding efficient algorithms for computationally intensive tasks posed by molecular biology and many interesting and hard problems remain in this area. The reader can refer to [8] for a lucid presentation of the computational needs that future advances in molecular biology pose. Here we limit ourselves to mentioning a few open problems that are tightly related to the topic of this paper.

- Can the $O(nm\alpha(n))$ bound of the convex sequence alignment problem be improved? Can a practical algorithm achieve this goal?

- Can the space for convex or concave sequence alignment be reduced, similarly to Hirschberg's reduction for linear sequence alignment? Galil and Rabani [17] have shown that the current algorithms require space $O(nm)$, even using Hirschberg's technique, and so new methods would be needed.

- Can the bounds for the fragment alignment problem be reduced? In particular can we achieve the optimal time bounds of $O(M + n)$ for linear and/or convex (concave) cost functions?

- Can the times for the other RNA secondary structure computations be reduced? Can convexity or concavity of loop energy cost functions be used to speed up the computation of multiple-loop RNA secondary structure?

- Can the space of our algorithms, and the other algorithms for single loop RNA structure, be reduced below the current $O(n^2)$ bound? Can the space for efficient computation of multiple loop RNA structure with general (or convex or concave) cost functions be reduced below $O(n^3)$?

- Is dynamic programming strictly necessary to solve sequence alignment problems? Notice that algorithms based on dynamic programming will take at least $O(mn)$ time in aligning two sequences of length m and n.

Acknowledgments

We would like to thank Stuart Haber, Hugo Martinez, Peter Sellers, Temple Smith, and Michael Waterman for useful comments.

References

[1] A. Aggarwal, M. M. Klawe, S. Moran, P. Shor, and R. Wilber, Geometric Applications of a Matrix-Searching Algorithm, Algorithmica 2, 1987, pp. 209–233.

[2] A. Aggarwal and J. Park, Searching in Multidimensional Monotone Matrices, 29th IEEE Symp. Found. Comput. Sci., 1988, pp. 497–512.

[3] A. V. Aho, J. E. Hopcroft, and J. D. Ullman, The Design and Analysis of Computer Algorithms, Addison-Wesley, 1974.

[4] A. V. Aho, J. E. Hopcroft, and J. D. Ullman, Data Structures and Algorithms, Addison-Wesley, 1983.

[5] A. Apostolico and C. Guerra, The Longest Common Subsequence Problem Revisited, Algorithmica 2, 1987, pp. 315–336.

[6] J. L. Bentley and J. B. Saxe, Decomposable Searching Problems I: Static-to-Dynamic Transformation. J. Algorithms 1(4), December 1980, pp. 301–358.

[7] H. S. Bilofsky, C. Burks, J. W. Fickett, W. B. Goad, F. I. Lewitter, W. P. Rindone, C. D. Swindel, and C. S. Tung, The GenBank Genetic Sequence Databank, Nucl. Acids Res. 14, 1986, pp. 1–4.

[8] C. DeLisi, Computers in Molecular Biology: Current Applications and Emerging Trends, Science, 240, 1988, pp. 47–52.

[9] D. Eppstein, Sequence Comparison with Mixed Convex and Concave Costs, J. of Algorithms, 11, 1990, pp. 85–101.

[10] D. Eppstein, Z. Galil, and R. Giancarlo, Speeding Up Dynamic Programming, 29th IEEE Symp. Found. Comput. Sci., 1988, pp. 488–496.

[11] D. Eppstein, Z. Galil, R. Giancarlo, and G. F. Italiano, Sparse Dynamic Programming I: Linear Cost Functions, J. ACM, to appear.

[12] D. Eppstein, Z. Galil, R. Giancarlo, and G. F. Italiano, Sparse Dynamic Programming II: Convex and Concave Cost Functions, J. ACM, to appear.

[13] M. J. Fischer and R. Wagner, The String to String Correction Problem, J. ACM 21, 1974, pp. 168–178.

[14] W. M. Fitch, Weighted Parsimony, Workshop on Algorithms for Molecular Genetics, Washington D.C., 1988.

[15] W. M. Fitch and T. F. Smith, Optimal Sequence Alignment, Proc. Nat. Acad. Sci. USA 80, 1983, pp. 1382–1385.

[16] Z. Galil and R. Giancarlo, Speeding Up Dynamic Programming with Applications to Molecular Biology, Theor. Comput. Sci., 64, 1989, pp. 107–118.

[17] Z. Galil and Y. Rabani, On the Space Requirement for Computing Edit Distances with Convex or Concave Gap Costs, Theor. Comp. Sci., to appear.

[18] M. R. Garey and D. S. Johnson, Computers and Intractability: A Guide to the Theory of NP-Completeness, W.H. Freeman, 1979.

[19] O. Gotoh, An Improved Algorithm for Matching Biological Sequences, J. Mol. Biol. 162, 1982, pp. 705–708.

[20] G. H. Hamm and G. N. Cameron, The EMBL Data Library, Nucl. Acids Res. 14, 1986, pp. 5–9.

[21] J. P. Haton, Practical Application of a Real-Time Isolated-Word Recognition System using Syntactic Constraints, IEEE Trans. Acoustics, Speech and Signal Proc. ASSP-22(6), 1974, pp. 416–419.

[22] D. S. Hirschberg, A Linear Space Algorithm for Computing Maximal Common Subsequences, Comm. ACM 18, 1975, pp. 341–343.

[23] D. S. Hirschberg, Algorithms for the Longest Common Subsequence Problem, J. ACM 24, 1977, pp. 664–675.

[24] D. S. Hirschberg and L. L. Larmore, The Least Weight Subsequence Problem, 26th IEEE Symp. Found. Comput. Sci., 1985, 137–143, and SIAM J. Comput. 16, 1987, pp. 628–638.

241

[25] M. K. Hobish, The Role of the Computer in Estimates of DNA Nucleotide Sequence Divergence, in S. K. Dutta, ed., DNA Systematics, Volume I: Evolution, CRC Press, 1986.

[26] J. W. Hunt and T. G. Szymanski, A Fast Algorithm for Computing Longest Common Subsequences, C. ACM 20(5), 1977, pp. 350–353.

[27] D. B. Johnson, A Priority Queue in Which Initialization and Queue Operations Take $O(\log\log D)$ Time, Math. Sys. Th. 15, 1982, pp. 295–309.

[28] M. I. Kanehisi and W. B. Goad, Pattern Recognition in Nucleic Acid Sequences II: An Efficient Method for Finding Locally Stable Secondary Structures, Nucl. Acids Res. 10(1), 1982, pp. 265–277.

[29] Z. M. Kedem and H. Fuchs, On Finding Several Shortest Paths in Certain Graphs, 18th Allerton Conf., 1980, pp. 677–686.

[30] M. M. Klawe and D. Kleitman, An Almost Linear Algorithm for Generalized Matrix Searching, Tech. Rep. IBM Almaden Research Center, 1988.

[31] D. E. Knuth and M. F. Plass, Breaking Paragraphs into Lines, Software Practice and Experience 11, 1981, pp. 1119–1184.

[32] A. G. Ivanov, Distinguishing an approximate word's inclusion on Turing machine in real time, Izv. Acad. Nauk USSR Ser. Mat. 48, 1984, pp. 520–568.

[33] L. L. Larmore and B. Schieber, On-Line Dynamic Programming with Applications to the Prediction of RNA Secondary Structure, J. Algorithms, to appear.

[34] V. I. Levenshtein, Binary Codes Capable of Correcting Deletions, Insertions and Reversals, Sov. Phys. Dokl. 10, 1966, pp. 707–710.

[35] D. Maier, The Complexity of Some Problems on Subsequences and Supersequences, J. ACM 25, 1978, pp. 322–336.

[36] T. Maniatis, Recombinant DNA, in D.M. Prescott, ed., Cell Biology, Academic Press, New York, 1980.

[37] H. Martinez, Extending RNA Secondary Structure Predictions to Include Pseudoknots, Workshop on Algorithms for Molecular Genetics, Washington D.C., 1988.

[38] W. J. Masek and M. S. Paterson, A Faster Algorithm Computing String Edit Distances, J. Comp. Sys. Sci. 20, 1980, pp. 18–31.

[39] A. M. Maxam and W. Gilbert, Sequencing End-Labeled DNA with Base Specific Chemical Cleavages, Meth. Enzymol. 65, 1980, p. 499.

[40] W. Miller and E. W. Myers, Sequence Comparison with Concave Weighting Functions, Bull. Math. Biol., 50(2), 1988, pp. 97–120.

[41] S. B. Needleman and C. D. Wunsch, A General Method applicable to the Search for Similarities in the Amino Acid Sequence of Two Proteins, J. Mol. Biol. 48, 1970, p. 443.

[42] R. Nussinov, G. Pieczenik, J. R. Griggs, and D. J. Kleitman, Algorithms for Loop Matchings, SIAM J. Appl. Math. 35(1), 1978, pp. 68–82.

[43] R. Nussinov and A. Jacobson, Fast Algorithm for Predicting the Secondary Structure of Single-Stranded RNA, Proc. Nat. Acad. Sci. USA 77, 1980, pp. 6309–6313.

[44] G. N. Reeke, Protein Folding: Computational Approaches to an Exponential-Time Problem, Ann. Rev. Comput. Sci. 3, 1988, pp. 59–84.

[45] T. A. Reichert, D. N. Cohen, and A. K. C. Wong, An Application of Information Theory to Genetic Mutations and the Matching of Polypeptide Sequences, J. Theor. Biol. 42, 1973, pp. 245–261.

[46] H. Sakoe and S. Chiba, A Dynamic-Programming Approach to Continuous Speech Recognition, Proc. Int. Cong. Acoustics, Budapest, 1971, Paper 20 C 13.

[47] F. Sanger, S. Nicklen, and A. R. Coulson, Chain Sequencing with Chain-Terminating Inhibitors, Proc. Nat. Acad. Sci. USA 74, 1977, 5463.

[48] David Sankoff, Matching Sequences under Deletion-Insertion Constraints, Proc. Nat. Acad. Sci. USA 69, 1972, pp. 4–6.

[49] D. Sankoff, Simultaneous Solution of the RNA Folding, Alignment and Protosequence Problems, SIAM J. Appl. Math. 45(5), 1985, pp. 810–825.

[50] D. Sankoff, J. B. Kruskal, S. Mainville, and R. J. Cedergren, Fast Algorithms to Determine RNA Secondary Structures Containing Multiple Loops, in D. Sankoff and J. B. Kruskal, editors, Time Warps, String Edits, and Macromolecules: The Theory and Practice of Sequence Comparison, Addison-Wesley, 1983, pp. 93–120.

[51] D. Sankoff and J. B. Kruskal, editors, Time Warps, String Edits, and Macromolecules: The Theory and Practice of Sequence Comparison, Addison-Wesley, 1983.

[52] P. H. Sellers, On the Theory and Computation of Evolutionary Distance, SIAM J. Appl. Math. 26, 1974, pp. 787–793.

[53] P. H. Sellers, Personal Communication, 1989.

[54] T. Smith and M. S. Waterman, Identification of Common Molecular Subsequences, J. Mol. Biol. 147 (1981), pp. 195–197.

[55] E. Ukkonen, On approximate string matching, J. of Algorithms, 6, 1985, pp. 132–137.

[56] V. M. Velichko and N. G. Zagoruyko, Automatic Recognition of 200 Words, Int. J. Man-Machine Studies 2, 1970, pp. 223–234.

[57] T. K. Vintsyuk, Speech Discrimination by Dynamic Programming, Cybernetics 4(1), 1968, 52–57; Russian Kibernetika 4(1), 1968, pp. 81–88.

[58] R. A. Wagner, On the Complexity of the Extended String-to-String Correction Problem, 7th ACM Symp. Theory of Computing, 1975, pp. 218–223.

[59] M. S. Waterman, Sequence alignments in the neighborhood of the optimum with general applications to dynamic programming, Proc. Natl. Acad. Sci. USA, 80, 1983, pp. 3123–3124.

[60] M. S. Waterman, Efficient Sequence Alignment Algorithms, J. of Theor. Biol., 108, 1984, pp. 333.

[61] M. S. Waterman, General Methods of Sequence Comparison, Bull. Math. Biol. 46, 1984, pp. 473–501.

[62] M. S. Waterman Editor, Mathematical Methods for DNA Sequences, CRC Press, Inc., 1988.

[63] M. S. Waterman and T. F. Smith, RNA Secondary Structure: A Complete Mathematical Analysis, Math. Biosciences 42, 1978, pp. 257–266.

[64] M. S. Waterman and T. F. Smith, New Stratigraphic Correlation Techniques, J. Geol. 88, 1980, pp. 451–457.

[65] M. S. Waterman and T. F. Smith, Rapid Dynamic Programming Algorithms for RNA Secondary Structure, Adv. Appl. Math. 7, 1986, pp. 455–464.

[66] M. S. Waterman, T. F. Smith, and W. A. Beyer, Some Biological Sequence Metrics, Adv. Math. 20, 1976, pp. 367–387.

[67] Robert Wilber, The Concave Least Weight Subsequence Problem Revisited, J. Algorithms 9(3), 1988, pp. 418–425.

[68] W. J. Wilbur and D. J. Lipman, Rapid Similarity Searches of Nucleic Acid and Protein Data Banks, Proc. Nat. Acad. Sci. USA 80, 1983, pp. 726–730.

[69] W. J. Wilbur and D. J. Lipman, The Context Dependent Comparison of Biological Sequences, SIAM J. Appl. Math. 44(3), 1984, pp. 557–567.

[70] M. Zucker, The Use of Dynamic Programming Algorithms in RNA Secondary Structure Prediction, in M. S. Waterman editor, Mathematical Methods for DNA Sequences, CRC Press, 1988, pp. 159–184.

[71] M. Zuker, and P. Stiegler, Optimal Computer Folding of Large RNA Sequences using Thermodynamics and Auxiliary Information, Nucl. Acids Res. 9, 1981, pp. 133.

244

Coding Trees as Strings for Approximate Tree Matching *

Roberto Grossi   Fabrizio Luccio   Linda Pagli

Dipartimento di Informatica, Università di Pisa
Corso Italia 40, I-56125 Pisa, Italy

Abstract

In this paper we consider matching problems on arbitrary ordered labelled trees and ranked trees, which have important applications in many fields such as molecular biology, term rewriting systems and language processing. Given a text tree T and a pattern tree P, we derive an algorithm to find all occurrences of P in T with bounded *distance* k, in time $O(k|T| + |P|)$. The distance refers to the number of subtrees to be inserted or deleted from T to obtain P. This problem is an extension of the *tree pattern matching* problem where deletions of subtrees occur only in T, and of the *approximate string matching* problem applied to trees. Extensions of the algorithm to solve other relevant problems, such as ranked trees matching, as well as their parallel versions are then devised.

Keywords: Analysis of algorithms, ordered trees, ranked trees, approximate matching, strings.

1   Introduction

Many papers have been devoted to the study of trees in various forms. A minor part of this literature is directed towards the problem of *tree isomorphism* and *matching*, both for its mathematical interest, and for motivations in different fields such as theory of programming [7, 1] and molecular biology [15, 21].

---

*This work has been partially supported by MURST of Italy

Given a pattern tree P and a text tree T, the problem is generally the one of determining all the (exact, or approximate) occurrences of P as a part of T. In particular, the trees are rooted and ordered. Labels may be assigned to the tree nodes; and ranks (number of children of a labelled node) may be assigned to the labels. Different alternatives lead to problems of different complexities.

A classical contribution is due to Hoffmann and O'Donnell [7]. They propose several algorithms for the subtree replacement problem in a ranked tree, whose worst case complexity is quadratic in the size of the trees. A linear expected complexity has been proved by Steyaert and Flajolet for several of these algorithms [18]. Recently, the complexity of a relevant restricted case has been lowered to $O(|T| \log |P|)$ [2].

Two recent papers are particularly innovative by a computational point of view. In [9] Kosaraju solves the so called *tree pattern matching* problem for labelled trees, where P matches with subtrees of T after the removal of some subtrees from T (see next section). The time required by the algorithm of [9] is $O(|T||P|^{0.75} \text{polylog}(|P|))$ against the $O(|T||P|)$ of the naive bound. The result of [9] has been improved by Dubiner, Galil and Magen [3], to $O(|T||P|^{0.5} \text{polylog}(|P|))$.

Tai [20] has defined the *edit distance* between ordered trees, as the weighted number of *insert*, *delete* and *modify* operations to transform one tree into another. In fact, under a delete operation an internal node u is removed, and the children of u are attached to its father. Insert is the complement operation. Modify is to change one node label. Zhang and Shasha [21] have given efficient means of measuring the edit distance, under various hypotheses. Due to the intrinsic difficulty of their problems, all their algorithms run in time more than linear in $|T|$ and $|P|$. In a subsequent paper the same authors have approached a restricted problem with unit weights, sequentially and in parallel [17]. An even more restricted version has been solved in [8], where only delete operations are considered. In both cases the total work is at least quadratic, in the worst case.

An efficient way of approaching matching problems in trees is to code trees with strings, and solve the problems on such codings. With such a technique, Ramesh and Ramakrishnan [14] solve the problem of ranked tree matching with $\leq k$ subtree replacements in time $O(k|T|)$. Mäkinen [13] shows how to determine all the exact occurrences of P as subtree of T, in time $O(|P| + |T|)$, applying a fast string matching algorithm on the coded trees. Shasha and Zhang [17] note that approximate string matching may be used to detect approximate matching on trees. Luccio and Pagli [11, 12] use the suffix tree data structure to support fast algorithms on string coded trees, for different approximate tree matching problems. Grossi [5] uses a similar technique for the matching of labelled trees, where label mismatches are allowed.

In this paper we consider arbitrary ordered trees and ranked trees, that are important in many fields. They appear, for example, in biochemistry as RNA secondary structures [15, 19], in term rewriting systems, in language processing and so on (e.g., see [7]).

General ordered trees are studied first. For two such trees T, T' we define a (non metric) distance as a function of the number of subtrees to be inserted or deleted in T to obtain T', and introduce an approximate tree matching problem (Problem 1), that consists of finding all the occurrences of P in T, with bounded distance k. On one side, this problem is an extension of the tree pattern matching problem treated in [3, 9]; in our case, subtrees can be removed from P besides from T. On another side, it is an extension of the approximate string matching problem [4, 10] to ordered trees. We give an algorithm to solve our problem in time $O(|P| + k|T|)$, which obviously applies to the tree pattern matching problem. If k is chosen independently of $|P|$, as it is normally assumed in approximate string matching, the algorithm compares favorably with the previous ones.

The matching of ranked trees is then considered (Problem 2), where P and T are labelled, and a special label ν in the leaves of P stands for any subtree in T. An extension of the previous algorithm allows to solve this problem in time complexity $O(|P| + k|T|)$, where k is the number of labels ν in P. In fact this is the same result of [14], obtained with a simpler technique. We also solve Problems 1 and 2 with a bounded amount of label mismatchings. For a particular restriction (Problem 3) where only label mismatchings are allowed (i.e., P occurs exactly in T except for a bounded number k of labels), we provide an algorithm which runs in $O(|T|)$. Note that k may be arbitrarily large, in this case without affecting complexity. For all our problems we also give efficient parallel algorithms in the PRAM model.

2   The general problem

The main approximate tree matching problem treated in this paper is related to ordered trees whose nodes are labelled with the symbols of an alphabet Λ.

For one such tree T, and a node $N \in T$, we denote by $\lambda(N)$ the label of N, and by $T[N]$ the subtree of T rooted at N. If T is non empty, it is coded as a string W_T, recursively defined as follows. Let $R \in T$ be the root of T, and let $T_1, \ldots, T_r$ be the subtrees of T rooted at the sons of R. We pose:

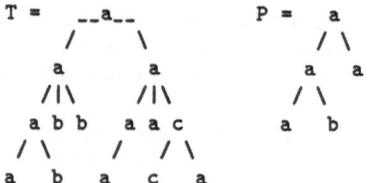

```
T =    __a__              P =     a
      /     \                    / \
     a       a                  a   a
    /|\     /|\                / \
   a b b   a a c              a   b
  / \     / / \
 a   b   a c   a
```

```
pos.  1  2  3  4  5  6  7  8  9 10 11 12 13 14 15 16 17 18 19 20 21 22 23 24 25 26 27 28

W     a  a  a  a  0  b  0  0  b  0  b  0  0  a  a  a  0  0  a  0  c  c  0  a  0  0  0  0
LST  28 13  8  5  0  7  0  0 10  0 12  0  0 27 18 17  0  0 20  0 26 23  0 25  0  0  0  0
PROX  2  3  4  6  0  9  0  0 11  0 14  0  0 15 16 19  0  0 21  0 22 24  0 29  0  0  0  0

W     a  a  a  0  b  0  0  a  0  0
LST  10  7  4  0  6  0  0  9  0  0
```

Figure 1: String representation of two trees T, P.

$$W_T = \begin{cases} \lambda(R)0 & \text{for } |T| = 1, \\ \lambda(R)W_{T_1} \ldots W_{T_r}0 & \text{for } |T| > 1, \end{cases}$$

where 0 is a new symbol, $0 \notin \Lambda$.

W_T can be built in time $O(|T|)$ by traversing T in preorder, and entering $\lambda(N)$ for each node N encountered during the traversal and 0 for each return to the previous level. Clearly W_T consists of $|T|$ labels and $|T|$ zeroes, and all proper prefixes of W_T have a number of labels greater than the number of zeroes. An easy inductive argument shows that there is a one to one correspondence between ordered trees T and sequences W_T.

For example, the sequences W_T, W_P for the two trees T, P are shown in fig. 1. We now define the two basic operations of PRUNE and GRAFT to transform a tree T into another tree T'. Let $N \in T$, q be a non negative integer, and $T_1, \ldots, T_r$ be the subtrees of T rooted at the sons of N. We pose:

$$\text{PRUNE}(T, N, q) \longrightarrow T'$$

where T' is obtained by removing the subtrees $T_{q+1}, \ldots, T_r$ from T. For $q = 0$, N

becomes a leaf in T', while for $q \geq r$ we have $T' = T$. See the examples of fig. 2. Let now $V = \{V_1, \ldots, V_s\}$ be an ordered forest of trees labelled over Λ. We pose:

$$\text{GRAFT}(T, N, V) \longrightarrow T'$$

where T' is obtained by inserting $V_1, \ldots, V_s$ in T, as a rightmost addition to the family of subtrees $T_1, \ldots, T_r$ at N. Then, after GRAFT node N has $r + s$ sons in T. PRUNE and GRAFT are inverse operations, in the sense that $T = \text{GRAFT}(\text{PRUNE}(T, N, q), N, \{T_{q+1}, \ldots, T_r\})$. See the example of fig. 2.

For two trees T, T' we define the transformation from T to T' as a sequence $S(T, T')$ of PRUNE or GRAFT operations on T to obtain T'. In this transformation we establish a correspondence between all pairs of matching nodes $N \in T, N' \in T'$, denoted by $N \equiv N'$. For this purpose, we require that $\lambda(N) = \lambda(N')$, otherwise T can not be transformed to T', and we pose $S(T, T') = S_\infty$. Let R and R' be the roots of T and T'; $T_1, \ldots, T_r$ and $T'_1, \ldots, T'_{r'}$ be the subtrees rooted at the sons of R and R', respectively. The sequence $S(T, T')$ is recursively built as follows:

Definition 1
if $\lambda(R) \neq \lambda(R')$ **then** $S(T, T') := \text{MISMATCH}$ **else**
  case
    $|T| = |T'| = 1 : S(T, T') := \Phi;$
    $|T| = 1, |T'| > 1 : S(T, T') := \text{GRAFT}(T, R, \{T'_1, \ldots, T'_{r'}\});$
    $|T| > 1, |T'| = 1 : S(T, T') := \text{PRUNE}(T, R, 0);$
    $|T| > 1, |T'| > 1 :$
      case
        $r > r' : S(T, T') := S(T_1, T'_1) \ldots S(T_{r'}, T'_{r'}) \, \text{PRUNE}(T, R, r');$
        $r = r' : S(T, T') := S(T_1, T'_1) \ldots S(T_r, T'_r);$
        $r < r' : S(T, T') := S(T_1, T'_1) \ldots S(T_r, T'_r) \text{GRAFT}(T, R, \{T'_{r+1}, \ldots, T'_{r'}\});$
if MISMATCH occurs in $S(T, T')$ **then** $S(T, T') := S_\infty$.

Note that, for any given pair T, T', the sequence $S(T, T')$ is uniquely defined (possibly, as S_∞). Definition 1 induces the correspondence $R \equiv R'$, recursively repeated for all the nodes of the matching positions of T, T', or does not induce any correspondence if $S(T, T') = S_\infty$. If $S(T, T') \neq S_\infty$, the transformation from T to T' is resolved in a (possibly empty) sequence of PRUNE and GRAFT operations. $|S(T, T')|$ is the number of such operations. If $S(T, T') = S_\infty$ we set $|S(T, T')| = \infty$.

Definition 2 The *subtree distance* $d(T, T')$ between two ordered labelled trees is given by $|S(T, T')|$.

249

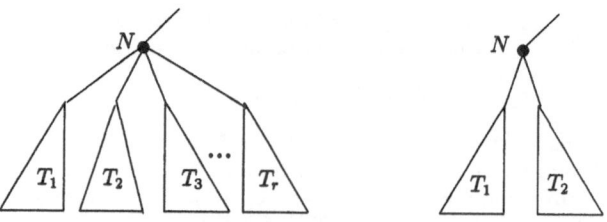

$$\text{PRUNE}(T, N, 2) \longrightarrow T'; \text{GRAFT}(T', N, \{T_3, \ldots, T_r\}) \longrightarrow T$$

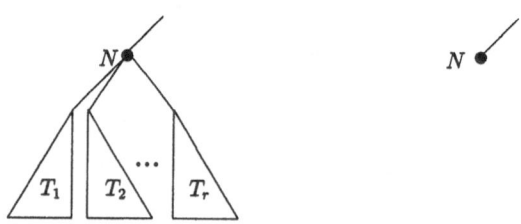

$$\text{PRUNE}(T, N, 0) \longrightarrow T'; \text{GRAFT}(T', N, \{T_1, \ldots, T_r\}) \longrightarrow T$$

Figure 2: Two examples for the operations PRUNE and GRAFT.

In the sample trees T, P of fig. 1 we have $d(T, P) = 3$, $d(T[14], P) = 2$, $d(T[2], P) = \infty$ (node numbering corresponds to node positions in W_T). Based on definition 2 we pose:

Problem 1 Given two ordered labelled trees T, P, and an integer $k \geq 0$, find all the nodes $N \in T$ such that $d(T[N], P) \leq k$.

Observation 1 If we compute $d(T[N], P)$ under a restriction of definition 1, where we put $S(T, T') := \text{MISMATCH}$ for $|T| = 1$, $|T'| > 1$, or for $T| > 1, |T'| > 1, r < r'$ (i.e., we do not accept to graft in T), and take $k = |P|$, problem 1 becomes the tree pattern matching problem of [3, 9].

We solve problem 1 using the W representation of the trees, and a suffix tree built on it. Recall that, for a given sequence X of n elements, the suffix tree on X and the lowest common ancestor algorithm allow to compute, in time $O(1)$, the length PREFIX(X, i, j) of the longest common prefix between the subsequences $X(i), ..., X(n)$ and $X(j), ..., X(n)$ (e.g., see [4]).

Let $W(i)$ denote the i-th element of the sequence W and, for $W(i) \neq 0$, let N_i be the tree node corresponding to $W(i)$. Define the new sequence LAST, with $|\text{LAST}| = |W|$, as follows. For $W(i) = 0$, LAST$(i) = 0$. For $W(i) \neq 0$ ($W(i)$ is the label of N_i), LAST$(i) = j$, where j is the last position in W of the subsequence $W_{T[N_i]}$ corresponding to the subtree rooted at N_i. For example, the sequences LAST$_T$, LAST$_P$ for the trees T, P are shown in fig. 1. Similarly define the sequence PROX, with $|\text{PROX}| = |W|$, as follows. If $W(i) = 0$, then PROX$(i) = 0$; if $W(i) \neq 0$, then PROX$(i) = j$, where $j > i$ is the minimum value for which $W(j) \neq 0$ (if $W(j) = 0$ for all $j > i$, then PROX$(i) = |W| + 1$). Note that PROX contains pointers to the successive nonvoid subtrees, in preorder. Finally define FTHR(i) as the position in W of the father of N_i.

Problem 1 can be solved with the method reported below, consisting of the preliminary phase followed by algorithm 1.

Theorem 1 *Problem 1 is solved in time $O(|P| + k|T|)$, using the preliminary phase and algorithm 1.*

Proof: *Correctness.* The inner **while** loop of algorithm 2 checks the matchings of P with a subtree of T. A point of mismatch is reached via a PREFIX or a LAST computation. We have three cases: 1) $W_T(i) \neq 0$ and $W_P(j) \neq 0$. We have a label mismatch, hence P is not found (defined := **false**). 2) $W_T(i) \neq 0$ and $W_P(j) = 0$. A PRUNE is performed in T, from the father of N_i. 3) $W_T(i) = 0$ and $W_P(j) \neq 0$. Symmetrical to case 2.

Complexity. In the preprocessing phase, W_T, W_P, LAST$_T$, LAST$_P$, FTHR$_T$, FTHR$_P$ and PROX can be built with a visit of the two trees; the suffix tree for X can be built in linear time in the size of X, as already pointed out [4]. Therefore, the preliminary phase requires time $O(|P|+|T|)$. To analyze algorithm 1 note that the internal nodes of T can be detected in time $O(|T|)$, scanning W_T. Consider the comparison between $T[N_i]$ and P, for a given t. The worst case occurs when defined is always **true** (i. e., the **then** branch in statement 2 never occurs). The inner **while** loop is executed at most $k + 1$ times (at this point the procedure stops with failure). When step 1 is executed, the function PREFIX brings the indices i, j to two values for which $W_T(i) \neq 0$ and $W_P(j) = 0$ or $W_T(i) = 0$ and $W_P(j) \neq 0$. That is, at the next iteration, step 3 is again executed. Therefore, step

1 is executed at most as many times as step 3. Since PREFIX can be computed in time $O(1)$, the comparison between $T[N_i]$ and P is performed in time $O(k)$, for $k > 0$. The procedure is repeated for all the internal nodes of T, that is $|T|$ times. Therefore the overall time complexity of the algorithm is $O(k|T|)$. $\qquad\square$

Preliminary phase

1. build the sequences W, LAST and FTHR for T and P, and PROX for T;

2. build the suffix tree for the string $X = W_T \$ W_P \&$ ($\$, \&$ are markers $\notin \Lambda \cup \{0\}$).

Algorithm 1

```
t := 1;
while t ≤ |W_T| do
   begin
   j := 1; i := t; d := 0; defined:=true;
   while i ≤ LAST_T(t) and j ≤ |W_P| and d ≤ k and defined do
1:   if W_T(i) = W_P(j)
        then begin r := PREFIX(X,i,j + |W_T| + 1); i := i + r; j := j + r end
        else
2:        if W_T(i) ≠ 0 and W_P(j) ≠ 0 then defined:=false {label mismatch}
3:        else if W_T(i) ≠ 0 and W_P(j) = 0
              then begin i := LAST_T(FTHR_T(i)) + 1; j := j + 1; d := d + 1 end
              else begin j := LAST_P(FTHR_P(j)) + 1; i := i + 1; d := d + 1 end;
   if d ≤ k and defined then PRINT(t); {we have d(T[N_i], P) ≤ k }
   t := PROX(t)
   end
end algorithm.
```

For each occurrence of P in T, a simple variation of the algorithm allows to report the sequence of PRUNE and GRAFT operations to transform $T[N_i]$ to P. Note that the time complexity stated in theorem 1 refers to a bounded Λ. Otherwise, such a complexity becomes $O(|P| + k|T| + (|P| + |T|) \log \min(|\Lambda|, |P| + |T|))$ as well known in approximate string matching.

Observation 2 If we compare the complexity result of theorem 1 with the ones of [3, 9], we see that our method is superior for $k < |P|^{0.5}$ (in particular, if k is chosen independently of $|P|$). Still, we solve a more general problem (observation 1).

Our algorithm 1 is based on known techniques for approximate string matching. Along the same lines, we can prove the following

Theorem 2 *Problem 1 can be solved in* $O(\max(k, \log|T|))$ *parallel time using a CRCW-PRAM with* $O(|T|)$ *processors.*

Proof: Build the W sequences for T, P in parallel as two arrays of size $2|P|$ and $2|T|$. This can be done in a EREW-PRAM model, in the following way. Assign a processor to each node N of T and P. Each processor works on two records A_N, B_N of two elements each, where $A_N[1], B_N[1]$ will contain elements of W, and $A_N[2], B_N[2]$ will contain pointers. Start with $A_N[1] = \lambda(N), B_N[1] = 0, A_N[2] = B_N[2] = 0$. All A's and B's are linked in parallel with respect to the tree preorder and without conflicts. If N is a leaf, then $A_N[2]$ is linked to B_N; otherwise, $A_N[2]$ is linked to $A_{N'}$, where N' is the first son of N. Next, if N is the last son of N'' then $B_N[2]$ is linked to $A_{N''}$; otherwise, $B_N[2]$ is linked to $A_{N'''}$, where N''' is the next sibling of N. If N is the tree root, then $B_N[2]$ is set to an end-of-list marker. This procedure generates two linked lists of A's and B's, for T and P respectively. By applying a list-ranking algorithm, the two lists are stored in the corresponding arrays in $O(\log|T|)$ parallel time in the EREW model.

Now, build the suffix tree in parallel time $O(\log|T|)$ with $O(|T|)$ processors, with a CRCW-PRAM (see [4]). Finally apply algorithm 1 in parallel at all the nodes of T (time $O(k)$). □

3 Variations and extensions

We can extend problem 1 by allowing a bounded number of label mismatches. For this purpose the variable MISMATCH introduced in definition 1 is used as a counter of mismatches. Initially MISMATCH is set to 0. The first line of the definition is replaced by:

if $\lambda(R) \neq \lambda(R')$ **then** MISMATCH := MISMATCH+1;

and the last line is removed (the **case** statement remains unchanged).

Definition 3 *The mismatch distance* $d_m(T, T')$ *(number of mismatches) between two ordered labelled trees is given by the final value of MISMATCH.*

We have:

Problem 1' Given two ordered labelled trees T, P, and two integers $k, k_m \geq 0$, find all the nodes $N \in T$ such that $d(T[N], P) \leq k, d_m(T[N], P) \leq k_m$.

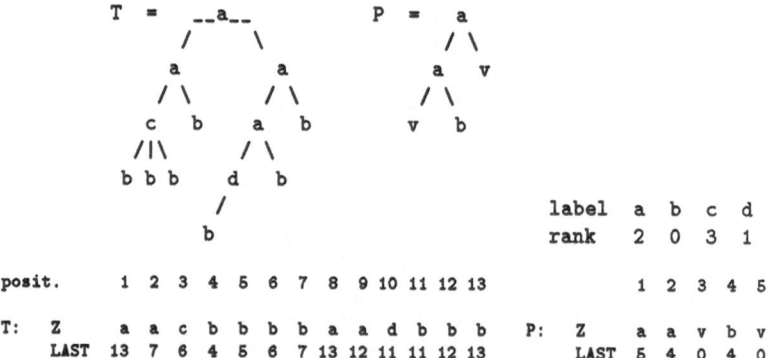

posit.	1	2	3	4	5	6	7	8	9	10	11	12	13				1	2	3	4	5
T: Z	a	a	c	b	b	b	b	b	a	a	d	b	b	P:	Z	a	a	v	b	v
LAST	13	7	6	4	5	6	7	13	12	11	11	12	13		LAST	5	4	0	4	0

Figure 3: Two ranked trees and their string representation.

Problem 1' can be solved with immediate extensions of the sequential and parallel algorithms for problem 1. We have:

Corollary 1 *Problem 1' can be solved in $O(|P| + \max(k, k_m)|T|)$ sequential time, and in $O(\max(k, k_m, \log|T|)$ parallel time with $O(|T|)$ processors.*

A relevant application of algorithm 1 is in the area of ranked tree matching, that is ordered trees where the number of sons of each node depends on the node label. In fact, a non-negative integer rank $\rho(\lambda)$ is associated to each label $\lambda \in \Lambda$; $\rho(\lambda(N))$ gives the number of sons of node N. A leaf L has $\rho(\lambda(L)) = 0$. The string representation of a ranked tree T is easier than in the general case, because T can be described by the sequence Z_T of *the labels encountered in a preorder traversal of* T.

An important problem on ranked trees comes from the field of automatic theorem proving. Consider a special label $\nu \notin \Lambda$ that can be placed on the leaves, to indicate any possible ranked subtree. For two ranked trees T, T', where T has k labels ν (and T' has no label ν), we say that T *matches with* T' if there exist k subtrees $T'_1, \ldots, T'_k$ of T', such that T coincides with T' if $T'_1, \ldots, T'_k$ are substituted with the k leaves ν of T. For example, consider the two ranked trees T, P of fig. 3, with the relative Z sequences, and rank values (LAST has the same meaning of problem 1; we set $\text{LAST}(i) = 0$ if the corresponding node has label ν). P matches with the subtrees $T[1]$ and $T[8]$. We have from [7]:

254

Problem 2 Given two ranked trees T, P where P contains $k > 0$ labels ν, find all the nodes $N \in T$ such that P matches with $T[N]$.

Problem 2 can be solved with the following extension of algorithm 1, applied to the sequences Z_T, Z_P.

Preliminary phase

1. build the sequences Z and LAST for T and P;
2. build the suffix tree on the Z sequences.

Algorithm 2

Similar to algorithm 1, except that the distance d is not computed, and the **if** statement 2, occurring for $Z_T(i) \neq Z_P(j)$, yields the two cases:

- $Z_P(j) \neq \nu$ (label mismatch): the current tree comparison is stopped with failure;
- $Z_P(j) = \nu$: skip ν in P, and the corresponding subtree in T.

We easily derive:

Corollary 2 *Problem 2 is solved in sequential time $O(|P|+k|T|)$ using the preliminary phase and algorithm 2, and in parallel time $O(\max(k, \log|T|))$ with $O(|T|)$ processors.*

Among others, the following variations of problem 2 are significant, and can be solved with easy variants of algorithm 2:

1. Different labels of type ν, namely $\nu_1, \nu_2, \ldots$, may appear in P, with the intention that all the occurrences of each ν_i must correspond to the same subtree in T (for the significance of this problem in term rewriting systems see [14]). If k is the total number of ν occurrences, this case can also be solved in $O(k|T| + |P|)$ time, by detecting identical subtree occurrences for the same ν_i with the PREFIX function.

2. The presence of label ν is allowed in P and T, with obvious meaning. If the total number of ν's in the two trees is k, the algorithm complexity remains unchanged.

3. Label mismatches are also allowed. Letting k, k_m respectively denote the total number of ν's, and the maximum allowed number of mismatches, the algorithm complexity becomes $O(|P| + \max(k, k_m)|T|)$.

255

4 An interesting restriction

Let us now consider a restriction of problem 1', where PRUNE and GRAFT operations are not allowed, but label mismatches are. We have:

Problem 3 Given two ordered labelled trees T, P, and an integer $k_m \geq 0$, find all the nodes $N \in T$ such that $d(T[N], P) = 0$ and $d_m(T[N], P) \leq k_m$.

A sequential solution of problem 3 is obtained with the following algorithm, that can be easily completed by the reader.

Preliminary phase

   1. build the sequences W_T, W_P;

   2. compute the size of all the subtrees of T, and maintain a list of the positions i in W_T such that $|T[N_i]| = |P|$.

Algorithm 3

   for each node N such that $|T[N]| = |P|$ **do**
   check $d(T[N], P) = 0, d_m(T[N], P) \leq k_m$
      by comparing W_P with the portion $W_{T[N]}$ of W_T.

We can now state:

Corollary 3 *Problem 3 is solved in $O(|T|)$ sequential time using the preliminary phase and algorithm 3, and in $O(\log |T|)$ parallel time using an EREW-PRAM with $O(|T|/\log |T|)$ processors.*

Proof: Let a *candidate subtree* $T[N]$ of T be one such that $|T[N]| = |P|$. For any pair of such subtrees $T[N], T[N']$ we have $W_{T[N]}$ and $W_{T[N']}$ are disjoint subsequences of W_T.

In the sequential case, the preliminary phase can be easily performed in $O(|T|)$ time by a preorder traversal of T and P. To analyze algorithm 3 note that there are at most $|T|/|P|$ candidate subtrees $T[N]$ in T. The verification of $d(T[N], P) = 0, d_m(T[N], P) \leq k_m$ requires $O(|P|)$ time by scanning $W_P, W_{T[N]}$. Hence the overall complexity is $O(|T|)$.

In the parallel case W_T, W_P can be built as in the proof of theorem 2. However, one processor is now used to work on $2 \log |T|$ nodes, to fill their records A's and B's, with a total of $O(|T|/\log |T|)$ processors. In this way the construction of the list of A's and B's requires $O(\log |T|)$ instead of $O(1)$, without affecting the overall time complexity. (Note that the succeeding list ranking can also be performed within the same bounds). Then, the subsequences $W_{T[N]}$ of the candidate subtrees are detected. Let the root of each such subtree be called the *leader* of the subtree. Since $|T[N]| = (i_A - i_B + 1)/2$, where i_A, i_B are the positions in W_T of A_N, B_N, the leaders can be found in time $O(\log |T|)$ with $O(|T|/\log |T|)$ processors. For any element $W_T(i)$ such that N_i is not a leader, detect the closest leader N_j to the left in W_T, within the same complexity bounds, by a prefix-sum-like procedure. The *displacement* $q = i - j$ between $W_T(i)$ and $W_T(j)$ is also computed, and associated to $W_T(i)$. If $q < 2|P|$, then $W_T(i)$ belongs to the substring $W_{T[N_j]}$; otherwise, $W_T(i)$ is not a member of a candidate subtree.

An element $W_T(i)$ with displacement $q < 2|P|$ needs also to receive the $(q+1)$-th symbol of W_P. This can be done by a multiple broadcast of all the $2|P|$ symbols of W_P. For this purpose the string W_T is condensed into a new string W'_T, obtained by removing from W_T all symbols with $q \geq 2|P|$ with a prefix-sum-like computation. Now, the $(q+1)$-th symbols of the $W'_{T[N]}$'s are $2|P|$ distant from each other and the corresponding $(q+1)$-th symbols of W_P are sent to them with independent broadcasts. All these steps are performed in optimal parallel time without read and write conflicts.

Now the conditions $d(T[N], P) = 0, d_m(T[N], P) \leq k_m$ for a candidate subtree $T[N]$ can be checked in a tree-like fashion by comparing the two strings $W_{T[N]}$ and W_P in $O(\log |P|)$ time with $O(|P|/\log |P|)$ processors. However, since only $O(|T|/\log |T|)$ processors are to be used, the above computation must be performed as follows. *Case 1:* $|P| \geq \log |T|$. Since $|T|/|P| \leq |T|/\log |T|$, that is the number of candidate subtrees is smaller than the number of processors, assign $O(|P|/\log |P|)$ processors to each candidate subtree. By simulating the $O(|P|/\log |P|)$ needed processors, a time slowdown of $O(\frac{\log |T|}{\log |P|})$ per step is introduced. Globally, the simulation requires $O(\frac{\log |T|}{\log |P|} \log |P|) = O(\log |T|)$ time. *Case 2:* $|P| < \log |T|$. The candidate subtrees are more than the processors, thus assign to each processor $O(\frac{\log |T|}{|P|})$ candidate subtrees of size $O(|P|)$. Each processor checks sequentially the condition on d, d_m for each assigned subtree, requiring overall $O(\frac{\log |T|}{|P|} |P|) = O(\log |T|)$ time and using $O(|T|/\log |T|)$ processors. □

Note that, due to the independence (i.e. non overlap) of the candidate subtrees, the complexity of the sequential and parallel solutions are optimal and independent of k_m and Λ.

257

References

[1] A.V. Aho, M. Ganapathi and S.W.K. Tjiang, Code generation using tree matching and dynamic programming, *ACM Trans. on Prog. Lang. Sys.* **11** (1989) 491-516.

[2] J. Cai, R. Paige and R. Tarjan, More efficient bottom-up tree pattern matching, *Proc. CAAP '90*, in: LNCS **431** (1990) 72-86.

[3] M. Dubiner, Z. Galil and E. Magen, Faster tree pattern matching, *Proc. 31-st IEEE Symp. on Found. of Comp. Sc.*, (1990) 145-150.

[4] Z. Galil and R. Giancarlo, Data structures and algorithms for approximate string matching, *J. Complexity* **4** (1988) 33-72.

[5] R. Grossi, A note on the subtree isomorphism for ordered trees and related problems, *Inf. Proc. Let.* (to appear).

[6] D. Harel and R.E. Tarjan, Fast algorithms for finding nearest common ancestors, *SIAM J. Comp.* **13** (1984) 338-355.

[7] C.M. Hoffmann and M.J. O'Donnel, Pattern matching in trees, *J. ACM* **29** (1982) 68-95.

[8] P. Kilpelänen and H. Mannila, The tree inclusion problem, *Proc. CAAP '91*, in: LNCS **493**, 1 (1991) 202-214.

[9] S.R. Kosaraju, Efficient tree pattern matching, *Proc. 30-th IEEE Symp. on found. of Comp. Sc.*, (1989) 178-183.

[10] G.M. Landau and U. Vishkin, Fast parallel and serial approximate string matching, *J. of Algorithms* **10** (1989) 157-169.

[11] F. Luccio and L. Pagli, Simple solutions for approximate tree matching problems, *Proc. CAAP '91*, in: LNCS **493**, 1 (1991) 193-201.

[12] F. Luccio and L. Pagli, An efficient algorithm for some tree matching problems, *Inf. Proc. Let.* (to appear).

[13] E. Mäkinen, On the subtree isomorphism problem for ordered trees, *Inf. Proc. Let.* **32** (1989) 271-273.

[14] R. Ramesh and I.V. Ramakrishnan, Nonlinear pattern matching, *Proc. ICALP '88*, in: LNCS **317** (1988) 473-488.

[15] D. Sankoff and J.B. Kruskal, Eds., *Time warps, string edits and macro-molecules: the theory and practice of sequence comparison*, Addison-Wesley, Reading, MA 1983.

[16] B.Schieber and U.Vishkin, On finding lowest common ancestor: simplification and parallelization, *SIAM J. Comp.* **17** (1988) 1253-1262.

[17] D. Shasha and K. Zhang, Fast algorithms for the unit cost editing distance between trees, *J. of Algorithms* **11** (1990) 581-621.

[18] J.M. Steyaert and P. Flajolet, Patterns and pattern-matching in trees: an analysis, *Information and Control* **58** (1983) 19-58.

[19] L. Stryer, *Biochemestry*, 3-rd edition, W.H.Freeman and Co., New York, NY 1988.

[20] K.-C. Tai, The tree-to-tree correction problem, *J. ACM* **26** (1979) 422-433.

[21] K. Zhang and D. Shasha, Simple fast algorithms for the editing distance between trees and related problems, *SIAM J. Comp.* **18** (1989) 1245-1262.

259

DECIDING CODE RELATED PROPERTIES BY MEANS OF FINITE TRANSDUCERS

Tom Head, Mathematical Sciences, State University of New York at Binghamton, Binghamton, New York 13902-6000, U.S.A.

Andreas Weber[*], Fachbereich Informatik, J.W. Goethe-Universität, Postfach 111 932, D-6000 Frankfurt am Main 11, West Germany

Abstract: Algorithms that run in time bounded by polynomials of low degree are given for deciding five code related properties. These algorithms apply to rational languages presented by means of non-deterministic finite automata. The algorithms avoid the construction of equivalent unambiguous automata. Instead they involve the construction of non-deterministic finite transducers. In each case the property being decided holds if and only if the associated transducer is single-valued.

Keywords: Codes, sequential machines, transducers, coproducts, semigroups, homophonic codes, multivalued encoding, free monoids, rational languages

1. Introduction

When one or more regular languages are specified by means of non-deterministic finite automata, it is often possible to

---

[*] This author's research was supported in part by a Postdoctoral Fellowship of the Japan Society for the Promotion of Science.

decide properties of the language, or the set of languages, without resorting to the construction of unambiguous (or deterministic) automata recognizing the languages. The general procedure is to convert the given non-deterministic automaton, or set of automata, into a non-deterministic finite transducer in such a way that the transducer will be single-valued if and only if the language, or set of languages, has the property in question. Five such reductions for code-related properties are given here.

Single-valuedness of finite transducers is decidable [Sch76] [Bl&He77] [An&Li78], and in polynomial time [Gu&Ib83] [We90]. For transducers which are non-deterministic sequential machines (NSMs) an elementary algorithm is available for deciding single-valuedness that requires only quadratic time. Since this algorithm suffices for the first three of our five applications, it is presented in Section 2. In Sections 3, 4 & 5 procedures are given for deciding: (1) whether the language recognized by a non-deterministic finite automaton (NFA) is a code; (2) whether the submonoids recognized by a finite collection of NFAs constitute a coproduct decomposition of the submonoid they generate; and (3) whether the code recognized by an NFA is immutable.

In Section 6 the problem of deciding single-valuedness of finite transducers that are not NSMs is discussed. In Sections 7 & 8 procedures are given for deciding: (4) whether the languages recognized by a finite set of NFAs are the homophone classes of a homophonic code; and (5) whether an encoding function is uniquely decipherable relative to the language recognized by an NFA.

2. Single-valuedness for Non-deterministic Sequential Machines

A _finite transducer_, T, is a 6-tuple $T = (Q,A,B,E,q_0,F)$ where Q is a finite set of _states_, A & B are finite sets called the _input & output alphabets_, E is a finite subset of $Q \times A^* \times B^* \times Q$ consisting of _edges_, q_0 is in Q and is called _the initial state_, and F is a subset of Q consisting of _final states_.

Each such T defines a relation $Rel(T) \subseteq A^* \times B^*$ as follows: The pair (u,v) in $A^* \times B^*$ is in $Rel(T)$ if there is a sequence of edges (q_0,u_1,v_1,q_1) $(q_1,u_2,v_2,q_2)\ldots(q_{n-2},u_{n-1},v_{n-1},q_{n-1})$ (q_{n-1},u_n,v_n,q_n) with q_n in F for which $u = u_1\ldots u_n$ and $v = v_1\ldots v_n$. The pair (λ,λ), where λ is the null string, belongs to $Rel(T)$ if the initial state q_0 is also final. The <u>domain</u> of the relation $Rel(T)$ is $D(T) = \{u$ in A^*: there is a v in B^* for which (u,v) is in $Rel(T)\}$. The <u>range</u> of $Rel(T)$ is $R(T) = \{v$ in B^*: there is a u in A^* for which (u,v) is in $Rel(T)\}$. We say that T is <u>single-valued</u> if, for each u in $D(T)$, the v for which (u,v) is in $Rel(T)$ is unique. Thus T is single-valued if and only if $Rel(T)$ defines a function from $D(T)$ onto $R(T)$.

By a <u>transducer</u> we will always mean a finite transducer as defined above. For asserting the time requirements of algorithms in "Big Oh" form it will be adequate to define the <u>size</u> of a transducer $M = (Q,A,B,E,q_0,F)$ to be the number of occurrences of symbols of Q, A & B in a list consisting of all the elements of the sets Q, A, B, E, $\{q_0\}$, and F.

Only two restricted types of transducer will be of concern in Sections 3, 4 & 5. A <u>non-deterministic sequential machine</u>, NSM, is a transducer M for which each edge is in $Q \times A \times B \times Q$. A <u>non-deterministic finite automaton</u>, NFA, is a transducer G for which each edge is in $Q \times A \times \{\lambda\} \times Q$. This unusual way of defining an NFA is clearly equivalent to the more usual definitions. In specifying an NFA G we do not list the unnecessary alphabet of output symbols, nor do we list λ as the third component in the edges of G. A subset of A^* is called a <u>language</u>. The language <u>recognized</u> by an NFA G is the domain $D(G)$. A language is <u>rational</u> if there is an NFA that recognizes it.

Let M be an NSM. By an elementary procedure that requires only time $O(N^2)$, where N is the size of M, it is decidable whether or not M is single-valued. Although the technique is not novel, we present this procedure because it is foundational to the algorithms of Sections 3, 4 & 5.

Algorithm 0. Decide whether an NSM $M = (Q,A,B,E,q_0,F)$ is single-valued as follows. Let G' be the NFA $G' = (Q \times Q, \{0,1\}, E', (q_0,q_0), F \times F)$ where E' is constructed as follows. Each pair of pairs (p,p') & (q,q') will be connected by at most one edge in G'. The pair will be connected by an edge $((p,p'),bit,(q,q'))$ in E' if and only if there is a symbol a in A for which there are edges (p,a,b,q) & (p',a,b',q') in M. If, for every such a in A, $b = b'$ then bit = 0, otherwise bit = 1. Let G be the NFA that is obtained by trimming and cotrimming G'. M is single-valued if and only if 1 does not appear as the middle term of any edge of G.

Remark 0. The algorithm above is easily adapted to produce one for deciding whether an NFA $G = (Q,A,E,q_0,F)$ is unambiguous. Let M be the NSM $M = (Q,A,E,E',q_0,F)$ where E' is constructed as follows. For each (p,a,q) in E, place $(p,a,(p,a,q),q)$ in E'. G is unambiguous if and only if M is single-valued.

3. Rational Codes

Algorithms for deciding whether a given rational language is a code are given in [La79] and [BePe85]. The complexity of making this decision is discussed in [Bo85] and [BePe86]. Polynomial time algorithms for this decision are given in these references, but each starts from an *unambiguous* NFA recognizing the language. When the language is specified by means of an NFA that is ambiguous then one must first construct an equivalent unambiguous NFA. In the worst case the construction of this unambiguous equivalent requires an exponential explosion in the number of states. The following algorithm avoids this explosion and provides the decision in time $O(N^2)$ where N is the size of the recognizing NFA.

Algorithm 1. Decide whether the language L recognized by the NFA $G = (Q,A,E,q_0,F)$ is a code as follows. If q_0 is in F then λ is in L and L is not a code, so continue only if q_0 is not

in F. Let M be the NSM M = $(Q,A,\{0,1\}, E',q_0,\{q_0\})$, where E'
is constructed as follows. For each edge (p,a,q) in E, place
$(p,a,0,q)$ in E' and, if q is in F, place $(p,a,1,q_0)$ in E'
also. L is a code if and only if M is single-valued.

Remark 1. B. Litow showed [Li90] that the problem of
deciding whether the language recognized by an NFA is a code
is in the parallel complexity class NC and consequently that
it is decidable in sequential polynomial time.

4. Coproduct Decompositions

Let S be a monoid. Let $\{S_i: \text{i in I}\}$ be an indexed family of
non-trivial submonoids of S, where I is an arbitrary index
set. We say that S is the coproduct of this indexed family of
submonoids if: (1) this family generates S and (2) for each
monoid X and each family of monoid homomorphisms $h_i: S_i \text{-->X}$ (i
in I), there is a unique homomorphism h:S-->X for which each
of the composite homomorphisms $S_i \subseteq \text{S-->X}$ (i in I) coincides
with h_i.

The concept of a coproduct as defined here is in accord
with its use in the theory of categories [MacL72]. This
concept generalizes the concept of a code in the following
sense. For an indexed family $\{s_i: \text{i in I}\}$ of non-null strings
in a submonoid S of A^*, S is the coproduct of the indexed
family of submonoids $\{s_i^*: \text{i in I}\}$ if and only if the subset
$\{s_i: \text{i in I}\}$ of S is a code that generates S and has the
property that $s_i=s_j$ holds only when i=j.

Now let $S_1,\ldots,S_n$ be an indexed family of non-trivial
rational submonoids of A^* for which, for each i ($1 \leq i \leq n$), $S_i =$
$L(G_i)$ for an NFA $G_i = (Q_i,A,E_i,q_{0i},F_i)$. Without loss of
generality we may assume that $F_i = \{q_{0i}\}$ and that Q_i and Q_j are
disjoint unless i=j. Let S be the submonoid of A^* generated
by the S_i ($1 \leq i \leq n$). The following algorithm requires at most
time $O(N^2)$ where N is the sum of the sizes of the NFAs G_i
($1 \leq i \leq n$).

<u>Algorithm 2</u>. With the above settings, decide whether S is the coproduct of the indexed family of submonoids S_i ($1 \leq i \leq n$) as follows. Let $B = \{b_1, \ldots, b_n\}$. Let $Q = \{q_0\} \cup (\cup\{Q_i : 1 \leq i \leq n\})$ where q_0 is a symbol not in $\cup\{Q_i : 1 \leq i \leq n\}$. Let M be the NSM M $= (Q, A, B, E, q_0, \{q_0\})$ where E is constructed as follows. For each i ($1 \leq i \leq n$) and each edge (p, a, q) in G_i, place (p, a, b_i, q) in E. If $p = q_{0i}$ then place (q_0, a, b_i, q) in E also. If $q = q_{0i}$ then place (p, a, b_i, q_0) in E also. If $p = q = q_{0i}$ then place (q_0, a, b_i, q_0) in E also. Then S is the coproduct of the indexed family S_i ($1 \leq i \leq n$) if and only if M is single-valued.

Remark 2. Call a monoid <u>decomposable</u> if it is the coproduct of two of its proper submonoids. From Algorithm 2 it follows that the problem of deciding whether a finitely generated submonoid of a free monoid is decomposable is in the complexity class NP. We have not investigated the following natural questions: Is this decomposability problem in Co-NP? Is it NP-complete?

5. Immutability of Codes and Languages

E.L. Leiss introduced the concept of immutability of codes in [Le84] and provided algorithms for deciding whether a finite code is immutable in [Le87]. For the variable length case the algorithm given requires exponential time. R. Capocelli, L. Gargano & U. Vaccaro provided an algorithm that requires only time $O(N^2)$ where N is the sum of the lengths of the code words [CaGaVa89,90]. Here we give a polynomial time algorithm for deciding the immutability of a rational language L.

Let A be a finite non-empty set and let $R \subseteq A \times A$ be a reflexive relation. A language $L \subseteq A^*$ is <u>immutable with respect to</u> R if whenever $u = a_1 \ldots a_n$ and $v = b_1 \ldots b_n$ are in L, with each a_i & b_i in A and each (a_i, b_i) in R, we have u=v.

The concept of the immutability of a code C is subsumed under the concept of the immutability of a language L by letting $L = C^*$. In [Le87] & [CaGaVa89,90] a directed graph with vertex set A, called a <u>subversion graph</u> SG_A, is

265

specified. The directed graph SG_A provides the reflexive relation $R = R(SG_A)$ appearing in the definition above. In certain naturally occurring situations, as with optical disks or punched cards [Le87], SG_A is acyclic and in this case the associated relation R is anti-symmetric. Since anti-symmetry plays no part in our decision procedure we assume only that R is reflexive. We summarize: <u>A code C is immutable with respect to a subversion graph</u> SG_A <u>if and only if the language</u> C^* <u>is immutable with respect to the reflexive relation</u> $R(SG_A)$.

By the product G X M of an NFA G = (Q,A,E,q_0,F) and an NSM M = (Q',A,B,E',q_0',F') we mean the NSM G X M = (Q X $Q',A,B,E'',(q_0, q_0'),F X F')$ where E" is constructed as follows. For each pair of edges (p,a,q) in G and (p',a,b,q') in M, place ((p,p'),a,b,(q,q')) in E".

<u>Algorithm 3</u>. Decide whether the language L recognized by the NFA G = (Q,A,E,q_0,F) is immutable with respect to the reflexive relation R ⊆ A X A as follows. Let M' be the NSM M' = (Q,A,A,E', q_0,F) where E' is constructed as follows. For each edge (p,a,q) in E and each b for which (a,b) is in R, place (p,b,a,q) in E'. Let M be the NSM M = G X M'. L is immutable with respect to R if and only if M is single-valued.

Remark 3. Immutability of a rational language L with respect to an attack by a specified finite transducer T, which need not be length preserving, is in general harsher than a subversion as specified by a binary relation on the alphabet. Immutability with respect to such an attack is still decidable [He89], but the final test of single-valuedness will often require the algorithm of Section 6, rather than the simpler Algorithm 0.

6. Single-valuedness for Finite Transducers
The best algorithm we have obtained for deciding single-valuedness for arbitrary finite transducers applies recent work [KaRyJa] that was kindly communicated to us by

J.Karhumaki in advance of publication. We also acknowledge valuable communication from P.Turkainen. The presentation of this algorithm will require a separate publication. Here we only state as a theorem the time requirements of the algorithm. By a _random access machine_ (RAM) we mean a deterministic RAM without multiplications and divisions using the uniform cost criterion. See [Pa78] or [AhHoUl74] as RAM references.

Theorem. Let K be a positive integer. For each T in the class of all finite transducers $T = (Q,A,B,E,q_0,F)$ for which (1) cardinal A $\leq$ K; (2) cardinal B $\leq$ K; and (3) for each (p,u,v,q) in E, length v $\leq$ K: it is decidable on a random access machine in time $O(N^2 \log N)$, where N is the size of M, whether M is single-valued.

Three decades ago S.Even gave in [Ev62] an $O(N^2)$ algorithm for deciding whether a generalized automaton is information lossless. Since [Ev62] is essentially transducer theoretic it is reviewed in the present context. A _generalized automaton_ G is a finite transducer $G = (Q,A,B,E,q_0,F)$ that satisfies two conditions: (1) $E \subseteq Q \times A \times B^+ \times Q$; and (2) G is deterministic, i.e. (p,a,u,q) & (p,a,u',q') in E implies that u=u' & q=q'. Let $G^{\cdot}$ be the transducer $G^{\cdot} = (Q,B,A,E^{\cdot},q_0,F)$ where $E^{\cdot} = \{(p,u,a,q): (p,a,u,q)$ is in E'}. The following statement may be taken as the definition of information losslessness in the present context. G is _information lossless_ if $G^{\cdot}$ is single-valued. Thus information losslessness can be decided by the algorithm referred to in the Theorem above, but it can also be decided by Algorithm 0, after an adaptation of $G^{\cdot}$ is made as indicated below.

Let M be the NSM $M = (Q',B,C,E',q_0,F)$ where from each edge $e=(p,b_1...b_n,a,q)$ of $G^{\cdot}$ is made the sequence of n abutted edges of M: $(q_{e(i-1)},b_i,e_i,q_{e(i)})$ where $1 \leq i \leq n$, $q_{e(0)}=p$, $q_{e(n)}=q$, and no $q_{e(i)}$ for which $1 \leq i \leq n-1$ occurs in any other edge of M. This defines E' and implies the appropriate definitions of Q' and C. Note that the output that M associates with each string w in B* is the set of accepting paths in G⁻ traversed as w is consumed. From the determinism of G it follows that G is information lossless if and only if M is single-valued.

7. Homophonic Codes

A homophonic code with alphabet A consists of a subset H of A* and a partition of H into a finite family of subsets $H_1, \ldots, H_n$ with the property that whenever an equation of the form $u_1 \ldots u_p = v_1 \ldots v_q$ holds with each u_i & each v_j in H it follows that p=q and, for each i ($1 \leq i \leq p$), u_i & v_i lie in the same member of the partition of H. Each member H_i of the partition is said to be a homophone class. Let H be a homophonic code with n such homophone classes and let B = $\{b_1, \ldots, b_n\}$ be a second alphabet with n symbols. For each i ($1 \leq i \leq n$), regard each string in H_i as a code word for the symbol b_i. With this understanding each string in H* can then be uniquely decoded into a string in B*. If H is a subset of A* which is a code then any partition of H provides a homophonic code. An example of such a homophonic cipher that was used four centuries ago is given in [Ka67,p.113]. Discussions of homophonic codes often deal with the even more restricted case in which H is a uniform code [St73] [De82] [Sa85]. Partitioning the set H = {a,ab,ba,bc,cb} into H_1 = {a,ab,ba} & H_2 = {bc,cb} yields a homophonic code with alphabet A = {a,b,c} for which H is not a code. Note that in this example a(bc)(ba) = (ab)(cb)a yet the decoding based on either factorization is $b_1 b_2 b_1$.

Homophonic codes have also been studied under the name: multivalued encodings. K. Sato gave an algorithm for deciding whether a finite collection of finite sets of strings forms a homophonic code in [Sat79]. A surprisingly efficient algorithm for this purpose has just recently been given by Capocelli, Gargano and Vaccaro [CaGaVa] who have also investigated decoding procedures for multivalued encodings in [CaVa89] and [CaGaVa91].

268

Now let $L_1, \ldots, L_n$ be an indexed family of rational languages in A^* for which, for each i ($1 \leq i \leq n$), $L_i = L(G_i)$ for an NFA $G_i = (Q_i, A, E_i, q_{0i}, F_i)$. Without loss of generality we may assume that Q_i and Q_j are disjoint unless i=j. We wish to decide whether the n languages L_i can be used as the n homophone classes of a homophonic code. Since a language containing the null string cannot be a homophone class we assume, for each i ($1 \leq i \leq n$), that q_{0i} is not in F_i. With n and cardinal A bounded by a constant K, the following algorithm requires time at most $O(N^2 \log N)$ where N is the sum of the sizes of the NFAs G_i ($1 \leq i \leq n$).

Algorithm 4. With the above settings, decide whether the L_i are the homophone classes of a homophonic code of size n as follows. Let $B = \{b_1, \ldots, b_n\}$. Let $Q = \{q_0\} \cup (\cup\{Q_i : 1 \leq i \leq n\})$ where q_0 is a symbol not in $\cup\{Q_i : 1 \leq i \leq n\}$. Let T be the transducer $T = (Q, A, B, E, q_0, \{q_0\})$ where E is constructed as follows. For each i ($1 \leq i \leq n$) and each edge (p, a, q) in G_i: Place (p, a, λ, q) in E. If $p = q_{0i}$ then place (q_0, a, λ, q) in E also. If q is in F_i then place (p, a, b_i, q_0) in E also. If $p = q_{0i}$ and q is in F_i, then place (q_0, a, b_i, q_0) in E also. The indexed family L_i ($1 \leq i \leq n$) can be used as the homophone classes of a homophonic code of size n if and only if T is single-valued.

Remark 4. Let $L_1, \ldots, L_n$ be the homophone classes of a homophonic code $H = \cup\{L_i : 1 \leq i \leq n\}$. Although H need not be a code, it must be a __precode__ where, by a precode we mean a subset P of a free monoid A^* for which, whenever an equation $u_1 \ldots u_m = v_1 \ldots v_n$ holds with each u_i & v_j in P, we have m=n. To decide whether a rational language P is a precode apply Algorithm 4 to the case in which P is the unique homophone class. In this case the time complexity of Algorithm 4 can be improved to $O(N^2)$ by using the fact that B is a singleton. One can show that for each precode there is a unique finest homophonic partition that refines every other homophonic partition. A precode is a code if and only if its finest homophonic partition consists of singletons.

8. Unique decipherability relative to a language

Let L be the language recognized by the NFA $G = (Q,A,E,q_0,F)$ and let $f:A \to B^*$ be a function that encodes each letter a in A as the string $f(a)$ of symbols of a second alphabet B. If the set $C = \{f(a): a \text{ in } A\}$ is not a code then there exist strings in C^* that decode into more than one string in A^*. However, it may be that each string in C^* decodes into at most one string in L. When this is the case we say that the encoding f is uniquely decipherable relative to L. The following algorithm requires time at most $O((KN)^2 \log(KN))$ where K is the maximum of the lengths of the strings in C, N is the size of the NFA G, and the alphabets A & B are of constant size.

Algorithm 5. With the above settings of L, G, f & B, let T be the transducer $T = (Q,A,B,E',q_0,F)$ where E' is constructed as follows. For each (p,a,q) in E, place $(p,f(a),a,q)$ in E'. The encoding function f is uniquely decipherable with respect to L if and only if T is single-valued.

Remark 5. Unique decipherability relative to a rational language can be obtained for more general encoding relations than the functional one used above [He80]. Unique decipherability relative to linear context free languages is undecidable even for the functional encoding scheme discussed above [He80].

Acknowledgement We thank R. Capocelli, S. Even, and U. Vaccaro for their remarks and correspondence which have resulted in improvements in the content and bibliography of this article.

References

[AhHoUl74] Aho,A., J.Hopcroft and J.Ullman: The Design and Analysis of Computer Algorithms. Reading, Mass.: Addison-Wesley 1974

[AnLi78] Anisimov,A.V. and L.P.Lisovik: Equivalence problems for finite-automaton mappings into free and commutative semigroups. Cybernetics, 14, 321-327(1978)

[BePe85] Berstel,J. and D.Perrin: Theory of Codes. Orlando, Florida: Academic Press 1985

[BePe86] Berstel,J. and D.Perrin: Trends in the theory of codes. Bull. European Assoc. Theor. Comput. Sci., 29, 84-95(1986)

[BlHe77] Blattner,M. and T.Head: Single-valued a-Transducers. J. Comput. Syst. Sci., 15, 310-327(1977)

[Bo85] Book,R.: The base of the intersection of two free monoids. Discrete Appl. Math., 12, 13-20(1985)

[CaGaVa89] Capocelli,R.M., L.Gargano and U.Vaccaro: An efficient algorithm for testing immutability of variable-length codes. IEEE Trans. Inform. Theory, IT-35, 1310-1314(1989)

[CaGaVa90] Capocelli,R.M., L.Gargano and U.Vaccaro: Immutable codes. In: Sequences: Combinatorics, Compression, Security and Transmission, R.Capocelli, Ed. New York: Springer-Verlag 1990

[CaGaVa91] Capocelli,R.M., L.Gargano and U.Vaccaro: Decoders with initial state invariance for multivalued encodings. Theor. Comput. Sci., to appear, (1991)

[CaGaVa] Capocelli,R.M., L.Gargano and U.Vaccaro: A test for the unique decipherability of multivalued encodings. Preprint, (1990)

[CaVa89] Capocelli,R.M. and U.Vaccaro: Structure of decoders for multivalued encodings. Discrete Appl. Math., 23, 55-71(1989)

[De82] Denning,D.E.R.: Cryptography and Security. Reading, Mass.: Addison-Wesley 1982

[Ev62] Even,S.: Generalized automata and their information losslessness. Proc. Third Annual Symp. on Switching Circuit Theory and Logical Design, 3, 144-147(1962)

[GuIb83] Gurari and O.Ibarra: A note on finite-valued and finitely ambiguous transducers. Math. Systems Theory, 16, 61-66(1983)

[He80] Head,T.: Unique decipherability relative to a language. Tamkang J. Math., 11, 59-66(1980)

[He89] Head,T.: Deciding the immutability of regular codes and languages under finite transductions. Inform. Process. Letters, 31, 239-241(1989)

[Ka67] Kahn,D.: The Codebreakers. New York: Macmillan 1967

[KaRyJa] Karhumäki,J., W.Rytter and S.Jarominek: Efficient constructions of test sets for regular and context-free languages. Manuscript, (1991)

[La79] Lallement,G.: Semigroups and Combinatorial Theory. New York: Wiley 1979

[Le84] Leiss,E.L.: Data integrity on optical disks. IEEE Trans. Comput., C-33, 818-827(1984)

[Le87] Leiss,E.L.: On testing immutability of codes. IEEE Trans. Inform. Theory, IT-33, 934-938(1987)

[Li90] Litow,B.E.: Parallel complexity of the regular code problem. Inform. and Comput., 86, 107-114(1990)

[MacL72] MacLane,S.: Categories for the Working Mathematician. New York: Springer-Verlag 1972

[Pa78] Paul,W.: Komplexitätstheorie. Stuttgart: Teubner 1978

[Sa85] Salomaa,A.: Computation and Automata. Cambridge: Cambridge Univ. Press 1985

[Sat79] Sato,K.: A decision procedure for the unique decipherability of multivalued encodings. IEEE Trans. Inform. Theory, IT-25, 356-360(1979)

[Sc76] Schützenberger,M.P.: Sur les relations rationnelles entre monoïdes libres. Theor. Comput. Sci., 3, 243-259(1976)

[St73] Stahl,F.A.: A homophonic cipher for computational cryptography. Proc. National Comput. Confer., 42, 565-568(1973)

[We90] Weber,A.: On the valuedness of finite transducers. Acta Inform., 27, 749-780(1990)

On the Derivation of Spline Bases

(Extended Summary)

Abraham Lempel[†] *and Gadiel Seroussi*

Hewlett-Packard Laboratories
1501 Page Mill Road
Palo Alto, CA 94304

Abstract

In this paper we present an explicit derivation of general spline bases over function spaces closed under differentiation. We determine the necessary and sufficient conditions for the existence of B-splines in such a function space and prove that whenever such a basis exists it is essentially unique. Rather than following the common practice of presenting a mathematical definition of B-splines and then proceeding to prove some of their desired properties, we begin by stipulating a set of design objectives and then proceed to derive functions that meet these objectives. We stipulate the following standard design objectives: a given degree of continuity, least feasible order for the given continuity requirement, and shape invariance under translation. As it turns out, these objectives form a complete set in the sense that no other requirement can be imposed without it being already implied by the ones listed. In other words, when the listed objectives are translated into algebraic constraints, the resulting equations have an essentially unique solution with no remaining degrees of freedom. Clearly, when other desirable objectives such as convexity and the variation diminishing property are attainable, they follow as a by-product by virtue of uniqueness.

Keywords: B-splines, Tchebycheff systems.

[†] This work was done while A. Lempel was on leave from Technion - Israel Institute of Technology, Haifa, Israel.

1. Introduction

The mathematical properties of B-splines and their applications have been the subject of increasing interest in recent years [2]-[5]. Although the subject has been studied in great depth and generality by the mathematics community, most of the attention in certain application areas such as computer graphics has been focused on the relatively simple class of polynomial B-splines and iterative methods of construction.

In this paper we present an explicit derivation of the more general Tchebycheffian spline bases over function spaces closed under differentiation. These splines have been studied in the literature, in the context of approximation theory [3, Chs. 9,10], [6]. We determine the necessary and sufficient conditions for the existence of B-splines in such a function space and prove that whenever such a basis exists it is essentially unique. The derivation is constructive, and leads to explicit expressions for the computation of the basis functions. Our method of derivation is diametrically opposite to the conventional approach. Instead of presenting a mathematical definition of B-splines and then proceeding to prove some of their desired properties, we begin by stipulating a set of design objectives and then proceed to derive functions that meet these objectives. While the proofs of existence and uniqueness are based on some elements of the theory of Tchebycheffian systems, the expressions for the spline bases are derived in an elementary manner by explicitly solving a system of linear equations derived from the design objectives, and thus, readily lend themselves to algorithmic implementation.

More specifically, we stipulate the following standard design objectives: a given degree of continuity, least feasible order for the given continuity requirement, and shape invariance under translation. As it turns out, these objectives form a complete set in the sense that no other requirement can be imposed without it being already implied by the ones listed. In other words, when the listed objectives are translated into algebraic constraints, the resulting equations have an essentially unique solution with no remaining degrees of freedom. Clearly, when other desirable objectives such as convexity and the variation diminishing property are attainable, they follow as a by-product by virtue of uniqueness.

The geometric model and the design objectives are precisely defined in Section 2. In Section 3 we translate the design objectives into a set of corresponding basic constraints. Next, in Section 4, we impose the requirement that the solution space be closed under differentiation, but allow complex-valued functions. In Section 5, the basic constraints are adjusted accordingly, resulting in a set of simple matrix equations. The general explicit solution is derived in Section 6, where we also present the necessary and sufficient conditions for the existence of a spline basis, and prove uniqueness when one exists. In Section 7 we make the connection with

Tchebycheff systems, thereby establishing the existence of unique spline bases in a large subclass of real function spaces closed under differentiation. Here, we also discuss the derivation of real-valued B-splines from complex-valued function spaces. Finally, in Section 8 we present a complete characterization of the uniform B-spline case.

This extended summary contains no proofs of the stated results. For proofs, the reader is referred to [8].

2. The Geometric Model and Design Objectives

Let $P_0, P_1, \cdots, P_m, \cdots, P_n$ be points in the p-dimensional Euclidean space $\mathbb{R}^p$. Let t be a nonnegative integer, and let $u_0 < u_1 < \cdots < u_{n-t+1}$ be real numbers. A p-dimensional *spline curve* S of *order* t, with *control points* $\{P_m\}$, and *knot vector* $(u_0 u_1 \cdots u_{n-t+1})$ is defined as the concatenation of the $n-t+1$ curve segments $S_0, S_1, \cdots, S_m, \cdots, S_{n-t}$, as follows:

$$S(u) = S_m(u), \quad u_m \le u < u_{m+1},$$

$$S_m(u) = \sum_{k=0}^{t} P_{m+k} f_{m,k}(u - u_m), \quad u_m \le u \le u_{m+1}, \ 0 \le m \le n-t. \tag{1}$$

Here, the $f_{m,k}(u)$ are real valued, linearly independent, continuous functions defined on the interval $0 \le u \le u_{m+1} - u_m$. The derivation of such functions, guaranteeing a given degree of continuity at every knot, and the determination of the conditions under which they exist is the subject of this paper. A portion of a typical spline curve S with its control points is illustrated in Figure 1.

One of the main advantages of this model is its *local control* property: each curve segment is controlled by $t+1$ control points (the segment's *support*), and, conversely, each control point supports $t+1$ curve segments. In addition, the functions $f_{m,k}(u)$ must satisfy the following design objectives:

Continuity. For a given nonnegative integer d, the curve S must be differentiable d times with respect to u, and the curve and its first d derivatives must be continuous for all values of u in the range $u_0 \le u < u_{n-t+1}$. In particular, we must have

$$S_m^{(j)}(u_{m+1}) = S_{m+1}^{(j)}(u_{m+1}), \quad 0 \le m < n-t, \ 0 \le j \le d. \tag{2}$$

Here, $S^{(j)}(u)$ denotes the j-th derivative of $S(u)$ with respect to u.

275

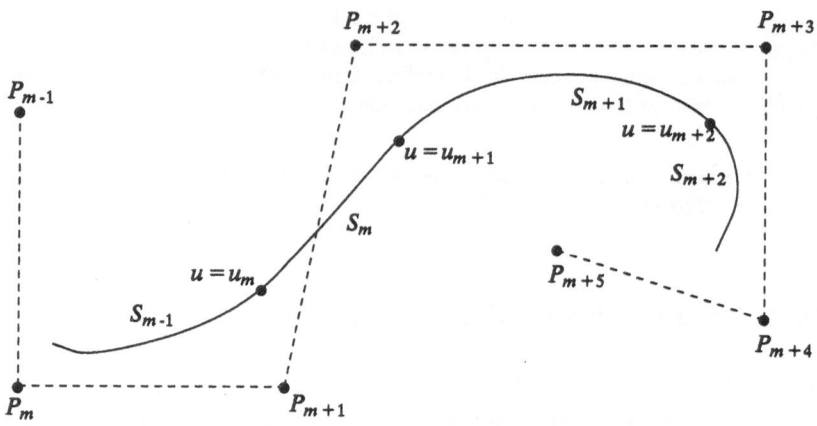

Figure 1: A typical spline curve.

Maximum local control. The support parameter t should be the least integer for which the given continuity requirement can be attained.

Shape preservation under translation. For any fixed vector $Q \in \mathbf{R}^p$ we must have

$$Q + S_m(u) = \sum_{k=0}^{t} (Q + P_{m+k}) f_{m,k}(u - u_m), \quad u_m \leq u \leq u_{m+1}, \quad 0 \leq m \leq n-t. \quad (3)$$

Functions $f_{m,k}(u)$ satisfying these design objectives will be called *segment basis functions* (in short, *SB functions*). These functions are concatenated to form the *spline basis functions* (*B-splines*) $N_m(u)$, $0 \leq m \leq n$, defined over the interval $u_0 \leq u < u_{n-t+1}$ as follows:

$$N_m(u) = \begin{cases} 0 & u < u_{m-t} \text{ or } u \geq u_{m+1} \\ & \qquad\qquad\qquad\qquad\qquad\qquad 0 \leq m \leq n. \\ f_{m-k,k}(u - u_{m-k}) & u_{m-k} \leq u < u_{m-k+1}, \quad 0 \leq k \leq t, \end{cases}$$

It follows from (1) and from the definition of $N_m(u)$ above that the spline curve $S(u)$ is given by

276

$$S(u) = \sum_{m=0}^{n} P_m N_m(u).$$

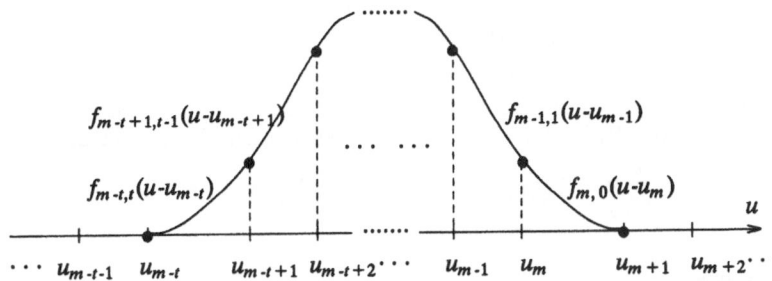

Figure 2: Spline basis function $N_m(u)$.

The relation between the SB functions $f_{m,k}(u)$ and the spline basis functions $N_m(u)$ is shown in Figure 2. We will use the term *solution space* to refer to the linear space of functions spanned by the functions $f_{m,k}$ for a fixed m and $0 \le k \le t$. Notice that, while the functions $f_{m,0}, f_{m,1}, \cdots, f_{m,t}$ are segments of the B-splines $N_m, N_{m+1}, \cdots, N_{m+t}$ (respectively), they do form a basis for the solution space they span. Also, defining

$$\mathbf{f}_m(u) = \begin{bmatrix} f_{m,0}(u) \\ f_{m,1}(u) \\ \cdot \\ \cdot \\ \cdot \\ f_{m,t}(u) \end{bmatrix},$$

we notice that the sets $\{\,\mathbf{f}_m(u),\ 0 \le m \le n-t\,\}$ and $\{\,N_m(u),\ 0 \le m \le n\,\}$ represent essentially the same mathematical objects (the SB functions), arranged in two different ways. We shall loosely use the term *spline basis* to refer to both arrangements. In this paper, we concentrate on the derivation of the SB functions $f_{m,k}(u)$. Clearly, this is equivalent to the derivation of the B-splines $N_m(u)$.

Much of the extensive literature on B-splines deals with two important special cases: *polynomial* B-splines, where the functions $f_{m,k}(u)$ are polynomials, and *uniform* B-splines, where $u_i = u_0+i$ (equally spaced knots). These two special cases will be addressed in Sections 7 and 8 as specific examples of our derivation.

3. The Basic Constraints (I)

We now proceed to translate the above requirements to constraints on the SB functions $f_{m,k}(u)$. Define $\delta_m = u_{m+1}-u_m$. It follows from (1) and (2) that

$$\sum_{k=0}^{t} P_{m+k} f_{m,k}^{(j)}(\delta_m) = \sum_{k=0}^{t} P_{m+1+k} f_{m+1,k}^{(j)}(0), \quad 0 \le j \le d.$$

This can be rewritten as

$$f_{m,0}^{(j)}(\delta_m)P_m + \sum_{k=1}^{t} [f_{m,k}^{(j)}(\delta_m)-f_{m+1,k-1}^{(j)}(0)]P_{m+k} + f_{m+1,t}^{(j)}(0)P_{m+t+1} = 0, \quad 0 \le j \le d.$$

Since the above equality must hold for every choice of control points, the coefficient of each P_k must be zero. Hence, we must have

$$f_{m,0}^{(j)}(\delta_m) = 0, \tag{4.1}$$

$$f_{m,k}^{(j)}(\delta_m)-f_{m+1,k-1}^{(j)}(0) = 0, \quad 1 \le k \le t, \tag{4.2}$$

$$f_{m+1,t}^{(j)}(0) = 0, \tag{4.3}$$

$$0 \le j \le d, \quad 0 \le m \le n\text{-}t.$$

Expanding the right hand side of (3) and using the definition of $S_m(u)$ from (1), the invariance-under-translation property translates into

$$Q+S_m(u) = Q \sum_{k=0}^{t} f_{m,k}(u-u_m) + S_m(u), \quad u_m \le u \le u_{m+1}.$$

Since this must hold for all $Q \in \mathbf{R}^p$, we obtain

$$\sum_{k=0}^{t} f_{m,k}(u) = 1, \quad 0 \le u \le \delta_m. \tag{4.4}$$

The system of equations (4.1) through (4.4) form the basic set of constraints on the SB functions.

278

4. The Solution Space

So far, we have imposed no restriction on the search space for the functions $f_{m,k}(u)$ satisfying our design objectives. We shall now restrict the search space to the class of SB functions that span their own derivatives and, thus, also become more amenable to analysis and explicit determination from the basic constraints. More specifically, we shall search for solutions $f_m(u)$ satisfying (4.1)-(4.4) which span a $(t+1)$-dimensional space and obey a differential equation

$$\frac{d}{du}f_m(u) = Df_m(u), \tag{5}$$

where the matrix D is independent of u. Notice that the conventional polynomial B-splines fall into the category of the proposed solution space.

As is well known [7], the general solution to (5) is of the form

$$f_m(u) = F_m b(u), \tag{6}$$

where F_m is a $(t+1)\times(t+1)$ matrix independent of u, and $b(u)$ is of the form

$$b(u) = \begin{bmatrix} b_0(u) \\ b_1(u) \\ \cdot \\ \cdot \\ \cdot \\ b_g(u) \end{bmatrix},$$

with

$$b_k(u) = \begin{bmatrix} e^{\lambda_k u} \\ u e^{\lambda_k u} \\ \cdot \\ \cdot \\ \cdot \\ u^{t_k} e^{\lambda_k u} \end{bmatrix}$$

$$0 \le k \le g, \quad t_k \ge 0, \quad g + \sum_{k=0}^{g} t_i = t.$$

The λ_k are, in general, complex numbers such that if $g>0$ and $j \ne k$, then $\lambda_j \ne \lambda_k$.

279

The elements of the vector $\mathbf{b}(u)$ form the *canonical basis* of the space defined by (5). The special case of $g=0$ and $\lambda_0=0$ reduces to the canonical polynomial basis $[\,1\ u\ u^2\ \cdots\ u^t\,]'$, where the prime denotes transposition.

It can be readily verified that $\mathbf{b}(u)$ satisfies the differential equation

$$\frac{d}{du}\mathbf{b}(u) = \Delta\mathbf{b}(u), \tag{7}$$

where Δ is the block-diagonal matrix

$$\Delta = \begin{bmatrix} \Delta_0 & & & & \\ & \Delta_1 & & & \\ & & \cdot & & \\ & & & \cdot & \\ & & & & \Delta_g \end{bmatrix} \triangleq \mathrm{diag}\left(\Delta_0\ \Delta_1\ \cdots\ \Delta_g\right)$$

with

$$\Delta_k = \begin{bmatrix} \lambda_k & & & & \\ 1 & \lambda_k & & & \\ & 2 & \cdot & & \\ & & \cdot & & \\ & & & t_k & \lambda_k \end{bmatrix} \triangleq J_k + \lambda_k I. \tag{8}$$

Here, I denotes an identity matrix of order t_k+1, and J_k denotes a sub-diagonal matrix with the entries $1,2,...,t_k$ in its sub-diagonal.

From (6) and (7), we obtain

$$f_m^{(j)}(u) = F_m\,\mathbf{b}^{(j)}(u) = F_m\Delta\mathbf{b}^{(j-1)}(u) = \cdots = F_m\Delta^j\mathbf{b}(u), \quad j\geq 0. \tag{9}$$

Consider now the *state matrix* $\mathbf{B}(u)$, defined by

$$\mathbf{B}(u) = \left[\mathbf{b}(u)\ \mathbf{b}^{(1)}(u)\ \mathbf{b}^{(2)}(u)\ \cdots\ \mathbf{b}^{(t)}(u)\right]$$

$$= \left[\mathbf{b}(u)\ \Delta\mathbf{b}(u)\ \Delta^2\mathbf{b}(u)\ \cdots\ \Delta^t\mathbf{b}(u)\right]. \tag{10}$$

Using this matrix and (9), we obtain

$$F_m B(u) = \left[\; \mathbf{f}_m(u) \; \mathbf{f}_m^{(1)}(u) \; \mathbf{f}_m^{(2)}(u) \; \cdots \; \mathbf{f}_m^{(t)}(u) \; \right] \tag{11}$$

In terms of the initial state vector $\mathbf{b}(0)$, the general solution of the differential equation (7) can also be written as [7]

$$\mathbf{b}(u) = e^{u\Delta} \mathbf{b}(0), \tag{12}$$

where

$$e^{u\Delta} = \sum_{k=0}^{\infty} \frac{u^k \Delta^k}{k!} = \text{diag} \left(e^{u\Delta_0} \; e^{u\Delta_1} \; \cdots \; e^{u\Delta_s} \right).$$

An explicit computation of $e^{u\Delta_k}$ yields

$$e^{u\Delta_k} = \sum_{j=0}^{\infty} \frac{u^j}{j!} (J_k + \lambda_k I)^j = \sum_{j=0}^{\infty} \frac{u^j}{j!} \sum_{l=0}^{j} \binom{j}{l} \lambda_k^{j-l} J_k^l$$

$$= \sum_{l=0}^{t_k} \frac{u^l J_k^l}{l!} \sum_{j=l}^{\infty} \frac{u^{j-l} \lambda_k^{j-l}}{(j-l)!} \triangleq M_k(u) \, e^{\lambda_k u}, \tag{13}$$

where

$$M_k(u) = \begin{pmatrix} \binom{0}{0} & & & & & \\ u\binom{1}{0} & \binom{1}{1} & & & & \\ u^2\binom{2}{0} & u\binom{2}{1} & \binom{2}{2} & & & \\ \cdot & \cdot & \cdot & \cdot & & \\ \cdot & \cdot & \cdot & \cdot & \cdot & \\ \cdot & \cdot & \cdot & \cdot & \cdot & \cdot \\ u^{t_k}\binom{t_k}{0} & u^{t_k-1}\binom{t_k}{1} & \cdot & \cdot & \cdot \, u\binom{t_k}{t_k-1} & \binom{t_k}{t_k} \end{pmatrix}$$

Lemma 1: The state matrix $B(u)$ is nonsingular for all u.

5. The Basic Constraints (II)

We are ready now to address the "maximum-local-control" objective of Section 2. Let $F_{m,k}$ denote the k-th row of F_m, $0 \le k \le t$.

Theorem 1: Given the continuity degree d, the spline order t must satisfy $t \ge d + 1$.

In the sequel, we shall assume $d = t + 1$, thus meeting the "maximum-local-control" requirement. Another item that can be cleared up at this stage concerns constraint (4.4). Let **1** denote a row vector of $t + 1$ ones. It follows from (4.4) that

$$(1F_m) \, b(u) = 1 \, (F_m \, b(u)) = 1.$$

Applying differentiation, we obtain

$$(1F_m) \, \Delta \, B(u) = 0,$$

which implies that the matrix Δ must be singular or, equivalently, that one of the λ_i must equal zero. Without loss of generality, we shall set $\lambda_0 = 0$. This implies that $(1\,0\,0\,\cdots\,0)$ is the unique representation of the constant 1 with respect to the basis $b(u)$. Hence, we must have $1F_m = (1\,0\,0\,\cdots\,0)$.

In view of the above, we can now rewrite the basic constraints (4.1)-(4.4) as follows:

$$F_{m,0} B(\delta_m) = (0\,0\,\cdots\,0\,\phi_{m,0}), \tag{20.1}$$

$$F_{m,k} B(\delta_m) - F_{m+1,k-1} B(0) = (0\,0\,\cdots\,0\,\phi_{m,k}), \quad 1 \le k \le t, \tag{20.2}$$

$$- F_{m+1,t} B(0) = (0\,0\,\cdots\,0\,\phi_{m,t+1}), \tag{20.3}$$

$$1 F_m = (1\,0\,0\,\cdots\,0), \tag{20.4}$$

$$0 \le m \le n - t.$$

The row vectors at the right hand sides of equations (20.1)-(20.4) are all of dimension $t+1$. The entries $\phi_{m,k}$, $0 \le k \le t+1$ are parameters indicating the "discontinuity" of the $(d+1)$-st derivatives at the knots and yet remain to be determined. As we shall see, whenever a solution to the basic constraints exists, it uniquely determines these $\phi_{m,k}$ parameters.

6. The General Solution

We proceed now to solve equations (20.1)-(20.4) for the matrix F_m. Let $\mathbf{a}$ denote the last row of $\mathbf{B}(0)^{-1}$, and let

$$\sigma_{m,r} \overset{\Delta}{=} \sum_{j=0}^{r} \delta_{m+j} = u_{m+r+1} - u_m, \quad -1 \le r \le t, \; 0 \le m \le n-t.$$

Notice that, for a given m, we have $0 = \sigma_{m,-1} < \sigma_{m,0} < \sigma_{m,1} < \cdots < \sigma_{m,t}$.

It follows from (14) that $\mathbf{B}(u)^{-1} = \mathbf{B}(0)^{-1} e^{-u\Delta}$. This and the definition of $\mathbf{a}$ imply

$$(0 \; 0 \; \cdots \; 0 \; \phi_{m,k}) \mathbf{B}(\delta_m)^{-1} = \phi_{m,k} \mathbf{a} \, e^{-\delta_m \Delta}. \tag{21}$$

Hence, using (21) to solve for $F_{m,0}$ in (20.1), we obtain

$$F_{m,0} = (0 \; 0 \; \cdots \; 0 \; \phi_{m,0}) \mathbf{B}(\delta_m)^{-1} = \phi_{m,0} \mathbf{a} \, e^{-\delta_m \Delta}. \tag{22}$$

Similarly, we can solve for $F_{m,k}$ in (20.2) to obtain

$$F_{m,k} = F_{m+1,k-1} e^{-\delta_m \Delta} + \phi_{m,k} \mathbf{a} \, e^{-\delta_m \Delta}, \quad 1 \le k \le t. \tag{23}$$

Equation (23) can be applied recursively to $F_{m+1,k-1}$, yielding

$$F_{m,k} = \left(F_{m+2,k-2} e^{-\delta_{m+1}\Delta} + \phi_{m+1,k-1} \mathbf{a} \, e^{-\delta_{m+1}\Delta} \right) e^{-\delta_m \Delta} + \phi_{m,k} \mathbf{a} \, e^{-\delta_m \Delta}$$

$$= F_{m+2,k-2} e^{-\sigma_{m,1}\Delta} + \begin{bmatrix} \phi_{m,k} & \phi_{m+1,k-1} \end{bmatrix} \begin{bmatrix} \mathbf{a} \, e^{-\sigma_{m,0}\Delta} \\ \mathbf{a} \, e^{-\sigma_{m,1}\Delta} \end{bmatrix}.$$

Applying the recursion a total of k-1 times, and using (22) to obtain the value of $F_{m+k,0}$, we obtain

$$F_{m,k} = \left[\phi_{m,k} \ \phi_{m+1,k-1} \ \cdots \ \phi_{m+k,0} \right] \begin{bmatrix} a \, e^{-\sigma_{m,0}\Delta} \\ a \, e^{-\sigma_{m,1}\Delta} \\ \cdot \\ \cdot \\ \cdot \\ a \, e^{-\sigma_{m,k}\Delta} \end{bmatrix}, \quad 0 \le k \le t. \qquad (24)$$

Define the lower triangular matrix

$$\Phi_m = \begin{bmatrix} \phi_{m,0} & & & & \\ \phi_{m,1} & \phi_{m+1,0} & & & \\ \cdot & \cdot & \cdot & & \\ \cdot & \cdot & \cdot & \cdot & \\ \cdot & \cdot & \cdot & \cdot & \cdot \\ \phi_{m,t} & \phi_{m+1,t-1} & \cdot & \cdot & \cdot & \phi_{m+t,0} \end{bmatrix} \qquad (25)$$

and the matrix

$$E_m = \begin{bmatrix} a \, e^{-\sigma_{m,0}\Delta} \\ a \, e^{-\sigma_{m,1}\Delta} \\ \cdot \\ \cdot \\ \cdot \\ a \, e^{-\sigma_{m,t}\Delta} \end{bmatrix}. \qquad (26)$$

It follows from (24) that

$$F_m = \Phi_m \, E_m. \qquad (27.1)$$

Also, (20.4) can be rewritten as

$$1 \, \Phi_m \, E_m = (1 \ 0 \ 0 \ \cdots \ 0). \qquad (27.2)$$

In order for the functions $F_{m,k}(u)$ to be linearly independent, the matrix F_m has to be nonsingular. Therefore, we must have

$$| \, E_m \, | \ne 0, \qquad (27.3)$$

and

284

$$\phi_{m+j,0} \neq 0, \quad 0 \leq j \leq t. \tag{27.4}$$

Equations (27.1)-(27.4) were derived from the basic constraints (20.1),(20.2), and (20.4). It remains to be shown that every solution for F_m, satisfying (27.1)-(27.4), is also consistent with (20.3). To this end it suffices to show that there exists a choice for $\phi_{m,t+1}$ (actually, a unique one) under which (20.3) is implied by (20.1), (20.2), and (20.4).

By (20.4), we can write

$$1\, F_m = 1\, F_{m+1} = (1\ 0\ 0\ \cdots\ 0).$$

Hence, summing both sides of (20.1) and (20.2) yields

$$(1\ 0\ 0\ \cdots\ 0)\, \mathbf{B}(\delta_m) - [(1\ 0\ 0\ \cdots\ 0) - F_{m+1,t}]\, \mathbf{B}(0) = (0\ 0\ \cdots\ 0\ \sum_{k=0}^{t} \phi_{m,k}),$$

or,

$$F_{m+1,t}\mathbf{B}(0) = (0\ 0\ \cdots\ 0\ \sum_{k=0}^{t} \phi_{m,k}) - (1\ 0\ 0\ \cdots\ 0)[\mathbf{B}(\delta_m) - \mathbf{B}(0)].$$

Since $\lambda_0 = 0$, the first row of $[\mathbf{B}(\delta_m) - \mathbf{B}(0)]$ is all zeros. Therefore, setting $\phi_{m,t+1} = -\sum_{k=0}^{t} \phi_{m,k}$, we obtain (20.3) and establish consistency.

A brief summary at this point will serve to clarify where we stand at the moment. Our design objectives imply the set of basic constraints (4.1)-(4.4). After imposing closure under differentiation on the solution space, we arrive at a stricter set of constraints (20.1)-(20.4), with (as yet to be determined) parameters $\phi_{m,k}$, $0 \leq k \leq t+1$. Consistency of the latter constraints imposes the relation $\sum_{k=0}^{t+1} \phi_{m,k} = 0$, and thus establishes the dependence of (20.3) on (20.1), (20.2), and (20.4). The remaining three constraints yield an explicit expression (27.1) for F_m in terms of E_m and Φ_m, and equation (27.2) to be satisfied by these matrices. In order for F_m to be a valid solution, both E_m and Φ_m must be nonsingular, and it is so imposed by (27.3) and (27.4).

We will show in the sequel that condition (27.4) is, actually, implied by (27.3). But, first, we will derive a more explicit condition, equivalent to (27.3). To this end, we need the following two lemmas.

Lemma 2: Write $\mathbf{a} = [\mathbf{a}_0 \ \mathbf{a}_1 \ \cdots \ \mathbf{a}_g]$, where $\mathbf{a}_k = [a_{k,0} \ a_{k,1} \ \cdots \ a_{k,t_k}]$, $0 \le k \le g$. Then, $a_{k,t_k} \ne 0$ for all $0 \le k \le g$.

Now, let $\mathbf{v} = (v_0 \ v_1 \ \cdots \ v_t)$ denote a row vector of complex numbers. Then,

$$\mathbf{v} \, E_m = \sum_{k=0}^{t} v_k \, \mathbf{a} \, e^{-\sigma_{m,k}\Delta} = \mathbf{a} \left[\sum_{k=0}^{t} v_k \, e^{-\sigma_{m,k}\Delta} \right] \tag{28}$$

Recalling the block-diagonal structure of $e^{-\sigma_{m,k}\Delta}$,

$$e^{-\sigma_{m,k}\Delta} = \mathrm{diag}(\, e^{-\sigma_{m,k}\Delta_0} \ e^{-\sigma_{m,k}\Delta_1} \ \cdots \ e^{-\sigma_{m,k}\Delta_s} \,), \quad 0 \le k \le t,$$

(28) can be rewritten as

$$\mathbf{v} \, E_m = [\, \mathbf{z}_0 \ \mathbf{z}_1 \ \cdots \ \mathbf{z}_g \,], \tag{29}$$

where

$$\mathbf{z}_j \triangleq \mathbf{a}_j \sum_{k=0}^{t} v_k e^{-\sigma_{m,k}\Delta_j}, \quad 0 \le j \le g.$$

Thus, $\mathbf{v} \, E_m = 0$ if and only if $\mathbf{z}_j = 0$ for all $0 \le j \le g$.

By (13), the entry $z_{j,c}$, in column c of $\mathbf{z}_j$, is given by

$$z_{j,c} = \mathbf{a}_j \sum_{k=0}^{t} v_k \left(e^{-\sigma_{m,k}\Delta_j} \right)_c = \sum_{k=0}^{t} v_k e^{-\lambda_j \sigma_{m,k}} \mathbf{a}_j \left(M_j(-\sigma_{m,k}) \right)_c, \quad 0 \le j \le g, \quad 0 \le c \le t_j.$$

Therefore, we have

$$z_{j,c} = \sum_{r=0}^{t_j} a_{j,r} \binom{r}{c} \sum_{k=0}^{t} v_k (-\sigma_{m,k})^{r-c} e^{-\lambda_j \sigma_{m,k}} = \sum_{l=0}^{t_j-c} \binom{c+l}{c} a_{j,c+l} \, V_j(l), \tag{30}$$

where

$$V_j(l) \triangleq \sum_{k=0}^{t} v_k (-\sigma_{m,k})^l e^{-\lambda_j \sigma_{m,k}}, \quad 0 \le l \le t_j. \tag{31}$$

Lemma 3: $z_j = 0$ if and only if $V_j(c) = 0$ for all $0 \le c \le t_j$, $0 \le j \le g$.

Define the $(t+1) \times (t+1)$ matrix

$$T_m = \left[\mathbf{b}(-\sigma_{m,0}) \; \mathbf{b}(-\sigma_{m,1}) \; \cdots \; \mathbf{b}(-\sigma_{m,t}) \right]' . \tag{32}$$

The rows (note the transposition) of T_m are simply the evaluations of the canonical basis elements at $-\sigma_{m,0}$ through $-\sigma_{m,t}$.

Theorem 2: E_m is nonsingular if and only if T_m is nonsingular.

Thus, Theorem 2 replaces condition (27.3) on E_m with a similar condition on the simpler and more explicit matrix T_m. Notice that, unlike E_m, T_m is independent of **a**.

Consider now equation (27.2). Assuming $|T_m| \ne 0$, we can write

$$(\psi_{m,0} \; \psi_{m,1} \; \cdots \; \psi_{m,t}) \triangleq 1\Phi_m = (1 \; 0 \; 0 \; \cdots \; 0) E_m^{-1}. \tag{34.1}$$

By the structure (25) of Φ_m, we then have

$$\phi_{m+t,0} = \psi_{m,t} \tag{34.2}$$

and

$$\phi_{m+k,t-k} = \psi_{m,k} - \psi_{m-1,k+1}, \quad 0 \le k < t. \tag{34.3}$$

Consequently, except for some end-effects, which may result from unspecified δ_m for $m < 0$ and $m > n$-t, equations (34.1)-(34.3) uniquely determine the last row of Φ_m. Since the first t rows of Φ_m appear in Φ_{m-1}, this determines all of Φ_m and, thus, F_m.

In order to satisfy condition (27.4), equation (34.2) must yield a nonzero value for $\phi_{m+t,0}$. By (34.1), the value of $\phi_{m+t,0}$ will be nonzero if and only if the matrix $\hat{E}_m$, obtained by deleting the first column and last row from E_m, is nonsingular. Since the i-th row of E_m is given by

$$E_{m,i} = \mathbf{a} \, \text{diag} \left(M_0(-\sigma_{m,i}) \; e^{-\lambda_1 \sigma_{m,i}} M_1(-\sigma_{m,i}) \; \cdots \; e^{-\lambda_g \sigma_{m,i}} M_g(-\sigma_{m,i}) \right) ,$$

the i-th row of $\hat{E}_m$ is obtained from $E_{m,i}$, $0 \le i \le t$-1, by deleting the first column of $M_0(-\sigma_{m,i})$. An examination of equations (30) and (31) indicates that this change from E_m to $\hat{E}_m$ affects only the component V_0 of V. Since, for $j = 0$, the range of c in (30) is now restricted to $1 \le c \le t_0$, V_0 is replaced by $\hat{V}_0 = [V_0(0) \; V_0(1) \; \cdots \; V_0(t_0\text{-}1)]$. That is, the last entry of V_0, $V_0(t_0)$, is deleted

287

to obtain $\hat{V}_0$. By the same arguments as in the proofs of Lemma 3 and Theorem 2, it follows that $\hat{E}_m$ is nonsingular if and only if the $t \times t$ matrix

$$\hat{T}_m = \left[\hat{b}(-\sigma_{m,0}) \; \hat{b}(-\sigma_{m,1}) \; \cdots \; \hat{b}(-\sigma_{m,t-1})\right]' \tag{35}$$

is nonsingular. Here, $\hat{b}(u)$ is obtained from $b(u)$ by deleting the last component u^{t_0} from the first block $b_0(u)$.

We summarize the above results in the following theorem.

Theorem 3: Given a $(t+1)$-dimensional canonical basis $b(u)$, and a knot vector $(u_0 u_1 \cdots u_{n-t+1})$, there exists a spline basis $f_m(u)$ for the span of $b(u)$ if and only if the matrix T_m of (32) and the $t+1$ matrices $\hat{T}_{m-j}$, $0 \le j \le t$, of (35) are nonsingular. Moreover, when $f_m(u)$ exists, it is essentially unique and given by (27.1) with (25)-(26) and (34.1)-(34.3).

In the next section we will see that for practically every case of interest, the nonsingularity of $\hat{T}_m$ is implied by that of T_m, with the latter condition being independent of m and solely determined by the components of $b(u)$.

7. Tchebycheff Systems

The entries of the matrices T_m, (32), and $\hat{T}_m$, (35), are functions of the λ_j and the knot-vector segment lengths δ_k. The λ_j are part of the definition of the space spanned by $b(u)$, or $\hat{b}(u)$, while the δ_k determine the points in the parameter space at which the canonical basis is evaluated to form the rows of T_m, or $\hat{T}_m$. For all practical purposes, in the context of non-uniform splines, the validity of a spline basis must not depend on the chosen knot-vector. Our goal, therefore, is to obtain conditions on the λ_i under which T_m and $\hat{T}_m$ remain nonsingular for every choice of positive δ_k, independent of the segment index m, and of the number n of control points. For real λ_i, these conditions are fully accounted for by the classical theory of Tchebycheff systems (in short, T-systems) [1], [3].

A set $\beta = \{ \beta_k(u) \mid 0 \le k \le t \}$ of $t+1$ continuous functions over the reals is called a *T-system* if for all $x_0 < x_1 < \cdots < x_t$, the determinant of the matrix

$$\| \beta \| \triangleq \left[\beta_j(x_i) \right], \quad 0 \le i,j \le t, \tag{36}$$

is positive. A T-system β is called *complete* (*CT-system*) if for all $0 \le r \le t$, $\{ \beta_k(u) \mid 0 \le k \le r \}$ is a T-system. Finally, a T-system β of functions of sufficient continuity is called an *extended* T-system (*ET-system*) if for all

$x_0 \leq x_1 \leq \cdots \leq x_t$, and

$$x_{i-1} < x_i = x_{i+1} = \cdots = x_{i+s} < x_{i+s+1},$$

the determinant of the matrix $\|\beta^*\|$ obtained from $\|\beta\|$ of (36) by replacing $\beta_j(x_{i+l})$ with the l-th derivative $\beta^{(l)}(x_i)$, $0 \leq l \leq s$, is positive. A T-system is said to be an *ECT-system* if it is both complete and extended. A space of functions is called an *ECT-space* if it has a basis forming an ECT-system.

Theorem 4: Let $b_k(u)$ denote the k-th component of the canonical basis $\mathbf{b}(u)$, where all the λ_i are real numbers. Then, there exists a sequence $(i_0 i_1 \cdots i_t) \in \{0,1\}^{t+1}$ such that the $t+1$ functions

$$\beta_k(u) = (-1)^{i_k} b_k(u), \quad 0 \leq k \leq t, \tag{37}$$

form an ECT-system.

The following is an immediate corollary of Theorem 4.

Corollary 1: For every set of real, pairwise distinct λ_j, $0 \leq j \leq g$, the matrix T_m of (32) is nonsingular for all m.

Since the reduced basis $\hat{\mathbf{b}}(u)$ of (35) shares all the relevant properties of $\mathbf{b}(u)$, Theorem 4 also applies to $\hat{\mathbf{b}}(u)$, and we have

Corollary 2: For every set of real, pairwise distinct λ_j, $0 \leq j \leq g$, the matrix $\hat{T}_m$ of (35) is nonsingular for all m.

Actually, since the order among the g blocks of $\mathbf{b}(u)$ can be changed arbitrarily without affecting any of the above results, the ECT property of the space spanned by $\mathbf{b}(u)$ guarantees that the nonsingularity of $\hat{T}_m$ is implied by that of T_m.

Corollaries 1 and 2, combined with Theorem 3, yield

Theorem 5: Given a $(t+1)$-dimensional canonical basis $\mathbf{b}(u)$, where all the λ_j are pairwise distinct real numbers, and a knot vector $(u_0 u_1 \cdots u_{n-t+1})$, there exists an essentially unique spline basis $\mathbf{f}_m(u)$ for the span of $\mathbf{b}(u)$.

Notice that the special case of polynomial splines $(g=0, \lambda_0=0)$ is fully covered by Theorem 5.

Real function spaces closed under differentiation can arise in two ways: Either all the λ_j in the canonical basis $\mathbf{b}(u)$ are real, or, if some of them are complex, the set of the λ_j is closed under complex conjugation. In the first case, the existence of a spline basis is guaranteed by Theorem 5. In the second case the general existence question remains open, though in several important cases it can be answered one way or another.

First, we observe that if a spline basis exists, it is all-real. This is due to the proven uniqueness of the spline basis when one exists. With both $b(u)$ and its complex conjugate $b^*(u)$ spanning the same space, they must give rise to the same spline basis which, therefore, must be real.

Second, by inspection of T_m, one can readily verify that, without restricting the δ_k, it is necessary that no pair of distinct λ_j have equal real parts. This, of course, excludes the presence of complex conjugate pairs and, necessarily, requires a restriction on the choice of knot-vectors. Restricting the interval in which the knot-vector may exist leads to a class of trigonometric splines [3, p. 452]. By imposing uniformity, a complete characterization can be obtained and, as shown in Section 8, the necessary and sufficient condition for the existence of a uniform spline basis (i.e., all δ_k equal to 1) is that no two distinct λ_j differ in an integer multiple of $2\pi i$.

Finally, it should be noted that real splines can also be obtained by taking the real part of a proper solution with unrestricted complex λ_j. However, if the set of λ_j is not closed under complex conjugation then the space spanned by the resulting splines is not necessarily closed under differentiation.

8. Uniform Splines

In the case of uniform splines we can set $\delta_k = 1$ for all k, resulting in $\sigma_{m,j} = j + 1$, $0 \leq j \leq t$. In this case, we can apply the minimal polynomial of e^{Δ} to fully characterize the λ_j for which there exists a spline basis. It can be readily verified that, due to the special form of e^{Δ}, the minimal polynomial of e^{Δ} equals its characteristic polynomial, and is given by

$$\mu(x) = \prod_{j=0}^{g} \left(x - e^{\lambda_j} \right)^{t_j + 1}$$

if and only if

$$j \neq k \implies \lambda_j \not\equiv \lambda_k \ (\text{mod } 2\pi i). \tag{40}$$

Consequently, since the degree of $\mu(x)$ equals $t + 1$, Lemma 2 implies that vE_m cannot equal 0 unless $v = 0$. It is also easy to verify that (40) is a necessary condition for E_m to be nonsingular. Similar arguments apply to $\hat{E}_m$, and we have

Theorem 6: The space spanned by $b(u)$ has a uniform spline basis if and only if the λ_j satisfy (40).

References

[1] S. Karlin and W. J. Studden, *Tchebycheff Systems: With Applications in Analysis and Statistics*, John Wiley, New York, NY, 1966.

[2] R.F. Riesenfeld, *Application of B-spline Approximation to Geometric Problems*, Ph.D. dissertation, Department of Systems and Information Science, Syracuse University, Syracuse, NY, May 1973.

[3] L.L. Schumaker, *Spline Functions: Basic Theory*, John Wiley, New York, NY, 1981.

[4] R.H. Bartels, J.C. Beatty, and B.A. Barsky, *An Introduction to Splines for Use in Computer Graphics and Geometric Modeling*, Morgan Kaufman Publishers, Los Altos, CA, 1987.

[5] C. de Boor, *A Practical Guide to Splines*, Springer-Verlag, New York, 1978.

[6] K. Scherer and L.L. Schumaker, "A Dual Basis for L-Splines and Applications", J. Approximation Theory, Vol. 29, 1980, pp. 151-169.

[7] E.A. Coddington and N. Levinson, *Theory of Ordinary Differential Equations*, Robert E. Krieger Publishing Co., Malabar, FL, 1987.

[8] A. Lempel and G. Seroussi, "Systematic Derivation of Spline Bases", HP Laboratories Technical Report, HPL-90-34, May 1990.

Optimal Parallel Pattern Matching
Through Randomization*

Michael O. Rabin

Aiken Computation Lab., Harvard University
Institute of Mathematics, the
Hebrew University of Jerusalem

Abstract

We present an optimal parallel pattern matching algorithm for d-dimensional patterns. Namely, for two d-dimensional arrays A and B, $|A| = m^d = M$, $|B| = n^d = N$, and $k \leq N/\log_2 m$ processors, all occurrences of A in B can be found in time $d \cdot cN/k + \log_2 k$ on an exclusive read exclusive write (EREW) PRAM, or in time $d \cdot cN/k$ on a concurrent read exclusive write (CREW) PRAM. In the CREW algorithm, concurrent read is invoked just once. The main tools are parallel prefix computation and randomization. All previous results in this area [Vishkin, 85; Kedem, Landau, Palem, 89] employed a concurrent read concurrent write (CRCW) PRAM model. As a demonstration of the power of this method, we provide a simple optimally efficient algorithm for the suffix-prefix matching problem of [Kedem, Landau Palem, 89] again for an EREW rather than for a CRCW machine.

1 Refinement of the Parallel Prefix Algorithm

Let S be a semi-group, i.e., a set of elements with an associative multiplication. Assume that the elements of $x \in S$ are represented by words of our parallel machine, and that the operation $x \bullet y$ can be performed by a fixed number of instructions of any one of the processors of the parallel machine. For example, let $S = \{0, 1, \ldots, p-1\}$, where p is a prime which when written in binary fits into a machine word. Let the operation $x \bullet y = (x \cdot y) \bmod p$.

It is well known that for an EREW PRAM with $n/2$ processors, given $x_1, \ldots, x_n \in S$, all the prefix products

$$x_1, \; x_1 \bullet x_2, \; \ldots, \; x_1 \bullet x_2 \bullet \ldots \bullet x_n \tag{1}$$

can be computed in time $2\log_2 n$, performing $2n$ operations.

*Research supported by research contracts ONR-N0001491-J-1981 and NSF-CCR-90-07677 at Harvard University.

We wish to refine this result to the case of bounded parallelism. To do this we use a proposition due to R. Brent.

Proposition 1. *Let C be a computation which can be performed by some number of processors, executing a total number N of instructions, in parallel time T. Then C can be performed by k processors in parallel time T_k where*

$$T_k \leq N/k + T \ .$$

Corollary 1. *Let S be as above and let $k \leq n/\log_2 n$. The prefix products (1) can be computed by k processors in parallel time*

$$T_k \leq 4n/k \ .$$

Proof. Combining the parallel-prefix result with Brent's Proposition we have

$$T_k \leq 2n/k + 2\log_2 n \ .$$

Now, $k \leq n/\log_2 n$ implies $2\log_2 n \leq 2n/k$. □

Assume further that S is a *group* and that the operation $x^{-1} \bullet y$, where x^{-1} is the group-inverse of $x \in S$, can be performed by a fixed number of processor instructions.

Let $x[1 \mathinner{.\,.} n]$ be an array of length n of elements of S. Denote by $x(i, j)$, where $1 \leq i < j \leq n$, the product

$$x(i, j) = x[i] \bullet \ldots \bullet x[j] \ .$$

Lemma 1. *Let $2 \leq m \leq n$ and $k \leq n/\log_2 m$, then all products of length m*

$$x(i, i + m - 1) \ , \qquad 1 \leq i \leq n - m + 1 \tag{2}$$

can be computed by k processors in parallel time $5n/k$.

Proof. Let us first consider the case $n_0 = 2m$, $k_0 \leq n_0/\log_2 m$. It follows from Corollary 1 that all prefix products $x(1, i)$, $2 \leq i \leq n_0$, can be computed by k_0 processors in time $4m/k_0$. Now,

$$x(i, i + m - 1) = x(1, i - 1)^{-1} \bullet x(1, i + m - 1) \ , \qquad 1 \leq i \leq m + 1 \ . \tag{3}$$

So k_0 processors will require less than time m/k_0 for computing all the left-hand side of (3). Thus, altogether, k_0 processors will evaluate the products in question in time $5m/k_0$.

Consider now the general case $x[1 \mathinner{.\,.} n]$ with $k \leq n/\log_2 m$. Let $\ell = n/m$. Cover the array $x[1 \mathinner{.\,.} n]$ by $\ell - 1$ *overlapping* arrays $x[1 \mathinner{.\,.} 2m], x[m+1 \mathinner{.\,.} 3m], \ldots, x[(\ell-2)m+1 \mathinner{.\,.} n]$. Assign $k_0 = k/\ell - 1$ processors to each of these arrays of length $n_0 = 2m$. Now,

$$2m/\log_2 m = 2n/(\ell \cdot \log_2 m) \geq n/((\ell - 1) \cdot \log_2 m) \geq k/(\ell - 1) = k_0 \ ,$$

so that the condition on k_0 in the special case $n_0 = 2m$ holds. Thus for every $x[jm + 1 \mathinner{.\,.} (j + 2)m]$, all products

$$x(i, i + m - 1) \ , \qquad jm + 1 \leq i \leq (j + 1)m + 1 \ ,$$

can be computed in time

$$5m/k_0 = 5m/(k/\ell - 1) = 5m(\ell - 1)/k \le 5n/k .$$

Since every product (2) is a subproduct of one of the above arrays of length $2m$, the Lemma is proved. $\qquad\qquad\square$

2 Fingerprinting of Strings

The material in this section is essentially taken from [Karp, Rabin, 87] and is presented here for the sake of completeness.

Consider the matrices

$$K(0) = \begin{bmatrix} 1 & 0 \\ 1 & 1 \end{bmatrix}, \qquad K(1) = \begin{bmatrix} 1 & 1 \\ 0 & 1 \end{bmatrix} .$$

Given a string $x = x_1 x_2 \ldots x_M \in \{0, 1\}^M$, we define

$$K(x) = K(x_1) \bullet K(x_2) \bullet \ldots \bullet K(x_M) = \begin{bmatrix} a & b \\ c & d \end{bmatrix} .$$

It is readily seen that if x and y are strings and $x \ne y$, then $K(x) \ne K(y)$. Let now p be a prime. We define the *fingerprint* $K_p(x)$ of x by

$$K_p(x) = \begin{bmatrix} a \bmod p & b \bmod p \\ c \bmod p & d \bmod p \end{bmatrix} .$$

It is clear that for strings x and y, $K_p(xy) = K_p(x) \bullet K_p(y)$, where $\bullet$ is the multiplication of matrices $\bmod p$. Whereas the mapping $x \mapsto K(x)$ was one to one, the mapping $x \mapsto K_p(x)$ cannot, in general, be one to one. Thus trying to decide if $x \ne y$ by testing whether $K_p(x) \ne K_p(y)$ can lead to a so-called *false match* where $x \ne y$ but $K_p(x) = K_p(y)$ holds. It will, however, turn out that if we choose p *randomly*, then the probability for a false match can be made small.

Let $R = \{(x(i), y(i)) : x(i), y(i) \in \{0, 1\}^M, 1 \le i \le N\}$ be a set of N pairs of 0–1 strings of length M.

Proposition 2. *If a prime $p \le MN^2$ is randomly chosen, then*

$$\Pr\left(\exists i \; x(i) \ne y(i) \land K_p(x(i)) = K_p(y(i))\right) \le .697/N .$$

This is, in essence, Corollary 10 of [Karp, Rabin, 87].

In the following applications of fingerprints and parallel-prefix computations, the group S will consist of all 2×2 matrices with elements in $\{0, 1, \ldots, p-1\}$ and with determinant 1. Note that for $A \in S$, since $|A| = 1$, we have

$$A = \begin{bmatrix} a & b \\ c & d \end{bmatrix} \rightarrow A^{-1} = \begin{bmatrix} a & p-c \\ p-b & d \end{bmatrix} .$$

Thus the condition concerning the computability by a fixed number of instructions of $A^{-1}B$, for $A, B \in S$, holds true. Note that $K_p(x) \in S$ for all $x \in \{0, 1\}^*$.

294

3 Pattern Matching for d-Dimensional Arrays

Let $A[1 \mathinner{\ldotp\ldotp} m, \ldots, 1 \mathinner{\ldotp\ldotp} m]$ be a d-dimensional 0–1 array of size $m^d = M$. We shall refer to A as the *pattern*. Let $B[1 \mathinner{\ldotp\ldotp} n, \ldots, 1 \mathinner{\ldotp\ldotp} n]$, where $m \leq n$, be a d-dimensional 0–1 array of size $n^d = N$. We shall refer to B as the *text*. For any $1 \leq i_1, \ldots, i_d \leq n{-}m{+}1$ we define the sub-array $C(i_1, \ldots, i_d)$ of size M by

$$C(i_1, \ldots, i_d) = B[i_1 \mathinner{\ldotp\ldotp} i_1 + m - 1, \ldots, i_d \mathinner{\ldotp\ldotp} i_d + m - 1] \, .$$

Strictly speaking, we shall view $C(i_1, \ldots, i_d)$ as a d-dimensional array where the d indices again range on $1 \mathinner{\ldotp\ldotp} m$. Thus for $1 \leq j_1, \ldots, j_d \leq m$,

$$C(i_1, \ldots, i_d)[j_1, \ldots, j_d] = B[i_1 + j_1 - 1, \ldots, i_d + j_d - 1] \, .$$

Definition 1. We shall say that the pattern A has a *match* at $(i_1, \ldots, i_d)$ in the text B if $A = C(i_1, \ldots, i_d)$.

In depicting d-dimensional arrays we shall use the convention that, say for $d = 3$, the indices i_1, i_2, i_3 correspond to the x, y, and z coordinate directions. So that for $d = 2$ and an array $C[1 \mathinner{\ldotp\ldotp} 3, 1 \mathinner{\ldotp\ldotp} 3]$, $C[1, 1]$ will be at the left-lower corner, and $C[3, 3]$ will be at the right-upper corner of the 3×3 square. With this convention

$$A = \begin{matrix} 0 & 1 \\ 1 & 1 \end{matrix}$$

has a match at $(2, 2)$ in

$$B = \begin{matrix} 0 & 0 & 1 \\ 1 & 1 & 1 \\ 0 & 1 & 0 \end{matrix}$$

The d-dimensional pattern matching problem is: Given a d-dimensional pattern A and a d-dimensional text B, find all the matches of A in B.

The strategy of our solution is as follows.

i. Arrange the symbols (bits) of A and of each $C(i_1, \ldots, i_d)$ as strings Str A and Str $C(i_1, \ldots, i_d)$, according to the *post-lexicographic* ordering of the indices. Thus, a 3-dimensional array $A[1 \mathinner{\ldotp\ldotp} 2, 1 \mathinner{\ldotp\ldotp} 2, 1 \mathinner{\ldotp\ldotp} 2]$ will be transformed into

Str $A = A[1,1,1]A[2,1,1]A[1,2,1]A[2,2,1]A[1,1,2]A[2,1,2]A[1,2,2]A[2,2,2] \, .$

The above 2-dimensional array B will be transformed into

Str $B = 010\,111\,001 \, .$

ii. Choose a random prime $p \leq MN^2$, where $|A| = m^d = M$, $|B| = n^d = N$, and compute $K_p(\text{Str } A)$ and $K_p(\text{Str } C(i_1, \ldots, i_d))$ for $1 \leq i_1, \ldots, i_d \leq n - m + 1$.

iii. For every $(i_1, \ldots, i_d)$, $1 \leq i_1, \ldots, i_d \leq n{-}m{+}1$, if $K_p(\text{Str } A) = K_p(\text{Str } C(i_1, \ldots, i_d))$, then print "match at $(i_1, \ldots, i_d)$."

Lemma 2. *Given a d-dimensional array B, $|B| = n^d = N$, and an $m \leq n$. Let $M = m^d$ and $k \leq N/\log_2 m$. All fingerprints $K_p(\text{Str } C(i_1, \ldots, i_d))$ can be computed by a k processor EREW parallel machine in parallel time $d \cdot 5 \cdot N/k$.*

295

Proof. For ease of visualization let us consider the "typical" case of $d = 3$. The parallel computation proceeds in 3 epochs.

In the first epoch we compute for every

$$(i_1, i_2, i_3), \qquad 1 \leq i_1, i_2, i_3 \leq n - m + 1, \tag{4}$$

the product

$$K_p(\text{Str } B[i_1 .. i_1 + m - 1, i_2, i_3]) = K_p(B[i_1, i_2, i_3]) \bullet \ldots \bullet K_p(B[i_1 + m - 1, i_2, i_3]). \tag{5}$$

Keeping i_1, i_2 fixed, the products (5) are all the subproducts of length m of the array $\text{MATX}_1 B[1 .. n, i_2, i_3]$ of matrices $K_p(B[j, i_2, i_3])$.

Assign to each of the n^2 linear arrays $\text{MATX}_1 B[1 .. n, i_2, i_3]$, $k_0 = k/n^2$ processors. We have

$$k_0 = k/n^2 \leq \left(n^3 / \log_2 m\right) / n^2 = n / \log_2 m.$$

So the conditions of Lemma 2 hold and all products (5), for $1 \leq i_1 \leq n - m + 1$, can be computed in parallel time $5n/k_0 = 5n^3/k = 5N/k$.

We now have, for every (i_1, i_2, i_3) in (4), the fingerprint, call it $L(i_1, i_2, i_3)$,

$$L(i_1, i_2, i_3) = K_p(\text{Str } B[i_1 .. i_1 + m - 1, i_2, i_3]).$$

Note that for every (i_1, i_2, i_3) the fingerprint of the 2-dimensional $m \times m$ array attached at (i_1, i_2, i_3) is the product

$$K_p(\text{Str } B[i_1 .. i_1 + m - 1, i_2 .. i_2 + m - 1, i_3]) =$$
$$L(i_1, i_2, i_3) \bullet L(i_1, i_2 + 1, i_3) \bullet \ldots \bullet L(i_1, i_2 + m - 1, i_3). \tag{6}$$

So we define, for every $1 \leq i_1, i_3 \leq n - m + 1$, an array of matrices $\text{MATX}_2 B[i_1, 1 .. n, i_3]$ by

$$\text{MATX}_2 B[i_1, j, i_3] = L(i_1, j, i_3), \qquad 1 \leq j \leq n.$$

In epoch 2 we employ, for every i_1, i_3, k/n^2 processors to compute the products (6). This again requires time $5 \cdot N/k$. Denote the left-hand side of (6) by $SQ(i_1, i_2, i_3)$. Then for $C(i_1, i_2, i_3)$ we have

$$K_p(\text{Str } C(i_1, i_2, i_3)) = SQ(i_1, i_2, i_3) \bullet \ldots \bullet SQ(i_1, i_2, i_3 + m - 1). \tag{7}$$

Thus we compute, in epoch 3, all products (7) in time $5 \cdot N/k$. Altogether we required parallel time $3 \cdot 5N/k$ for the computation of all fingerprints (7). □

Theorem 3. *Let A and B be d-dimensional arrays, $|A| = m^d = M$, $|B| = n^d = N$ and let $k \leq N/\log_2 m$. All matches of A in B can be found by a k-processor EREW parallel machine in time $d \cdot 11N/k + \log_2 k$, and with a probability of having even a single error smaller than $.697/N$. Using a CREW parallel machine, all matches can be found in time $d \cdot 11N/k$, with the same probability of error.*

Proof. Choose a random prime $p \leq MN^2$. Compute all fingerprints $K_p(\text{Str } C(i_1, \ldots, i_d))$ in parallel time $d \cdot 5N/k$.

296

Compute $K_p(\text{Str } A)$ in parallel time $d \cdot 5N/k$. This is done by padding the array A, which is of size M, into an array D of size N by adding 1's. We again use Lemma 2 and note that

$$K_p(\text{Str } A) = K_p(\text{Str } D(1, 1, \ldots, 1)) .$$

Assume that we have a concurrent read exclusive write (CREW) parallel computer. The k processors start by reading $K_p(\text{Str } A)$. This is the *only* time that concurrent read is required. Now, for $1 \leq i_1, \ldots, i_d \leq n - 1m + 1$ do

if $K_p(\text{Str } A) = K_p(\text{Str } C(i_1, \ldots, i_d))$ then MATCH $[i_1, \ldots, i_d] :=$ "yes"
else MATCH $[i_1, \ldots, i_d] :=$ "no".

Here MATCH is a d-dimensional array holding the matching information. The time required is N/k. Thus total time is

$$d \cdot 10N/k + N/k \leq d \cdot 11N/k .$$

If we have an EREW parallel computer, then we start by making k copies of the fingerprint $K_p(\text{Str } A)$. This is done by doubling the number of copies in each cycle, and requires time $\log_2 k$. We then proceed as before. Thus, the total time is $d \cdot 11N/k + \log_2 k$.

The statement concerning probability of even one error is Proposition 2 in Section 2. Note that all actual matches are correctly identified. The only possible error is for a match being declared where there is none. □

As usual, if we want to decrease the probability of error we can randomly choose several primes $p_1, \ldots, p_r \leq MN^2$ and do r of the above computations in time $r \cdot d \cdot cN/k$. The probability of even a single error (false match) becomes smaller than $(.697/N)^r$.

The result in Theorem 3 is best possible for a CREW parallel computer. Consider the case $k = N/\log_2 m$. To compute MATCH $[1, \ldots, 1]$ we must determine whether $A = C(1, \ldots, 1)$. Thus MATCH $[1, \ldots, 1]$ is the conjunction of the results of the $M = m^d$ tests $A[i_1, \ldots, i_d] = B[i_1, \ldots, i_d], 1 \leq i_1, \ldots, i_d \leq m$. In [Cook, Dwork, Reischuck, 86] it is shown, in Theorem 7, that on an *exclusive write* parallel computer with any number of processors, the computation of the conjunction of M boolean values will require time at least $\log_{4.79} M \geq d \cdot 1/3 \cdot \log_2 m$. This holds true even if we assume that every processor has a local memory and can store all past values it read in shared memory. Now, for our k we have $N/k = \log_2 m$ so that we require at least time $d \cdot 1/3 \cdot N/k$ for computing MATCH $[1, \ldots, 1]$.

Thus the result of Theorem 3 is optimal, up to a constant independent of d, and cannot be improved for a *deterministic* algorithm on a CREW parallel machine. We do not know whether it is also optimal for randomized algorithms, but conjecture that it is so. To the best of our knowledge there is no proof that the $d \cdot cN/k$ result [Kedem, Landau, Palem, 89] is the best possible for deterministic algorithms on a CRCW parallel machine, i.e. that the factor d (the dimension) cannot be improved.

4 The Suffix-Prefix Matching Problem

Let $A[1..n]$ and $B[1..n]$ be two 0–1 arrays. We can also assume that $A[i], B[i] \in \Sigma$, $|\Sigma| \leq n$. The S-P matching problem is to find all $1 \leq i \leq n$ such that

$$A[1..n] = B[n-i+1..n] . \tag{8}$$

The result should be written in an array $\delta[1..n]$ so that $\delta[i] = 1$ if (8) holds and $\delta[i] = 0$ otherwise.

In [Kedem, Landau, Palem, 89] it is shown that for $k \leq n/\log_2 n$ processors, the S-P matching problem can be solved in parallel time $c \cdot n/k$ on a CRCW parallel machine.

Theorem 4. *With the above notations, the S-P matching problem can be solved in time $10n/k$ on an EREW parallel machine, with a probability of having even a single error smaller than $.697/n$.*

Proof. Choose randomly a prime $p \leq n^3$. Using Corollary (1) with respect to the prefixes of $A[1..n]$ and of $B[1..n]$, compute $K_p(\text{Str } A[1..i])$, $K_p(\text{Str } B[1..i])$, $1 \leq i \leq n$. This will require parallel time $2 \cdot 4n/k$. Next compute

$$K_p(\text{Str } B[n-i+1..n]) = K_p(\text{Str } B[1..n-i])^{-1} \bullet K_p(\text{Str } B[1..n]) , \qquad 1 \leq i \leq n .$$

This requires time n/k.

For $1 \leq i \leq n$ do
if $K_p(\text{Str } A[1..i]) = K_p(\text{Str } B[n-i+1..i])$, then $\delta[i] := 1$,
else $\delta[i] := 0$.

This again requires time n/k.

The statement concerning the probability of error follows from Proposition 2, observing that it remains true if we assume $\ell(x(i)), \ell(y(i)) \leq M$. □

5 Conclusions

We have seen that the method of randomization produces simple and fast algorithms for a variety of exact pattern-matching problems. In fact, the present author has yet to see a problem in this area not involving pre-processed data, which was solved by a deterministic algorithm, and which is not amenable to the randomized approach.

6 Acknowledgement

The author thanks Z. Galil and L. Valiant for bringing the [Cook, Dwork, Reischuck, 86] reference to his attention.

7 References

[Cook, Dwork, Reischuck, 86] Cook, S., Dwork, C., Reischuck, R., "Upper and lower time bounds for parallel random access machines without simultaneous writes," SIAM Jour. on Comp., vol. 15 (1986), pp. 87–98.

[Karp, Rabin, 87] Karp, R.M., Rabin, M.O., "Efficient randomized pattern-matching algorithms," IBM Jour. of Res. and Dev., vol. 31 (1987), pp. 249–260.

[Kedem, Landau, Palem, 89] Kedem, Z.M., Landau, G.M., Palem, K.V., "Optimal parallel suffix-prefix matching algorithm and applications," Proc. of the 1989 ACM Symp. on Parallel Algorithms and Architectures, pp. 388–399.

[Vishkin, 85] Vishkin, U., "Optimal parallel matching in strings," Info. and Control, vol. 67 (1985), pp. 91–113.

Approximate string-matching and the q-gram distance *

Esko Ukkonen

Department of Computer Science, University of Helsinki

Teollisuuskatu 23, SF-00510 Helsinki, Finland

Abstract. Some results are summarized on approximate string-matching with a string distance function that is computable in linear time and is based on the so-called q-grams ('n-grams'). An algorithm is given for the associated string matching problem that finds the locally best approximate occurrences of pattern P, $|P| = m$, in text T, $|T| = n$, in time $O(n \log(m - q))$. The occurrences with distance $\leq k$ can be found in time $O(n \log k)$. This should be compared to the edit distance based k-differences problem for which the best algorithm currently known needs $O(kn)$. The q-gram distance yields a lower bound for the unit cost edit distance, which leads to a fast hybrid algorithm for the k-differences problem.

1 Introduction

The *approximate string-matching* problem is to find the approximate occurrences of a pattern string P in a text string T [6, 7]. The approximation quality can be measured with different string distance functions. Recently, the version of the problem that is based on the edit distance has received lot of attention [2, 8, 9, 13, 14, 20, 21, 24].

In this paper we study the approximate string-matching in connection with another distance measure, based on the so-called q-grams. Our main motivation is to find alternatives to the edit distance because it leads to dynamic programming that is often relatively slow.

The q-grams ('n-grams') are simply substrings of length q; the concept dates back to Shannon [19]. They have been applied, in many variations, e.g. in different spelling correction methods. Such systems typically preprocess the text (which represents a static dictionary) to make the subsequent searches for 'correct' words faster [12, 17]. In Section 2 we study some properties of a q-gram based string distance measure. In Section 3 we give algorithms for different string matching problems that are based on this measure. For finding the approximate occurrences of P in T that are locally the best ones according to the q-gram distance, we give a solution with running time $O(|T| \log(|P| - q))$. For the threshold version of the problem, in which one wants to

---

*This work was supported by the Academy of Finland and by the Alexander von Humboldt Foundation (Germany). The work was done during a visit to the Institut für Informatik, University of Freiburg.

find the occurrences with a distance $\leq k$, an algorithm with running time $O(|T|\log k)$ is given. We only deal with the case *without* text preprocessing.

The final section points out an important connection to the edit distance based string matching. The q-gram distance provides a non-trivial lower bound on the unit cost edit distance of approximate occurrences of P in T. This leads to the following scheme for solving the k differences problem (the problem of finding the substrings of T such that the unit cost edit distance between the substring and P is $\leq k$ where k is a given threshold value): Compute at each text location i the lower bound for the edit distance between P and the potential occurrence of P that ends at i. At the locations where the bound is $\leq k$, check by some dynamic programming method whether or not there really is an occurrence with at most k differences ending at that location. The dynamic programming can skip over the locations with bound $> k$. Other algorithms with similar hybrid structure have recently been proposed in [2, 9, 20].

The full paper [23] gives the details and some additional results missing in the present summary.

2 The q-gram distance

Let Σ be a finite alphabet, and let Σ^* denote the set of all strings over Σ and Σ^q all strings of length q over Σ, for $q = 1, 2, \ldots$. A q-gram is any string $v = a_1 a_2 \cdots a_q$ in Σ^q.

Definition. Let $x = a_1 a_2 \cdots a_n$ be a string in Σ^*, and let v in Σ^q be a q-gram. If $a_i a_{i+1} \cdots a_{i+q-1} = v$ for some i, then x has an *occurrence* of v. Let $G(x)[v]$ denote the total number of the occurrences of v in x. The *q-gram profile* of x is the vector $G_q(x) = (G(x)[v]), v \in \Sigma^q$.

The distance between two strings is defined as the L_1 norm of the difference of their q-gram profiles.

Definition. Let x, y be strings in Σ^*, and let $q > 0$ be an integer. The *q-gram distance* between x and y is

$$D_q(x, y) = \sum_{v \in \Sigma^q} |G(x)[v] - G(y)[v]| \tag{1}$$

Example. Let $x = 01000$ and $y = 001111$ be strings in the binary alphabet. Their 2-gram profiles are, listed in the lexicographical order of the 2-grams, $(2, 1, 1, 0)$ and $(1, 1, 0, 3)$, and their 2-gram distance is 5.

It is an easy exercise to prove the following properties of the q-gram distance. The length of a string x is denoted $|x|$.

Theorem 2.1 *For all x, y, z in Σ^*,*

(i) $D_q(x,y) = D_q(y,x)$;

(ii) $D_q(x,y) \leq D_q(x,z) + D_q(z,y)$;

(iii) $|(|x| \dot{-} (q-1)) - (|y| \dot{-} (q-1))| \leq D_q(x,y) \leq (|x| \dot{-} (q-1)) + (|y| \dot{-} (q-1))$
where $r \dot{-} s = \max(0, r-s)$;

(iv) $D_q(x_1 x_2, y_1 y_2) \leq D_q(x_1, y_1) + D_q(x_2, y_2) + 2(q-1)$;

(v) *If h is a non-length-increasing homomorphism on Σ^*, then*
$D_q(h(x), h(y)) \leq D_q(x,y)$.

By the properties (i) and (ii), the q-gram distance is a pseudometric. It is not a metric as $D_q(x,y)$ can be 0 even if $x \neq y$. This is the case if x and y have identical q-gram profiles.

It is not difficult to see that $D_q(x,y)$ can be evaluated in time linear in $|x|$ and $|y|$. The main task is to compute fast the non-zero part of the q-gram profiles of x and y. We will present two alternative methods, both based on well-known techniques.

First method. In general we can not assume that a q-gram v as such could serve as an index. Rather, we need an encoding of v as an integer. A natural encoding is to interprete v directly as a c-ary integer, where $c = |\Sigma|$.

Let $v = b_1 b_2 \cdots b_q$, and let $\Sigma = \{A_0, A_1, \ldots, A_{c-1}\}$. Then the *integer code* of the q-gram v is

$$\tilde{v} = \tilde{b}_1 c^{q-1} + \tilde{b}_2 c^{q-2} + \ldots + \tilde{b}_q c^0$$

where $\tilde{b}_i = j$ if $b_i = A_j$. Then, let $x = a_1 \cdots a_n$, and let $v_i = a_i \cdots a_{i+q-1}$, $1 \leq i \leq n - q + 1$, be the q-grams of x. Obviously,

$$\tilde{v}_{i+1} = (\tilde{v}_i - \tilde{a}_i \cdot c^{q-1}) \cdot c + \tilde{a}_{i+q}. \tag{2}$$

By setting $\tilde{v}_1 = \sum_{i=1}^{q} \tilde{a}_i c^{q-i}$ and then applying Eq. (2) for $1 \leq i \leq n - q$ we get the integer codes for all q-grams of x (c.f. [11]). Simultaneously we count the number of the occurences of each q-gram in an array $G[0 : c^q - 1]$ by setting $G[\tilde{v}_i] \leftarrow G[\tilde{v}_i] + 1$ for all i. At the end $G[\tilde{v}] = G(x)[v]$ for each q-gram v. Moreover, we create a list L of codes that really occur in x. Assuming that each application of Eq. (2) takes constant time (this holds true at least for small c and q), the total time for computing G and L is $O(|x|)$. Let us denote these G and L as G_1 and L_1.

Similarly we get $G = G_2$ and $L = L_2$ for a string y in time $O(|y|)$.

Now Eq. (1) gets the form $D_q(x,y) = \sum_{\tilde{v} \in L_1 \cup L_2} |G_1[\tilde{v}] - G_2[\tilde{v}]|$ which obviously can be evaluated in time $O(|x| + |y|)$.

Theorem 2.2 *The q-gram distance $D_q(x,y)$ can be evaluated in time $O(|x| + |y|)$ and in space $O(|\Sigma|^q + |x| + |y|)$.*

The $O(|\Sigma|^q)$ space requirement for tables G_1 and G_2 can for large Σ and q limit the applicability of this very simple algorithm. As only at most $|x| + |y| - 2(q-1)$ elements of tables G_1 and G_2 are active and hence the tables are often sparse, their space requirement can be reduced by standard hashing techniques without increasing the (expected) running time of the method.

Second method. Next we represent a space-efficient but more complicated method that does not use hashing or integer arithmetic. The method codes only the relevant q-grams with small numbers, that are found using suffix automata [1, 4]; suffix trees [26, 16] could be used as well.

Let $Gr(x)$ be the set of different q-grams of x. Eq. (1) can be written as

$$D_q(x,y) = \sum_{v \in Gr(x)} |G(x)[v] - G(y)[v]| + \sum_{v \in \Sigma^q - Gr(x)} G(y)[v]. \tag{3}$$

For evaluating $D_q(x,y)$ it therefore suffices to know the q-gram profile of x and y restricted to $Gr(x)$, and the total number of q-gram occurrences of y such that the corresponding q-gram does not occur in x.

Let $Gr(x) = \{v_1, \ldots, v_r\}$. For each q-gram v in Σ^q, the *code of v with respect to x* is $\bar{v} = i$, if $v = v_i$, and $\bar{v} = 0$ otherwise. Hence there are $r + 1 \leq |x| - q + 2$ different codes.

The codes with respect to x for the q-grams of any string can be found by scanning the string with a modified suffix automaton for x.

Linear time algorithms for constructing the suffix automaton are given in [1, 5].

We use a modification of the standard automaton such that when it scans a string $b_1 b_2 \cdots b_m$, it outputs the codes $\bar{w}_1, \bar{w}_2, \ldots, \bar{w}_{m-q+1}$ for the q-grams $w_i = b_i b_{i+1} \cdots b_{i+q-1}$. Such a modified suffix automaton is denoted $SA_q(x)$.

Theorem 2.3 *The q-gram distance $D_q(x,y)$ can be evaluated using automaton $SA_q(x)$ in time $O(|x||\Sigma| + |y|)$ and in space $O(|x||\Sigma|)$.*

3   String matching with the q-gram distance

Definition Let $T = t_1 t_2 \cdots t_n$ be the *text* and $P = p_1 p_2 \cdots p_m$ the *pattern*, and let q be an integer, $0 < q \leq m$. Both T and P are strings in the alphabet Σ. Let d_i be the minimum q-gram distance between P and the substrings of T ending at t_i, that is, $d_i = \min_{1 \leq j \leq i+1} d_i(j)$ where $d_i(j) = D_q(P, t_j \cdots t_i)$. Moreover, let s_i be the starting location of the *longest* substring of T that gives d_i, that is, s_i is the smallest j such that $d_i = d_i(j)$, $1 \leq j \leq i+1$. (The requirement of the longest string is only to make the problem well-defined; we could require the shortest string as well.) The *approximate string matching problem with the q-gram distance* is to find (d_i, s_i) for $1 \leq i \leq n$.

The following simple properties of d_i and s_i are useful. First, we have

$$0 \leq d_i \leq m - q + 1 \tag{4}$$

as the q-gram distance is nonnegative by definition, and $d_i(i - q + 2) = m - q + 1$ because $t_{i-q+2} \cdots t_i$ contains no q-grams but P contains $m - q + 1$ of them. Moreover,

$$i - 2m + q \leq s_i \leq i - q + 2. \tag{5}$$

The upper bound holds true because always $d_i(j) = m - q + 1$ for $i - q + 2 \leq j \leq i+1$, and s_i is the *smallest* starting point of a substring giving d_i. For the lower bound, we note first that $t_j \cdots t_i$ contains more than $2m - 2q + 2$ q-grams when $j < i - 2m + q$. More than $m - q + 1$ of them can not occur in P because P has only $m - q + 1$ q-grams. Therefore $d_i(j) > m - q + 1$ when $j < i - 2m + q$. On the other hand, $d_i(i - q + 2) = m - q + 1$, and the lower bound follows.

To evaluate d_i it suffices by (5) to find the minimum of $d_i(j)$ for $i - 2m + q \leq j \leq i - q + 2$. These values $d_i(j)$ clearly satisfy the recursion

$$
d_i(j-1) = \begin{cases} d_i(j) - 1 & \text{if the number of the occurrences of the } q\text{-gram} \\ & \quad v = t_{j-1} \cdots t_{j+q-2} \text{ in } t_j \cdots t_i \text{ is } < G(P)[v] \\ d_i(j) + 1 & \text{otherwise} \end{cases} \tag{6}
$$

where the starting value is given by $d_i(i - q + 2) = m - q + 1$. This rule simply says that the q-gram v at t_{j-1} makes the distance $d_i(j - 1)$ smaller (compared to $d_i(j)$) if the number of the occurrences of v in $t_j \cdots t_i$ is not as large as in P; otherwise v makes the distance larger.

In (6) we need to know for every q-gram v of T whether or not v occurs in P. This information is provided by scanning T with the automaton $SA_q(P)$, whose construction was described in Section 2.

Let the different q-grams of P be $w_1, w_2, \ldots, w_M$, in the order of the first occurrence in P. When $SA_q(P)$ has scanned w_h, it outputs code h. A q-gram not in P gets code 0. We denote by τ_j the code of the q-gram $t_j t_{j+1} \cdots t_{j+q-1}$ (the jth q-gram of T), $1 \leq j \leq n - q + 1$.

The q-gram profile $G_q(P)$ of P is represented as a table $G[0 : M]$ such that $G[0] = 0$ and for $h > 0$, $G[h] = G(P)[w_h]$. The construction of $SA_q(P)$ also produces table G.

Algorithm 3.1 is a straightforward implementation of recursion (6) for computing (d_i, s_i). The algorithm uses table $C[0 : M]$ for counting the occurrences of the q-grams of P in $t_j \cdots t_i$ for $j = i - q + 1, \ldots, i - 2m + q$. Function $Scan$ makes $SA_q(P)$ to scan the next text symbol t_i and to return the code of the q-gram ending at t_i; if $i < q$, then the automaton returns 0.

Algorithm 3.1 Evaluation of (d_i, s_i), $1 \leq i \leq n$, for text $T = t_1 \cdots t_n$ and pattern $P = p_1 \cdots p_m$.

1. Construct automaton $SA_q(P)$ and q-gram profile $G[0 : M]$ $(= G_q(P))$
2. for $i \leftarrow 1 - q$ downto $1 - 2m + q$ do $\tau_i \leftarrow 0$
3. for $i \leftarrow 1, \ldots, n$ do
4. $d \leftarrow D \leftarrow m - q + 1$
5. $C[0 : M] \leftarrow 0$; $S \leftarrow 0$
6. $\tau_{i-q+1} \leftarrow Scan(SA_q(P), t_i)$
7.   for $j \leftarrow i - q + 1$ downto $i - 2m + q$ do
8.   if $C[\tau_j] < G[\tau_j]$ then
9. $d \leftarrow d - 1$; $C[\tau_j] \leftarrow C[\tau_j] + 1$
10.   else $d \leftarrow d + 1$
11.   if $d \leq D$ then

12. $S \leftarrow j; D \leftarrow d$
13.   enddo
14. $(d_i, s_i) \leftarrow (D, S)$
15. enddo

Theorem 3.1 *Given a pattern P of length m, a text T of length n, and an integer $q > 0$, Algorithm 3.1 solves the approximate string-matching problem with the q-gram distance function in time $O(m|\Sigma|)$ for preprocessing P, and in time $O((m-q)n)$ for processing T. The algorithm needs working space $O(m|\Sigma|)$.*

Next we develop a method, based on balanced search trees, that implements the innermost minimization loop (lines 7 - 13) of Algorithm 3.1 in time $O(\log(m - q))$.

We write (6) as
$$d_i(j-1) = d_i(j) + h_i(j-1).$$
where $h_i(j-1)$ is $+1$ or -1, as explained in (6). Then solving (6) gives $d_i(j) = m - q + 1 + H_i(j)$ where $H_i(j) = \sum_{k=j}^{i-q+1} h_i(k)$. To find d_i we have to find the minimum of the values $H_i(j)$, $i - 2m + q \le j \le i - q + 1$, for $i = 1, 2, \ldots, n$. We will do this by maintaining a balanced binary search tree.

At the moment when the minimal $H_i(j)$ can be read from the tree for some i, the tree has $2(m - q + 1)$ leaves, that represent, from left-to-right, the numbers $h_i(j)$, $i - 2m + q \le j \le i - q + 1$. Hence the leftmost leaf stores $h_i(i - 2m + q)$, the rightmost leaf stores $h_i(i - q + 1)$, and the leaves of the subtree rooted at some node ν store an interval of values $h_i(j)$.

Each node ν has the normal *llink*, *rlink* and *father* fields. An explicit search key is not needed; logically the key for a leaf λ storing $h_i(j)$ is j. A direct access to λ is given by *Leaf(j)*.

Node ν has also three special fields: *sum*, *min*, and *minindex*. They are defined as follows. Assume that the leaves of the subtree rooted at ν are nodes *Leaf(j)*, $j_1 \le j \le j_2$ for some $j_1 \le j_2$. Then
$$\begin{cases} sum(\nu) = \sum_{j=j_1}^{j_2} h_i(j) \\ min(\nu) = \min_{j_1 \le k \le j_2+1} \sum_{j=k}^{j_2} h_i(j) \\ minindex(\nu) = \text{smallest } k, \ j_1 \le k \le j_2 + 1, \text{ that gives } min(\nu) \text{ above.} \end{cases} \tag{7}$$

Hence *sum* stores the total sum of the values $h_i(j)$ in the leaves of the subtree, *min* stores the smallest partial sum of these values when summed up from right to left, and *minindex* stores the left end point of the longest interval giving the smallest sum.

Given the *sum*, *min*, and *minindex* values for nodes $\nu_r = rlink(\nu)$ and $\nu_l = llink(\nu)$ that satisfy (7), it is immediate that the following rules give these values for ν such that (7) is again satisfied.
$$\begin{cases} sum(\nu) & = \ sum(\nu_l) + sum(\nu_r) \\ min(\nu) & = \ \min(sum(\nu_r) + min(\nu_l), min(\nu_r)) \\ minindex(\nu) & = \ \text{if } sum(\nu_r) + min(\nu_l) \le min(\nu_r) \text{ then} \\ & \quad minindex(\nu_l) \text{ else } minindex(\nu_r). \end{cases} \tag{8}$$

With rule (8) bottom-up building and updating of the tree is possible in constant time per node.

The root ($Root$) of the tree satisfying (7) gives the information we need, because $min(Root) = \min\{H_i(j) : 1-2m+q \leq j \leq i-q+1\}$. Hence $d_i = m-q+1+min(Root)$ and $s_i = minindex(Root)$.

Let B_i be the tree described above and $Root(B_i)$ its root; we call B_i the *tail sum tree* for sequence $(h_i(j))$, $i - 2m + q \leq j \leq i - q + 1$.

Tree B_{i+1} is obtained from B_i by a rather simple transformation. As the leaves of B_{i+1} represent sequence $(h_{i+1}(j))$, $i - 2m + q + 1 \leq j \leq i - q + 2$, we have to find out how $h_{i+1}(j)$ differs from $h_i(j)$.

We call j the *change point* for $i + 1$, denoted $j = cp(i + 1)$, if the following three conditions are met (recall that τ_j is the code of the q-gram starting at t_j and $G[0 : M]$ represents the profile $G_q(P)$).

1. $\tau_j > 0$;

2. there are exactly $G[\tau_j]$ occurrences of the q-gram $t_j \cdots t_{j+q-1}$ in $t_j \cdots t_i$;

3. $\tau_j = \tau_{i-q+2}$ (hence there are $G[\tau_j] + 1$ occurrences of $t_j \cdots t_{j+q-1}$ in $t_j \cdots t_{i+1}$).

Note that the change point is not always defined.

Numbers $h_{i+1}(j)$ and $h_i(j)$ differ only at the change point as we have

$$h_{i+1}(j) = \begin{cases} +1 & (= -h_i(j)) \text{ if } j = cp(i + 1) \\ h_i(j) & \text{otherwise} \end{cases} \qquad (9)$$

Rule (9) shows how the contribution $h_i(j)$ of τ_j to $d_i(j')$, $j' \leq j$, can differ from its contribution $h_{i+1}(j)$ to $d_{i+1}(j')$. If there are $G[\tau_j]$ occurrences of the q-gram $t_j \cdots t_{j+q-1}$ in $t_j \cdots t_i$ but $G[\tau_j] + 1$ occurrences in $t_j \cdots t_{i+1}$, then τ_j contributes -1 to $d_i(j')$ but $+1$ to $d_{i+1}(j')$; only the $G[\tau_j]$ rightmost occurrences of $t_j \cdots t_{j+q-1}$ in $t_j \cdots t_{i+1}$ can decrease $d_{i+1}(j')$.

Change point $j = cp(i+1)$ can be found in constant time by using queues $L[1 : M]$. Queue $L[\tau]$ contains indexes r, in increasing order, such that $\tau_r = \tau$, $h_i(r) = -1$, and tree B_i has a leaf representing $h_i(r)$. The size of $L[\tau]$ is given by $S[\tau]$. Obviously, if $\tau_{i-q+2} > 0$ and $S[\tau_{i-q+2}] = G[\tau_{i-q+2}]$ then $cp(i + 1)$ is the head of $L[\tau_{i-q+2}]$ (i.e., the smallest index stored in this list); otherwise $cp(i + 1)$ is not defined. When $j = cp(i + 1)$ is defined, we update the *sum* field of *Leaf(j)* to $+1$ and remove j from $L[\tau_{i-q+2}]$.

The other changes to B_i include deleting the leftmost leaf $Leaf(i-2m+q)$, because it does not belong to B_{i+1}. The rightmost leaf of B_{i+1} should represent

$$h_{i+1}(i - q + 2) = \begin{cases} -1 & \text{if } \tau_{i-q+2} > 0 \\ +1 & \text{otherwise.} \end{cases}$$

This has no predecessor in B_i. Therefore we insert a new rightmost leaf, pointed by $Leaf(i - q + 2)$.

Algorithm 3.2 summarizes the transformation of B_i into B_{i+1}.

Algorithm 3.2 Updating a tail sum tree B_i into B_{i+1}.

$Delete(\lambda)$ removes leaf λ from B_i, rebalances the tree, and updates the nodes with (8);
$Change(\lambda, h)$, where λ is a leaf, sets $sum(\lambda) \leftarrow h$, and updates the nodes on the path from λ to $Root$ with (8);
$Insert(j, h)$ creates a new rightmost leaf λ such that $Leaf(j) = \lambda$, sets $sum(\lambda) \leftarrow h$, rebalances the tree, and updates the nodes with (8);
$Enqueue$ and $Dequeue$ are the standard queue operations of adding a tail element and removing the head element.

    1. $Delete(Leaf(i - 2m + q))$
    2. if $Head(L[\tau_{i-2m+q}]) = i - 2m + q$ then
    3. $Dequeue(L[\tau_{i-2m+q}])$
    4. $S[\tau_{i-2m+q}] \leftarrow * - 1$
    5. if $\tau_{i-q+2} > 0$ and $S[\tau_{i-q+2}] = G[\tau_{i-q+2}]$ then
    6. $j \leftarrow Dequeue(L[\tau_{i-q+2}])$ % $j = cp(i+1)$
    7. $S[\tau_{i-q+2}] \leftarrow * - 1$
    8. $Change(Leaf(j), 1)$
    9. $Insert(i - q + 2, \text{if } \tau_{i-q+2} > 0 \text{ then } -1 \text{ else } 1)$
    10. $Enqueue(L[\tau_{i-q+2}], i - q + 2)$
    11. $S[\tau_{i-q+2}] \leftarrow * + 1$

The approximate string-matching problem can now be solved with Algorithm 3.3 that uses Algorithm 3.2 as a subroutine.

Algorithm 3.3 Evaluation of (d_i, s_i), $1 \leq i \leq n$, for text $T = t_1 \cdots t_n$ and pattern $P = p_1 \cdots p_m$ using a tail sum tree.

  1. Construct $SA_q(P)$ and $G[0 : M]$
  2. Construct the initial tail sum tree B_0, representing
      $h_0(j) = 1$ for $-2m + q \leq j \leq -q + 1$
  3. for $i \leftarrow 0, \ldots, n - 1$ do
  4. $\tau_{i-q+2} \leftarrow Scan(SA_q(P), t_{i+1})$
  5.   Update B_i to B_{i+1} with Algorithm 3.2
  6. $(d_{i+1}, s_{i+1}) \leftarrow (min(Root(B_{i+1})) + m - q + 1, minindex(Root(B_{i+1})))$
     enddo

Theorem 3.2 *Given a pattern P of length m, a text T of length n, and an integer $q > 0$, Algorithm 3.3 solves the approximate string-matching problem with the q-gram distance function in time $O(m|\Sigma|)$ for preprocessing P and in time $O(n \log(m - q))$ for processing T. The algorithm needs working space $O(m|\Sigma|)$.*

Proof. The preprocessing phase is the same as for Alg. 3.1 and hence needs time and space $O(m|\Sigma|)$. The initial tail sum tree (a balanced binary tree with $2m - 2q + 1$ leaves) can be constructed by standard methods in time and space $O(m - q)$.

The main loop makes n calls to Alg. 3.2. Each call includes one deletion, at most one change, and one insertion to a balanced binary tree of height $O(\log(m-q))$. Each operation also includes re-establishing conditions (8). All this can be performed in $O(\log(m - q))$ time if our tail sum tree is implemented by augmenting some standard balanced tree scheme such as the red-black trees (see e.g. [3, Theorem 15.1]). The operations on queues L in Alg. 3.2 clearly take time $O(1)$. The total time for scanning T (line 4) is again $O(n)$. Hence the main loop takes time $O(n \log(m - q))$.

A working space of $O(m)$ suffices for the tail sum tree, the lists L, and the relevant values τ_i. $\square$

The balanced tree in Algorithm 3.3 creates a considerable overhead compared to the simple Algorithm 3.1. However, the tree is used in a very restricted fashion: all deletions remove the leftmost leaf, all insertions create a rightmost leaf, and the size of the tree remains unchanged.

Therefore it is rather easy to design a tailor-made tree structure with smaller overhead for this particular application. In this structure, the tail sum tree B is represented as a subtree of a larger balanced binary tree C that has twice as many leaves as B has. Tree B circularly glides over C when new leaves are inserted and old ones are deleted. The shape of C can be kept unchanged which means that no time-consuming rebalancing operations are needed. We leave the details of the construction as an exercise to the interested reader.

Next we consider a natural variation of the approximate string matching problem.

Definition The *threshold problem* is to find, for a given integer $k \geq 0$, all text locations i such that the q-gram distance d_i is $\leq k$.

Lemma 3.1 *If $d_i \leq k$ then $i - m + 1 - k \leq s_i \leq i - m + 1 + k$.*

By the lemma, one can test whether or not $d_i \leq k$ by computing $d = \min_{a(i)-k \leq j \leq a(i)+k} d_i(j)$ where $a(i)$ denotes $i - m + 1$. If $d \leq k$, then we know that $d_i \leq k$ because $d_i = d$. If $d > k$, then also $d_i > k$. As d is the minimum of only $2k + 1$ elements, it turns out that it can be found for each i in time $O(\log k)$. Only a minor modification to Algorithm 3.3 is needed.

Theorem 3.3 *Given pattern P of length m and text T of length n in alphabet Σ, and $k \geq 0$, the threshold problem can be solved in time $O(m|\Sigma|)$ for preprocessing P and $O(n \log k)$ for processing T. The working space requirement is $O(m|\Sigma|)$.*

Another variation of interest is to find the distances between P and substrings of T of a fixed length, that is, we want to evaluate values $d_i(i - \delta)$ for some fixed $\delta \geq 0$.

A linear time method, based on straightforward bookkeeping, is easy to develop. We formulate it as Algorithm 3.4; a similar method in the special case $q = 1$ is given in [9].

308

Algorithm 3.4 Evaluation of $D_i = d_i(i - \delta)$, $\delta + 1 \leq i \leq n$, for text $T = t_1 \cdots t_n$, and pattern $P = p_1 \cdots p_m$, and integer $\delta \geq 0$. We assume that $\delta + 1 \geq q$; otherwise the problem is trivial as then each $D_i = m - q + 1$.

1. Construct $SA_q(P)$ and $G[0:M]$
2. $Scan(SA_q(P), t_1 \cdots t_{q-1})$
3. **for** $i \leftarrow q, \ldots, \delta$ **do**
4. $\tau_{i-q+1} \leftarrow Scan(SA_q(P), t_i)$
5. $G[\tau_{i-q+1}] \leftarrow * - 1$
   enddo
6. $D \leftarrow |G[0]| + \cdots + |G[M]|$
7. **for** $i \leftarrow \delta + 1, \ldots, n$ **do**
8. $\tau_{i-q+1} \leftarrow Scan(SA_q(P), t_i)$
9. $G[\tau_{i-q+1}] \leftarrow * - 1$
10.   **if** $G[\tau_{i-q+1}] \geq 0$ **then** $D \leftarrow D - 1$ **else** $D \leftarrow D + 1$
11. $D_i \leftarrow D$ % $D_i = d_i(i - \delta)$
12. $G[\tau_{i-\delta}] \leftarrow * + 1$
13.   **if** $G[\tau_{i-\delta}] \leq 0$ **then** $D \leftarrow D - 1$ **else** $D \leftarrow D + 1$
   enddo

In Algorithm 3.4, array G is initialized (line 1), as before, to represent the q-gram profile of P. During the main loop at line 11 G satisfies the invariant $G[\bar{v}] = G_q(P)[v] - G_q(t_{i-\delta} \cdots t_i)[v]$ for the q-grams v of P, and $G[0]$ counts (in negative) the number of other q-grams in $t_{i-\delta} \cdots t_i$. Moreover, D satisfies the invariant $D = \sum_{j=0}^{M} |G[j]|$. Hence $D = D_q(t_{i-\delta} \cdots t_i, P)$, that is, the algorithm is correct. Clearly, it needs time $O(n)$ for processing T.

Theorem 3.4 *Algorithm 3.4 evaluates the q-gram distances $d_i(i-\delta)$ for $\delta+1 \leq i \leq n$ in time $O(n)$ for processing T. The preprocessing time and working space are as in Theorem 3.3.*

4   A hybrid method for the k-differences problem

The *edit distance $DE(x,y)$* between strings x and y is defined as the minimum possible number of editing steps that convert x into y [15, 25]. We restrict our consideration to the case where each editing step is a rewriting rule of the form $a \rightarrow \varepsilon$ (a deletion), $\varepsilon \rightarrow a$ (an insertion), or $a \rightarrow b$ (a change) where a, b in Σ are any symbols, $a \neq b$, and ε is the empty string. Each symbol of x is allowed to be rewritten by some editing operation at most once.

The associated approximate pattern matching problem with threshold is, given pattern P and text T as before, and an integer $k \geq 0$, to find all i such that the minimum of the edit distances between P and the substrings of T ending at t_i is $\leq k$; the problem is also known as the *k differences problem*. Hence, if we let $d_i = \min_{1 \leq j \leq i} DE(P, t_j \cdots t_i)$, the problem is to find all i such that $d_i \leq k$.

Now, let $d_i(j) = D_q(P, t_j \cdots t_i)$ be as in Section 3, for some $q > 0$.

Theorem 4.1 *For $1 \le i \le n$, $d_i(i - m + 1)/(2q) \le de_i$.*

Proof. Let $d = de_i$. The theorem follows if we show that at most dq of the q-grams of P are missing in $t_{i-m+1} \cdots t_i$ (here $t_j = \varepsilon$ if $j < 1$), as then $d_i(i - m + 1) \le 2dq$.

Let P' be the substring of T ending at t_i such that $DE(P, P') = d$. String P' can be obtained from P with at most d insertions, deletions and changes. A deletion or a change at character p_i of P destroys at most q q-grams of P, namely those that contain p_i. An insertion between p_i and p_{i+1} destroys at most $q - 1$ q-grams of P, namely those that contain both p_i and p_{i+1}. Hence at most $d_1 q + d_2(q - 1)$ q-grams of P are missing in P' where d_1 is the total number of deletions and changes, and d_2 is the total number of insertions. As $|P'| \le m + d_2$, string $t_{i-m+1} \cdots t_i$ contains all q-grams of P' except for at most d_2. Hence at most $d_1 q + d_2(q - 1) + d_2 = dq$ of the q-grams of P are not present in $t_{i-m+1} \cdots t_i$, which proves the theorem. $\square$

A similar proof shows that we also have $d_i/(2q) \le de_i$ where $d_i = \min_{j \le i} d_i(j)$, and that for all strings x, y, we have $D_q(x, y)/(2q) \le DE(x, y)$.

Values de_i can be found by evaluating table (D_{ji}), $0 \le i \le n$, $0 \le j \le m$, where $D_{ji} = \min_{1 \le i' \le i} DE(t_{i'} \cdots t_i, p_1 \cdots p_j)$ from recursion

$$D_{ji} = \min\{D_{j-1,i} + 1, D_{j,i-1} + 1, \text{if } p_j = t_i \text{ then } D_{j-1,i-1} \text{ else } D_{j-1,i-1} + 1\}$$

with initial conditions $D_{0,i} = 0$ for $0 \le i \le n$ and $D_{j,0} = j$ for $1 \le j \le m$. This dynamic programming method solves the problem as $de_i = D_{mi}$, see [18, 21]. In particular, the k differences problem can be solved in $O(kn)$ time by a 'diagonal' modification of this method [8, 24]. Such algorithms can be speeded-up by the following hybrid method.

First, evaluate by Algorithm 3.4 the distances $d_i(i - m + 1)$ for $1 \le i \le n$. Mark all i such that $d_i(i - m + 1)/(2q) \le k$. Then evaluate de_i only for the marked i. As only such de_i can be $\le k$ by Theorem 5.1, we get the solution of the k differences problem. By Theorem 3.4, the marking phase takes time $O(n)$, with $O(m|\Sigma|)$ time for preprocessing P.

A *diagonal* e of (D_{ji}) consists of entries D_{ji} such that $i - j = e$. If $de_i \le k$, it is easy to see that to find $de_i = D_{mi}$ correctly, it suffices to restrict the evaluation of (D_{ji}) to $2k + 1$ successive diagonals with the diagonal of D_{mi} in the middle. The simplest way to do this is just to apply the above recursion for (D_{ji}), restricted to these diagonals. This takes time $O(km)$.

An asymptotically faster method is obtained by restricting the algorithm of [8] or [24] to the $2k + 1$ diagonals. This gives an method requiring time $O(k^2)$ for the evaluation of the diagonals. It also needs time $O(m^2)$ for preprocessing P (this can be improved to $O(m)$ by using lowest common ancestor algorithms). Moreover, the method scans $t_{i-m-k+1} \cdots t_i$. However, we have to scan over this string anyway when marking i.

If the relevant diagonals for different marked i overlap, their evaluation can be combined such that each diagonal is evaluated at most once. As the method spends time $O(k)$ per diagonal, the total time for evaluating the marked values de_i is always

$O(kn)$. The marking and dynamic programming phases can be combined so that repeated scanning of T is not necessary.

Summarizing, we have:

Theorem 4.2 *The k differences problem can be solved in time $O(\min(n + rk^2, kn))$ where r is the number of indexes i such that $d_i(i - m + 1)/(2q) \leq k$. The method needs time $O(m|\Sigma| + m^2)$ for the preprocessing of P.*

Grossi and Luccio [9] propose a similar method for the special case where $q = 1$ and the only editing operation used in the definition of the edit distance is the change (the k *mismatches problem*).

References

[1] A. Blumer, J. Blumer, D. Haussler, A. Ehrenfeucht, T. Chen and J. Seiferas: The smallest automaton recognizing the subwords of a text. *Theor. Comp. Sci.* 40 (1985), 31–55.

[2] W. I. Chang and E. L. Lawler: Approximate string matching in sublinear expected time. In: *Proc. IEEE 1990 Ann. Symposium of Foundations of Computer Science*, pp. 116–124.

[3] T. H. Cormen, C. E. Leiserson and R. L. Rivest: *Introduction to Algorithms.* (The MIT Press 1990.)

[4] M. Crochemore: Transducers and repetitions. *Theor. Comp. Sci.* 45 (1986), 63–89.

[5] M. Crochemore: String matching with constraints. In: *Proc. MFCS'88 Symposium.* Lect. Notes in Computer Science 324, (Springer-Verlag 198), 44–58.

[6] G. R. Dowling and P. Hall: Approximate string matching. *ACM Computing Surveys* 12 (1980), 381–402.

[7] Z. Galil and R. Giancarlo: Data structures and algorithms for approximate string matching. *J. Complexity* 4 (1988), 33–72.

[8] Z. Galil and K. Park: An improved algorithm for approximate string matching. In: *Automata, Languages, and Programming (ICALP'89).* Lect. Notes in Computer Science 372 (Springer-Verlag 1989), 394–404.

[9] R. Grossi and F. Luccio: Simple and efficient string matching with k mismatches. *Inf. Proc. Letters* 33 (1989), 113–120.

[10] P. Jokinen, J. Tarhio, and E. Ukkonen: A comparison of approximate string matching algorithms. Submitted.

[11] R. M. Karp and M. O. Rabin: Efficient randomized pattern matching. *IBM J. Res. Dev.* 31 (1987), 249–260.

[12] T. Kohonen and E. Reuhkala: A very fast associative method for the recognition and correction of misspellt words, based on redundant hash-addressing. In: *Proc. 4th Joint Conf. on Pattern Recognition*, 1978, Kyoto, Japan, pp. 807–809.

[13] G. Landau and U. Vishkin: Fast string matching with k differences. *J. Comp. Syst. Sci.* 37 (1988), 63–78.

[14] G. Landau and U. Vishkin: Fast parallel and serial approximate string matching. *J. Algorithms* 10 (1989), 157–169.

[15] V. I. Levenshtein: Binary codes of correcting deletions, insertions and reversals. *Sov. Phys.-Dokl* 10 (1966), 707–710.

[16] E. M. McCreight: A space-economical suffix tree construction algorithm. *J. ACM* 23 (1976), 262–272.

[17] O. Owolabi and D. R. McGregor: Fast approximate string matching. *Software – Practice and Experience* 18 (1988), 387–393.

[18] P. H. Sellers: The theory and computation of evolutionary distances: pattern recognition. *J. Algorithms* 1 (1980), 359–373.

[19] C. E. Shannon: A mathematical theory of communications. *The Bell Systems Techn. Journal* 27 (1948), 379–423.

[20] J. Tarhio and E. Ukkonen: Boyer–Moore approach to approximate string matching. In: *Proc. 2nd Scand. Workshop on Algorithm Theory (SWAT'90)*, Lect. Notes in Computer Science 447 (Springer-Verlag 1990), 348–359.

[21] E. Ukkonen: Finding approximate patterns in strings. *J. Algorithms* 6 (1985), 132–137.

[22] E. Ukkonen: Algorithms for approximate string matching. *Information and Control* 64 (1985), 100–118.

[23] E. Ukkonen: Approximate string-matching with q-grams and maximal matches.

[24] E. Ukkonen and D. Wood: Approximate string matching with suffix automata. Submitted. Report A-1990-4, Department of Computer Science, University of Helsinki, April 1990.

[25] R. E. Wagner and M. J. Fisher: The string-to-string correction problem. *J. ACM* 21 (1974), 168–173.

[26] P. Weiner: Linear pattern matching algorithms. In: *Proc. 14th IEEE Ann. Symp. on Switching and Automata Theory*, 1973, pp. 1–11.

Universal Discrimination of Individual
Sequences via Finite-State Classifiers

Jacob Ziv and Neri Merhav

Department of Electrical Engineering
Technion - Israel Institute of Technology
Haifa 32000, ISRAEL

Abstract

The problem of classifying individual sequences into a finite number of classes with respect to a given set of training sequences using finite-state (FS) classifiers, is investigated. FS classifiers are sought that discriminate between two given distinct sources of sequences. In practice, the sources are not known and hence such a classifier is not implementable. We propose a simple classification algorithm which is universal in the sense of being independent of the unknown sources. The proposed algorithm discriminates between the sources whenever they are distinguishable by some finite-memory classifier, for almost every given training sets from these sources. This algorithm is based on the Lempel-Ziv data compression algorithm and is associated with a new notion of empirical informational divergence between two individual sequences.

I. Introduction

The problem of classifying information sources is traditionally posed in a probabilistic framework, where the goal is normally to minimize the probability of error or some other related performance criterion. In the classical theory of hypothesis testing (see, e.g., [1],[2]), complete knowledge of the underlying probability measures is assumed. Since this assumption is rarely met in practice, considerable efforts have been

This research was supported by the U.S - Israel Binational Science Foundation.

313

made in recent years to relax its necessity and to develop, for certain classes of sources, universal classification rules that are independent of the unknown underlying measures and yet perform asymptotically as well as the optimal likelihood ratio test (see, e.g., [3]-[8]).

In this presentation, we attempt to make an additional step towards universality and relaxation of statistical modeling assumptions. Similarly to the approach taken in [9], rather than modeling the data generation mechanism by an underlying probability measure, we allow the data to be arbitrary but we limit the class of permissible schemes to consist only of these that can be implemented by finite-state machines (FSM's). This setup often reflects a realistic situation, where we have no faithful statistical model on one hand, and we are equipped with limited resources of computation power and storage, on the other. Furthermore, this approach is more general than that of both assuming a statistical model and limiting the class of allowed schemes (see, e.g., [10]-[12] and references therein).

Specifically, we seek a classifier with a rejection option, which is trained by a given set of training sequences from each class, and that assigns to each class a small set of vectors such that the sources will be distinguishable. This is done subject to an axiomatic consistency constraint that if a test sequence appears in the training set, it will be classified correctly. The rationale behind this objective of creating small classification sets (clusters) is that small clusters correspond to "high resolution" classifiers that can distinguish well between sources which are relatively "close" to each other. Analogously, in the probabilistic framework, the atoms of classification cells are equivalence classes induced by the possible values of the test statistics. The more detailed statistics are extracted from the data, the smaller are these atoms and the resolution (or refinement) of the classifier improves. On the other hand, one would like to avoid a situation of over-resolution since in the case, there might be an insufficient number of available training sequences to cover all atoms. We shall elaborate further on this point later on.

314

We propose a computationally attractive universal classification rule, based on the Lempel-Ziv (LZ) data compression algorithm, which fulfills the above requirements for every possible training set except for a vanishingly small fraction of training sequences from each class. This classification rule introduces an interesting generalized notion of informational divergence (or relative entropy) between two individual sequences.

II. Preliminaries and Problem Formulation

For the sake of simplicity and convenience, we shall consider a two-class problem and assume that all observation sequences are of the same length n. The results will extend straightforwardly to the more general situation.

The following generic notation will be used. A sequence z is an ordered collection $z_1, z_2, ..., z_n$ of observations, where z_i takes on values in a finite set A with $|A| = A$ elements. The space of all A^n possible sequences of length n will be denoted by A^n. Let $\phi_i = \{x_1^i, x_2^i, \cdots x_k^i\}$, $i = 1, 2$, denote two disjoint given collections of k arbitrary (nonrandom) vectors in A^n. The sequences x^j, $j = 1, 2, ..., k$, will be referred to as *training sequences* and ϕ_i, $i = 1, 2$, are the *training sets* corresponding to "sources" σ_i, $i = 1, 2$. Here a source represents an abstract entity which governs the respective training set in the sense of possessing certain characteristic features which are shared by vectors of one training set but not by vectors of the other. Therefore, it is convenient to think of a source as a large hidden collection of vectors which is partly seen via the training set, i.e., $\sigma_i \supseteq \phi_i$. Normally, each training set contains only a small fraction of vectors from the source. For example, the source may include an exponentially large number of sequences (as a function of n) while the training set size k is only polynomial or even fixed.

The classification problem is as follows. Upon observing a test vector $y \in A^n$ to be classified and given the training sets ϕ_i, $i = 1, 2$, decide whether $y \in \sigma_i$, $i = 1, 2$ or reject y, namely, decide that $y \in \sigma_0 \triangleq \sigma_1^c \cap \sigma_2^c$, where the superscript c denotes the

315

complementary set. A classification rule M is a partition of A^n into three disjoint sets M_i, $i = 0, 1, 2$, where M_0 is the rejection region, and M_1 and M_2 are decision regions corresponding to the two sources σ_1 and σ_2. We allow only classification rules that are *consistent* with the training sets, i.e., $M_i \supseteq \phi_i$, $i = 1, 2$, and seek a consistent classification rule M with small classification sets M_1 and M_2. Ideally, one wishes to find a classification rule such that M_1 and M_2 are so small that $M_1 \cap \sigma_2$ and $M_2 \cap \sigma_1$ are empty sets, namely, classification errors are avoided.

The permissible family $\mathbf{M}$ in this paper consists of all classifiers that can be realized by FSM's followed by modulu-n counters. Specifically, an S-state classifier is a triple $C = (S, g, \Omega)$, where S is a finite set of states with S elements, $g: A \times S \to S$ is a *next-state function*, and Ω is a partition of the space of empirical probability measures on $A \times S$, into three disjoint regions Ω_i, $i = 0, 1, 2$, depending on the training sets, where Ω_0 is the rejection region and Ω_1 and Ω_2 are acceptance regions of σ_1 and σ_2, respectively. When a test sequence $\mathbf{y} = y_1, y_2, ..., y_n$ is fed into C, which in turn is initialized with $s_0 \in S$, a state sequence $\mathbf{s} = s_1, s_2, ..., s_n, s_i \in S$, is generated by

$$s_t = g(y_t, s_{t-1}), \qquad t = 1, 2, ..., n. \tag{1}$$

Let $n_{\mathbf{y}}^g(a, s)$, $a \in A$, $s \in S$, denote the joint count of $y_t = a$ and $s_{t-1} = s$ along the pair sequence $(\mathbf{y}, \mathbf{s})$, and let $q_{\mathbf{y}}^g(a, s) = n_{\mathbf{y}}^g(a, s)/n$ denote the empirical joint probability of a and s with respect to g. The empirical joint probability distribution $Q_{\mathbf{y}}^g = \{q_{\mathbf{y}}^g(a, s),$ $a \in A, s \in S\}$ serves as test statistics for classifying $\mathbf{y}$, that is, the classification rule $M = \{M_i\}_{i=0}^2$, associated with a FS classifier C is given by

$$M_i = \{\mathbf{y} \in A^n : Q_{\mathbf{y}}^g \in \Omega_i\}, \qquad i = 0, 1, 2. \tag{2}$$

where the partition Ω (and hence also M) in turn depends on the training sequences via the empirical distributions $Q_{\xi_j}^g$, $j = 1, 2, ..., k$, $i = 1, 2$. These empirical distributions are precomputed in the training phase.

Two sequences $\mathbf{y}$ and $\mathbf{z}$ are said to be of the same *type with respect to g*, or, of the same g-*type* if $Q_y^g = Q_z^g$. The g —type of z is defined as

$$T_g(z) \triangleq \{y \in A^n: Q_y^g = Q_z^g\}. \tag{3}$$

It should be pointed out that by confining our interest to FS classifiers, we exclude uninteresting trivialities, e.g., the classification rule $M_i = \phi_i$, $i = 1, 2$, which requires an exponential number of states as a function of n. Furthermore, we avoid the need for an exponentially large number of training sequences in each set. Generally speaking, a good training set need not contain more than one representative from each type with respect to every possible next-state function g, because two training sequences of the same g-type carry exactly the same information accessible to a classifier C that employs g as a next-state function. Since there are less than $(n + 1)^{AS}$ different types with respect to every g [12], there is no point in sampling more than $(n + 1)^{AS}$ good training sequences for a given g from each source. Observe that any family M of classifiers that requires a relatively small (i.e., subexponential) number of training sequences and hence a small number of equivalence classes, can be a reasonable choice for a class of practical classification schemes. Thus, the family of FS classifiers is a good choice in that respect.

III. Classification Rules

We first observe that for a given next-state function g, the smallest acceptance regions associated with a consistent FS classifier must include the entire g-type of each training sequence. Suppose that the g-types corresponding to every two training sequences from different sources are disjoint. Thus, the acceptance regions

$$M_i = \bigcup_{j=1}^{k} T_g(x_j^i), \qquad i = 1, 2 \tag{4}$$

are the best possible for a given g in the above defined sense.

We next confine our interest to the case where g is the next-state function of an lth order *Markovian* machine (i.e., finite-memory machine), where the state at time t, s_t is

317

the string of l preceding letters $(x_{t-l}, x_{t-l+1}, \ldots, x_{t-1})$. It will be assumed that the sources are distinguishable by this machine for some l but the value(s) of such an l is unknown.

Let l be a fixed integer and consider the empirical probabilities of vectors of length $(l + 1)$ defined as

$$q_z(a^{l+1}) = \frac{1}{n} \sum_{i=1}^{n} 1\{z_i^{i+l} = a^{l+1}\}, \qquad a^{l+1} \in A^{l+1} \tag{5}$$

where z_j^i ($j \le i$) denotes the string $(z_j, z_{j+1}, \ldots, z_i)$ with the cyclic convention that $z_{-i} = z_{n-i}$, and $1\{D\}$ is the indicator function of an event D. Let $q_z(a_{l+1}|a^l) \triangleq q_z(a^{l+1})/q_z(a^l)$ denote the empirical transition probability from a string $a^l \in A^l$ to a letter $a \in A$ where a^{l+1} denotes the concatenation of a^l and a_{l+1}. Let $Q_z = \{q_z(a^{l+1}), a^{l+1} \in A^{l+1}\}$ denote the empirical distribution of $(l+1)$–vectors, and define $T^l(z)$ as the type of z with respect to Q_z. Define the distance between the empirical distributions associated with two sequences y and z as

$$d(y,z) \triangleq \sum_{a^{l+1} \in A^{l+1}} |q_y(a^{l+1}) - q_z(a^{l+1})| - \sum_{a^l \in A^l} |q_y(a^l) - q_z(a^l)|. \tag{6}$$

It is easy to see that $d(y,z)$ is strictly positive iff the empirical conditional probabilities $q_y(\cdot|a^l)$ and $q_z(\cdot|a^l)$ are distinct for some $a^l \in A^l$. Thus, in fact, $d(\cdot,\cdot)$ can be regarded as a distance measure between the conditional measures although these do not appear explicitly in (6). The motivation of this distance definition is that the conditional relative entropy between two lth order Markov processes

$$D(Q_y|Q_z) = \sum_{a^{l+1} \in A^{l+1}} q_y(a^{l+1}) \log \frac{q_y(a_{l+1}|a^l)}{q_z(a_{l+1}|a^l)} \tag{7}$$

can be lower bounded in terms of the distance d between them (see Appendix), but unlike the relative entropy, the distance measure (6) is not too sensitive near singularity points of the reference measure Q_z. Thus, it will be useful as a measure of distinguishability between two sources which is defined as follows.

Definition: A pair of sources (σ_1, σ_2) is said to be (ε, l) *- separable* for some small $\varepsilon > 0$ and some positive integer l if $d(x_i^1, x_j^2) \geq \varepsilon$ for any pair of training sets $\phi_i \subseteq \sigma_i$, $i = 1, 2$ and for any $i, j \in \{1, 2, \ldots, k\}$.

The underlying assumption is that the two sources are distinguishable (separable) if there exists some value of l such that the l-th order conditional measures are sufficiently far apart. Define next the δ-neighborhood type $T_\delta^l(z)$ of a vector z as the set of all sequences y such that $D(Q_y | Q_z) \leq \delta$. The classification rule M^* is defined as follows.

$$M_1^* = \bigcup_{i=1}^{k} T_\delta^l(x_i^1) - \bigcup_{i=1}^{k} T_\delta^l(x_i^2) \tag{8.a}$$

and

$$M_2^* = \bigcup_{i=1}^{k} T_\delta^l(x_i^2) - \bigcup_{i=1}^{k} T_\delta^l(x_i^1), \tag{8.b}$$

where $\delta > 0$ is a parameter which can be made arbitrarily small. To avoid classification errors, the training sequences of σ_2 and their $\delta-$ neighborhoods are excluded from M_1^*, and vice versa. Note, that M^* is a version of (4) where g is Markovian and the types are slightly extended to include their δ-neighborhoods. This extension, which is made for technical reasons, does not affect significantly the discrimination ability of the classifier but it makes M^* a convenient reference for our universal classification rule which will be defined later.

It can be shown (see Appendix) that if δ is chosen smaller than $\varepsilon^2/(8\ln 2)$ in (8), then the training sequences of σ_i and their δ-neighborhoods are all included in M_i^*, $(i = 1, 2)$ provided that the pair (σ_1, σ_2) is (ε, l)-separable. This means that in this case M^* is consistent with the training sets and it discriminates between the sources without classification errors. Furthermore, it is easy to show (see, e.g., [18]) that if there exists a value of l much larger than $\log S$ for which the sources are (ε, l)-separable, then the cardinality of M_i^* is exponentially not larger than that of M_i of (4) (up to a small factor depending on δ, l, S, and n), for any S-state classifier. Following the discussion at the end of Sect. II, the required number of training sequences for M^* grows double-exponentially with l.

In the implementation of M^* there is a tradeoff between the computational complexity and the storage requirement where either of these quantities grows at least as fast as $n \log n$. A more crucial difficulty with M^*, however, is that the value of l for which the sources are (ε, l)-separable is normally unknown. Furthermore, one *cannot* guarantee (ε, l)-separability by selecting a "large enough" l because $d(y, z)$ is not necessarily a monotonic function of l and thus (ε, l)-separability does not imply $(\varepsilon, l + 1)$- separability.

We next present an alternative classification scheme, based on the Lempel-Ziv (LZ) data compression algorithm, which alleviates this difficulty and where both computational complexity and memory size grow linearly with n [16],[17]. To this end, we first describe an incremental parsing procedure of a vector y with respect to another vector z.

First, find the longest prefix of y that appears as a string in z, i.e., the largest integer p such that $(y_1, y_2, \ldots, y_p) = (z_i, z_{i+1}, \ldots, z_{i+p-1})$ for some i. The string $(y_1, y_2, \ldots, y_p)$ is defined as the first phrase of y with respect to z. Next, start from y_{p+1} and find the longest prefix of $y_{p+1}, y_{p+2}, \ldots, y_n$, which appears in z in a similar manner, and so on. The procedure is terminated once the entire vector y has been parsed with respect to z. Let $c(y|z)$ denote the number of phrases of y with respect to z, and let $c(y)$ denote the number of phrases in y resulting from "self-parsing" [9], i.e., sequential incremental parsing of y into distinct phrases such that each phrase is the shortest string which is not a previously parsed phrase. For example, let A = $\{0,1\}$, $n = 11$, y = (01111000110), and z = (10010100110). Then, parsing y with respect to z yields (011,110,00110), that is, $c(y|z) = 3$, and self-parsing of y results in (0,1,11,10,00,110), namely, $c(y) = 6$.

For two sequences y and z, define the function

$$\Delta(y|z) = \frac{1}{n}[c(y|z) \log n - c(y) \log c(y)]. \tag{9}$$

320

The function $\Delta(y|z)$ can be interpreted as a generalized notion of the relative entropy between y and z. The analogy is as follows. Similarly to the relative entropy between two probability distributions $D(P|Q)$, which expresses the expected redundancy when encoding a source P using a codebook that best matches Q, the function $\Delta(y|z)$ carries a similar meaning for universal coding of individual sequences. The first term in (8) is essentially the number of bits required for encoding y given z, where each phrase is encoded by the index of its first location in z. The second term is approximately the LZ codeword length of y. In fact, it is shown in [15] that $\Delta(y|z)$ and the relative entropy between the respective empirical distributions are strongly related.

Fix $\xi > 0$ and define the classification rule M^{**} by

$$M_1^{**} = \bigcup_{i=1}^{k} \left\{ y : \Delta(y|x_i^1) \le \xi \right\} - \bigcup_{i=1}^{k} \left\{ y : 0 < \Delta(y|x_i^2) \le \xi \right\} \qquad (10.\text{a})$$

and

$$M_2^{**} = \bigcup_{i=1}^{k} \left\{ y : \Delta(y|x_i^2) \le \xi \right\} - \bigcup_{i=1}^{k} \left\{ y : 0 < \Delta(y|x_i^1) \le \xi \right\}. \qquad (10.\text{b})$$

The next theorem summarizes some properties of interest associated with M^{**}.

Theorem: Let ϕ_1 and ϕ_2 be training sets sampled from (ε, l)–separable sources σ_1 and σ_2. Then there exists a sufficiently small value of ξ, depending on ε, l, S and A such that the following hold.

a) $M_i^{**} \supseteq \phi_i$, $i = 1,2$

b) If the (σ_1, σ_2) are (ε, l)-separable for some $l \gg \log S$ then the cardinality of M_i^{**} is exponentially smaller than that of M_i $(i = 1, 2)$ up to a small factor depending on ξ.

c) For $i, j = 1, 2, i \ne j$, $y \in \sigma_i$ implies $y \in (M_j^{**})^c$ and $y \in M_j^{**}$ implies $y \in \sigma_i^c$ for every pair of training sets $\phi_i \in \times_{j=1}^{k} T_\delta^l(x_j^i)$, $i = 1, 2$, from a given collection of $\delta-$ neighborhood types $T_\delta^l(x_j^i)$ except for training sets that contain vectors from a bad set $B_y \subset A^n$ depending on y, where

$$\lim_{n \to \infty} \frac{|T_\delta^l(x_j^j) \cap B_y|}{|T_\delta^l(x_j^j)|} = 0. \tag{11}$$

d) $y \in (M_i^*)^c$ implies $y \in (M_i^{**})^c$ for almost every pair of training sets in the sense of (11).

e) $y \in M_i^*$ implies $y \in M_i^{**}$ for almost every y and almost every pair of training sets in the sense of (11).

The proof appears in [15].

Part (a) of the theorem tells us that M^{**} is a consistent rule. Part (b) states that the rule is asymptotically optimal in the sense of yielding small acceptance regions provided that l is fairly large. In part (c) it is claimed that for almost every combination of training sequences from given δ– neighborhood types, there will be no classification errors for a given y. It should be pointed out that part (c) does not imply that $M_1^{**} \cap \sigma_2$ and $M_2^{**} \cap \sigma_1$ are empty sets as the bad set B_y depends on y. However, it follows from part (c) and the (ε, l)-separability that for almost every $y \in \sigma_i$ and $\phi_i \in X_{j=1}^k T_\delta^l(x_j^j)$, $i = 1, 2$, there will be no classification errors. Finally, parts (d) and (e) tell us that M^{**} is almost always a good approximation to the optimal but unrealizable rule M^* as it accepts and rejects essentially the same vectors. Another implication of parts (d) and (e) is associated with the learning ability of M^{**} as compared to M^*. Both rules require about the same number of training sequences in the sense explained at the end of Sect. II. Again, it should be stressed that unlike M^*, the rule M^{**} does not depend on the unknown l.

Appendix

A lower bound on the relative entropy in terms of $d(\cdot, \cdot)$.

$$d(y,z) = \sum_{a^{l+1}} |q_y(a^{l+1}) - q_z(a^{l+1})| - \sum_{a^l} |q_y(a^l) - q_z(a^l)|$$

$$= \sum_{a^{l+1}} |q_y(a^{l+1}) - q_z(a^{l+1})| - \sum_{a^l} |q_y(a^l) - q_z(a^l)| \sum_{a_{l+1}} |q_z(a_{l+1}|a^l)|$$

$$\leq \sum_{a^{l+1}} |q_y(a^{l+1}) - q_z(a^{l+1}) + q_z(a^{l+1}) - q_y(a^l)q_z(a_{l+1}|a^l)|$$

$$= \sum_{a^l} q_y(a^l) \sum_{a_{l+1}} |q_z(a_{l+1}|a^l) - q_y(a_{l+1}|a^l)|$$

$$\leq \sum_{a^l} q_y(a^l)\sqrt{(2\ln 2) \sum_{a_{l+1}} q_y(a_{l+1}|a^l)\log \frac{q_y(a_{l+1}|a^l)}{q_z(a_{l+1}|a^l)}}$$

$$\leq \sqrt{(2\ln 2) \sum_{a^l} q_y(a^l) \sum_{a_{l+1}} q_y(a_{l+1}|a^l)\log \frac{q_y(a_{l+1}|a^l)}{q_z(a_{l+1}|a^l)}}$$

$$= \sqrt{(2\ln 2)D(Q_y\|Q_z)} ,$$

where the last two inequalities follow from Pinsker's inequality [13. Chap. 3, problem 17] and Jensen's inequality, respectively.

References

[1] H. Van Trees, *Detection, Estimation, and Modulation Theory.* Wiley, New York, 1968.

[2] R. E. Blahut, *Principles and Practice of Information Theory,* Addison-Wesley, 1987.

[3] J. Ziv, "On Classification with Empirically Observed Statistics and Universal Data Compression," *IEEE Trans. Inform. Theory,* Vol. IT-34, No. 2, pp. 278-286, March 1988.

[4] M. Gutman, "Asymptotically Optimal Classification for Multiple Tests with

Empirically Observed Statistics," *IEEE Trans. Inform. Theory,* Vol. IT-35, No. 2, pp. 401-408, March 1989.

[5] M. Gutman, "On Tests for Independence, Tests for Randomness and Universal Data Compression," submitted for publication.

[6] N. Merhav, M. Gutman, and J. Ziv, "On the Estimation of the Order of a Markov Chain and Universal Data Compression," *IEEE Trans. Inform. Theory,* Vol. IT-35, No. 5, pp. 1014-1019, September 1989.

[7] O. Zeitouni and M. Gutman, "On Universal Hypotheses Testing via Large Deviations," *IEEE Trans. Inform. Theory,* Vol. IT-37, No. 2, pp. 285-290, March 1991.

[8] O. Zeitouni, J. Ziv, and N. Merhav, "When is the Generalized Likelihood Ratio Test Optimal?" submitted to *IEEE Trans. Inform. Theory.*

[9] J. Ziv and A. Lempel, "Compression of Individual Sequences via Variable-Rate Coding," *IEEE Trans. Inform. Theory,* Vol. IT-24, No. 5, pp. 530-536, September 1978.

[10] J. Pearl, "Capacity and Error Estimates for Boolean Classifiers with Limited Complexity," *IEEE Trans. on Pattern Analysis and Machine Intelligence,* Vol. PAMI-1, No. 4, pp. 350-355, October 1979.

[11] L. Devroye, "Automatic Pattern Recognition: A Study of the Probability of Error," *IEEE Trans. on Pattern Analysis and Machine Intelligence,* Vol. PAMI-10, No. 4, pp. 530-543, July 1988.

[12] T. M. Cover, "Geometrical and Statistical Properties of Systems of Linear Inequalities with Applications in Pattern Recognition," *IEEE Trans. on Electronic Computers,* pp. 326-334, June 1965.

[13] I. Csiszár and J. Korner, *Information Theory: Coding Theorems for Discrete Memoryless Systems.* New York: Academic, 1981.

[14] L. D. Davisson, G. Longo, and A. Sgarro, "The Error Exponent for the Noiseless

Encoding of Finite Ergodic Markov Sources," *IEEE Trans. Inform. Theory*, Vol. IT-27, No. 4, pp. 431-438, July 1981.

[15] J. Ziv and N. Merhav, "A Measure of Relative Entropy Between Individual Sequences with Application to Universal Classification," in preparation.

[16] M. Rodeh, V. R. Pratt and S. Even, "Linear Algorithms for Data Compression via String Matching," *J. Assoc. Comput. Mach.*, Vol. 28, pp. 16-24, January 1981.

[17] A. D. Wyner and J. Ziv, "Some Asymptotic Properties of the Entropy of a Stationary Ergodic Data Source with Applications to Data Compression," *IEEE Trans. on Information Theory*, pp. 1250-1258, Nov. 1989.

[18] M. Feder, N. Merhav, and M. Gutman, "Universal Prediction of Individual Sequences," submitted to *IEEE Trans. Inform. Theory*.

Security

Improving the Efficiency and Reliability of Digital Time-Stamping

Dave Bayer*
Barnard College
Columbia University
New York, N.Y. 10027 U.S.A.
dab@math.columbia.edu

Stuart Haber
Bellcore
445 South Street
Morristown, N.J. 07960 U.S.A.
stuart@bellcore.com

W. Scott Stornetta
Bellcore
445 South Street
Morristown, N.J. 07960 U.S.A.
stornetta@bellcore.com

March 1992

Abstract

To establish that a document was created after a given moment in time, it is necessary to report events that could not have been predicted before they happened. To establish that a document was created before a given moment in time, it is necessary to cause an event based on the document, which can be observed by others. Cryptographic hash functions can be used both to report events succinctly, and to cause events based on documents without revealing their contents. Haber and Stornetta have proposed two schemes for digital time-stamping which rely on these principles [HaSt 91].

We reexamine one of those protocols, addressing the resource constraint required for storage and verification of time-stamp certificates. By using trees, we show how to achieve an exponential increase in the publicity obtained for each time-stamping event, while reducing the storage and the computation required in order to validate a given certificate.

We show how time-stamping can be used in certain circumstances to extend the useful lifetime of different kinds of cryptographic certifications of authenticity, in the event that the certifying protocol is compromised. This can be applied to digital signatures, or to time-stamping itself, making the digital time-stamping process renewable.

*Partially supported by NSF grant DMS-90-06116.

329

1   Introduction

Causality fixes events in time. If an event was determined by certain earlier events, and determines certain subsequent events, then the event is sandwiched securely into its place in history. Fundamentally, this is why paper documents have forensic qualities allowing them to be accurately dated and examined for signs of after-the-fact tampering. However, documents kept in digital form need not be closely tied to any physical medium, and tampering may not leave any tell-tale signs in the medium.

Could an analogous notion of causality be applied to digital documents to correctly date them, and to make undetected tampering infeasible? Any solution would have to time-stamp the data itself, without any reliance on the properties of a physical medium, and would be especially useful and trustworthy if the date and time of the time-stamp could not be forged.

In [HaSt 91], Haber and Stornetta posed this problem, and proposed two solutions. Both involve the use of cryptographic hash functions (discussed in §2 below), whose outputs are processed in lieu of the actual documents. In the *linking* solution, the hash values of documents submitted to a time-stamping service are chained together in a linear list into which nothing can feasibly be inserted or substituted and from which nothing can feasibly be deleted. This latter property is insured by a further use of cryptographic hashing. In the *random-witness* solution, several members of the client pool must date and sign the hash value; their signatures form a composite certification that the time-stamp request was witnessed. These members are chosen by means of a pseudorandom generator that uses the hash of the document itself as a seed. This makes it infeasible to deliberately choose which clients should and should not act as witnesses.

In both of these solutions, the record-keeping requirements per time-stamping request are proportional to the number of (implicit) observers of the event. In §3 below we address the following problem: What if an immense flood of banal transactions want their time-stamps to become part of the historical record, but history just isn't interested? We propose to merge many unnoteworthy time-stamping events into one noteworthy event, using a tournament run by its participants. The winner can be easily and widely publicized. Each player, by remembering a short list of opponents, can establish participation in the tournament. We do this by building trees in place of the linked list of the linking solution, thus achieving an exponential increase in the number of observers. Such hash trees were previously used by Merkle [Merk 80] for a different purpose, to produce authentication certificates for a directory of public enciphering keys.

There are several ways in which a cryptographic system can be compromised. For example, users' private keys may be revealed; imprudent choice of key-lengths may be overtaken by an increase in computing power; and improved algorithmic techniques may render feasible the heretofore intractable computational problem on which the system is based. In §4 below we show how time-stamping can be used in certain circumstances to extend the useful lifetime of digital signatures. Applying the same technique to time-stamping itself, we demonstrate that digital time-stamps can be renewed.

Finally, in §5 we discuss the relationships between the different methods of digital time-stamping that have been proposed.

2 Hash functions

The principal tool we use in specifying digital time-stamping schemes, here as in [HaSt 91], is the idea of a cryptographic hash function. This is a function compressing digital documents of arbitrary length to bit-strings of a fixed length, for which it is computationally infeasible to find two different documents that are mapped by the function to the same *hash value*. (Such a pair is called a *collision* for the hash function.) Hence it is infeasible to fabricate a document with a given hash value. In particular, a fragment of a document cannot be extended to a complete document with a given hash value, unless the fragment was known before the hash value was created. In brief, a hash value must follow its associated document in time.

There are practical implementations of hash functions, for example those of Rivest [Riv 90] and of Brachtl, *et al.* [BC$^+$ 88], which seem to be reasonably secure.

In a more theoretical vein, Damgård defined a family of *collision-free hash functions* to be a family of functions $h : \{0,1\}^* \to \{0,1\}^l$ compressing bit-strings of arbitrary length to bit-strings of a fixed length l, with the following properties:

1. The functions h are easy to compute, and it is easy to pick a member of the family at random.

2. It is computationally infeasible, given a random choice of one of these functions h, to find a pair of distinct strings x, x' satisfying $h(x) = h(x')$.

He gave a constructive proof of their existence, on the assumption that there exist one-way "claw-free" permutations [Dam 87]. For further discussion of theoretical questions relating to the existence of families of cryptographic hash functions (variously defined) see [HaSt 91] and the references contained therein.

In the rest of this paper, we will assume that a cryptographic hash function h is given: either a particular practical implementation, or one that has been chosen at random from a collision-free family.

3 Trees

In the linking scheme, the challenger of a time-stamp is satisfied by following the linked chain from the document in question to a time-stamp certificate that the challenger considers trustworthy. If a trustworthy certificate occurs about every N documents, say, then the verification process may require as many as N steps. We may reduce this cost from N to $\log N$, as follows.

Suppose we combine the hash values of two users' documents into one new hash value, and publicize only the combined hash value. (We will consider a "publicized" value to be trustworthy.) Either participant, by saving his or her own document as well as the other contributing hash value, can later establish that the document existed before the time when the combined hash value was publicized.

More generally, suppose that N hash values are combined into one via a binary tree, and the resulting single hash value is widely publicized. To later establish priority, a participant need only record his own document, as well as the $\lceil \log_2 N \rceil$ hash values that were directly combined with the document's hash value along the path to the root of the tree. In addition, along with each combining hash value, the user needs to record its "handedness," indicating whether the newly computed value was placed before or after the combining hash value. Verification consists simply of recomputing the root of the tree from this data.

Once hashing functions are chosen, such a scheme could be carried out like a world championship tournament: Heterogeneous local networks could govern local subtrees under the scrutiny of local participants, and regional "winners" could be combined into global winners under the scrutiny of all interested parties. Global communication facilities are required, and a broadcast protocol must be agreed upon, but no centralized service bureau need administer or profit from this system. For example, given any protocol acceptable separately to the western and eastern hemispheres for establishing winners for a given one-hour time period, the winners can be broadcast by various members of the respective hemispheres, and anyone who wants to can carry out the computations to determine the unique global winner for that time period. Winners for shorter time periods can similarly be combined into unique winners for longer time periods, by any interested party.

At a minimum, daily global winners could be recorded in newspaper advertisements, to end up indefinitely on library microfilm. The newspaper functions as a widely available public record whose long-term preservation at many locations makes tampering very difficult. An individual who retains the set of values tracing the path between his document and the hash value appearing in the newspaper could establish the time of his document, without any reliance on other records. Anyone who wishes to be able to resolve time-stamps to greater accuracy needs only to record time-stamp broadcasts to greater accuracy.

4 Using time-stamping to extend the lifetime of a threatened cryptographic operation

The valid lifetime of a digitally signed document can be extended with digital time-stamping, in the following way. Imagine an implementation of a particular digital signature scheme, with a particular choice of key lengths, and consider a plaintext document D and its digital signature σ by a particular user. Now let the pair (D, σ) be time-stamped. Some time later the signature may become invalid, for any of a variety of reasons, including the compromise of the user's private key, an increase in available computing power making signatures with keys of that length unsafe, or the discovery of a basic flaw in the signature scheme. At that point, the document-signature pair becomes questionable, because it may be possible for someone other than the original signer to create valid signatures.

However, if the pair (D, σ) was time-stamped at a time before the signature was compromised, then the pair still constitutes a valid signature. This is because it is known to have been created at a time when only legitimate users could have produced it. Its validity is not in question even though new signatures generated by the compromised method might no longer be trustworthy.

The same technique applies to other instances of cryptographic protocols. In particular, the technique can be used to renew the time-stamping process itself. Once again, imagine an implementation of a particular time-stamping scheme, and consider the pair (D, C), where C is a valid time-stamp certificate (in this implementation) for the document D. If (D, C) is time-stamped by an improved time-stamping method before the original method is compromised, then one has evidence not only that the document existed prior to the time of the new time-stamp, but that it existed at the time stated in the original certificate. Prior to the compromise of the old implementation, the only way to create a certificate was by legitimate means. (The ability to renew time-stamps was mentioned in [HaSt 91] but an incorrect method was given. The mistake of the previous work was in assuming that it is sufficient to renew the certificate alone, and not the document-certificate pair. This fails, of course, if the compromise in question is a method of computing hash collisions for the hash function used in submitting time-stamp requests.)

5 Different methods of time-stamping

To date, three different digital time-stamping techniques have been proposed: linear linking, random witness and linking into trees. What is the relationship between them? Does one supersede the others? Initially, one might think that trees satisfy time-stamping requirements better than the two previously proposed methods, because the tree protocol seems to reduce storage requirements while increasing the number of interested parties who serve as witnesses. But there are other tradeoffs to consider.

First we consider the linking protocol. In certain applications, such as a laboratory notebook, it is crucial not only to have a trustworthy date for each entry but also to establish in a trustworthy manner the exact sequence in which all entries were made. Linear linking of one entry to the next provides the most straightforward means of achieving this.

Next we consider the difference between the random-witness method and the tree method. While trees increase the number of witnesses to a given time-stamping event in proportion to the number of documents time-stamped, they do not guarantee a minimum number of witnesses. Neither do they guarantee that witnesses will retain their records. In contrast, in random witness the effective number of witnesses is the entire population, though only a small fraction are actually involved in any given time-stamping event. Furthermore, the set of signatures computed by the random-witness protocol explicitly creates a certificate which is evidence that a time-stamping event was widely witnessed. Thus, the protocol does not depend for its final validity on witnesses keeping records. Random witness is somewhat analogous to placing an advertisement in the newspaper, as discussed earlier, but with an additional refinement. Like the newspaper ad, it is effectively a widely witnessed event, but in addition it creates a record of the witnessing.

Given these tradeoffs, we imagine that the three methods may be used in a complementary fashion, as the following example illustrates. An individual or company might use linear linking to time-stamp its own accounting records, sending the final summary value for a given time period to a service maintained by a group of individuals or parties. This service constructs linked trees at regular intervals. The root of each tree is then certified as a widely viewed event by using the random-witness protocol among the participants. In this way, individual and group storage needs can be minimized, and the number of events which require an official record of witnessing can be greatly reduced.

References

[BC+ 88] B. O. Brachtl, D. Coppersmith, M. M. Hyden, S. M. Matyas, Jr., C. H. W. Meyer, J. Oseas, Sh. Pilpel, and M. Shilling. Data authentication using modification detection codes based on a public one way encryption function. U.S. Patent No. 4,908,861, issued March 13, 1990. (Cf. C. H. Meyer and M. Shilling, Secure program load with modification detection code. In *Securicom 88: 6ème Congrès mondial de la protection et de la sécurité informatique et des communications*, pp. 111–130 (Paris, 1988).)

[Dam 87] I. Damgård. Collision-free hash functions and public-key signature schemes. In *Advances in Cryptology—Eurocrypt '87*, Lecture Notes in Computer Science, Vol. 304, pp. 203–217, Springer-Verlag (Berlin, 1988).

[HaSt 91] S. Haber, W. S. Stornetta, How to time-stamp a digital document, *Journal of Cryptography*, Vol. 3, No. 2, pp. 99–111 (1991). (Presented at Crypto '90.)

[Merk 80] R. C. Merkle, Protocols for public key cryptosystems. In *Proc. 1980 Symp. on Security and Privacy*, IEEE Computer Society, pp. 122–133 (Apr. 1980).

[Riv 90] R. L. Rivest. The MD4 message digest algorithm. In *Advances in Cryptology—Crypto '90*, Lecture Notes in Computer Science, Vol. 537 (ed. A. J. Menezes, S. A. Vanstone), pp. 303–311, Springer-Verlag (Berlin, 1991).

A Note on Secret Sharing Schemes*

R. M. Capocelli[1], A. De Santis[2], L. Gargano[2], U. Vaccaro[2]

[1] Dipartimento di Matematica, Università di Roma, 00185 Roma, Italy

[2] Dipartimento di Informatica, Università di Salerno, 84081 Baronissi (SA), Italy

Abstract

A secret sharing scheme is a method for dividing a secret key k among a set $\mathcal{P}$ of participants in such a way that: if the participants in $A \subseteq \mathcal{P}$ are qualified to know the secret they can reconstruct the secret key k; but any set $A \subseteq \mathcal{P}$, which is not qualified to know the secret, has absolutely no information on k.

In this paper we give further evidence that Information Theory is source of valuable tools to analyze and design efficient secret sharing schemes.

1 Introduction

A secret sharing scheme is a method for dividing a secret key k among a finite set $\mathcal{P}$ of participants in such a way that only certain specified subsets of $\mathcal{P}$ can compute k.

The first secret sharing schemes that have been studied are (t, n) threshold schemes [Bl], [Sh]. A (t, n) threshold scheme allows a secret to be shared among n participants in such a way that any t of them can recover the secret, but any $t - 1$, or fewer, have absolutely no information on the secret (see [Si] for a comprehensive survey and extensive bibliography on secret sharing schemes).

Ito, Saito, and Nishizeki [ItSaNi] gave a more general method of secret sharing. An *access structure* Γ is a specification of all the subsets of participants who can recover the secret and is said *monotone* if $A \in \Gamma$ and $A \subseteq B \subseteq \mathcal{P}$ implies $B \in \Gamma$. Ito, Saito, and Nishizeki gave a methodology to realize secret sharing schemes for arbitrary monotone access structures. Subsequently, Benaloh and Leichter [BeLe] gave a simpler and more efficient way to realize secret sharing schemes for monotone access structures; see also [Br] for related results.

An important issue in the implementation of secret sharing schemes is the size of shares given to participants, since the security of a system degrades as the amount of the information that must be kept secret increases. In all secret sharing schemes the size of the

*This work was partially supported by the National Council of Research (C.N.R.) under grant 91.02326.CT12 and by M.U.R.S.T. in the framework of Project: "Algoritmi, Sistemi di Calcolo e Strutture Informative".

335

shares cannot be less than the size of the secret. Moreover, there are access structures for which any corresponding secret sharing scheme must give to some participant a share of size strictly bigger than the secret size [BeLe], [BrSt]. Capocelli *et al.* [CaDeGaVa] proved that there exists an access structure for which at least one participant must receive a share of size 50% bigger than the secret size. For more on this subject see also [BlDeStVa], [BlDeGaVa].

In this paper we continue the study of secret sharing schemes from an information–theoretic point of view and derive new and old results in an unified manner.

2 Preliminaries

In this section we present the basic information theoretic measures. For a complete treatment of the subject, see [Ga], [CsKo].

Given a probability distribution $\{p(x)\}_{x \in X}$ on a finite set X, define the *entropy* of X, $H(X)$, as

$$H(X) = - \sum_{x \in X} p(x) \log p(x)^1.$$

The entropy $H(X)$ satisfies the following inequality:

$$0 \le H(X) \le \log |X|, \tag{1}$$

where $H(X) = 0$ if and only if there exists $x_0 \in X$ such that $p(x_0) = 1$; $H(X) = \log |X|$ if and only if $p(x) = 1/|X|$, for all $x \in X$.

Given two sets X and Y and a joint probability distribution $\{p(x,y)\}_{x \in X, y \in Y}$ on their cartesian product, the *conditional entropy* $H(X|Y)$ of X given Y is defined as

$$H(X|Y) = - \sum_{y \in Y} \sum_{x \in X} p(y)p(x|y) \log p(x|y).$$

It is easy to see that

$$H(X|Y) \ge 0, \tag{2}$$

and the joint entropy $H(XY)$ is equal to

$$H(XY) = H(X) + H(Y|X). \tag{3}$$

The *mutual information* between X and Y is defined by

$$I(X;Y) = H(X) - H(X|Y), \tag{4}$$

and enjoys the following properties:

$$I(X;Y) = I(Y;X), \tag{5}$$

---

[1] All logarithms in this paper are of base 2

and

$$I(X;Y) \geq 0. \tag{6}$$

From inequality (6) one gets the following relation between the entropy of X and the conditional entropy of X given Y

$$H(X) \geq H(X|Y). \tag{7}$$

The *conditional mutual information* between X and Y given Z is defined by

$$I(X;Y|Z) = H(X|Z) - H(X|YZ). \tag{8}$$

The conditional mutual information $I(X;Y|Z)$ satisfies the important properties

$$I(X;Y|Z) \geq 0, \tag{9}$$

$$I(X;Y|Z) = I(Y;X|Z), \tag{10}$$

and

$$I(X;YZ) = I(X;Z) + I(X;Y|Z).$$

Formulæ (8) and (9) imply the following generalization of inequality (7)

$$H(X|Z) \geq H(X|YZ). \tag{11}$$

3 Secret Sharing Schemes

We recall some general definitions and notations about secret sharing schemes. Suppose that $\mathcal{P} = \{P_1, \ldots, P_{|\mathcal{P}|}\}$ is the set of participants. Denote by Γ the set of subsets of participants which we desire to be able to determine the key; hence $\Gamma \subseteq 2^{\mathcal{P}}$. Γ is called the *access structure* of the secret sharing scheme. A natural property for an access structure Γ is that of being *monotone*, i.e., if $B \in \Gamma$ and $B \subseteq C \subseteq \mathcal{P}$, then $C \in \Gamma$.

Let $\mathcal{K}$ be a set of q elements called *keys*. For every participant $P \in \mathcal{P}$, let S_P be a set of s_P elements. Elements of the sets S_P are called *shares*. Suppose a *dealer* D wants to share the secret key $k \in \mathcal{K}$ among the participants in $\mathcal{P}$ (we will assume that $D \notin \mathcal{P}$). He does this by giving each participant $P \in \mathcal{P}$ a share from S_P. We assume that on the set of keys $\mathcal{K}$ is defined a probability distribution $\{p(k)\}_{k \in \mathcal{K}}$ according to which the dealer chooses the secret key. Any secret sharing scheme for keys in $\mathcal{K}$ and the probability distribution $\{p(k)\}_{k \in \mathcal{K}}$ naturally induce a probability distribution on the joint space[2] $S_{P_1} \times \ldots \times S_{P_{|\mathcal{P}|}}$ that define the probability that participants receive given shares. We say that the scheme is *perfect* (with respect to access structure Γ) if the following two properties are satisfied:

1. For any set of participants $A = \{P_1, \ldots, P_{|A|}\} \notin \Gamma$, for any set of possible values of their shares $(s_1, \ldots, s_{|A|}) \in S_{P_1} \times \ldots \times S_{P_{|A|}}$ and for any value $k \in \mathcal{K}$ it holds $p(k|\, s_1, \ldots, s_{|A|}) = p(k)$;

---

[2]To maintain notation simpler, we shall denote random variables with the same symbol of the set of their possible values.

337

2. For any set of participants $A = \{P_1, \ldots, P_{|A|}\} \in \Gamma$, for any set of possible values of their shares $(s_1, \ldots, s_{|A|}) \in S_{P_1} \times \ldots \times S_{P_{|A|}}$ there exists an unique value $k \in \mathcal{K}$ such that $p(k|\, s_1 \ldots s_{|A|}) = 1$.

Condition 1. implies that any non qualified subset of participants cannot gain any information whatsoever on the possible value of the secret key, whereas condition 2. implies that all qualified sets of participants can uniquely determine the secret key. Using the information measure introduced in the previous section, we can restate above two conditions in the following way (cf. also [KaGrHe] and [Ko])[3]

1'. For any set of participants $A = \{P_1, \ldots, P_{|A|}\} \notin \Gamma$ it holds $H(\mathcal{K}|\, P_1, \ldots, P_{|A|}) = H(\mathcal{K})$;

2'. For any set of participants $A = \{P_1, \ldots, P_{|A|}\} \in \Gamma$ it holds $H(\mathcal{K}|\, P_1, \ldots, P_{|A|}) = 0$.

The first secret sharing schemes to be studied were *threshold schemes* [Bl], [Sh] in which the access structure defining the qualified sets contains all subsets of the set of participant $\mathcal{P}$ of cardinality greater than, or equal to, a fixed number t. Karnin, Greene, and Hellman [KaGrHe] proved that in any threshold scheme any set X_i from which the i-th share is taken satisfies $H(X_i) \geq H(\mathcal{K})$. We shall derive the same bound for general access structures. Assume a set of participants Y cannot determine the secret, but they could if another participant (or group of participants) X would be willing to pool its own share. Intuitively, the uncertainty on the shares given to X is at least as big as that on the secret itself, from the point of view of Y, otherwise, the set of participants Y would have some information on the secret and could decrease their uncertainty on $\mathcal{K}$. This is formally stated in the next lemma by Capocelli *et al.* [CaDeGaVa] which constitutes an extension and a sharpening on Theorem 1 of Karnin, Greene, and Hellman [KaGrHe]. The proof of the lemma is so simple that we present it here for reader's convenience.

Lemma 3.1 *Let* $\Gamma \subseteq 2^{\mathcal{P}}$ *be an access structure and* $X, Y \subset \mathcal{P}$ *such that* $Y \notin \Gamma$ *and* $X \cup Y \in \Gamma$. *Then* $H(X|Y) = H(\mathcal{K}) + H(X|Y\mathcal{K})$.

Proof. By (4) and (5) we have $I(X; \mathcal{K}|Y) = H(X|Y) - H(X|Y\mathcal{K}) = H(\mathcal{K}|Y) - H(\mathcal{K}|XY)$. From 1'. we have $H(\mathcal{K}|Y) = H(\mathcal{K})$ and from 2'. we have $H(\mathcal{K}|XY) = 0$. Therefore $H(X|Y) = H(\mathcal{K}) + H(X|Y\mathcal{K})$. □

4 Bounds on the size of shares

One of the basic problems in the field of secret sharing schemes is to derive bounds on the amount of information that must be kept secret. This is important from the practical point of view since the security of any system degrades as the amount of secret information

---

[3]In order to further simplify the notation, we shall denote with the same symbol a (set of) participant(s), the set of associated shares, and the random variable(s) defined on this set.

increases. The first non trivial lower bound on the size of each share that is to be given to participants in threshold schemes derives immediately from the entropy bound by Karnin, Greene, and Hellman [KaGrHe] and implies that each share must be at least as big as the secret. This bound is tight as Shamir scheme [Sh] shows.

Benaloh and Leichter [BeLe] gave the first example of an access structure (not representable as a threshold scheme) for which any secret sharing scheme must give to some participant shares which are from a domain larger than that of the secret. The access structure they considered is $\mathcal{AS} = closure\{\{A,B\},\{B,C\},\{C,D\}\}$. Recently, Brickell and Stinson [BrSt] showed that there are only two access structures with 4 participants which are the closure of a graph (i.e., the closure of a family whose elements are pairs of participants), satisfying above limitation. Such access structures are $\mathcal{AS}$ and $\mathcal{AS}2 = closure\{\{A,B\},\{B,C\},\{C,D\},\{B,D\}\}$. Finally, Capocelli et al. [CaDeGaVa] showed that either the share for B or that for C must be at least 50% bigger than the secret. More precisely, they proved that either the size $|B|$ of the set from which the shares for B are taken or the size $|C|$ of the set from which the shares for C are taken must be at least $|\mathcal{K}|^{1.5}$. Moreover, they showed that this bound is best possible for $\mathcal{AS}$ and $\mathcal{AS}2$. In [BlDeStVa] and [BlDeGaVa] several general bounds on the size of shares are provided.

Bounds on the size of sets from which shares are taken have been provided also in a different model of secret sharing schemes (see [ScSt], [StVa]). Briefly, the two main differences between the model considered so far and the model studied in [ScSt] and [StVa] are as follows.

H1 In [ScSt] and [StVa] different participants must receive different shares.

H2 In [ScSt] and [StVa] a key is determined as a function of the shares held by a set of participants. In the model considered in [Sh] and in the previous part of this paper, the key is determined as a function of the shares held by participants and the *identity* of the participants themselves.

Let n be the number of participants in the scheme, t be the threshold, $\mathcal{K}$ be the set of possible secret and V be the set from which the shares are taken. Under above hypothesis **H1** and **H2** Stinson and Vanstone proved that

$$|V| \geq (n - t + 1)|\mathcal{K}| + t - 1. \tag{12}$$

In the rest of this section we present a simple proof of (12) using the information theoretic characterization of secret sharing schemes.

We first recall the following condition of *regularity* that in [ScSt] and [StVa] has been made on the distribution of shares of the participants.

H3 There exists a positive integer ℓ such that to any $k \in \mathcal{K}$ it is associated a subset $\phi(k) \subset V^n$ consisting of ℓ elements. To share a secret $k \in \mathcal{K}$, an element from $\phi(k)$ is choosen with uniform probability and the components of the chosen vector are given to participants as shares.

In order to prove (12) denote by X_i the random variable assuming as values the shares given to the i-th participant. Obviously, conditions 1'. and 2'. translate as

a) $H(\mathcal{K}|X_{i_1} \ldots X_{i_t}) = 0$ for any choice of indices $i_1 \ldots i_t$;

b) $H(\mathcal{K}|X_{i_1} \ldots X_{i_{t-1}}) = H(\mathcal{K})$ for any choice of indices $i_1 \ldots i_{t-1}$.

From Lemma 3.1 one has

$$H(\mathcal{K}) = H(X_{i_t}|X_1 \ldots X_{i_{t-1}}) - H(X_{i_t}|X_{i_1} \ldots X_{i_{t-1}}\mathcal{K}).$$

Supposing $p(k) = 1/|\mathcal{K}| \;\; \forall k \in \mathcal{K}$, one has $H(\mathcal{K}) = \log|\mathcal{K}|$. From hypothesis **H2** one has that the set of all possible values the random variable X_{i_t} can have, once known the value of the random variables $X_{i_1} \ldots X_{i_{t-1}}$, is at most $|V| - t + 1$. From inequality (1) one gets $H(X_{i_t}|X_{i_1} \ldots X_{i_{t-1}}) \leq \log(|V| - t + 1)$. On the other hand, under the hypothesis of this model, once the values of $X_{i_1}, \ldots, X_{i_{t-1}}$, and $\mathcal{K}$ are known, the random variable X_{i_t} can assume *at least* $n - t + 1$ values. By definition we have

$$H(X_{i_t}|X_{i_1} \ldots X_{i_{t-1}}\mathcal{K}) = - \sum_{x_{i_1},\ldots,x_{i_{t-1}},x_{i_t},k} p(x_{i_1} \ldots x_{i_{t-1}}x_{i_t}k) \log p(x_{i_t}|x_{i_1} \ldots x_{i_{t-1}}k)$$

$$= \sum_{x_{i_1},\ldots,x_{i_{t-1}},k} p(x_{i_1} \ldots x_{i_{t-1}}k) H(X_{i_t}|X_{i_1} = x_{i_1},\ldots,X_{i_{t-1}} = x_{i_{t-1}},\mathcal{K} = k)$$

Given that $X_{i_1} = x_{i_1},\ldots,X_{i_{t-1}} = x_{i_{t-1}},\mathcal{K} = k$, let z be the number of elements in $\phi(k) \subset V^n$ which have $x_{i_1},\ldots,x_{i_{t-1}}$ as $t-1$ of the n components. By hypotesis **H1** and **H3** it follows that $p(x_{i_t}|x_{i_1},\ldots,x_{i_{t-1}},k)$ is a multiple of $1/z(n - t + 1)$, that is $p(x_{i_t}|x_{i_1},\ldots,x_{i_{t-1}},k) = r/z(n - t + 1)$, where $r \in \{1, 2, \ldots, z\}$. Therefore,

$$H(X_{i_t}|X_{i_1} = x_{i_1},\ldots,X_{i_{t-1}} = x_{i_{t-1}},\mathcal{K} = k) \geq \min_{Q \in \mathcal{Q}} H(Q)$$

where $\mathcal{Q}$ is the set of probability distributions $Q = (q_1,\ldots,q_{|Q|})$ such that $q_i = r_i/z(n - t + 1)$, where $r_i \in \{1, 2, \ldots, z\}$.

The minimum above is reached by $Q^* = (1/(n - t + 1),\ldots,1/(n - t + 1))$. (Indeed, if the minimum were reached by Q' i with $q_1' = r_1/z(n - t + 1)$ and $q_2' = r_2/z(n - t + 1)$ where $z > r_1 \geq r_2 > 1$, then the distribution Q'' with $q_1'' = (r_1 + 1)/z(n - t + 1)$, $q_2'' = (r_2 - 1)/z(n - t + 1)$, and $q_i'' = q_i'$, if $i \geq 3$, satisfies $H(Q') > H(Q'')$. If the minimum were reached by Q' and $q_1' = q_2' = 1/z(n - t + 1)$, then the distribution Q'' with $|Q'| - 1$ probabilities $(q_1' + q_2', q_3', \ldots, q_{|Q'|}')$ satisfies $H(Q') > H(Q'')$.)

Therefore, it holds $H(X_{i_t}|X_{i_1} \ldots X_{i_{t-1}}\mathcal{K}) \geq H(Q^*) = \log(n - t + 1)$. Hence

$$\log|\mathcal{K}| \leq \log(|V| - t + 1) - \log(n - t + 1),$$

that proves (12).

5 Ramp schemes

Let $\mathcal{K}$ be the set of possible secrets, t and n positive integers, $t \leq n$. A (t, n) threshold scheme is a method of dividing a secret $k \in \mathcal{K}$ among a set $\mathcal{P}$ of n participants in such a way that (cf. Section 3)

S1. *Any set of t participants can reconstruct the secret:*
   For all $X_{i_1}, \ldots, X_{i_t} \in \mathcal{P}$, $H(\mathcal{K}|X_{i_1} \ldots X_{i_t}) = 0$.

S2. *Any set of less that t participants has no information on the secret:*
   For all $X_{i_1}, \ldots, X_{i_r} \in \mathcal{P}$, with $r < t$, $H(\mathcal{K}|X_{i_1} \ldots X_{i_r}) = H(\mathcal{K})$.

From Lemma 3.1 it follows that $H(X_i) \geq H(X_i|X_{i_1} \ldots X_{i_{t-1}}) \geq H(\mathcal{K})$. When only uniform distributions are involved, this implies that the size of the share given to a participant, i.e., $\log|X_i|$, is greater than or equal to $\log|\mathcal{K}|$, the size of the secret.

Since the size of shares is an important issue in the implementation of secret sharing schemes, a natural problem is to study schemes that distribute to participants shares of size smaller than the secret size. Therefore, let us suppose that we have to construct a scheme which satisfies the following two requirements:

R1. For all participant $X_i \in \mathcal{P}$, $H(X_i) \leq (1/m)H(\mathcal{K})$

R2. For all participants $X_{i_1}, \ldots, X_{i_t} \in \mathcal{P}$, $H(\mathcal{K}|X_{i_1} \ldots X_{i_t}) = 0$

for fixed values of m and t. When only uniform probability distributions are involved, condition R1. implies that the size of each share is at most $1/m$ times the size of the secret.

Obviously, in view of the previous discussion, we cannot expect that our schemes has maximum security, that is, it cannot happen that any set of participants not allowed to reconstruct the secret does not have absolutely no information on the secret. We want now to derive bounds on the "level" of security such a scheme can provide.

Let $r < t$ and consider the mutual information $I(\mathcal{K}; X_{i_{r+1}} \ldots X_{i_t}|X_{i_1} \ldots X_{i_r})$ that can be written either as $H(\mathcal{K}|X_{i_1} \ldots X_{i_r}) - H(\mathcal{K}|X_{i_1} \ldots X_{i_t}) = H(\mathcal{K}|X_{i_1} \ldots X_{i_r})$, or as $H(X_{i_{r+1}} \ldots X_{i_t}|X_{i_1} \ldots X_{i_r}) - H(X_{i_{r+1}} \ldots X_{i_t}|\mathcal{K}X_{i_1} \ldots X_{i_r})$. Therefore, we have

$$
\begin{aligned}
H(\mathcal{K}|X_{i_1} \ldots X_{i_r}) &= H(X_{i_{r+1}} \ldots X_{i_t}|X_{i_1} \ldots X_{i_r}) - H(X_{i_{r+1}} \ldots X_{i_t}|\mathcal{K}X_{i_1} \ldots X_{i_r}) \\
&\leq H(X_{i_{r+1}} \ldots X_{i_t}|X_{i_1} \ldots X_{i_r}) \text{ (from (2))} \\
&\leq H(X_{i_{r+1}} \ldots X_{i_t}) \text{ (from(7))} \\
&\leq \sum_{j=r+1}^{t} H(X_{i_j}) \text{ (from(3) and(7))} \\
&\leq \frac{t-r}{m} H(\mathcal{K}) \text{ (from } R1.).
\end{aligned}
$$

Thus, it is natural to consider (t, m) *ramp schemes*, that is secret sharing schemes which satisfy the following property, referred to as the ramp property:

$$H(\mathcal{K}|X_{i_1} \ldots X_{i_r}) = \left[\frac{t-r}{m}\right]_* H(\mathcal{K}), \tag{13}$$

where $1 \leq r \leq t$ and $[x]_* = \min\{1, x\}$.

Essentially, in ramp schemes the uncertainty that a coalition of participants have on the secret decreases "gradually" from the maximum, i.e., $H(\mathcal{K})$, to 0 as a function of the number of participants in the coalition. Next theorem tells us that in a (t, m) ramp scheme the information held by any r participants is at least a fraction r/m of the secret entropy. This constitutes an extension to ramp schemes of Karnin's *et al.* bound and it can be used to show that the ramp schemes provided in [McSa] or in [BlMe] are optimal with respect to the sizes of shares given to participants.

Theorem 5.1 *The joint entropy of any r participant's shares $X_{i_1}, \ldots, X_{i_r}$ in a (t, m) ramp scheme satisfies*

$$H(X_{i_1} \ldots X_{i_r}) \geq \frac{r}{m} H(\mathcal{K}).$$

Proof. Let $X_{i_1}, \ldots, X_{i_r}, X_{i_{r+1}}, \ldots, X_{i_t}$ be t participants and $r \leq t$. The mutual information $I(\mathcal{K}; X_{i_1} \ldots X_{i_r} | X_{i_{r+1}} \ldots X_{i_t})$ can be written either as

$$H(\mathcal{K}|X_{i_{r+1}} \ldots X_{i_t}) - H(\mathcal{K}|X_{i_1} \ldots X_{i_t}) = H(\mathcal{K}|X_{i_{r+1}} \ldots X_{i_t}) = (r/m)H(\mathcal{K}),$$

or as

$$H(X_{i_1} \ldots X_{i_r} | X_{i_{r+1}} \ldots X_{i_t}) - H(X_{i_1} \ldots X_{i_r} | \mathcal{K} X_{i_{r+1}} \ldots X_{i_t}).$$

Hence, it follows that

$$H(X_{i_1} \ldots X_{i_r} | X_{i_{r+1}} \ldots X_{i_t}) \geq (r/m)H(\mathcal{K}).$$

The theorem follows from above inequality and formula (7). □

When keys in $\mathcal{K}$ are chosen with uniform probability, above theorem implies that in any secret sharing scheme satisfying (13), the entropy of the sample space from which shares to participants X_i are taken obey to the requirement

$$H(X_i) \geq \frac{1}{m} \log |\mathcal{K}|.$$

Therefore, by (1), the size $\log |X_i|$ of the shares given to X_i is at least $1/m$ times the size of the secret K, that is

$$\log |X_i| \geq \frac{1}{m} \log |\mathcal{K}|. \tag{14}$$

The algorithms presented in [BlMe] and [McSa] satisfy above bound with equality and therefore are optimal with respect to the size of shares given to participants.

References

[BeLe] J. C. Benaloh and J. Leichter, Generalized Secret Sharing and Monotone Functions, Proceedings of Crypto '88, *Advances in Cryptology*, Lecture Notes in Computer Science, vol. 403, S. Goldwasser, Ed., Springer–Verlag, Berlin, 1990, pp. 27–35.

[Bl] G. R. Blakley, Safeguarding Cryptographic Keys, *Proceedings of AFIPS 1979 National Computer Conference*, vol. 48, New York, NY, pp. 313-317, June 1979.

[BlMe] G. R. Blakley and C. Meadows, Security of Ramp Schemes, Proceedings of Crypto '84, *Advances in Cryptology*, Lecture Notes in Computer Science, vol. 196, G. R. Blakley and D. Chaum, Eds., Springer–Verlag, Berlin, 1985, pp. 411–431.

[BlDeStVa] C. Blundo, A. De Santis, D. R. Stinson, and U. Vaccaro, Graph Decomposition and Secret Sharing Schemes, *Eurocrypt '92.*

[BlDeGaVa] C. Blundo, A. De Santis, L. Gargano, and U. Vaccaro, On the Information Rate of Secret Sharing Schemes, *preprint.*

[Br] E. F. Brickell, Some Ideal Secret Sharing Schemes, *J. Combin. Math. and Comb. Comput.*, vol. 9, 105–113, 1989.

[BrDa] E. F. Brickell and D. M. Davenport, On the Classification of Ideal Secret Sharing Schemes, *J. Cryptology*, vol. 4, 123–143, 1991.

[BrSt] E. F. Brickell and D. R. Stinson, Some Improved Bounds on the Information Rate of Perfect Secret Sharing Schemes, in " Advances in Cryptology - CRYPTO '90", S. A. Vanstone, Ed., Lectures Notes in Computer Science, Springer–Verlag, Berlin.

[CaDeGaVa] R. M. Capocelli, A. De Santis, L. Gargano, and U. Vaccaro, On the Size of Shares in Secret Sharing Schemes, in "Advances in Cryptology - CRYPTO 91", J. Feigenbaum, Ed., Lecture Notes in Computer Science, Springer-Verlag, Berlin.

[CsKo] I. Csiszár and J. Körner, *Information Theory. Coding theorems for discrete memoryless systems,* Academic Press, 1981.

[De] D. Denning, *Cryptography and Data Security,* Addison–Wesley, Reading, MA, 1983.

[Ga] R. G. Gallager, *Information Theory and Reliable Communications,* John Wiley & Sons, New York, NY, 1968.

[GoMiWi] O. Goldreich, S. Micali, and A. Wigderson, How to Play Any Mental Game, Proceedings of the 19th Annual ACM Symposium on Theory of Computing, 1987, New York, pp. 218–229.

[ItSaNi] M. Ito, A. Saito, and T. Nishizeki, Secret Sharing Scheme Realizing General Access Structure, Proceedings of IEEE Global Telecommunications Conference, Globecom 87, Tokyo, Japan, 1987, pp. 99–102.

[KaGrHe] E. D. Karnin, J. W. Greene, and M. E. Hellman, On Secret Sharing Systems, *IEEE Transactions on Information Theory*, vol. IT-29, no. 1, Jan. 1983, pp. 35-41.

[Ko] S. C. Kothari, Generalized Linear Threshold Schemes, Proceedings of Crypto '84, *Advances in Cryptology*, Lecture Notes in Computer Science, vol. 196, G. R. Blakley and D. Chaum, Eds., Springer–Verlag, Berlin, 1985, pp. 231-241.

[McSa] R. J. McEliece and D. Sarwate, On Sharing Secrets and Reed–Solomon Codes, *Communications of the ACM*, vol. 24, n. 9, pp. 583-584, September 1981.

[ScSt] P. J. Schellenberg and D. R. Stinson, Threshold Schemes from Combinatorial Design, *J. Combin. Math. and Combin. Computing*, vol. 5, pp. 143–160, 1989.

[Sh] A. Shamir, How to Share a Secret, *Communications of the ACM*, vol. 22, n. 11, pp. 612-613, Nov. 1979.

[Si] G.J. Simmons, *An Introduction to Shared Secret and/or Shared Control Schemes and Their Application*, Contemporary Cryptology, IEEE Press, pp. 441–497, 1991.

[StVa] D. R. Stinson and A. Vanstone, A combinatorial approach to threshold schemes, *SIAM J. Disc. Math.*, vol. 1, May 1988, pp.230–236.

Privacy of Dense Symmetric Functions*
(extended abstract)

Benny Chor[†] Netta Shani[‡]
Department of Computer Science
Technion, Haifa 32000, Israel

Abstract

An n argument function, f, is called t – private if there exists a distributed protocol for computing f, so that no coalition of $\leq t$ processors can infer any additional information from the execution of the protocol. It is known that every function defined over a finite domain is $\lfloor \frac{n-1}{2} \rfloor$-private. The general question of t – privacy (for $t \geq \lceil \frac{n}{2} \rceil$) is still unresolved.

In this work we relate the question of $\lceil \frac{n}{2} \rceil$-privacy for the class of *symmetric* functions of Boolean arguments $f : \{0,1\}^n \to \{0,1,\ldots,n\}$ to the structure of weights in $f^{-1}(b)$ ($b \in \{0,1,\ldots,n\}$). We show that if f is $\lceil \frac{n}{2} \rceil$-private, then every set of weights $f^{-1}(b)$ must be an *arithmetic progression*. For the class of *dense* symmetric functions (defined in the sequel), we refine this to the following necessary and sufficient condition for $\lceil \frac{n}{2} \rceil$ – privacy of f: Every collection of such arithmetic progressions must yield distinct remainders, when computed modulo the greatest common divisor of their differences. This condition is used to show that for dense symmetric functions, $\lceil \frac{n}{2} \rceil$-privacy implies n–privacy.

1 Introduction

An n-argument function $f(x_1,\ldots,x_n)$ is called t – *private* if there exists a protocol for distributively computing f, so that no coalition of $\leq t$ processors can infer any *additional information* (in the information-theoretic sense) from the execution of the protocol. Additional information is any information on inputs of non-coalition members

---

*Research supported by US-Israel Binational Science Foundation grant 88-00282.
[†]Contact author, e-mail: benny@techsel.bitnet , benny@cs.technion.ac.il .
[‡]Supported in part by the Julius and Dorothea Harbrand and by the Edwards fellowships.

which does not follow from inputs of coalition members and the value $f(x_1, \ldots, x_n)$. Ben-Or, Goldwasser and Wigderson [1] and Chaum, Crepeau and Damgard [4] have shown that over *finite* domains, every function can be computed $\left\lfloor \frac{n-1}{2} \right\rfloor$ – privately. Some functions, like modular addition [3], are even n – private, while others, like Boolean OR, are $\left\lfloor \frac{n-1}{2} \right\rfloor$ – private but not $\left\lceil \frac{n}{2} \right\rceil$ – private [1]. No general characterization for t – privacy, for $t \geq \left\lceil \frac{n}{2} \right\rceil$, has been found yet.

For the special case of Boolean functions $f : A_1 \times A_2 \times \ldots \times A_n \rightarrow \{0,1\}$, Chor and Kushilevitz [6] showed that $\left\lceil \frac{n}{2} \right\rceil$ – privacy implies n – privacy, by proving that every $\left\lceil \frac{n}{2} \right\rceil$ – private Boolean function can be expressed as the exclusive-or of n Boolean functions, each depending on a single variable $f(x_1, x_2, \ldots, x_n) = f_1(x_1) \oplus f_2(x_2) \oplus \ldots \oplus f_n(x_n)$. For two argument functions $f(x_1, x_2)$, Kushilevitz [7] and Beaver [2] gave a combinatorial characterization which determines 1 – privacy of f. This characterization provides a necessary condition for t – privacy of n argument functions $f(x_1, \ldots, x_n)$. If f is t – private ($t \geq \left\lceil \frac{n}{2} \right\rceil$), then for every partition of $\{1, 2, \ldots, n\}$ to two sets $S, \bar{S}$ with $|S| = t$, the induced two argument function must be 1 – private.

In this paper, we study the question of t – privacy (for $t \geq \left\lceil \frac{n}{2} \right\rceil$) of *symmetric* functions $f : \{0,1\} \times \ldots \times \{0,1\} \rightarrow \{0, 1, \ldots, n\}$ (without loss of generality, this is the range of f). Using partition arguments, we show that if f is $\left\lceil \frac{n}{2} \right\rceil$-private, then every set of weights $f^{-1}(b)$ must be an *arithmetic progression*. That is, if $f^{-1}(b)$ is not empty, then the set of weights in it is of the form $L(a, d) \stackrel{\text{def}}{=} \{a + id \mid i = 0, 1, \ldots\} \cap \{0, \ldots, n\}$. If $f^{-1}(b)$ contains a single weight, then the difference d in the corresponding arithmetic progression is 0. Otherwise, this d is at least 1 and at most n. We say that an $\left\lceil \frac{n}{2} \right\rceil$ – private symmetric function is *dense*, if for every $f^{-1}(b)$ which contains more than a single weight, the difference d satisfies $d \leq \left\lceil \frac{n}{4} \right\rceil$. (Density is more generally defined – see Section 5.) For the class of dense symmetric functions, we prove the following necessary condition for $\left\lceil \frac{n}{2} \right\rceil$ – privacy of f: Every collection of arithmetic progressions which correspond to $f^{-1}(b)$, for different b's, must yield non-identical remainders, when computed modulo the greatest common divisor of their differences. That is let $L(a_1, d_1), \ldots, L(a_k, d_k) \subseteq \{0, \ldots, n\}$ be a collection of $k > 1$ disjoint arithmetic progressions which correspond to $f^{-1}(b_1), \ldots, f^{-1}(b_k)$ with $0 < d_i \leq \left\lceil \frac{n}{4} \right\rceil$, and $e \stackrel{\text{def}}{=} \gcd(d_1, \ldots, d_k)$. If f is a dense $\left\lceil \frac{n}{2} \right\rceil$-private function and the above sequences correspond to some (not necessarily all) of its values, then there exist $1 \leq j < i \leq k$ satisfying $a_j \not\equiv a_i \pmod{e}$. When this condition is met, we construct an n-private protocol for computing f. This protocol "converges" to the right value by performing a sequence of n-private modular additions, where each modulus equals the greatest common divisor of the differences of "relevant" arithmetic progressions.

The remaining of this paper is organized as follows. In Section 2 we describe the model and give formal definition of privacy. In Section 3 we quote known, relevant results. In Section 4 we prove the arithmetic progression condition. In Section 5 we prove the necessary condition for $\left\lceil \frac{n}{2} \right\rceil$ – privacy of dense functions, while in Section 6 we give the protocol which proves the sufficiency of this condition. Finally, Section 7 contains some concluding remarks and open problems.

2   Model and Definitions

In this section we describe the distributed model which is used, and present the formal definition of t – privacy. We consider a synchronous distributed network which connects n computationally unbounded processors $P_1, P_2, \ldots, P_n$. These processors are honest, namely they follow their specified protocols. There is a reliable and secure communication line between each pair of processors, so that messages sent are received with no modification, and eavesdropping by a third party is not possible. At the beginning of an execution, each processor P_i has an input x_i (no probability space is associated with the inputs). In addition, each processor has a random input r_i taken from a source of randomness R_i (the random inputs are independent). The processors wish to compute the value of a function $f(x_1, x_2, \ldots, x_n)$. To this end, they exchange messages as prescribed by a randomized protocol $\mathcal{F}$. Messages are sent in rounds, where in each round every processor can send a message to every other processor. Each message a processor P_i sends in the k-th round is determined by its input x_i, its random input r_i, the messages it received so far, and the identities of the sender and the receiver. We say that the protocol $\mathcal{F}$ computes f if in every execution, the last message in the protocol contains the value of the function $f(x_1, x_2, \ldots, x_n)$.

We say that a coalition (i.e. a set of processors) T *does not learn any additional information* (other than what follows from its input and the function value) from the execution of a randomized protocol $\mathcal{F}$, which computes f, if the following holds: For every two input vectors $\vec{x}, \vec{y}$ that agree in their T entries (i.e. $\forall i \in T : x_i = y_i$) and satisfy $f(\vec{x}) = f(\vec{y})$, and for every choice of random inputs $\{r_i\}_{i \in T}$, the messages passed between T and $\bar{T}$ are identically distributed. That is, for every communication S,

$$Pr(S|\vec{x}, \{r_i\}_{i \in T}) = Pr(S|\vec{y}, \{r_i\}_{i \in T}) ,$$

where the probability space is over the random inputs of all processors in $\bar{T}$.

We say that a protocol $\mathcal{F}$ for computing f is *t-private* if every coalition T of size $\leq t$ does not learn any additional information from the execution of the protocol. We say that a function f is *t-private* if there exists a t-private protocol that computes it. (It is possible to relax these privacy and correctness requirements, as in [6, 7], without effecting the results of this paper.) For two argument functions, we use the term "private" to mean 1–private.

3 Related Results

In this section we summarize some known results about privately computable functions, which will be used in the sequel. To make the exposition self contained, we include the following definition:

Let M be a matrix with rows from a set C and columns from a set D. The relation $\sim_M$ is defined on rows of the matrix – $x_1, x_2 \in C$ satisfy $x_1 \sim_M x_2$ if there exists $y \in D$ such that $M_{x_1,y} = M_{x_2,y}$. The relation $\equiv_M$ is defined as the transitive closure of $\sim_M$. In a similar way, we define the relations $\sim_M$ and $\equiv_M$ on the matrix columns. We say that a matrix is *forbidden* if it contains at least two entries with different values, all its rows are in the same equivalence class (with respect to $\equiv_M$), and all its columns are in the same equivalence class. Given a function $f : C \times D \to B$, we associate with it a matrix M_f with rows from C, columns from D, such that the entry $M_{x,y}$ contains the value $f(x, y)$. Using these definitions, the privately computable functions of two arguments have the following characterization.

Two Party Characterization Theorem [7, 2]: A function $f : C \times D \to B$ is private if and only if the matrix which represents it, M_f, does not contain any forbidden submatrix.

The following special case of the Privacy Characterization Theorem deals with 2-by-2 submatrices, and yields a useful necessary condition.

Corners Lemma [6]: Suppose f is private. For every $x_1, x_2, y_1, y_2 \in \{0,1\}^n$, if $f(x_1, y_1) = f(x_1, y_2) = f(x_2, y_1) = b$, then $f(x_2, y_2) = b$.

Finally, the following lemma states a necessary condition for t – privacy ($t \geq \lceil \frac{n}{2} \rceil$) of f, in terms of 1 – privacy of a related two-argument function.

Reduction Lemma [6]: Let $A_1, A_2, \ldots, A_n$ and B be non-empty sets, $t \geq \lceil \frac{n}{2} \rceil$, and $f : A_1 \times A_2 \times \ldots \times A_n \to B$ be t – private. Let $S \subseteq \{1, 2, \ldots, n\}$ be any subset of size t. Denote by C (resp. D) the Cartesian product of the A_i with $i \in S$ (resp. $i \in \bar{S}$). Then, viewing f as a two argument function $f' : C \times D \to B$, f' is 1–private.

4   The Arithmetic Progression Condition

In this section we develop necessary conditions for $\lceil \frac{n}{2} \rceil$ privacy of symmetric functions $f : \{0,1\}^n \to \{0, \ldots, n\}$. This condition, termed the arithmetic progression condition, states that for every b in the range of f, the weights of the vectors in $f^{-1}(b)$ form an arithmetic progression in $\{0, \ldots, n\}$.

Let $f : \{0, 1\}^n \to \{0, \ldots, n\}$ be a symmetric function. For $\vec{x} \in \{0, 1\}^n$, let $weight(\vec{x})$ denote the Hamming weight of $\vec{x}$ (the number of 1's in $\vec{x}$). We define the functions h and g, which are related to f:

- $h : \{0, \ldots, n\} \to \{0, \ldots, n\}$ is defined by $h(weight\,(\vec{x})) = f(\vec{x})\,(\vec{x} \in \{0,1\}^n)$.

- $g : \{0, \ldots, \lceil \frac{n}{2} \rceil\} \times \{0, \ldots, \lfloor \frac{n}{2} \rfloor\} \to \{0, \ldots, n\}$ is defined by $g\,(x, y) = h(x + y)$.

For brevity, we will often denote $B \stackrel{\text{def}}{=} \{0, \ldots, n\}$, $C_1 \stackrel{\text{def}}{=} \{0, \ldots, \lceil \frac{n}{2} \rceil \}$, and $C_2 \stackrel{\text{def}}{=} \{0, \ldots, \lfloor \frac{n}{2} \rfloor \}$. By the reduction Lemma we have the following:

Property 1: If f is $\lceil \frac{n}{2} \rceil$-private, then g is 1-private.

Lemma 1: [Arithmetic Progression Lemma] Let $f : \{0, 1\}^n \to \{0, \ldots, n\}$ be an $\lceil \frac{n}{2} \rceil$-private symmetric function. Let $b \in \{0, \ldots, n\}$ be an element in f's range with a non-empty preimage $f^{-1}(b)$. Then there are $a, d \in \{0, \ldots, n\}$ such that

$$h^{-1}(b) = L(a, d) \stackrel{\text{def}}{=} \{a \pm id \mid i = 0, 1, \ldots\} \cap \{0, \ldots, n\} .$$

349

Proof: If $h^{-1}(b)$ is a singleton $\{a\}$, then simply take $d = 0$. Otherwise, $h^{-1}(b)$ contains at least two distinct elements. Let u, v be two elements in $h^{-1}(b)$ such that their difference $d = u - v > 0$ is of minimal value (without loss of generality $v < u$). Let a be the minimal element in $\{v - id \mid i = 0, 1 \ldots, n\} \cap \{0, \ldots, n\}$. We first show that for these values of a, d and for every $i \geq 0$ with $a + i \cdot d \leq n$, the equality $h(a \pm i \cdot d) = b$ holds. That is, $L(a, d) \subseteq h^{-1}(b)$. We then show the reverse inclusion, namely $h^{-1}(b) \subseteq L(a, d)$. The proof uses Property 1 and the Corners Lemma. We omit further details. □

Examples

1. For any integer $d \geq 1$, the function $\sum_{i=1}^{n} x_i \pmod{d}$ is $\left\lceil \frac{n}{2} \right\rceil$-private. For every $0 \leq b \leq d - 1$, $h^{-1}(b) = L(b, d)$.

2. For the Boolean OR function, $h^{-1}(0) = 0$, while $h^{-1}(1) = \{1, \ldots, n\}$ which is not an arithmetic progression in $\{0, \ldots, n\}$. Indeed, this symmetric function is not $\left\lceil \frac{n}{2} \right\rceil$-private [1, 6].

By Lemma 1, we can identify every $\left\lceil \frac{n}{2} \right\rceil$-private symmetric function with a sequence of arithmetic progressions $L(a_1, d_1), \ldots, L(a_m, d_m)$. These progressions partition $\{0, \ldots, n\}$ to disjoint subsets.

5   Dense Symmetric Functions

In this section we define the class of *dense* symmetric functions of Boolean variables. This class includes functions where, if the set $h^{-1}(b)$ is not a singleton, then it is "fairly large". We develop a necessary condition for $\left\lceil \frac{n}{2} \right\rceil$ privacy of such dense symmetric functions. This condition will be used in the next section to show that the n-private protocol, defined there, always terminates. We start with the definition of density. It is stated for any symmetric function of Boolean variables, in terms of the matrix of the two variable function $g : C_1 \times C_2 \to B$. An equivalent condition, for $\left\lceil \frac{n}{2} \right\rceil$-private dense function will follow.

Definition 1: Let $f : \{0, 1\}^n \to \{0, \ldots, n\}$ be a symmetric function. We call f *dense* if $\forall b$ with $|h^{-1}(b)| > 1$ and $\forall x \in C_1, \exists y \in C_1$ and $\exists z \in C_2$ such that $y \neq x$ and $g(x, z) = g(y, z) = b$.

Remark: If f is dense, then $\forall\, b$ with $|h^{-1}(b)| > 1$, and $\forall\, z \in C_2 \,\exists\, x \in C_1$ such that $g(x,z) = b$.

Examples

1. For $2 \le v \le 8$, the function $\sum_{i=1}^{n} x_i \pmod 3$ is $\left\lceil \frac{n}{2} \right\rceil$-private, but is not dense.

2. For $n \ge 5$, the Boolean XOR function is both $\left\lceil \frac{n}{2} \right\rceil$-private and dense.

3. For $n \ge 2$, the Boolean OR function is not $\left\lceil \frac{n}{2} \right\rceil$-private, but it is dense.

For $\left\lceil \frac{n}{2} \right\rceil$-private symmetric functions, the following claim states a simple condition which is equivalent to density.

Claim 1: Let $f : \{0,1\}^n \rightarrow \{0,\ldots,n\}$ be an $\left\lceil \frac{n}{2} \right\rceil$-private symmetric function. Let $L(a_1, d_1), \ldots, L(a_m, d_m)$ be the corresponding arithmetic progressions. Then f is dense iff for all $1 \le k \le m$, $d_k \le \left\lceil \frac{n}{4} \right\rceil$.

Proof: Omitted. $\qquad\qquad\qquad\qquad\qquad\qquad\qquad\qquad\qquad\qquad$ $\square$

Using the last claim, it is clear that if all $L(a_i, d_i)$ with $d_i > 0$ contain at least 5 elements, then necessarily $d_i \le \left\lceil \frac{n}{4} \right\rceil$ and therefore f is dense. We give two examples showing that 4 elements in $L(a,d)$ neither suffice nor are necessary to guarantee density.

Examples

1. Let $f : \{0,1\}^{12} \rightarrow \{0,\ldots,12\}$ have $h^{-1}(0) = \{0,4,8,12\}$, and all other weights are singletons. This f is $\left\lceil \frac{n}{2} \right\rceil$-private, but $d_0 = \frac{n}{3} > \left\lceil \frac{n}{4} \right\rceil$, and so f is not dense.

2. Let $f : \{0,1\}^{14} \rightarrow \{0,\ldots,14\}$ have $h^{-1}(0) = \{3,7,11\}$ all other weights singletons. This f is both $\left\lceil \frac{n}{2} \right\rceil$-private and dense ($d_0 = 4 = \left\lceil \frac{n}{4} \right\rceil$), despite the fact that $1 < |h^{-1}(0)| < 4$.

The arithmetic progression lemma gave us a necessary condition for $\left\lceil\frac{n}{2}\right\rceil$-privacy of symmetric functions of Boolean variables, namely for every b in f's range $h^{-1}(b) = L(a_b, d_b)$. The following condition on the relations between the sequences $L(a_b, d_b)$ is a necessary condition for $\left\lceil\frac{n}{2}\right\rceil$ privacy of *dense* functions. The proof of this theorem uses the Two Party Characterization Theorem, and Euclid's algorithm for computing the greatest common divisor.

Theorem 1: Let $L(a_1, d_1), \ldots, L(a_k, d_k) \subseteq \{0, \ldots, n\}$ be a collection of $k \geq 1$ disjoint arithmetic progressions, with $0 < d_i \leq \left\lceil\frac{n}{4}\right\rceil$. Let $e \stackrel{\text{def}}{=} \gcd(d_1, \ldots, d_k)$. If f is a dense $\left\lceil\frac{n}{2}\right\rceil$-private function and the above sequences correspond to some (not necessarily all) values in f's range, $b_1, \ldots, b_k$, then there exist $1 \leq j < i \leq k$ satisfying $a_j \not\equiv a_i \pmod{e}$.

Proof: Assuming that for all $1 \leq i < j \leq k$, $a_j \equiv a_i \pmod{e}$, we will show that the two variable function g contains a *forbidden submatrix*. By the Two Party Characterization Theorem, this implies that g is not 1-private, which, by property 1, implies that f is not $\left\lceil\frac{n}{2}\right\rceil$-private, a contradiction.

Without loss of generality $h(a_i) = a_i$ $(i = 1, \ldots, k)$. Define the submatrix $M = M_1 \times M_2 \subseteq C_1 \times C_2$ by

$$M_1 = L(a_1, e) \cap C_1$$

$$M_2 = L(0, e) \cap C_2$$

We first show that the assumption $a_2 \equiv a_1 \pmod{e}$ implies that g is non-constant over $M_1 \times M_2$. Since $0 \leq a_1 \leq d_1 - 1$, we have $a_1 \leq \left\lceil\frac{n}{2}\right\rceil$. Thus $(a_1, 0) \in M_1 \times M_2$ and thus g attains the value 1 ($= g(a_1, 0)$) in $M_1 \times M_2$. Similarly $0 \leq a_2 \leq d_2 - 1$, and so $a_2 \in M_1$. Since $a_2 \equiv a_1 \pmod{e}$, $a_1 \in M_1$ and thus g attains the value 2 ($= g(a_2, 0)$) in $M_1 \times M_2$.

We will now show that every two rows in M_1 (resp. every two columns in M_2) satisfy the $\equiv_M$ relation. The proof will be by a sequence of claims. In Claim 2, we show that for every $1 \leq j \leq k$, and every $x, y \in M_1$ (resp. every $x, y \in M_2$) if $x \equiv y$ $\pmod{d_j}$ then $x \equiv_M y$. In Claims 3-4, we show that for every $1 \leq j < i \leq k$, and every $x, y \in M_1$, ($x, y \in M_2$ respectively) if $x \equiv y$ $\pmod{\gcd(d_j, d_i)}$ then $x \equiv_M y$. Finally, in claim 5, we show that for every $x, y \in M_1$ ($x, y \in M_2$ respectively), if $x \equiv y$ $\pmod{e}$ then $x \equiv_M y$.

Claim 2: Let $1 \leq j \leq k$. If $x, y \in M_1$ satisfy $x \equiv y \pmod{d_j}$, then $x \equiv_M y$. If $x, y \in M_2$ satisfy $x \equiv y \pmod{d_j}$, then $x \equiv_M y$.

Proof: Omitted. $\qquad\square$

Claim 3: Let b, c be integers in the range $0 < b < c \leq \left\lceil \frac{n}{4} \right\rceil$, such that $e \mid b, e \mid c$, and $c = q \cdot b + r$, where $0 < r \leq b$. Suppose that for every $x, y \in M_1$ (resp. every $x, y \in M_2$) $x \equiv y \pmod{c}$ or $x \equiv y \pmod{b}$ implies that $x \equiv_M y$. Then for every $x, y \in M_1$ (resp. every $x, y \in M_2$), $x \equiv y \pmod{r}$ implies $x \equiv_M y$.

Proof: Omitted. $\qquad\square$

Claim 4: Let b, c be integers in the range $0 < b < c \leq \left\lceil \frac{n}{4} \right\rceil$, such that $e \mid b, e \mid c$. Suppose that for every $x, y \in M_1$ (resp. every $x, y \in M_2$) $x \equiv y \pmod{c}$ or $x \equiv y \pmod{b}$ implies that $x \equiv_M y$. Then for every $x, y \in M_1$ (resp. every $x_1, y \in M_2$), $x \equiv y \pmod{\gcd(b, c)}$ implies $x \equiv_M y$.

Proof: The proof uses Claim 3, employing the remainders $r_0, r_1, r_2 \ldots$ which are the partial results in Euclid's algorithm for computing the greatest common divisor of b, c. We omit further details. $\qquad\square$

Claim 5: Let $x, y \in M_1$ ($x, y \in M_2$ resp.) satisfy $x \equiv y \pmod{e}$. Then $x \equiv_M y$.

Proof:

$$
\begin{aligned}
\text{Let } e_1 &= \gcd(d_1, d_2) \\
e_2 &= \gcd(e_1, d_3) \\
&\vdots \\
e_{k-1} &= \gcd(e_{k-2}, d_k)
\end{aligned}
$$

Clearly $e_{k-1} = \gcd(d_1, \ldots, d_k)$, and $e \mid e_i$ for all $1 \leq i \leq k-1$. By Claim 2, if $x, y \in M_1$ and $x \equiv y \pmod{d_0}$ or $x \equiv y \pmod{d_1}$, then $x \equiv_M y$. Thus by Claim 4, if $x \equiv y \pmod{e_1}$ then $x \equiv_M y$. Repeated uses of Claim 4 imply the desired conclusion. $\qquad\square$

By M's definition, every $x, y \in M_1$ (resp. $x, y \in M_2$) satisfy $x \equiv y \pmod{e}$. Thus by Claim 5, every $x, y \in M_1$ (resp. $x, y \in M_2$) satisfy $x \equiv_M y$. We saw that the function g is non-constant over $M_1 \times M_2$. Thus $M = M_1 \times M_2$ is a forbidden submatrix of $C_1 \times C_2$. This completes the proof of Theorem 1. $\qquad\square$

6 An n-private Protocol

In this section we describe a protocol for computing any $\left\lceil \frac{n}{2} \right\rceil$-private dense symmetric function of Boolean variables. The protocol computes every such function n-privately. This proves the existence of a gap in the privacy hierarchy for the class of dense symmetric functions: No such function is t-private but not $t + 1$ private for any t in the range $\left\lceil \frac{n}{2} \right\rceil \leq t < n$. The correctness of the protocol is based on the condition developed in Theorem 1. This shows that the condition is not only necessary, but also sufficient for $\left\lceil \frac{n}{2} \right\rceil$- privacy.

Let $L(a_1, d_1), \ldots, L(a_r, d_r)$ be the collection of all arithmetic progressions corresponding to the function f (the collection includes the singletons, with $d_i = 0$). Finding $f(x_1, \ldots, x_n)$ is equivalent to finding a b with $\sum_{i=1}^{n} x_i \in L(a_b, d_b)$. If $d_b > 0$, this means finding the unique a_b such that $\sum_{i=1}^{n} x_i \equiv a_b \pmod{d_b}$)b. If $d_b = 0$, this means finding $\sum_{i=1}^{n} x_i$ itself.

Our protocol progresses in stages. In the j-th stage, a set D_j of potential differences $\{d_i\}$ is maintained. This set is updated at each stage by shrinking it until finally D_j either becomes a singleton containing exactly one $d_b > 0$ (which implies that $f(x_1, \ldots, x_n)$ equals the corresponding b) or empty, which implies that $d_b = 0$, and $f(x_1, \ldots, x_n)$ can be computed by finding $\sum_{i=1}^{n} x_i$. The protocol "converges" to the right value by performing a sequence of n-private modular additions, one per stage. We now give its formal description.

Let D_0 be the collection of all d_b's with $d_b > 0$. If D_0 is empty, then $h^{-1}(b)$ is a singleton for every b in f's range. In this case, finding f's value is equivalent to finding $\sum_{i=1}^{n} x_i$. For a set of positive integer A, let $\gcd(A)$ denote the gcd of all elements in A.

Stage 0: If $D_0 = \phi$ then compute $\sum_{i=1}^{n} x_i$ n-privately. Broadcast the corresponding value of f, and halt. Else ($D_0 \neq$) go to Stage 1.

Stage j: $(j \geq 1)$

1. Locally compute $e_j = \gcd(D_{j-1})$

2. Compute n-privately $c_j \overset{\text{def}}{=} \sum_{i=1}^n x_i \pmod{e_j}$.

3. Let $D_j = \{d_i | d_i \equiv c_j \pmod{e_j}\} \cap D_{j-1}$.

4. If $D_j = \phi$ then compute $\star$-privately $\sum_{i=1}^n x_i$, find f's value which correspond to this sum, broadcast it and halt.

5. Else, if $|D_j| = |D_{j-1}|$ then $|D_j| = 1$. Therefore there is a unique b satisfying $D_j = \{d_b\}$ and $a_b \equiv c_j \pmod{d_b}$. In this case, broadcast f's value which corresponds to $L(a_b, d_b)$, and halt.

6. Else $(0 < |D_j| < |D_{j-1}|)$ continue to stage $j + 1$.

We now prove that the protocol is correct, always terminates, and computes $f(x_1, \ldots, x_n)$ n-privately.

Theorem 2: The protocol computes $f(x_1, \ldots, x_n)$ n-privately.

Proof: We first show that the protocol always terminates. By the definition of D_j, clearly $D_j \subseteq D_{j-1}$. Suppose that for some input $\vec{x}$ and for some stage j, we have $|D_j| = |D_{j-1}| > 1$. This means that for $e_j = \gcd(D_j)$ and for all $d \in D_j$, $d \equiv b_j$ $\pmod{e_j}$. By Theorem 1, this constitutes a contradiction to the $\left\lceil \frac{n}{2} \right\rceil$-privacy of f.

We conclude that for all j with $|D_{j-1}| > 1$, D_j is a proper subset of D_{j-1}. As D_0 is finite, we conclude that after a finite number of stages, D_j will either be empty or a singleton, at which stage the protocol terminates.

Next, we show that upon termination, every processor holds the value $f(x_1, \ldots, x_n)$. Let $u = weight(x_1, \ldots, x_n)$, such that $u \in L(a, d)$. If $D_0 = \phi$, then for all arithmetic progressions $d = 0$. In this case u is computed at stage 0 of the protocol, and the corresponding value $g(u)$ is broadcast before termination. Otherwise, $D_0 \neq \phi$, and at least stage 1 is executed. By the termination conditions, the protocol terminates

355

when either $D_j = \phi$ or $|D_j| = |D_{j-1}| = 1$. We now show that if $d = 0$ then the protocol terminates with $D_j = \phi$, while if $d > 0$, the protocol terminates with $|D_j| = |D_{j-1}| = 1$.

For the case $u \in L(a, d)$ with $d = 0$, suppose the protocol terminates with $D_j = \{d_b\}$. As D_j contains only positive elements, $d_b > 0$. By definition, $e_j = \gcd(D_j) = d_b$, and thus, by c_j's definition $u \equiv c_j \pmod{d_b}$. But then we would have $u \in L(a_b, d_b)$ with $d_b > 0$, contradicting $d = 0$. Thus the final D_j must be empty, and so the protocol computes u and the corresponding value $f(\vec{x})$.

For the case $d > 0$, we show that for all stages j, the difference d is an element of D_j, and that for the final stage, $|D_j| = |D_{j-1}| = 1$. Thus, upon termination, $D_j = \{d\}$. The protocol computes $c_j = u \pmod{d}$, and the corresponding value $f(\vec{x})$. This implies that $D_j \neq \phi$ for all j.

The proof that $d \in D_j$ is by induction on the stage j. As $d > 0$, d is an element of D_0 by definition. For the induction step, suppose $d \in D_{j-1}$. Since $u \in L(a, d)$, $u \equiv a \pmod{d}$. By the protocol, $e_j = \gcd(D_{j-1})$, so $e_j | d$, and therefore $u \equiv a \pmod{e_j}$. As $c_j = u \pmod{e_j}$, we have $u \equiv c_j \pmod{e_j}$. By the protocol, this implies $d \in D_j$.

Finally we prove the n-privacy of our protocol. Let $\vec{x} = (x_1, \ldots, x_n)$ and $\vec{y} = (y_1, \ldots, y_n)$ be two inputs in $\{0, 1\}^n$, for which $f(\vec{x}) = f(\vec{y})$. Let $T \subseteq \{1, \ldots, n\}$ be a coalition, such that for all $i \in T$, $x_i = y_i$. Let $u = weight(\vec{x})$, $v = weight(\vec{y})$, and $u, v \in L(a, d)$. We distinguish between the cases $d = 0$ and $d > 0$.

If $d = 0$ then $u = v$. Therefore, for every stage j executed in the protocol,

$$c_j(\vec{x}) = u \bmod e_j = v \bmod e_j = c_j(\vec{y}) ,$$

namely c_j (the modular sum) gets the same values in the execution on input $\vec{x}$ and on input $\vec{y}$. Since the protocol for modular sum [3] is n-private, it follows that the messages sent by it, restricted to T, are identically distributed in the two cases $\vec{x}, \vec{y}$. The same holds for the final stage, when the sum itself is computed n-privately. As messages are exchanged only in the sum computations, this proves n-privacy for $d = 0$.

356

For the case $d > 0$, we have $u \bmod d = v \bmod d$. We saw that $d \in D_j$ for all stages j, thus $e_j | d$ and so $u \bmod e_j = v \bmod e_j$. Thus $c_j(\vec{x}) = u \bmod e_j$ and $c_j(\vec{y}) = v \bmod e_j$. As messages are exchanged only in the computations of the c_j's, the n-privacy of our protocol follows from the n-privacy of the modular sum protocol. This completes the proof of Theorem 2. $\qquad\square$

The proof of Theorem 2 implies that the gcd condition on the arithmetic progression is not only a necessary condition for $\lceil \frac{n}{2} \rceil$-privacy of dense symmetric functions, but also a sufficient condition, namely

Theorem 3: Let $f : \{0,1\}^n \to \{0,\ldots,n\}$ be a dense symmetric function. Then f is $\lceil \frac{n}{2} \rceil$-private if and only if

1. For every b in f's range there are a, d such that $h^{-1}(b) = L(a,d)$.

2. For any subset Q ($|Q| \geq 2$) of positive differences from the various progressions $L(a,d)$, there is at least one pair i, j with $d_i, d_j \in T$, for which the corresponding a_i, a_j satisfy $a_i \not\equiv a_j \pmod{\gcd(Q)}$.

In [1, 4] it was shown that every function, defined over a finite domain, is $\lfloor \frac{n-1}{2} \rfloor$-private. Our result states that dense symmetric functions which are $\lceil \frac{n}{2} \rceil$-private are also n-private. We conclude that there is a gap in the privacy hierarchy for this family of functions.

Theorem 4: Let $f : \{0,1\}^n \to \{0,\ldots,n\}$ be a dense symmetric function. Then f is either $\lfloor \frac{n-1}{2} \rfloor$ but not $\lceil \frac{n}{2} \rceil$-private, or it is n-private.

Finally, we remark that both levels of the privacy hierarchy for dense symmetric functions are non-empty: For example OR is a dense, $\lfloor \frac{n-1}{2} \rfloor$, but not $\lceil \frac{n}{2} \rceil$-private function, while XOR (for $n \geq 5$) is a dense, n-private symmetric function.

7 Concluding Remarks

One natural directions to extend our work is to consider domains with larger size ($\{0,1\ldots,k\}^n$ for some $k \geq 2$, instead of merely $\{0,1\}^n$), and to remove the density restriction.

All known proofs of non t – privacy for functions with finite domain $(t \geq \lceil \frac{n}{2} \rceil)$ are based on partition arguments. It is an open problem whether such argument always suffices. That is, whether non t – privacy can always be proved by a partition argument. Partition arguments are particularly convenient for use in discussing t – privacy of symmetric functions, as there is essentially only one partition for each t.

Symmetric functions of Boolean variables which satisfy the arithmetic progression condition but are not dense are potentially a good class of candidates to test the sufficiency of partition arguments for proofs of non t – privacy. As a specific example, consider the following function $f : \{0,1\}^6 \rightarrow \{0,\ldots,6\}$: The arithmetic progressions corresponding to f are

$$L(0,5) = \{0,5\}$$

$$L(1,3) = \{1,4\}$$

$$L(2,0) = \{2\}$$

$$L(3,0) = \{3\}$$

$$L(6,0) = \{6\} \ .$$

It is not hard to check that the induced two argument function, corresponding to any partition of the 6 arguments to two sets of size 3, is 1 – private. Therefore, if this function f is *not* 3 – private, the proof will not use partitions but rather will require a new argument.

Acknowledgments

We would like to thank David Peleg, who generalized our original statement of Claim 1, and simplified its proof. Thanks to David, Oded Goldreich, and Eyal Kushilevitz, for their comments on earlier versions of this manuscript.

References

[1] Ben-or M., S. Goldwasser, and A. Wigderson, "Completeness Theorems for Non-Cryptographic Fault-Tolerant Distributed Computation" *Proc. of 20th STOC*, 1988, pp. 1-10.

[2] Beaver, D., "Perfect Privacy for Two Party Protocols", Technical Report TR-11-89, Harvard University, 1989.

[3] Benaloh (Cohen), J.D., "Secret Sharing Homomorphisms: Keeping Shares of a Secret Secret", *Advances in Cryptography - Crypto86 (proceedings)*, A.M. Odlyzko (ed.), Springer-Verlag, Lecture Notes in Computer Science, Vol. 263, pp. 251-260, 1987.

[4] Chaum, D., C. Crepeau, and I. Damgard, "Multiparty Unconditionally Secure Protocols" *Proc. of 20th STOC*, 1988, pp. 11-19.

[5] Chor, B., M. Geréb-Graus, and E. Kushilevitz, "Private Computations Over the Integers", *31th IEEE Conference on the Foundations of Computer Science*, October 1990, pp. 335-344.

[6] Chor, B., and E. Kushilevitz, "A Zero-One Law for Boolean Privacy", *SIAM J. Discrete Math.*, Vol 4, No 1, 1991, pp. 36-47. Early version in *Proc. of 21th STOC*, 1989, pp. 62-72.

[7] Kushilevitz, E., "Privacy and Communication Complexity", *Proc. of 30th FOCS*, 1989, pp. 416–421. To appear in *SIAM Jour. Disc. Math.*

Efficient Reduction among Oblivious Transfer Protocols based on New Self-Intersecting Codes

Claude Crépeau *   Miklós Sántha †

Laboratoire de Recherche en Informatique
Université Paris-Sud
Bâtiment 490
91405 Orsay FRANCE

Abstract

A $\binom{2}{1}$-OT$_2$ (one-out-of-two Bit Oblivious Transfer) is a technique by which a party S owning two secret bits b_0, b_1, can transfer one of them b_c to another party $\mathcal{R}$, who chooses c. This is done in a way that does not release any bias about $b_{\bar{c}}$ to $\mathcal{R}$ nor any bias about c to S. One interesting extension of this transfer is the $\binom{2}{1}$-OT$_2^k$ (one-out-of-two String O.T.) in which the two secrets q_0, q_1 are elements of $GF^k(2)$ instead of bits. A reduction of $\binom{2}{1}$-OT$_2^k$ to $\binom{2}{1}$-OT$_2$ presented in [BCR86] uses $O(k^{\log_2 3})$ calls to $\binom{2}{1}$-OT$_2$ and thus raises an interesting combinatorial question: how many calls to $\binom{2}{1}$-OT$_2$ are necessary and sufficient to achieve a $\binom{2}{1}$-OT$_2^k$?

In the current paper we answer this question quite precisely. We accomplish this reduction using $\Theta(k)$ calls to $\binom{2}{1}$-OT$_2$. First, we show by probabilistic methods how to obtain such a reduction with probability essentially 1 and second, we give a deterministic polynomial time construction based on the algebraic codes of Goppa [Gop81].

1   Introduction

The equivalence between cryptographic primitives has recently become a major research topic as evidenced by many papers on the topic: [BCR86], [Cré88],[CK88],[Kil88],[Cré89]. So far a large number of cryptographic protocols have been shown equivalent to one another. Nevertheless very few of those reductions accomplish *perfectly* the task they are designed for. Generally the reduction will fail to achieve its goal with a small probability. An exception to this situation is the main contribution of [BCR86]. As an introduction to our work, we first summarize their result.

A $\binom{2}{1}$-OT$_2$ (one-out-of-two Bit Oblivious Transfer) is a technique by which a party S owning two secret bits b_0, b_1, can transfer one of them b_c to another party $\mathcal{R}$, who chooses c. This is done in a way that does not release any bias about $b_{\bar{c}}$ to $\mathcal{R}$ nor any bias about c to S. This primitive was first introduced in [EGL83] with application to contract signing protocols. A natural and interesting extension of this transfer is the $\binom{2}{1}$-OT$_2^k$ (one-out-of-two String Oblivious Transfer, know as ANNBP in [BCR86]) in which the two secrets q_0, q_1 are elements of $GF^k(2)$ instead of bits. The core of [BCR86] is a reduction of $\binom{2}{1}$-OT$_2^k$ to $\binom{2}{1}$-OT$_2$, i.e. an efficient two-party protocol to achieve $\binom{2}{1}$-OT$_2^k$ based on the assumption of the existence of a protocol for $\binom{2}{1}$-OT$_2$. The problem is solved using functions known as (n, k)-*zigzags* for historical reasons. They give an

---

*Supported in part by an NSERC Postdoctorate Scholarship. e-mail: crepeau@dmi.ens.fr

†Part of this research was supported by an Alexander von Humboldt Fellowship. e-mail: santha@dmi.ens.fr

efficient construction for a binary linear $(k^{\log_2 3}, k)$-zigzag. This was the best known construction of an (n, k)-zigzag so far. This construction raises an interesting combinatorial question: Minimize n as a function of k.

As pointed out in [BCR86], the problem of finding linear (n, k)-zigzags is equivalent to finding linear binary (n, k, d) codes with some special intersecting properties. Recently we have also found out that such codes had been extensively studied in the past, and are known to code theorists as *self-intersecting codes* [CL85]. Indeed the construction of the binary linear $(k^{\log_2 3}, k)$-zigzag of [BCR86] was known to coding theorists since 1984 by the work of Miklós [Mik84].

Denote by $zzl(n) = max\{k :$ there exists a binary linear (n, k)-zigzag $\}$. Katona and Srivastava [KS83] have shown that

n	$zzl(n)$
$3, 4, 5$	2
$6, 7, 8$	3
$9, 10, 11, 12$	4
$13, 14$	5

They also derived an upper bound on the asymptotic behavior of $\frac{zzl(n)}{n}$, whereas a lower bound was given by Komlós (reported in [Mik84]). These bounds are the following:

$$0.207 < \liminf_{n \to \infty} \frac{zzl(n)}{n} \le \limsup_{n \to \infty} \frac{zzl(n)}{n} < 0.283.$$

They imply that binary linear (ck, k)-zigzags exist for $c > 4.81$ and clearly not for $c < 3.53$. Retter [Ret86] showed that most (classical) Goppa codes (see [MS77]) of rate less than 0.0817 are self-intersecting.

The current paper focuses on the polynomial time constructibility of such zigzags. First we show in section 3 that for $\alpha = \log_{4/3} 4$ and any $\epsilon > 0$, a random $k \times (1 + \epsilon)\alpha k$ binary matrix (considered as a linear application) defines a $((1 + \epsilon)\alpha k, k)$-zigzag with probability asymptotically 1, and a random $k \times (1 - \epsilon)\alpha k$ binary matrix defines a $((1 - \epsilon)\alpha k, k)$-zigzag with probability asymptotically 0. In both cases the convergence is exponentially fast in k. Actually the first part of this result was also stated by Cohen and Lempel [CL85]. This leads to an obvious efficient probabilistic algorithm for the generation of binary linear $((1 + \epsilon)\alpha k, k)$-zigzags. Finally in section 4, we present a deterministic polynomial time construction for binary linear $(O(k), k)$-zigzags based on the algebraic codes of Goppa [Gop81].

2   Previous Work

Assume $\mathcal{S}$ and $\mathcal{R}$ have a mean of accomplishing $\binom{2}{1}$-OT$_2$ and that they wish to perform a $\binom{2}{1}$-OT$_2^k$ over the two k-bit strings q_0, q_1. For any string x, let x^i denote the ith bit of x. For a set of indices $I = \{i_1, i_2, ..., i_m\}$, we define x^I to be the concatenation $x^{i_1} x^{i_2} ... x^{i_m}$, the indices taken in increasing order.

First observe that performing $\binom{2}{1}$-OT$_2$ on each pair q_0^i, q_1^i, $1 \le i \le k$, in order to implement a $\binom{2}{1}$-OT$_2^k$ fails dramatically since a cheating $\mathcal{R}$ can get q_0^1 and q_1^2 for instance, i.e. partial information about q_0 and q_1. Assume instead that we find a function $f : GF^n(2) \to GF^k(2)$ with the nice property that for every two input strings x_0, x_1 and every disjoint sets I, J seeing the bits x_0^I and x_1^J releases information on at most one of $f(x_0)$ and $f(x_1)$. Let us be more precise about this.

361

Definition 2.1 A subset $I \subseteq \{1, 2, ..., n\}$ *biases* a function $f : GF^n(2) \to GF^k(2)$ if

$$\exists q_0, q_1, x \ \left[\#\{z | z^I = x^I, \ f(z) = q_0\} \neq \#\{z | z^I = x^I, \ f(z) = q_1\} \right] \tag{1}$$

such an x^I is said to *release information* about $f(x)$.

Definition 2.2 A (n, k)-*zigzag* is an easily computable function $f : GF^n(2) \to GF^k(2)$ for which random inverses can be easily computed for each input such that

$$\forall I \subseteq \{1, 2, ..., n\} \ \left[I \text{ or } \bar{I} \text{ does not bias } f \right] \tag{2}$$

The basic idea of [BCR86] is to find a zigzag f and use it as follows:

Protocol 2.1 ($\binom{2}{1}$-$\mathbf{OT}_2^k((q_0, q_1), c)$)

   1: S finds random x_0, x_1 such that $f(x_0) = q_0$ and $f(x_1) = q_1$.

   2: $\underset{i=1}{\overset{n}{\mathbf{DO}}}$ S transfers x_c^i out-of x_0^i, x_1^i to $\mathcal{R}$ using $\binom{2}{1}$-OT_2.

   3: $\mathcal{R}$ recovers q_c by computing $q_c = f(x_c)$.

This protocol works since an honest $\mathcal{R}$ gets $x_c^{\{1,...,n\}}$ and $x_{\bar{c}}^{\emptyset}$, leading him to q_c and no information about $q_{\bar{c}}$, while by the definition of f, a cheating $\mathcal{R}$ who gets x_0^I and $x_1^{\bar{I}}$ for some $I \neq \emptyset \neq \bar{I}$, only gets information about one of $f(x_0)$ or $f(x_1)$. Based on this protocol, our task is to build efficient f's.

Working under the assumption that $f(x)$ is a linear binary function, i.e. that $f(x) = Mx$ for a given $k \times n$ binary matrix M (we say that M *defines* f), [BCR86] shows that the characterization (2) of f

$$\forall I \subseteq \{1, 2, ..., n\} \ \left[I \text{ or } \bar{I} \text{ does not bias } Mx \right] \tag{3}$$

is equivalent to the two following characterizations (4) and (5).

Define for a matrix M, M^i to be the ith column of M. For $I = \{i_1, i_2, ..., i_m\}$, we define M^I to be the matrix obtained by the concatenation $M^{i_1} M^{i_2} ... M^{i_m}$. We say that two line vectors $\vec{v}_0$ and $\vec{v}_1$ of dimension n *intersect* if $\vec{v}_0 \times \vec{v}_1 \neq (0, 0, ..., 0) = 0^n$, where the product $\times$ is performed component by component.

Proposition 2.3 M *defines a linear zigzag if and only if*

$$\forall I \subseteq \{1, 2, ..., n\} \ \left[M^I \text{ or } M^{\bar{I}} \text{ has rank } k \right] \tag{4}$$

or

$$\forall \vec{a}, \vec{b} \in GF^k(2) - \left\{ 0^k \right\} \ \left[\vec{a}M \text{ and } \vec{b}M \text{ intersect } \right] . \tag{5}$$

Observations & Remarks

This last characterization turns out to be the most useful. Consider M as the generator matrix of a binary linear code. M defines an (n, k)-zigzag exactly if the (n, k, d) linear code generated by M does not contain two non-intersecting codewords $\vec{c}_1, \vec{c}_2$. A number of observations can be made about the code generated by M. The most important are:

362

Observation 2.1: If M defines a linear (n, k)-zigzag then it defines a linear (n, k, k) code. (see [BCR86])

Observation 2.2: If M defines a linear $(n, k, \frac{n}{2} + 1)$ code then it defines a linear (n, k)-zigzag. By a pigeon hole argument any two codewords of such a code must intersect.

Remark 2.1: Observation 2.2 combined with Plotkin's bound [MS77] implies that linear $(n, k, \frac{n}{2} + 1)$ codes cannot exist over $GF(2)$ when $n \in O(k)$. Nevertheless this observation will be crucial to the elaboration of the deterministic technique of section 4.

3 Random matrices for linear zigzags

In this section we determine very precisely the size of random matrices which define linear zigzags. If k denotes the number of lines, then there is a threshold function $t(k)$, which is linear in k, such that a random matrix with more than $t(k)$ columns defines a linear zigzag with high probability, whereas a random matrix with less than $t(k)$ columns does not define a linear zigzag, also with high probability. Komlós has already observed [Mik84] that some matrices with more than $t(k)$ columns define a linear zigzag.

Theorem 3.1 *Set* $\alpha = \log_{4/3} 4 \approx 4.81$, *and let* M *be a random* $k \times n$ *binary matrix. Then for every constant* $\epsilon > 0$, *there exists a constant* $0 < \eta < 1$ *such that we have the following two propositions:*

1. *If* $n \geq (1 + \epsilon)\alpha k$, *then* $Pr[M$ *does not define a linear zigzag*$] < \eta^k$,

2. *If* $n \leq (1 - \epsilon)\alpha k$, *then* $Pr[M$ *defines a linear zigzag*$] < \eta^k$.

Sketch of Proof: Let $\vec{a}$ and $\vec{b}$ be two non-zero binary line vectors of length k, and set $S = (\vec{a}, \vec{b})$. For every S, let D_S be the event that $\vec{a}M$ and $\vec{b}M$ do not intersect, and let X_S be the associated indicator random variable. As every bit in $\vec{a}M$ and $\vec{b}M$ is 1 with probability $1/2$, we have

$$E[X_S] = Pr[D_S] = \begin{cases} (\frac{3}{4})^n & \text{if } \vec{a} \neq \vec{b} \\ (\frac{1}{2})^n & \text{if } \vec{a} = \vec{b}. \end{cases}$$

Let us define the random variable $X = \sum_S X_S$, where the summation is done over all couples $S = (\vec{a}, \vec{b})$. Then M defines a linear zigzag if and only if $X = 0$. By the linearity of the expectation, $E[X] \approx 2^{2k}(\frac{3}{4})^n$.

Proposition 1.: $n \geq (1 + \epsilon)\alpha k$. Then we have:

$$Pr[X > 0] \leq E[X] \approx 2^{2k} \left(\frac{3}{4}\right)^n \leq \left(\frac{1}{4}\right)^{\epsilon k}.$$

Proposition 2.: $n \leq (1 - \epsilon)\alpha k$. Then using Chebishev's inequality, we have:

$$Pr[X = 0] \leq Pr[|X - E[X]| \geq E[X]] \leq \frac{Var(X)}{E[X]^2}.$$

We will show that $Var(X)$ is exponentially small compared to $E[X]^2$. We set $N = E[X]^2$. Let $\vec{a_1}, \vec{b_1}, \vec{a_2}$ and $\vec{b_2}$ be non-zero line vectors of length k, and set $S_1 = (\vec{a_1}, \vec{b_1}), S_2 = (\vec{a_2}, \vec{b_2})$. Since X_{S_1} and X_{S_2} are 0-1 random variables, we have:

$$Var(X) = \sum_{S_1, S_2} cov(X_{S_1}, X_{S_2}) = \sum_{S_1, S_2} E[X_{S_1}](E[X_{S_2}|X_{S_1} = 1] - E[X_{S_2}]).$$

If the random variables X_{S_2} and X_{S_1} are not independent, then we will bound $cov(X_{S_1}, X_{S_2})$ from above by $E[X_{S_1}]E[X_{S_2}|X_{S_1} = 1]$. If X_{S_2} and X_{S_1} are independent, then $cov(X_{S_1}, X_{S_2}) = 0$. The proof works out because for most S_1 and S_2 they are indeed independent. We will prove this with the help of the following Lemma:

Lemma: *If for some $m > 0$, the family of vectors $\{\vec{a_1}, \ldots \vec{a_m}\}$ is linearly independent then $\vec{a_1} M, \ldots, \vec{a_m} M$ are independent random variables.*
Proof of Lemma: It follows from Vazirani's Parity Lemma [Vaz87] when applied to the uniform distribution. $\quad\square$

It follows from the Lemma that the random variables X_{S_1} and X_{S_2} are independent when the family of four vectors $\{\vec{a_1}, \vec{b_1}, \vec{a_2}, \vec{b_2}\}$ is linearly independent. Thus we have to consider the covariances of only those random variables X_{S_1} and X_{S_2} where there is some linear dependence among the (not necessarily distinct) vectors of the family $\mathcal{U} = \{\vec{a_1}, \vec{b_1}, \vec{a_2}, \vec{b_2}\}$. We will distinguish several cases according to the rank of $\mathcal{U}$ and the possible dependencies. The analysis of all the cases is similar, we analyze here two of them.

Case 1: $rank(\mathcal{U}) = 3$ and $\vec{a_2} = \vec{a_1}$. The number of such families is less than 2^{3k}. Let us suppose that $\vec{a_1} M$ and $\vec{b_1} M$ do not intersect and let us compute the probability that $\vec{a_2} M$ and $\vec{b_2} M$ do not intersect either. Let a_1 and b_1 be the first bit of respectively $\vec{a_1} M$ and $\vec{b_1} M$. Similarly, let a_2 and b_2 be the first bit of $\vec{a_2} M$ and $\vec{b_2} M$. The string $a_1 b_1$ with probability $\frac{1}{3}$ takes each of the values 00, 01 and 10, since $\vec{a_1} M$ and $\vec{b_1} M$ do not intersect. Our hypotheses about the family $\mathcal{U}$ imply that $a_2 = a_1$ and b_2 is 1 with probability $\frac{1}{2}$, independently from the value of $a_1 b_1$. Thus we have

$$Pr[a_2 b_2 = 11 | a_1 b_1 \neq 11] = \frac{1}{3} \times 0 + \frac{1}{3} \times 0 + \frac{1}{3} \times \frac{1}{2} = \frac{1}{6}.$$

The analysis is identical in the other columns of the matrix, therefore the total contribution to the variance of these families is less than

$$2^{3k} \left(\frac{3}{4}\right)^n \left(\frac{5}{6}\right)^n \leq (0.83)^k N.$$

Case 2: $rank(\mathcal{U}) = 3$ and $\vec{a_1} + \vec{a_2} + \vec{b_1} + \vec{b_2} = \vec{0}$. The number of such families is once again less than 2^{3k}. We will use the same notation as previously. Then $a_2 = a_1 + z$ and $b_2 = b_1 + z$, where $z = a_1 + a_2 = b_1 + b_2$ is a random bit. Clearly, $a_2 b_2 = 11$ if and only if $a_1 b_1 = 00$ and $z = 1$. This happens with probability $\frac{1}{3} \times \frac{1}{2} = \frac{1}{6}$, and the total contribution of these families is less than $(0.83)^k N$. $\quad\square$

4 Deterministic Linear Size Construction

Although the result of section 3 implies that binary linear $(O(k), k)$-zigzags exist and can be obtained easily by picking one at random, it does not provide an efficient way of building a *guaranteed* zigzag of linear size. The problem of checking if a random matrix defines a zigzag seems rather hard (probably as hard as finding the minimal weight of a code, which is considered a hard problem [BET78]).

The solution we present now, is based on a new approach: using the theory of linear codes over $GF(q)$, for $q > 2$, in order to perform the reduction of $\binom{2}{1}$-OT_q^k to $\binom{2}{1}$-OT_q in the larger field $GF(q)$ ($\binom{2}{1}$-OT_q is the q-ary version of $\binom{2}{1}$-OT_2). We then use the construction of [BCR86]/[Mik84] to perform the reduction of $\binom{2}{1}$-OT_q to $\binom{2}{1}$-OT_2.

4.1 Reduction over $GF(q)$

Assume S and $\mathcal{R}$ have a mean of accomplishing $\binom{2}{1}$-OT_q for some fixed $q > 2$ and that they wish to perform a $\binom{2}{1}$-OT_q^k with the two (q-ary) strings q_0, q_1 of length k. As before, q_x^i defines the ith (q-ary) entry of q_x. In the same manner, for $I \subseteq \{1, 2, ..., n\}$ we define q_x^I as before but with respect to q-ary strings instead of binary. We also say that a subset I *biases* a function $f : GF^n(q) \rightarrow GF^k(q)$ exactly if the condition (1) is satisfied as in the case $q = 2$.

The basic idea of [BCR86], to find a function f with property (2) remains our approach to performing the reduction of $\binom{2}{1}$-OT_q^k to $\binom{2}{1}$-OT_q. Based on the assumption that f is a linear function (with operations in $GF(q)$ instead of $GF(2)$) it is possible to show that the characterizations (4) and (5) are also equivalent to (3) when interpreted over $GF(q)$ (the details are left out of this abstract). Therefore, in order to solve the problem over $GF(q)$, one possible approach is to find linear $(n, \Theta(n), \frac{n}{2} + 1)$ codes over $GF(q)$. Due to the work of Goppa [Gop81], Tsfasman, Vlădut and Zink [TVZ82], Katsman, Tsfasman and Vlădut [KTV84] and Manin and Vlădut [MV84] we actually get the codes that we need over $GF(q)$ where q is greater than 9 and is an even power of a prime.

More precisely, their result is as follows:

Proposition 4.1 ([MV84]) *If $q = p^{2m}$ for some prime p, it is possible to construct in polynomial time linear (n, k, d) codes over $GF(q)$ with parameters spreading over the line of equation*

$$\frac{k}{n} + \frac{d}{n} = 1 - \frac{1}{\sqrt{q} - 1}.$$

We use these codes to achieve our goal as follows.

4.2 Construction of a Linear zigzag over $GF(q)$

Let $q = p^{2m} > 9$ for some prime p, and let $\Gamma_q = \frac{1}{2} - \frac{1}{\sqrt{q}-1}$. According to proposition 4.1 it is possible to build in polynomial time a linear $(n, \Gamma_q n - 1, \frac{n}{2} + 1)$ code over $GF(q)$ for all sufficiently large n. This leads to the construction of a linear $(n, \Gamma_q n - 1)$-zigzag over $GF(q)$, i.e.

Proposition 4.2 *If $q = p^{2m} > 9$ for some prime p, it is possible to build in polynomial time a linear $(\frac{k+1}{\Gamma_q}, k)$-zigzag over $GF(q)$.*

Remark 4.1: An analysis similar to that of section 3 for the case over $GF(q)$ shows that for any $\epsilon > 0$, a random $GF(q)$-matrix of size $k \times (2 + \epsilon)k$ has asymptotically probability 1 of being a zigzag, as $q \rightarrow \infty$.

4.3 Construction of a Linear zigzag over $GF(2)$

Using the previous technique it is possible, for $m > 1$, to build a $\binom{2}{1}$-$OT_{2^{2m}}^k$ based on the existence of a $\binom{2}{1}$-$OT_{2^{2m}}$. First observe that a $\binom{2}{1}$-$OT_{2^{2m}}$ and a $\binom{2}{1}$-OT_2^{2m} are interchangeable since the base sets of the two structures are the same. On the other hand, building a $\binom{2}{1}$-OT_2^{2m} is very simple using the general construction of [BCR86]. Their technique is used purely to bootstrap our more efficient technique and is used in an ad-hoc fashion to solve the problem for this fixed length $2m$ over $GF(2)$. This technique is nothing else than building a concatenation code from the Goppa code and a fixed smaller binary linear code. The concatenation of these two codes will be a linear binary code.

Example 4.1 Set $q = 256$. Let q_0, q_1 be two binary strings of length k. Consider each of these two strings as $\frac{k}{8}$ bytes (i.e. elements of $GF(256)$). By proposition 4.2 one can build a linear $(\frac{k}{3}, \frac{k}{8})$-zigzag over $GF(256)$. Let M be the byte-matrix of size $\frac{k}{8} \times \frac{k}{3}$ that defines such a zigzag. Let z_0, z_1 be two random byte-strings of length $\frac{k}{3}$ such that $q_0 = Mz_0$ and $q_1 = Mz_1$ where the operations $+$ and $\times$ of $GF(256)$ are performed on bytes. Now let $\mathcal{M}$ be the 8×27 binary-matrix that defines the binary linear $(27, 8)$-zigzag of [BCR86]. For each byte z_0^i of z_0 and z_1^i of z_1, let y_0^i and y_1^i be random binary strings of length 27 such that $z_0^i = \mathcal{M}y_0^i$ and $z_1^i = \mathcal{M}y_1^i$ with the operations performed in $GF(2)$. Name $y_0 = y_0^1 y_0^2 ... y_0^{\frac{k}{3}}$ and $y_1 = y_1^1 y_1^2 ... y_1^{\frac{k}{3}}$ the two resulting binary strings of length $9k$.

Define $f : GF^{9k}(2) \rightarrow GF^k(2)$ to be

$$f(x) = G\mathcal{G}x$$

where $\mathcal{G}$ is the $\frac{8k}{3} \times 9k$ binary matrix defined from $\mathcal{M}$ as follows:

$$
\mathcal{G} = \begin{pmatrix}
\mathcal{M} & \begin{matrix} 0 & 0 & ... & 0 \\ \vdots & \vdots & \ddots & \vdots \\ 0 & 0 & ... & 0 \end{matrix} & ... & \begin{matrix} 0 & 0 & ... & 0 \\ \vdots & \vdots & \ddots & \vdots \\ 0 & 0 & ... & 0 \end{matrix} \\
\begin{matrix} 0 & 0 & ... & 0 \\ \vdots & \vdots & \ddots & \vdots \\ 0 & 0 & ... & 0 \end{matrix} & \mathcal{M} & ... & \begin{matrix} 0 & 0 & ... & 0 \\ \vdots & \vdots & \ddots & \vdots \\ 0 & 0 & ... & 0 \end{matrix} \\
\vdots & \vdots & \ddots & \vdots \\
\begin{matrix} 0 & 0 & ... & 0 \\ \vdots & \vdots & \ddots & \vdots \\ 0 & 0 & ... & 0 \end{matrix} & \begin{matrix} 0 & 0 & ... & 0 \\ \vdots & \vdots & \ddots & \vdots \\ 0 & 0 & ... & 0 \end{matrix} & ... & \mathcal{M}
\end{pmatrix}
$$

and G is the $k \times \frac{8k}{3}$ binary matrix constructed by replacing each line M_i of M ($\frac{k}{3}$ bytes per line) by 8 lines of $\frac{8k}{3}$ bits obtained by multiplying (in $GF(256)$) each component of M_i with the byte constants $00000001, 00000010, ..., 10000000$ and interpreting the result as binary strings, i.e.

$$
G = \begin{pmatrix}
M_1 \\
00000010 \times M_1 \\
\vdots \\
10000000 \times M_1 \\
\vdots \\
M_{\frac{k}{8}} \\
00000010 \times M_{\frac{k}{8}} \\
\vdots \\
10000000 \times M_{\frac{k}{8}}
\end{pmatrix}
$$

We then have that $q_0 = f(y_0)$, $q_1 = f(y_1)$ and the following proposition:

Proposition 4.3 *The above f is a binary linear $(9k, k)$-zigzag.*

Similar constructions can be obtained for any large enough field. For larger fields more efficient techniques than that of [BCR86]/[Mik84] should be used. Nevertheless, if their technique is used to bootstrap the process, then the field $GF(256)$ will lead to the family of zigzags with smallest expansion factor.

5 Conclusion and Open Problems

We have shown how to construct binary linear size zigzag functions both by probabilistic and deterministic polynomial time methods. The exact complexity of the decision problem "Given a matrix M, is it the generator of a zigzag?" still has to be determined (the best we can say is that it is in Co-NP). Another problem still open is to construct deterministically over $GF(q)$ some zigzag that will do better than the asymptotic bounds of section 3 and remark 4.1 . Finally we ask if non-linear functions can generate smaller zigzags than linear functions.

Acknowledgments

We would like to thank Paul Barbaroux, Gilles Brassard, Johannes Buchmann, W. Fernandez De La Vega, Antoine Joux, Joe Kilian, Jean-Marc Robert, and Jacques Stern for their help, comments, and support. Special thanks to Oded Goldreich and Silvio Micali for suggesting characterization (4) (c.f. [BCR86]). Claude wishes to thank Shafi Goldwasser for introducing him to Goppa codes, Peter Elias for providing an extensive bibliography on the topic and Gérard Cohen for extremely valuable information about his earlier research on intersecting codes. We are very greatful to Ivan Damgård for pointing out to us that our $(9k, k)$-zigzag was indeed linear.

References

[BCR86] G. Brassard, C. Crépeau, and J.–M. Robert. Information theoretic reductions among disclosure problems. In 27th Symp. of Found. of Computer Sci., pages 168–173, IEEE, 1986.

[BET78] E.R. Berlekamp, R.J. Mc Eliece, and H.C.A. Van Tilborg. On the inherent intractability of certain coding problems. IEEE Transaction on Information Theory, 384–386, 1978.

[CK88] C. Crépeau and J. Kilian. Achieving oblivious transfer using weakened security assumptions. In 28th Symp. on Found. of Computer Sci., pages 42–52, IEEE, 1988.

[CL85] G. Cohen and A. Lempel. Linear intersecting codes. Discrete Mathematics, 56:35–43, 1985.

[Cré88] C. Crépeau. Equivalence between two flavours of oblivious transfers (abstract). In C. Pomerance, editor, Advances in Cryptology: Proceedings of Crypto '87, pages 350–354, Springer-Verlag, 1988.

[Cré89] C. Crépeau. Verifiable disclosure of secrets and application. In Advances in Cryptology: Proceedings of Eurocrypt '89, Springer-Verlag, 1989.

[EGL83] S. Even, O. Goldreich, and A. Lempel. A randomized protocol for signing contracts. In R. L. Rivest, A. Sherman, and D. Chaum, editors, Proceedings CRYPTO 82, pages 205–210, Plenum Press, New York, 1983.

[Gop81] V.D. Goppa. Codes on algebraic curves. Soviet Mathematical Dokl., 24(1):170–172, 1981.

[Kil88] J. Kilian. Founding cryptography on oblivious transfer. In Proc. 20th ACM Symposium on Theory of Computing, pages 20–31, ACM, Chicago, 1988.

[KS83] G. Katona and J. Srivastava. Minimal 2-coverings of finite affine spaces based on $GF(2)$. Journal of Statist. Plann. Inference, 8:375–388, 1983.

[KTV84] G.L. Katsman, M.A. Tsfasman, and S.G. Vlădut. Modular curves and codes with a polynomial construction. *IEEE Transaction on Information Theory*, IT-30(2):353–355, 1984.

[Mik84] D. Miklós. Linear binary codes with intersecting properties. *Discrete Applied Mathematics*, 9(2):187–196, 1984.

[MS77] F.J. MacWilliams and N.J.A. Sloane. *The Theory of Error-Correcting Codes*. North-Holland, 1977.

[MV84] Yu.I. Manin and S.G. Vlădut. Linear codes and modular curves (in Russian). *Sovr. Prob. Mat.*, VINITI, 1984.

[Ret86] C.T. Retter. Intersecting goppa codes. 1986. manuscript.

[TVZ82] M.A. Tsfasman, S.G. Vlădut, and Th. Zink. Modular curves, Shimura curves, and Goppa codes, better than Varshamov-Gilbert bound. *Math. Nachr.*, 109:21–28, 1982.

[Vaz87] U. Vazirani. Efficiency considerations in using semi-random sources. In *Proc. 19th ACM Symposium on Theory of Computing*, pages 160–168, ACM, New York City, 1987.

Perfect Zero-Knowledge Sharing Schemes over any Finite Abelian Group[*]

Yvo Desmedt   Yair Frankel

Dept. EE & CS, Univ. of Wisconsin – Milwaukee
P.O. Box 784, WI 53201 Milwaukee, U.S.A.

Abstract

A secret sharing (threshold) scheme is an algorithm in which a distributor creates *shares* of a secret such that a minimum number of shares are needed to regenerate the secret. We propose a new homomorphic perfect secret sharing scheme over any finite Abelian group for which the group operation and inverses are computable in polynomial time. We introduce the concept of *zero-knowledge sharing scheme* to prove that the distributor does not reveal anything. A stronger condition than not revealing anything about the secret.

1   Introduction

Secret sharing schemes enable any subset of individuals (shareholders) specified by a predefined access structure to recompute a secret (key) k from *shares* which they have received at an earlier stage from a distributor. The first secret sharing schemes [3, 18] had an access structure which allowed any t out of l individuals to recompute the secret from their shares. These secret sharing schemes are often called *threshold* schemes. Although in the modern literature sharing schemes are often studied in a more general context than threshold schemes (due to [9]), we will consider them as synonyms. Only in our definition section (Section 2) we do not. Beside the above application, threshold schemes are used in multi-party protocols [7] and fault tolerant computing [16]. *Perfect* threshold schemes do not reveal anything about the secret when $t - 1$ shares are used. In Shamir's sharing scheme [18] the distributor generates a random polynomial, $f(x)$, of degree $t - 1$ over a field such that $f(0) = k$ and gives each individual i the share $f(i)$. Lagrange interpolation may be used by t shareholders, at a later stage, to regenerate k.

Benaloh [1] discussed homomorphic sharing (threshold) schemes as those having the property that when $s_{i,1}$ is i's share of k_1 and $s_{i,2}$ is i's share of k_2, then $s_{i,1} \cdot s_{i,2}$ is i's share of $k_1 * k_2$ and for such threshold schemes t shareholders can

---
[*]This research is being supported by NSF Grant NCR-9106327.

reconstruct $k_1 * k_2$ using their $s_{i,1} \cdot s_{i,2}$. All the schemes we discuss will have such property. When both operators "$*$" and "$\cdot$" are identical or when there is no confusion possible we will simply speak of *multiplicative* threshold schemes (when the operation is addition, we could speak about additive).

Existing sharing schemes are over fields or over some finite geometries (for an overview on current sharing schemes consult [20]). In this paper we present a method to create a multiplicative (homomorphic) perfect threshold scheme over any finite Abelian group. To be computationally feasible we only require that the group operation and inverses in the finite Abelian group can be computed in polynomial time.

At first glance it may seem trivial to make a multiplicative secret threshold scheme over a finite Abelian group due to the fundamental theorem of Abelian groups [10] and an appropriate homomorphism [1, p. 255]. Let e_G be the exponent (*i.e.* the smallest positive integer such that $x^{e_G} = 1$ for all x in G) and $\{g_1, \ldots, g_h\}$ be a set of generators for G. As in [18] the distributor will choose independently random polynomials $f_j(x)$ for $1 \le j \le h$ of degree $t-1$ such that the secret $k = g_1^{\gamma_1} \cdots g_h^{\gamma_h} \in G$ with $\gamma_j = f_j(0)$. Let the share for individual i be $s_i = g_1^{f_1(i)} \cdots g_h^{f_h(i)} \in G$. Then if e_G is known to the shareholders, any subgroup of t shareholders can easily recompute k using Lagrange interpolation. However, there are mathematical and computational problems with this solution [1, p. 255]. First, the distributor must solve the *discrete logarithm problem*, which is considered hard [4, 11, 13] for many groups even when generators are known. Second, this scheme is useless when $e_G = 2$ since l can only be one. Third, a multiple of e_G must be known by the distributor and shareholders to use Lagrange interpolation, but this reduces the usefulness of this scheme since multiples of e_G must be kept secret for many cryptographic algorithms [8, 12, 14, 17]. Finally, the distributor must know a set of generators. We solve all previous problems by requiring only that the group operation and inverses in the group can be calculated in polynomial time. Moreover, our scheme does not reveal anything new to any $t-1$ shareholders (*e.g.* e_G). To formally prove this we introduce a new concept called *zero-knowledge secret sharing*.

In Section 2 we will overview the concept of threshold scheme and define zero-knowledge sharing schemes. The foundations of this paper are laid by explaining multiplicative threshold schemes over any finite Abelian group which require the distributor to know a non-trivial multiple ($\neq 0$) of the exponent of the group (see Section 3). In Section 4 we present a multiplicative zero-knowledge threshold scheme over a (family of) finite Abelian groups. For this scheme it is not necessary to know the exponent. We provide an application of the group based multiplicative threshold scheme in Section 5.

2 Preliminaries and Notations

We now introduce formal definitions and notation used in this paper. When $\mathcal{A}$ is a set, $|\mathcal{A}|$ will denote the cardinality of the set. When a is a string, $|a|$ will denote the length of the string.

Definition 1 Let t and l be integers with $1 \le t \le l$. Let $\mathcal{K}$ be a set with elements called *keys*, or *secrets*. Let $\mathcal{S}$, $\mathcal{X}$ and $\mathcal{R}$ be non-empty sets and let $\mathcal{A} = \{1, \ldots, l\}$. We will use the set $\mathcal{A}$ to index the set of shareholders. Let $\rho : \mathcal{K} \times \mathcal{R} \to (\mathcal{S} \times \mathcal{X})^l : (k, r) \to ((s_1, x_1'), \ldots, (s_l, x_l'))$ and $\eta : (\mathcal{S} \times \mathcal{X})^t \to \mathcal{K}$. Let $\mathcal{P}_q(\mathcal{A}) = \{\mathcal{B} \mid \mathcal{B} \subset \mathcal{A} \text{ and } |\mathcal{B}| = q\}$. Let selection function $\sigma : (\mathcal{S} \times \mathcal{X})^l \times \mathcal{P}_t(\mathcal{A}) \to (\mathcal{S} \times \mathcal{X})^t$ for input $(((s_1, x_1'), \ldots, (s_l, x_l')), \mathcal{B})$ return tuples (s_i, x_i') for all $i \in \mathcal{B}$. For a given k, called the secret, (a machine) D during the *distribution phase* chooses an $r \in \mathcal{R}$ with a uniform probability distribution and computes $\rho(k, r)$. Using a private channel, D transmits the secret share, s_i, to each shareholder and publishes a public directory containing a description of all the strings x_i' for $i \in \mathcal{A}$. We say that $(\mathcal{K}, \mathcal{S}, \mathcal{X}, \mathcal{R}, D, t, l, \rho, \eta)$ is a (t, l)-threshold scheme when

1. $\forall \mathcal{B} \subset \mathcal{A}$ where $|\mathcal{B}| = t - 1$ holds: if $H(k) \ne 0$ then $0 < H(k|\mathcal{S}_\mathcal{B}, \mathcal{X}_\mathcal{A}) \le H(k)$ for H the entropy function [6], $\mathcal{X}_\mathcal{A} = \{x_i' \in \mathcal{X} \mid i \in \mathcal{A}\}$ and $\mathcal{S}_\mathcal{B} = \{s_i \in \mathcal{S} \mid i \in \mathcal{B} \text{ and } \mathcal{B} \subset \mathcal{A}\}$.

2. $\forall \mathcal{B} \subset \mathcal{A}$ where $|\mathcal{B}| = t$ holds then $\eta(\sigma(\rho(k, r), \mathcal{B})) = k$.

Schemes in which for any $\mathcal{B} \subset \mathcal{A}$ with $|\mathcal{B}| = t - 1$ holds that $H(k|\mathcal{S}_\mathcal{B}, \mathcal{X}_\mathcal{A}) = H(k)$ are called *perfect*. When a sharing scheme is perfect and $|\mathcal{K}|/|\mathcal{S}| = 1$, it is called an *ideal* sharing scheme.

A definition of threshold schemes based on design theory may be found in [21]. We call the phase in which η is calculated the *recomputation phase*.

Definition 2 Let $\mathcal{K}_n$ be a family of groups indexed by $n \in \mathcal{J}$, where $\mathcal{J}$ is an infinite subset of $\{0, 1\}^*$. We say that a family of groups $\mathcal{K}_n$ is polynomial time when the group operation and inverses may be computed in polynomial time in function of $|n|$. For a polynomial time family of groups, $\mathcal{K}_n$, we say that $(\mathcal{K}, \mathcal{S}, \mathcal{X}, \mathcal{R}, D, \rho, \eta)_{t,l,n}$ is a *practical* family of (t, l)-threshold schemes when the elements of $\mathcal{R}$ have a representation which is polynomial in function of $\max(l, |n|)$ and when ρ and η can be computed in polynomial time in function of $\max(l, |n|)$.

Without loss of generality, we assume that $\mathcal{K}(*)$ is a multiplicative group. We will usually write $a * b$ as ab when the context is clear. We also remark that for the above R's there exists a probabilistic polynomial time algorithm, given as input a random tape, generates an $r \in \mathcal{R}$ with a uniform probability. (A random tape is an infinite binary tape where the bits occur independently and with a uniform probability distribution). From now on when we say random, we mean random with a uniform probability distribution.

We now overview the definition of homomorphic threshold schemes [1].

Definition 3 Let $(\mathcal{K}, \mathcal{S}, \mathcal{X}, \mathcal{R}, D, t, l, \rho, \eta)$ be a threshold scheme. When $\mathcal{B} \subset \mathcal{A}$ with $|\mathcal{B}| = t$ is fixed η corresponds to some $\eta_B : \mathcal{S}^t \to \mathcal{K}$. If we have the operation "·" over $\mathcal{S}$ and for all $\mathcal{B} \subset \mathcal{A}$ with $|\mathcal{B}| = t$ holds that η_B is a homomorphism from $\mathcal{S}^t(\cdot)$ to $\mathcal{K}(*)$, then $(\mathcal{K}, \mathcal{S}, \mathcal{X}, \mathcal{R}, D, t, l, \rho, \eta)$ is a homomorphic threshold scheme. When both operations are identical we will just speak about multiplicative threshold scheme (or additive when appropriate).

When $H(k) = 0$ all sharing schemes are perfect, so perfect sharing looses its intuitive meaning. However often it is desirable that the shareholders do not receive extra information (*e.g.* the exponent of the group) from D. To formalize this we introduce zero-knowledge sharing schemes which broadens the concept of perfect sharing schemes. Informally a zero-knowledge threshold scheme satisfies the property that the key distributor is *not* giving *anything* new to any subset of $t - 1$ shareholders. Observe that the definition of perfect threshold scheme informally says that the key distributor does not give, to any subset of $t - 1$ shareholders, anything new from an *information theoretical viewpoint* about the *secret*. Our concept goes further. Any subset of $t - 1$ shareholders will not be able to calculate anything new about the secret. This distinction is important when the entropy of the secret is not maximal. Moreover the key distributor is not giving anything new (to any subset of $t - 1$ shareholders) about whatsoever. A formal definition will be given in the final paper.

The concept of zero-knowledge *threshold* schemes can easily be broadened towards sharing schemes with other access structures.

Because the key k could reveal something about the group that t shareholders could not calculate, we used $t - 1$ in the above. Indeed when the key k is a nontrivial square root of 4 modulo $n = p \cdot q$, p and q being primes, then t shareholders can factor n.

3 Foundation

Only in this section we assume that a non-trivial multiple of the exponent, $e'_{\mathcal{K}}$, is known to the distributor.

3.1 Exponent only contains large prime factors

We first propose a multiplicative threshold scheme over any finite Abelian group $\mathcal{K}(*)$ and l is less than the smallest prime factor of $e'_{\mathcal{K}}$. Let $\mathcal{S} = \mathcal{K}$, $\mathcal{R} = \mathcal{K}^{t-1}$ and $\mathcal{X} = \mathcal{A}$. For $\mathcal{B} \subset \mathcal{A}$ such that $|\mathcal{B}| = t$, we define

$$y_{i,B} = \prod_{\substack{j \notin B \\ j \in A}} (x_i - x_j) \prod_{\substack{j \in B \\ j \neq i}} (0 - x_j). \tag{1}$$

Distribution phase. For all $i \in \mathcal{A}$, let $x_i = i$. D chooses independently random (with uniform probability distribution) elements $s_i \in \mathcal{K}$ for $1 \leq i \leq t-1$. D then computes

$$s_j = \left(k * \prod_{i=1}^{t-1} s_i^{-y_{i,c_j}} \right)^{y_{j,c_j}^{-1}} \tag{2}$$

for $t \leq j \leq l$ where $C_j = \{1, 2, \ldots, t-1, j\}$. D privately transmits to shareholder i the share s_i.

Recomputation phase. The t shareholders in $\mathcal{B}$ compute $k = \Pi_{i \in B} s_i^{y_{i,B}}$.

Lemma 1 *For any finite Abelian group where l is less than the smallest prime factor of $e'_{\mathcal{K}}$, $(\mathcal{K}, \mathcal{S}, \mathcal{X}, \mathcal{R}, D, t, l, \rho, \eta)$ is a multiplicative ideal (t, l)-threshold scheme.*

Proof. See final paper. $\qquad\qquad\qquad\qquad\qquad\qquad\qquad\qquad\qquad\square$

3.2 Exponent is a small prime

Obviously, the above method does not work when l is large and the exponent is small (*e.g.* 2). We now propose a multiplicative perfect threshold scheme over a finite Abelian group $\mathcal{K}$ where the exponent is a small prime q and $q \leq l$. When the exponent of a (family of) groups is small, a probabilistic polynomial time algorithm may be used to calculate it, thus D does not need to keep the exponent secret. For the next threshold scheme we assume that the shareholders have not calculated $e_{\mathcal{K}}$ to stay consistent with the previous method, even though not doing so would simplify the scheme.

The difficulty with a small exponent is that the exponentiation operation in $\mathcal{K}$ is performed in Z_q. What we would like is to be able to perform the calculation in $GF(q^d)$ since the characteristic of this field is q. Although this is not possible, we will create a vector space which will behave in the way we want.

Let d be an integer such that $q^d - 1 \geq l$. Let u be the root to a monic irreducible polynomial $p(x)$ of degree d used to define $GF(q^d)$, so $u^d = a_0 + \cdots + a_{d-1}u^{d-1}$.

In our context we consider $\mathcal{K}^d = \mathcal{K} \times \cdots \times \mathcal{K}$, the direct product of $\mathcal{K}'s$, as a vector space over $GF(q^d)$ in the following manner. Elements of the additive group in $\mathcal{K}^d$ are written as $\vec{k} = [k_0, \ldots, k_{d-1}]$ and the identity element is $\vec{0} = [1, \ldots, 1]$. We remark that this vector space is *not* necessarily of dimension d and the notation $[k_0, \ldots, k_{d-1}]$ should not be confused with coordinates. Addition in $\mathcal{K}^d$ is defined as $[k_0, \ldots, k_{d-1}] + [k'_0, \ldots, k'_{d-1}] = [k_0 * k'_0, \ldots, k_{d-1} * k'_{d-1}]$. For $(b_0 + \cdots + b_{d-1}u^{d-1}) \in GF(q^d)$ and $[k_0, \ldots, k_{d-1}] \in \mathcal{K}^d$ the scalar operation $(\cdot)$ for the vector space is defined as $(b_0 + \cdots + b_{d-1}u^{d-1}) \cdot [k_0, \ldots, k_{d-1}] = \Sigma_{i=0}^{d-1} b_i u^i [k_0, \ldots, k_{d-1}]$ where $b \cdot [k_0, \ldots, k_{d-1}] = [k_0^b, \ldots, k_{d-1}^b]$ for $b \in GF(q)$ and $u \cdot [k_0, k_1, \ldots, k_{d-1}] = [1, k_0, \ldots, k_{d-2}] + [k_{d-1}^{a_0}, k_{d-1}^{a_1}, \ldots, k_{d-1}^{a_{d-1}}]$, recursively $bu^i \cdot [k_0, \ldots, k_{d-1}] = u \cdot (u^{i-1} \cdot (b \cdot [k_0, \ldots, k_{d-1}]))$. We leave it to the reader to prove that this is a vector space. Now let $S = \mathcal{K}^d, R = (\mathcal{K}^d)^{t-1}$.

Distribution phase. D chooses $p(x)$ as above. Then D chooses l different elements $x_i \in GF(q^d) \setminus 0$. and independently random elements $\vec{s}_i \in \mathcal{K}^d$ for $1 \leq i \leq t - 1$. Let k_0 be the secret in $\mathcal{K}$ and $\vec{k} = [k_0, 1, \ldots, 1]$ represent the secret in $\mathcal{K}^d$. Then D computes

$$\vec{s}_j = y_{j,\mathcal{C}_j}^{-1} \cdot (\vec{k} - (\sum_{\substack{i \neq j \\ i \in \mathcal{C}_j}} y_{i,\mathcal{C}_j} \cdot \vec{s}_i)) \tag{3}$$

for $t \leq j \leq l$ where $\mathcal{C}_j = \{1, 2, \ldots, t-1, j\}$ and $y_{i,B}$ is as in (1) except the computation is performed in $GF(q^d)$. D publishes all the $x'_i = \{(x_i, p(x))|i \in \mathcal{A}\}$ for all $i \in \mathcal{A}$ and privately transmits to shareholder i the share $\vec{s}_i = [s_{i,0}, \ldots, s_{i,d-1}]$.

Recomputation phase. The t shareholders in $\mathcal{B}$ compute $y_{i,B}$ in $Z[u]$ and then calculate $k_0 = F_0(\vec{k}) = \Pi_{i \in B} F_0(y_{i,B} \cdot \vec{s}_i)$ where $F_0 : \mathcal{K}^d \to \mathcal{K} : [k'_0, \ldots, k'_{d-1}] \to k'_0$.

Lemma 2 *For any finite Abelian group where $e_\mathcal{K} = q$, $(\mathcal{K}, \mathcal{S}, \mathcal{X}, \mathcal{R}, D, t, l, \rho, \eta)$ is a multiplicative perfect (t, l)-threshold scheme.*

Proof. See final paper. □

We say a vector space $\mathcal{K}$ is polynomial time if its additive group is polynomial time.

Corollary 1 *For any family of polynomial time finite vector spaces $\mathcal{K}_n$ over a finite field F_n, $(\mathcal{K}, \mathcal{S}, \mathcal{X}, \mathcal{R}, D, \rho, \eta)_{t,l,n}$ can be adapted to a practical additive ideal (t, l)-threshold scheme when $l < |F|$ and to a perfect one otherwise.*

Although many of the earlier proposed sharing schemes over finite geometries [20] can trivially be adapted to vector spaces, not all are practical for any polynomial time vector space. Indeed using the above it is not difficult to make vector spaces for which it is hard to find the coordinates. We are now able to discuss the general case.

3.3 General case

Let $e'_{\mathcal{K}} = q_1^{\delta_1} \cdots q_c^{\delta_c}$ where q_m is a prime $(1 \leq m \leq c)$, $q_m < q_{m+1}$ and $0 < \delta_m$, and all q_m are known to the distributor.

Distribution phase. Let d be an integer such that $q_1^d - 1 \geq l$. The distributor D chooses monic irreducible polynomials $p_m(x)$ of degree d over Z_{q_m} for all $m \leq c$ [2, 15]. Using the Chinese remainder theorem on the *coefficients* of the polynomial, D creates a new polynomial $p(x)$ with root u, so $p(x) \equiv p_m(x)$ over Z_{q_m}. For each $i \in \mathcal{A}$, D chooses a unique element $w_{i,m} = a_{i,m,0} + \cdots + a_{i,m,d-1}u^{d-1} \in Z_{q_m}[u]$ and then computes $x_i \equiv w_{i,m}$ over Z_{q_m} using the Chinese remainder theorem on the coefficients of $w_{i,m}$ for each $m \leq c$. To give the $\vec{s}_i$, the distributor proceeds as before except that $\vec{s}_i$ are elements of the module $\mathcal{K}^d$ over $Z_{e'_{\mathcal{K}}}$.

Recomputation phase. Similar as before, however working in a module over $Z[u]$.

Theorem 1 *For any finite Abelian group $\mathcal{K}$ and any l, the above is a multiplicative perfect (t, l)-threshold scheme.*

Proof. See final paper. $\qquad\qquad\qquad\qquad\qquad\qquad\qquad\qquad\qquad\qquad$ □

The disadvantage with the scheme is that the $p(x)$ can reveal information about $e_{\mathcal{K}}$ to the shareholders. Indeed if $p(x)$ is not irreducible mod q_i then the shareholders will know that q_i does not divide $e_{\mathcal{K}}$. Moreover to execute the algorithm in Theorem 1, the distributor needs to know the factorization of $e'_{\mathcal{K}}$ (although by modifying above knowing $e'_{\mathcal{K}}$ is sufficient). In the next section these problems will be resolved.

4 Zero-knowledge threshold scheme

In this section we *do not* assume that a multiple of the exponent, $e'_{\mathcal{K}}$, is known to the distributor. We now propose a *multiplicative perfect zero-knowledge (t, l)-threshold scheme for any family of finite Abelian groups $\mathcal{K}_n$ and any practical l without requiring that the exponent be known to anyone.* The concept of zero-knowledge sharing scheme will be used to prove that $t - 1$ shareholders will not obtain any more information about the group from the distributor.

Distribution phase. Let q be a prime greater than or equal to $l + 1$. Let u be a root of cyclotomic polynomial $p(x) = \sum_{j=0}^{q-1} x^j$ and $x_i = \sum_{j=0}^{i-1} u^j$. Let k_0 be the secret in $\mathcal{K}$ and $\vec{k} = [k_0, 1, \ldots, 1]$ represent the secret in $\mathcal{K}^{q-1}$. Then D computes

$\vec{s}_j$ as in (3) for $t \leq j \leq l$ except

$$y_{i,B} = \prod_{\substack{j \in B \\ j \neq i}} (x_i - x_j)^{-1}(0 - x_j) \in Z[u] \tag{4}$$

and the computation for $\vec{s}_j$ is performed in the module $\mathcal{K}^{q-1}$ over $Z[u]$. The distributor D privately transmits to shareholder i the share $\vec{s}_i = [s_{i,0}, \ldots, s_{i,q-2}]$. **Recomputation phase.** The t shareholders in B compute $y_{i,B}$ as in (4) in $Z[u]$ and then calculate $k_0 = F_0(\vec{k}) = \Pi_{i \in B} F_0(y_{i,B} \cdot \vec{s}_i)$.

Theorem 2 *The above is a multiplicative perfect zero-knowledge (t,l)-threshold scheme for any family of finite Abelian groups provided that $l = O(t|n|^c)$ for a constant c. The zero-knowledge threshold scheme takes $O(tl^2)$ group operations and $O(l)$ inverses for the shareholders. The distributor performs $O(tl^2(l-t))$ group operations and $O(l(l-t))$ inverse operations.*

Proof. See final paper. □

5 A cryptographic application

We briefly discuss an application of the group based threshold scheme. In their original context, sharing schemes *reveal* the secret to at least one person when the secret has been computed. Authentication protects against impersonation of the sender and substitution of the message [19]. In [5] a combination of threshold and authentication is presented such that any set of $t - 1$ can*not* impersonate or substitute authenticated messages, but any subset of t shareholders has the authority and capability to perform the authentication. This plays an important role in data security in cases that one does not want that one person has the sole capability of authenticating messages, such as wholesale financial transactions. The scheme in [5] however only works when n is the product of two safe primes, i.e., $n = p \cdot q$, with $(p-1)/2$ and $(q-1)/2$ primes. Because it is an open problem whether infinitely many such primes exist, we could not prove that the scheme in [5] is formally secure.

If we use above homomorphic zero-knowledge threshold scheme, we can formally prove that the scheme [5] can be adapted such that the scheme satisfies the following security property. If during the (distributed) authentication process somebody eavesdrops the local area network of the t shareholders, nothing will leak that allows the eavesdropper to perform an impersonation or substitution attack, as long as no set of t shareholders is willing to collaborate. Full details of this will be given in the final paper.

376

6 Conclusion

Existing perfect sharing scheme have been studied over finite fields or over some finite geometry. We have presented a homomorphic perfect zero-knowledge threshold scheme which works for any finite Abelian group provided that l is polynomial (to be more precise $l = O(t|n|^c)$). Whether such schemes exist for any l is a natural open problem.

Although not necessarily zero-knowledge, we have proposed homomorphic perfect threshold schemes over any finite Abelian group for any l. The distributor must know the exponent of the group for these schemes. When zero-knowledge is of little importance, e.g. when the exponent is public, theses schemes have certain advantages such as requiring less memory. This introduces the following open problems. Do homomorphic *ideal* sharing schemes exist over any finite Abelian group for *any* l? Can the efficiency of the above schemes be improved? Finally, what other algebraic structures allow for homomorphic perfect (zero-knowledge) sharing schemes?

Acknowledgments

The authors thank Alfredo De Santis for discussions related to this paper, which were helpful to prepare this text.

References

[1] J. C. Benaloh. Secret sharing homomorphisms: Keeping shares of a secret secret. In A. Odlyzko, editor, *Advances in Cryptology, Proc. of Crypto '86 (Lecture Notes in Computer Science 263)*, pp. 251–260. Springer-Verlag, 1987. Santa Barbara, California, U.S.A., August 11–15.

[2] E. R. Berlekamp. Factoring polynomials over large finite fields. *Mathematics of Computation*, 24(111), pp. 713–735, 1970.

[3] G. R. Blakley. Safeguarding cryptographic keys. In *Proc. Nat. Computer Conf. AFIPS Conf. Proc.*, pp. 313–317, 1979. vol.48.

[4] D. Coppersmith, A. Odlyzko, and R. Schroeppel. Discrete logarithms in $GF(p)$. *Algorithmica*, pp. 1–15, 1986.

[5] Y. Desmedt and Y. Frankel. Shared generation of authenticators and signatures. To be presented at Crypto '91, August 12–15, 1991, Santa Barbara, California, U.S.A., to appear in: Advances in Cryptology. Proc. of Crypto '90 (Lecture Notes in Computer Science), Springer-Verlag, 1991.

[6] R. G. Gallager. *Information Theory and Reliable Communications*. John Wiley and Sons, New York, 1968.

[7] O. Goldreich, S. Micali, and A. Wigderson. How to play any mental game. In *Proceedings of the Nineteenth annual ACM Symp. Theory of Computing, STOC*, pp. 218–229, May 25–27, 1987.

[8] S. Goldwasser and S. Micali. Probabilistic encryption. *Journal of Computer and System Sciences*, 28(2), pp. 270–299, April 1984.

[9] M. Ito, A. Saito, and T. Nishizeki. Secret sharing schemes realizing general access structures. In *Proc. IEEE Global Telecommunications Conf., Globecom'87*, pp. 99–102. IEEE Communications Soc. Press, 1987.

[10] N. Jacobson. *Basic Algebra I.* W. H. Freeman and Company, New York, 1985.

[11] A. Menezes, S. Vanstone, and T. Okamoto. Reducing elliptic curve logarithms to logarithms in a finite field. In *Proceedings of the Twenty third annual ACM Symp. Theory of Computing, STOC*, 1991.

[12] M. Naor and M. Yung. Public-key cryptosytems provably secure against chosen ciphertext attack. In *Proceedings of the twenty second annual ACM Symp. Theory of Computing, STOC*, pp. 427–437, May 14–16, 1990.

[13] A. M. Odlyzko. Discrete logs in a finite field and their cryptographic significance. In N. Cot T. Beth and I. Ingemarsson, editors, *Advances in Cryptology, Proc. of Eurocrypt 84 (Lecture Notes in Computer Science 209)*, pp. 224–314. Springer-Verlag, 1984. Paris, France April 1984.

[14] M. Rabin. Digitalized signatures and public-key functions as intractable as factorization. Technical report, Massachusetts Institute of Technology Technical Report MIT/LCS/TR–212, Cambridge, Massachusetts, January 1977.

[15] M. Rabin. Probabilistic algorithms in finite fields. *SIAM Journal on Computing*, 9(2), pp. 273–280, 1980.

[16] M. Rabin. Efficient dispersal of information for security, load balancing, and fault tolerance. *Journal of the ACM*, 36(2), pp. 335–348, April 1989.

[17] R. L. Rivest, A. Shamir, and L. Adleman. On digital signatures and pulickey cryptosystems. Technical report, Massachusetts Institute of Technology Technical Report LCS/TN–82, Cambridge, Massachusetts, April 1977.

[18] A. Shamir. How to share a secret. *Commun. ACM*, 22, pp. 612–613, November 1979.

[19] G. J. Simmons. A survey of information authentication. *Proc. IEEE*, 76(5), pp. 603–620, May 1988.

[20] G. J. Simmons. Robust shared secret schemes. *Congressus Numerantium*, 68, pp. 215–248, 1989.

[21] D. R. Stinson and S. A. Vanstone. A combinatorial approach to threshold schemes. *SIAM Journal on Discrete Mathematics*, 1(2), pp. 230–236, 1988. Extended abstract is in Advances in Cryptology, Proc. of Crypto '87 (Lecture Notes in Computer Science 293).

Some Comments on the Computation of n-th Roots in Z_N *

Michele Elia

Dipartimento di Elettronica

Politecnico di Torino - Italy

Abstract

Algorithms for the computation of n-th roots in the ring Z_N of integers modulo N are discussed with particular reference to computations in prime fields.

1 - Introduction

Gauss' proof of the quadratic reciprocity law [1] enhanced the importance of the equation

$$x^2 = a \bmod N \quad , \tag{1}$$

and definitively demonstrated its intrinsic non-triviality. He also observed that if the integer N factors as

$$N = \prod_{i=1}^{s} p_i^{\alpha_i} \qquad \alpha_i \geq 1 \quad ,$$

then all the roots of an algebraic equation in the ring Z_N may be obtained (*lifted*) by means of the Chinese reminder theorem from its roots in each ring $Z_{p_i^{\alpha_i}}$, which in turn are obtained from those in the field Z_{p_i} for all $i = 1, \ldots, s$.

In addition to their purely theoretical interest, algorithms for computing square and cube roots of elements in the ring Z_N, are of use in many places. Significant examples are some primality testings, Rabin's cryptographic public key scheme based on second-degree equations, [7] and the celebrated Lagrange-Galois' plan, which yields the roots of equations over fields, [4], in particular finite fields, [5, 6] in terms of the equation's coefficients.

*This work was financially supported in part by Politecnico di Torino under the internal grant n. POLI4169-87-Cap11205.

This paper discusses the computation of n-th roots of elements in Z_N, this being equivalent to the solution of the binomial equation

$$x^n = a \bmod N \quad , \tag{2}$$

where a is prime with N. It will be recalled that, whatever N may be, a is designated the n-th residue if the roots of (2) belong to Z_N, otherwise it is called non-residue. Moreover if a is an n-th residue and N is composite, then the roots in Z_N are obtained from the roots in subrings via the Chinese reminder theorem, namely

Theorem 1 (Chinese reminder theorem) *Let $N = \prod_{i=1}^{s} m_i$ product of s relatively prime integers. Any number $Y \in Z_N$ is then solely identified by the set of reminders*

$$Y = y_i \qquad \bmod m_i \qquad i = 1, \ldots, s$$

and the interpolation formula holds

$$Y = \sum_{i=1}^{s} y_i \psi_i \qquad \bmod N \quad , \tag{3}$$

where the interpolating coefficients

$$\psi_i = \left(\prod_{j \neq i} m_j \right) a_i$$

have been defined by choosing a_i such that $\psi_i = 1 \bmod m_i$ $\forall i = 1, \ldots, s$, and by satisfying orthonormality conditions in the sense that we have

$$\begin{cases} \psi_i \, \psi_j &= \quad 0 \qquad \bmod N \qquad \forall i \neq j \\ \psi_i^2 &= \quad \psi_i \qquad \bmod N \qquad \forall i \ . \end{cases}$$

$\square$

Using this notation, given any polynomial $\xi(x)$ in Z_N, it is immediately seen that:

$$\xi(Y) = \sum_{i=1}^{s} \xi(y_i) \, \psi_i \qquad \bmod N \ .$$

Therefore each root of the equation $\xi(x) = 0$ belonging to Z_N can be expressed by means of (3) as a linear combination of the roots of the same equation considered in every ring $Z_{p_i^{\alpha_i}}$. It follows that any equation in Z_N may have a number of roots greater than its degree.

The paper is organized as follows. In section 2 some aspects of arithmetical computations in cyclic groups are developed. Sections 3, 4 and 5 then discuss the computation of square roots, cube roots and q-th roots in prime fields Z_p, p being a prime. The final section offers a few comments and remarks.

2 - Arithmetic computations in cyclic groups

Despite their apparently simple structure, cyclic groups appear somewhere as the essential instrument characterizing the matter. As a meaningful example the cyclic group structure was exploited by Gauss to settle the problem of the construction of regular polygons with a rule and compass. A second important example is provided by the binomial equations whose Galois group, [4], is a cyclic group.

Cyclic groups of order N will be denoted by C_N. We say that an element $a \in C_N$ belongs to the exponent d, if such an integer is the minimum positive exponent for which

$$a^d = 1 \quad .$$

It is easy to verify that d must be a divisor of N. Conversely, we call primitive for d any element of C_N that belongs to the exponent d.

Some arithmetical computations in cyclic groups concern the evaluation of discrete logarithms, n-th powers and n-th roots. Moreover their complexity as a function of the parameters involved, i.e. the group order and exponent magnitude, is also an interesting issue. Since definitive measurements of computational complexity are still lacking, we will assume the complexity of certain operations as fundamental measurements in terms of which to express the complexity of our computations. Let p be a prime number. Complexity $\mathcal{L}(p)$ for computing the discrete logarithm in C_p, and complexity $\mathcal{P}(q)$ for computing the q-th power in C_N can then be taken as known.

2.1 - Discrete logarithm in cyclic groups

The discrete logarithm of $a \in C_N$ to base g, a group generator, is defined as the minimum positive integer $L_g(a)$ such that

$$a = g^{L_g(a)} \quad .$$

Computation of discrete logarithms in a cyclic group C_N has been well described in [10] and its complexity strongly depends on the prime factors of the group order N. In particular, the following result, useful for our purposes, is specialized from [10].

Proposition 1 *Let $N = p^m$ be a prime power. Computation of the discrete logarithm, given a generator $g \in C_N$, has complexity upper bounded by*

$$\frac{m(m-1)}{2}\mathcal{P}(p) + m\mathcal{L}(p) \quad .$$

PROOF. To compute the discrete logarithm, it is convenient to represent the integer $L_g(a)$ base p:

$$L_g(a) = \sum_{i=0}^{m-1} \ell_i p^i \quad ,$$

where the unknowns $\ell_i \in \{0, 1 \ldots, p-1\}$ will be recursively obtained. Setting

$$a_i = \frac{a_{i-1}}{g^{\ell_{i-1}p^{i-1}}}, \quad a_0 = a, \quad \forall i \geq 1 \quad ,$$

therefore ℓ_i is the minimum integer in $\{0, 1 \ldots, p-1\}$ such that

$$\left(\frac{a_i}{g^{\ell_i p^i}}\right)^{p^{m-i-1}} = 1 \quad .$$

Let $g_1 = g^{p^{m-1}}$ be a generator of the subgroup C_p, then ℓ_i may be computed as the logarithm to base g_1:

$$\ell_i = L_{g_1}(a_i^{p^{m-i-1}}) \quad .$$

The recursive technique just described requires evaluation of m discrete logarithms in C_p and computation of m powers with exponent p^{m-i-1}, $i = 0, \ldots, m-1$. The total complexity is found to be

$$\frac{m(m-1)}{2}\mathcal{P}(p) + m\mathcal{L}(p) \quad .$$

$\square$

2.2 - n-th root in cyclic groups

The computation of n-th roots in a cyclic group C_N is an easy task if n is relatively prime with N. In this case the following proposition holds.

Proposition 2 *If $(N, n) = 1$ then the n-th root of a is equal to the power a^x, with x satisfying the condition $xn = 1 \bmod N$, because any element in C_N is a n-th power.*

PROOF. If n and N are relatively prime, it is well known that two integers x and y exist such that

$$xn + yN = 1 \quad .$$

Let a be any element of C_N, therefore we have the following chain

$$a = a^{xn+yN} = a^{xn} = (a^x)^n \quad ,$$

from which we get

$$\sqrt[n]{a} = a^x$$

showing that the n-th root of a can be computed as a raised to power x.

□

The case in which n is a divisor of N can be addressed if $a \in C_N$ is the n-th power of a generator $g \in C_N$, i.e. $a = g^{\mu n}$. Assuming that $N = \lambda n$, we have $a^\lambda = 1$, and n different roots will result. Let $\vartheta = g^\lambda$ be a generator of the subgroup C_n, therefore the set of n-th roots $\sqrt[n]{a}$ is obtained as

$$g^\mu \vartheta^i \qquad i = 0, \ldots, n-1 \quad .$$

To get g^μ, let us compute the discrete logarithm $L_g(a) = \mu n$ of a to base g, thus

$$g^\mu = g^{L_g(a)/n} \quad .$$

3 - Square roots in Z_p

The practical solution of the congruence

$$x^2 = a \bmod p \quad , \tag{4}$$

when its possibility has been established is an interesting task, given that the form $\sqrt{a}$ could be hardly viewed as a proper answer. This problem is reported in [8], where it is also shown how to compute square roots in Z_p when p is of the form $4n+3$ or $8n+5$. Two iterative algorithms, due to Adleman, Manders and Miller [2], and Lehmer and Lucas respectively, which hold for all primes are reported in [7, 3]. Here we consider only direct methods, but the results may also throw some light on those of the iterative method. Moreover thereafter and whenever necessary we will assume, differently from [2], that a primitive element in Z_p is given. The question of finding such an element and evaluating its relative complexity will be not discussed. The proposed algorithms for computing square roots can be viewed as a completion of the method outlined in [8]. Because of their inherently different characteristic, following [8], it is convenient to consider separately primes of the two forms $4n + 3$ and $4n + 1$. Moreover the subclass of Fermat primes deserves particular attention.

Primes $p = 4n + 3$. This case is addressed by proposition 2 because quadratic residues form a cyclic group of odd order. Assuming that $a \in Z_p$ is a quadratic residue, the Euler-Fermat theorem implies

$$a^{2n+1} = 1 \bmod p \quad ,$$

which multiplied by a gives

$$a^{2n+2} = a \bmod p \quad ,$$

and the roots of equation (4) turn out to be

$$x = \pm a^{n+1} \bmod p \quad .$$

Fermat primes $p = 2^{2^t} + 1$. To be pragmatic, currently Fermat primes could be exhausted by working out three cases: 17, 257 *and* 65537. However, a general algorithm that also holds for any new Fermat prime which may be discovered can be viewed as a completion of the subject, the case highlight the whole discussion. The following facts are known:

1. the multiplicative group of non-zero elements in Z_p is a cyclic group of order 2^{2^t};

2. any primitive element is a quadratic non-residue and conversely;

3. primitive elements are odd powers of one another;

4. 3 is a primitive element for every t, a fact easily checked by means of the quadratic reciprocity law;

5. the discrete logarithm to base a primitive element can be computed by using proposition 1.

The square root of a quadratic residue a is computed as

$$\sqrt{a} = \pm 3^{L_3(a)/2} = \pm 3^{2^{u-1}(2\ell+1)} \quad ,$$

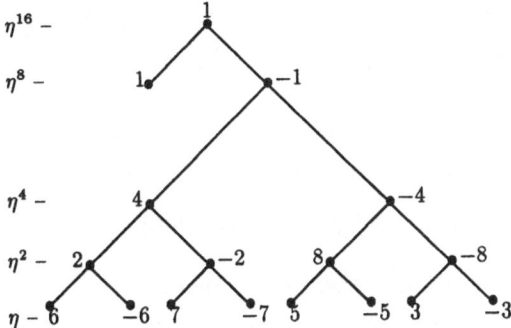

Figure 1: Tree of the even powers in Z_{17}

where the discrete logarithm, $L_3(a) = 2^u(2\ell + 1)$, *with* $u \geq 1$, is obtained by applying proposition 1.

A pictorial representation of the structure of $\mathcal{M}(p)$, the multiplicative group of Z_p, is obtained by associating each group element a at a node of a binary tree, with the square roots $\pm\sqrt{a}$ associated to the ending nodes of the outgoing branches from the node with label a. The identity 1 is associated to the root. Square roots of 1, i.e. 1 *and* -1, are associated at the next pair of nodes connected to the root. The remaining part of the tree extends under the node with label -1. Non-quadratic residues are associated to terminal nodes of this tree.

Every quadratic residue a can be considered the root of a tree with height $s < 2^t$.

The shape of the tree for $p = 17$ is reported as an example in figure 1.

Primes $p = 4n + 1$. Let $a \in Z_p$ be a quadratic residue. Computation of the square root of a now requires the knowledge of certain roots of unity in Z_p depending on a special representation of p in terms of powers of 2. A prime number p, not a Fermat prime, can always be written in the form

$$p = 2^q m + 2^{q-1} + 1 \quad with \quad q \geq 3 , \quad m > 0 \quad ,$$

This follows by considering that n will be either of the form $2^{q-3}(2m+1)$, with $q \geq 3$ and $m \geq 1$, or of the form 2^{2^t-2}. In the first instance, we have

$$p = 4 \ 2^{q-3}(2m+1) + 1 = 2^q m + 2^{q-1} + 1 \quad m \neq 0,$$

while in the second we have Fermat primes.

Let $b \in Z_p$ denote a primitive element, we have

$$b^{2^q m + 2^{q-1}} = 1 = b^{2^{q-1}(2m+1)} \quad ,$$

and on setting $\tilde{b} = b^{2m+1}$, it follows that $\tilde{b}$ belongs to the exponent 2^{q-1}. Since a is a quadratic residue, we find that

$$a^{2m+1} = \bar{a} \bmod p \tag{5}$$

is an element of order at most 2^{q-2}.

If we know a square root γ of $\bar{a}$, then we obtain the roots of equation (4) in a straightforward way. Multiplying equation (5) by a we get

$$a^{2m+2} = \bar{a} a \bmod p \quad ,$$

giving

$$\sqrt{\bar{a} a} = a^{m+1} \bmod p \quad ,$$

so that the roots of equation (4) are

$$x = \pm \frac{a^{m+1}}{\gamma} \quad .$$

To compute γ, we take the logarithm of $\bar{a}$ to base b, which can be evaluated according to proposition 1

$$L_b(\bar{a}) = 2^s(2\ell + 1) \qquad 1 \leq s \leq q - 2 \quad .$$

The square root of $\bar{a}$ is thus

$$\sqrt{\bar{a}} = b^{2^{s-1}(2\ell+1)} = \gamma \bmod p \quad ,$$

386

4 - Cube roots in Z_p

Solution of the congruence

$$x^3 = a \bmod p \quad , \tag{6}$$

when its possibility has been established, is similar to the square root evaluation. It is convenient to distinguish primes of two forms, $6n - 1$ and $6n + 1$.

Primes $p = 6n - 1$. This case is addressed by proposition 2, because the order $6n - 2$ of $\mathcal{M}(p)$ is prime with 3. Therefore every element is a cube. By multiplying term by term the two equalities

$$a^{6n-1} = a \quad ,$$

$$a^{6n-2} = 1 \quad ,$$

in fact, we obtain

$$a^{12n-3} = a^{3(4n-1)} = a \quad ,$$

which shows that the cube root of a is obtained as

$$\sqrt[3]{a} = a^{4n-1} \quad .$$

Note that the identity has only itself as a cube root in the field.

Primes $p = 6n + 1$. Let $a \in Z_p$ be a cubic residue, i.e. $a^{2n} = 1$, (cubic non-residues have no cube root in the field). Computation of the cube root of a now requires the knowledge of cube roots of unity in Z_p and depends on a special representation of p in terms of powers of 3.

Let b denote a primitive element in Z_p. If we set $\omega = b^{2n}$, then the 3 cube roots of the unity are 1, ω $\;and\;$ ω^2. It follows that any cubic residue a has three cube roots in the field

$$\sqrt[3]{a}, \quad \sqrt[3]{a}\,\omega \quad and \quad \sqrt[3]{a}\,\omega^2 \quad ,$$

which reduces the computational problem to the explicit evaluation of only one cube root determination $\sqrt[3]{a}$.

Assume first that $n = 3m + 1$, therefore

$$a^{2(3m+1)} = a^{6m+2} = 1 \quad .$$

Multiplying by a we get

$$a^{6m+3} = a \quad ,$$

from which it is evident that one determination of the cube root is

$$\sqrt[3]{a} = a^{2m+1} \quad .$$

Assume now that $n = 3m - 1$, therefore

$$a^{2(3m-1)} = a^{6m-2} = 1 \quad .$$

Squaring and multiplying by a we get

$$a^{12m-3} = a \quad ,$$

from which it is evident that one determination of the cube root is

$$a^{4m-1} = \sqrt[3]{a} \quad .$$

Let us now assume more in general that $n = 3^s(3m \pm 1)$, with $s \geq 1$, therefore

$$a^{2(3^s(3m\pm1))} = 1 \quad .$$

Setting $\tilde{b} = b^{2(3m\pm1)}$, it follows that $\tilde{b}$ belongs to the exponent 3^s. Since a is a cubic residue, we find that

$$a^{2(3m\pm1)} = \bar{a} \bmod p \tag{7}$$

is an element of order at most 3^{s-1}.

The next step is to find the discrete logarithm of $\bar{a}$ to base $\tilde{b}$, that is $L_{\tilde{b}}(\bar{a}) = 3^u(3\ell \pm 1)$ with $1 \leq u \leq s - 1$, so that

$$\sqrt[3]{\bar{a}} = \tilde{b}^{3^{u-1}(3\ell\pm1)} = \gamma \bmod p \quad .$$

Now from equation (7) we get

$$a^{2(3m\pm1)} a^{\pm1} = \bar{a} a^{\pm1} \bmod p \quad ,$$

so that by taking the cube root

$$\sqrt[3]{\bar{a} a^{\pm1}} = \gamma \sqrt[3]{a^{\pm1}} = a^{2m\pm1} \bmod p \quad ,$$

we finally obtain one determination of the cube root of a

$$\sqrt[3]{a} = \begin{cases} \dfrac{a^{2m+1}}{\gamma} & if \ \ n = 3^s(3m+1) \\[3mm] \dfrac{\gamma}{a^{2m-1}} & if \ \ n = 3^s(3m-1) \end{cases}$$

388

5 - q-th roots in Z_p

Let q be a prime number, therefore the congruence

$$x^q = a \bmod p \quad, \tag{8}$$

when its possibility has been established, presents a situation not very different from that analyzed in the previous sections. It is convenient to distinguish primes of two forms, $2qn + r$, $2 \le r \le (q-1)$ and $2qn + 1$ respectively.

Primes $p = 2qn + r$. This case is addressed by proposition 2 because the order $2qn + r - 1$ of $\mathcal{M}(p)$ is prime with q and every element is a q-th power. In particular, since q is prime with $r - 1$, there exists an integer k such that $k\,q = 1 \bmod (r-1)$ or equivalently $k\,q = 1 + (\lambda+1)(r-1) = r + \lambda(r-1)$ for some λ. By multiplying the equality

$$a^{2qn+r} = a \quad \bmod p \quad,$$

term by term with the equality

$$a^{2qn+r-1} = 1 \quad \bmod p \quad,$$

raised to power λ we obtain

$$a^{2qn(\lambda+1)+r+\lambda(r-1)} = a^{qB} = a \quad,$$

with $B = 2n(\lambda+1) + k$. It follows that the q-th root of a is obtained as

$$\sqrt[q]{a} = a^B \quad.$$

Primes $p = 2qn + 1$. Let $a \in Z_p$ be a q-residue. Now we have q roots of a. If ω denotes a q-th root of unity in Z_p, then all solutions of equation (8) are obtained as

$$x = \sqrt[q]{a}\,\omega^i \qquad i = 0,\ 1,\ \ldots,\ q-1 \quad,$$

therefore the computational problem is reduced to explicit evaluation of the only determination $\sqrt[q]{a}$.

Let $b \in Z_p$ denote a primitive element, therefore ω is obtained as the power

$$\omega = b^{2n} \quad.$$

To obtain a determination $\sqrt[q]{a}$, let us consider that in general $n = q^s(qm + r)$, with $s \geq 1$ and $1 \leq r \leq q - 1$, therefore

$$a^{2q^s(qm+r)} = 1 \quad .$$

Setting $\tilde{b} = b^{2(qm+r)}$, it follows that $\tilde{b}$ belongs to the exponent q^s. Since a is a q-th residue, we find that

$$a^{2(qm+r)} = \bar{a} \bmod p \tag{9}$$

is an element of order at most q^{s-1}.

The next step is to find the discrete logarithm of $\bar{a}$ to base $\tilde{b}$, that is $L_{\tilde{b}}(\bar{a}) = q^u(q\ell + r')$ with $1 \leq u \leq s - 1$ and $1 \leq r' \leq (q - 1)$, so that

$$\sqrt[q]{\bar{a}} = \tilde{b}^{q^{u-1}(q\ell + r')} = \gamma \bmod p \quad .$$

Now let us observe that a pair of integers s and t can be always found such that $2rs + qt = 1$, because $2r$ and q are relatively prime numbers. Therefore from equation (9) raised to the power s, we get

$$a^{2qms+2rs} = a^{2qms+1-qt} = \bar{a}^s \bmod p \quad .$$

In conclusion, by taking the q-th root

$$\sqrt[q]{\bar{a}^s} = \gamma^s = \sqrt[q]{a} \, a^{2sm-t} \bmod p \quad ,$$

we obtain a determination of the q-th root of a:

$$\sqrt[q]{a} = \frac{\gamma^s}{a^{2sm-t}} \quad .$$

6 - Final Observations

Let us conclude with Gauss's words showing how an old subject is a source of never-ending interest:

> The problem of distinguishing prime numbers from composite numbers and of resolving the latter into their prime factors is known to be one of the most important and useful in arithmetic. .... Further, the dignity of the science itself seems to require that every possible means be explored for the solution of a problem so elegant and celebrated.

Rabin, [7], in fact, has recently shown that computation of square roots in Z_N and factorization of N are equivalent problems. Therefore methods for the straightforward computation of square roots and, more in general, the n-th roots in finite rings Z_N, with N composite, deserve further investigations.

References

[1] **K.F. Gauss**, *Disquisitiones Arithmeticae*, Springer-Verlag, New York, 1986.

[2] **L. Adleman, K. Manders and G. Miller**, On taking roots in Finite Fields, *Proc. of the 20th Annual Symp. on the Foundation of Computer Science*, 1979, pp. 175-178.

[3] **D.M. Bressoud**, *Factorization and Primality Testing*, Springer-Verlag, New York, 1989.

[4] **E. Dehn**, *Algebraic Equations*, Dover, New York, 1960.

[5] **M. Elia**, Formulas for the Roots of Algebraic Equations over Finite Fields and the Lagrange-Galois' Plan, *Internal Report*, Politecnico di Torino, 1990.

[6] **J.W.P. Hirschfeld**, *Projective Geometries over Finite Fields*, Clarendon Press, Oxford, 1979.

[7] **E. Kranakis**, *Primality and Cryptography*, Wiley & Sons, New York, 1986.

[8] **G.B. Mathews**, *Theory of Numbers*, Chelsea, New York, 1990.

[9] **R.C. Peralta**, A Simple and Fast Probabilistic Algorithm for Computing Square Roots Modulo a Prime Number, *IEEE Transactions on Information Theory*, vol. IT-32, N.6, November 1986, pp. 846-847.

[10] **S.C. Pohling and M.E. Hellman**, An Improved Algorithm for Computing Logarithms over $GF(p)$ and its Cryptographic Significance, *IEEE Transactions on Information Theory*, vol. IT-24, N.1, January 1978, pp. 106-110.

The Varieties of Secure Distributed Computation

Matthew Franklin
Columbia University
Computer Science Department,
New York, NY 10027

Moti Yung
IBM Research Division,
T.J. Watson Center,
Yorktown Heights, NY 10598

Abstract

Secure distributed computing protocols allow a collection of processors, within some specific computational environment, to evaluate jointly the output of a function while maintaining the secrecy of privately held inputs. A specific example is secret ballot election, in which a final tally must be computed without revealing anything else about the individual votes. The secrecy constraints of more general protocol problems can be satisfied by incorporating secure distributed computing (either cryptographic or unconditional) as a basic building block. This paper surveys the extensive progress that has been made in the area of secure distributed computation over the past few years.

1 Introduction

Secure distributed computing (or "fault-tolerant distributed computing," or "oblivious circuit evaluation") is the problem of evaluating a function (or, equivalently, a circuit) to which each player has one secret input, such that the output becomes commonly known while the inputs remain secret. The problem would be trivial in the presence of a trusted agent. All players give their private inputs to the agent, who computes the function and distributes the output. The task of protocols for secure distributed computation can be considered to be the *simulation* of the presence of a trusted agent by a group of players who are mutually untrustworthy. The study of this problem was initiated by Yao [72] [73] and by Goldreich, Micali, and Wigderson [44].

Of course, untrustworthiness, for cryptographers, is a many-flavored thing. It might mean that all parties behave perfectly during the protocol, but later some group of gossippers compare notes to try to learn everyone else's secrets. A stronger meaning might be that some fixed fraction of the parties cheat during the protocol, but in such a way that the outcome (i.e., the value of the function) remains the same. Even stronger would be if some fixed fraction of the parties cheated arbitrarily during the protocol; yet stronger adversarial behavior can be imagined.

Three areas of cryptography are closely related to the problem of secure distributed computation. Secret-ballot election schemes [35] [17] are essentially a special-case of secure computation in which the function is a simple sum of ones and zeros. Instance-hiding schemes [1] [11] involve a weak computational agent exploiting one or more strong but untrusted computational agents to compute a function for secret inputs. When one player has a secret circuit, and the other player has non-secret data, then the problem of secure distributed computing reduces to a minimum-knowledge circuit simulation, solved in the general case by [43], and later improved by [55].

The next section of this paper gives short descriptions of cryptographic and other primitives that are used in the upcoming protocols, as well as the basic elements of one model for secure computation. Two-party cryptographic protocols are presented in Section 3, multi-party cryptographic protocols in Section 4, and multi-party non-cryptographic (sometimes called "unconditional") protocols in Section 5. Section 6 presents some of the complexity results and lower bounds that are known for these protocols.

2 Preliminaries

In this section, we give the basic models and definitions for secure distributed computation. The section begins with definitions of cryptographic and distributed computing primitives that will be used in upcoming protocols.

2.1 Cryptographic and Distributed Computing Primitives

2.1.1 Number Theory Basics

A quantity is "negligible" if it is smaller than the reciprocal of any polynomial of relevant parameters, and otherwise "non-negligible." When this term is used, the relevant parameters will usually not be stated explicitly. Negligibility can be used to formalize the "hardness" of a problem, a typical assumption made in cryptographic protocols. Factoring is often assumed to be hard, as is finding discrete logs; the latter assumption is formalized below.

Let p be a prime, let g be a generator of Z_p^* (i.e., $Z_p^* = \{g^i | i \in Z_p^*\}$), and let a be an integer between 0 and $p-1$. Define $DLP_{p,g}(a)$ to be i such that $g^i = x \bmod p$, $0 \leq i < p$. The **Discrete Log Assumption** (DLA) states that no probabilistic polynomial-time Turing Machine, can, on input p, g, and a, output $DLP_{p,g}(a)$ non-negligibly better than random guessing.

2.1.2 Indistinguishable Probability Distributions

The indistinguishability of probability distributions, due to Yao [72] and Goldwasser and Micali [47], combines computational assumptions with randomness in a condition that is useful for defining the security of cryptographic primitives. Informally, two distributions are indistinguishable if a guesser cannot tell them apart.

Let P and Q be probability distributions on k-bit strings. They are *distinguishable* by D if there is a constant $c > 0$ such that

$$\text{prob}(D(x) \neq D(y) | x \leftarrow P, y \leftarrow Q) \geq \frac{1}{2} + \frac{1}{k^c}.$$

[The notation $x \leftarrow P$ indicates that x is drawn from the distribution P.]

A "family" of distributions is an indexed collection $\{P_k\}$ of probability distributions, where P_k is a distribution on k-bit strings. Two families of distributions $\{P_k\}$ and $\{Q_k\}$ are "computationally indistinguishable" if, for every D which runs in time polynomial in k, there is a constant $K > 0$ such that P_k and Q_k are not distinguishable by D for all $k > K$.

Indistinguishability can be generalized to apply to "ensembles" of probability distributions. An ensemble is a family of *parametrized collections* of probability distributions, i.e., $\{P_k(z)\}$, where

$P_k(z)$ is a distribution on k-bit strings for each choice of $z \in L_k$. Fixing one z from each parameter set L_k "collapses" an ensemble into a family of distributions. Two ensembles are said to be computationally indistinguishable if the corresponding families of distributions are computationally indistinguishable no matter how they are collapsed.

Notice that there is a natural ensemble $\{M_k(z)\}$ associated with each probabilistic Turing Machine M and language L: $M_k(z)$ is the distribution of outputs of M on input $z \in L$ of length k. This will be useful in the upcoming definition of zero-knowledge proof systems.

2.1.3 Trapdoor Functions and Encryption Schemes

A function is "one-way" if it is easy to compute and hard to invert. More formally, suppose that $\{f_k\}$ is a family of functions defined on bit strings, where $f_k : \{0,1\}^k \rightarrow \{0,1\}^k$. The family $\{f_k\}$ is one-way if two requirements are satisfied: (a) there is a deterministic polynomial-time (in k) algorithm to compute each f_k; and (b) for every probabilistic polynomial-time algorithm A and for every c there exists a k_c such that

$$\text{prob}(f_k(A(f_k(x))) = f_k(x)|x \leftarrow \{0,1\}^k) < k^{-c}$$

for all $k > k_c$. The existence of one-way functions implies $P \neq NP$, so no definitive examples have been found; the discrete log functions from Section 2.1 is an example of a candidate that is believed by some researchers to be one-way.

A family of one-way functions is "trapdoor" if there exists some secret information (called the "trapdoor information" or "trapdoor key") for each function with which the inverse can be computed in probabilistic polynomial time with non-negligible chance of success. More formally, a randomized key-generating algorithm K takes as input a security parameter k and outputs a pair of functions E and D such that (1) E is one-way on $\{0,1\}$, and (2) the probability that $E(D(E(x))) \neq E(x)$ is non-negligible. If E is a permutation, then it can be the basis for a public-key encryption scheme: E is made public for encryption by anyone, while D is kept private so that no one else can decrypt.

A candidate for a trapdoor permutation is the RSA public-key encryption function [69]: $E(m) = m^e \bmod n$, where $n = pq$, and the trapdoor key is (p, q) (easily samplable given the family parameter $k = |p| = |q|$). The inverse permutation is $D(m) = m^d \bmod n$, where $ed = 1 \bmod (p-1)(q-1)$, which can be efficiently found from the trapdoor key.

2.1.4 Bit Commitment

Bit commitment is a protocol that simulates sealing a bit in an opaque envelope, so that the bit is unalterable by the hider (after sealing), and unreadable by the receiver (before opening). A bit commitment scheme can be considered to be a mapping from some large domain to $\{0,1\}$; a bit is committed by giving a random element in the pre-image of the mapping at that output value. The scheme is unalterable if the mapping is in fact a function. The scheme is unreadable if the distributions of elements in the pre-image of zero and elements in the pre-image of one are indistinguishable to the receiver.

The first bit commitment protocol was part of a protocol by Blum [22] for two-party coin flipping. A commitment by one party, followed by a guess by the second party, followed by a revelation by the first party, is equivalent to the flip of a coin. Naor [61] shows a general construction for basing bit commitment on any one-way function, based on earlier reductions [52] [50].

Basic bit commitment requires that the receiver be polynomially-bounded. Another flavor of bit commitment, called "strong bit commitment" [24] [26] [62] allows the receiver to be unbounded. In this case, unreadability requires that the two probability distributions be almost identical.

2.1.5 Interactive Proof Systems and Zero-Knowledge Proof

An interactive proof system is a two-party protocol in which a "prover" conveys a convincing argument to a polynomially-bounded probabilistic "verifier;" this idea was introduced independently by Goldwasser, Micali, and Rackoff [48] and Babai [3]. More precisely, given a language L, an interactive proof system for L consists of two probabilistic interactive machines called the prover and the verifier that have access to a common input tape containing a possible element x of L, and that send messages through a pair of communication tapes, but otherwise are inaccessible to one another. The verifier performs a number of steps polynomial in the length of the common input, ending in either an "accept" or a "reject" state. When the verifier behaves correctly, then (1) when $x \in L$, the probability that the verifier rejects is less than any inverse polynomial in the length of x; and (2) when $x \notin L$, the probability that the verifier accepts is also less than any inverse polynomial in the length of x, even when the prover behaves in an arbitrarily adversarial manner.

An interactive proof system for a language L is "computational zero-knowledge" [48] if the verifier learns nothing from the interaction except the validity of the proof, even if the verifier behaves in an arbitrarily adversarial (probabilistic polynomial-time) manner. This vague condition can be formalized in terms of the computational indistinguishability of ensembles of probability distributions. Consider the probability distribution generated by an adversarial verifier, called its "view," which covers everything seen by the verifier during an interactive proof: messages from and to the prover, plus verifier's coin tosses. For each adversarial verifier, there is an ensemble $View_k(z)$ of such distributions, parameterized over all $z \in L$ of length k. A proof system for membership in L is zero-knowledge if the view ensemble of any adversary V^* is computationally indistinguishable from the natural ensemble associated with L and some probabilistic polynomial time Turing Machine M_{V^*} (called the "simulator" for V^*).

2.1.6 Secret Sharing and Verifiable Secret Sharing

Underlying many recent results in secure distributed computing is a secret sharing scheme due to Shamir [70]. In this scheme, a secret s is shared by a "dealer" among n players by giving each player a point on an otherwise random degree t polynomial $p(x)$ such that $p(0) = s$ (e.g., give $p(i)$ to the ith player). Any $t+1$ players can interpolate to recover $p(x)$ and hence s. If all computation is performed over a finite field, then the secret is secure from any group of up to t players in the strongest possible (information-theoretic) sense. The parameter t is called the "threshold" of the secret sharing scheme.

When a secret s is shared among n players using a scheme with threshold t, then the piece given by the dealer to a single player may be called a "t-share" of s. Notice that this protocol has two phases. The dealer distributes t-shares during a "sharing phase," and the players interpolate to recover the secret during a "recovery phase."

Verifiable secret sharing (VSS), first introduced by Chor, Goldwasser, Micali, and Awerbuch [31], is a form of secret sharing in which cheating by the dealer and some of the players cannot prevent the honest players from receiving valid shares of some unique recoverable secret. More formally, a t-VSS scheme has three properties: (1) If the dealer is honest, then the secret is information-theoretically secure during the sharing phase from any group of up to t players; (2) The value that will be recovered by the honest players during the recovery phase is uniquely determined at the end of the sharing phase; and (3) If the dealer is honest during the sharing phase, then the recovered value will be the dealer's actual secret.

2.1.7 Instance Hiding

Instance hiding schemes, initiated by Abadi, Feigenbaum, and Kilian [1], are a means for computing with encrypted data. Assume that a computationally weak player wishes to compute a complicated function with the help of powerful but untrusted players (called "oracles"). The idea is for the weak player to ask questions of the powerful players from which the desired answer can be easily computed, but in such a way that sufficiently small subsets of the powerful players cannot determine anything about your intended computation from those questions. Every function has a multiple oracle instance hiding scheme that leaks no information to any single oracle [11] (but which allows any two oracles to determine the intended computation).

Instance hiding may be thought of as a close cousin to secure distributed computation. It is listed in this section on preliminaries because two interesting protocols for secure distributed computation are in fact based directly on instance hiding as a primitive [2] [12].

2.2 Models and Basic Protocols

In this subsection, the background definitions and concepts are presented for secure distributed computing.

2.2.1 Protocol Definitions

There are three fundamental components of a computation protocol: the parties that participate in the computation, the communication media over which messages are sent, and the adversaries who attempt to defeat the security of the protocol. Unless otherwise stated, all computation is assumed to be over some large finite field.

Each party (or "player" or "processor") in a protocol is modeled as an "interacting Turing machine." Each such machine has a private input tape (read-only), output tape (write-only), random tape (read-only), and work tape (read-write). All of the machines share a common input tape (read-only) and output tape (write-only). In addition, each machine may have a history tape, on which it may record any computation or commmunication in which it is involved. Other tapes may be associated with machines or with pairs of machines to model the communication media available for the protocol; this will be covered shortly.

For most of the protocols discussed in this paper, either $n = 2$ ("two-party") or $n \geq 3$ ("multi-party"). The ith processor, $1 \leq i \leq n$, has the secret value s_i (usually an element of a finite field) on its private input tape. A description of the function $f(x_1, \cdots, x_n)$ (e.g., the encoding of a circuit that evaluates the function) is on the common input tape. A protocol specifies a program for each processor, at the end of which each processor has the value $f(s_1, \cdots, s_n)$ on its output tape. Assume that the programs of the processors coordinate all messages to be sent and received

396

at the ends of "rounds" (i.e., at the ticks of a global clock, between which all local computation of all processors is done asynchronously).

If the protocol is "cryptographic," i.e., if it relies on some unproven intractability assumption, then the program for each interactive Turing Machine is assumed to run in probabilistic polynomial-time. Otherwise, the protocol is "non-cryptographic" (or "unconditional"), and there are no bounds on the running time of the programs. However, for most non-cryptographic protocols described in this paper (but not all [5] [12]), the programs for all players are polynomial-time.

In general, players communicate with one another through private channels or public channels (or both). A message sent through a private channel is inaccessible to anyone except the designated sender and receiver. A public channel, called a "broadcast" channel, sends a message to everyone. By labeling a message on a broadcast channel, it may be addressed to a single receiver, but its contents are accessible to all players.

If a pair of players are connected by a private channel, then the corresponding pair of interactive Turing Machines share two additional "communication" tapes. Each tape is exclusive-write for one of the two players (unerasable, with a unidirectional writing head), and is readable by both players. If there is a broadcast channel, then each interactive Turing Machine has an exclusive-write tape that is readable by all other players. The cryptographic protocols typically assume only a broadcast channel (in addition to whatever cryptographic assumptions are needed). The non-cryptographic protocols typically assume a complete (fully connected) network of private channels among the players, and possibly a broadcast channel as well.

A player is considered "correct" if it follows its program exactly, engaging in no additional communication or computation beyond what is specified by the protocol, and keeping all of its private tapes private. A player that is not correct is considered "faulty." To model worst-case behavior (maximally malicious coordination), the faulty players in a protocol are modeled as being under the control of a single adversary. In the cryptographic setting, the adversary is a probabilistic polynomial-time algorithm. In the unconditional setting, the adversary is an algorithm with no time or space bounds.

A "passive" adversary can read the private internal tapes and received messages of all faulty processors, but does not interfere with the programs of the faulty processors. This models the situation in which a group of dishonest players participate in the protocol properly, and afterwards pool their information in an attempt to learn more about the inputs of the honest players.

An "active" adversary can read private tapes and messages of faulty players, and can also specify the messages that are sent by faulty players. In other words, an active adversary can cause the faulty processors to violate the protocol in an arbitrary coordinated attack. An active adversary is "rushing" if it can read all messages sent to the faulty players in a round before deciding on the messages sent by those faulty players in that same round.

An adversary can also be either "static" or "dynamic." A static adversary controls the same set of faulty players throughout the protocol. A dynamic adversary can increase the set of faulty players under its control each round. Even within a single round, faulty players may be added dynamically, on the basis of messages and internal tapes of those players that have already been corrupted. For most of the protocols discussed in this paper (but not all [65] [38]), there is either a single static passive adversary or a single static active adversary.

2.2.2 Oblivious Transfer

Suppose that player A has a secret bit b that is unknown to player B. The two players engage in a protocol, at the end of which player B successfully receives bit b with probability $\frac{1}{2}$. B always knows with certainty whether or not the "transfer" of bit b was successful. By contrast, A cannot determine that the transfer was successful any better than random guessing (or non-negligibly better than random guessing, with respect to some security parameter). This elementary protocol describes the simplest version of Oblivious Transfer, due to Rabin [66] (abbreviated "OT"). It is analogous to mail service by an incompetent Postal Service; any sent letter arrives at the proper destination half of the time, and disappears the other half of the time.

There are many other types of Oblivious Transfer that have all been shown to be equivalent to the simple one [25] [36] [37]. One important alternative is called "1-2 Oblivious Transfer" [39] (abbreviated "1-2-OT"). In this version, player A begins with two secret bits b_0 and b_1. Player B can choose to receive exactly one of these bits, without letting A know which bit was chosen.

This Oblivious Transfer protocol is one of the most basic possible primitives that can break the "knowledge symmetry" between two players. In addition to other versions of Oblivious Transfer, more sophisticated protocols can be built from this primitive. In fact, secure distributed computation can be based directly on Oblivious Transfer [56] [13] [46], as will be discussed in later sections.

2.2.3 Byzantine Agreement

Suppose that there are n players, each player i has a single bit b_i, and some of the players are unreliable. The Byzantine Agreement problem is to devise a protocol at the end of which all of the non-faulty players agree on a bit b (called the "consensus value"). Moreover, we must have that $b = b_i$ if i is honest and all of the honest players had the same initial bit. This problem was initially considered by Lamport, Shostak, and Pease [58], who showed that it can be solved if no more than one-third of the players are faulty. As will be shown in the next subsection, Byzantine Agreement can be used to eliminate the need for a broadcast channel in some secure distributed computation protocols.

2.2.4 Security Definitions

There are two important security properties of computation protocols that are considered in this paper: "privacy" and "resilience." These terms capture the ability of a protocol to withstand some fraction of faulty players that are under the control of an adversary.

Privacy assumes a passive adversary. A protocol is said to be "t-private" if, given any set of at most t faulty processors, no passive adversary can learn anything more about the private inputs of the honest players than is otherwise revealed by knowing the output of the computation and knowing the private inputs of the faulty players. This can be formalized by considering whether there is an algorithm, of the same computational power as the adversary, that, given only the output and faulty players' inputs, can in some sense simulate anything achievable by the adversary throughout and at the end of the protocol.

Resilience assumes an active adversary. A protocol is said to be "t-resilient" if no active adversary, with at most t faulty processors under its control, can (1) learn anything more about the private inputs of the honest players than is otherwise revealed by knowing the output of the

398

computation and knowing the private inputs of the faulty players; or (2) prevent any honest player from having the correct result on its output tape at the end of the protocol.

Note that the notion of "correct result" in the definition of t-resiliency is not straightforward. Certainly the adversary could have one or more faulty players behave as though they held a different input than they actually did, while otherwise being in perfect accordance with the protocol. This behavior cannot be protected against by the honest players. What can be said is that if each player initially commits to an input (e.g., using a verifiable secret sharing scheme) then the honest players must learn the value of the function at those committed inputs; cheating players who fail to commit are eliminated from the computation. In addition to maintaining privacy, initial commitment also guarantees "independence of inputs," i.e., the faulty players' inputs cannot be related in any way to the honest players' inputs.

Notice also that under certain conditions a protocol that requires a broadcast channel is equivalent to a protocol that requires only a network of private channels. Obviously, if the adversary is passive, then each broadcast can be simulated by sending the same message privately to each player. More interestingly, if the adversary is active, then each broadcast can also be simulated on a private network if there are sufficiently few faulty players. Using Byzantine Agreement [58] [40], a t-resilient protocol that requires a broadcast channel can be converted into a t-resilient protocol that needs only a network of private channels, whenever $n > 3t$.

Rather than proving the above properties directly and individually, one may want to capture all of the "good" properties of a secure protocol in one crisp definition. Such definitions have been proposed very recently by Beaver [9] and by Micali and Rogoway [60].

3 Two-Party Cryptographic Secure Distributed Computation

3.1 Main Ideas

In this section, we present two-party computation protocols whose security rely on cryptographic assumptions. There are two types of protocols, both of which work from the circuit representation of the computed function. The first type has two largely non-interactive subprotocols; one player "scrambles" the circuit in some manner, then they interact, and then the second player "evaluates" the scrambled circuit. The second type is an interactive gate-by-gate evaluation of the circuit for encrypted inputs.

Note that secure two-party computation is not possible in general without some complexity assumption. As a simple example, Ben-Or, Goldwasser, and Wigderson [19] show that there cannot be an information-theoretically secure two-party protocol for computing the OR of two input bits. There are other impossibility results for unconditional two-party computation that are covered in a later section.

The first results in two-party secure computation are due to Yao [72]. He was interested in a setting in which both parties behaved honestly, but wished to maintain the secrecy of their inputs ("1-privacy" as defined previously). One of the problems he considered was called the "Millionaires' Problem." Two millionaires wish to know who is richer, without revealing any other information about the size of their respective fortunes. In this same paper, Yao indicated that any 2-ary function $f(x, y)$ could be computed privately by two players assuming the hardness of factoring.

399

3.2 Protocols

The first general two-party secure computation protocol result is due to Yao [73], who considers a task that is somewhat more general than the computation problem described earlier. Instead of computing a deterministic function, an "interactive computational problem" has inputs and outputs that are distributed according to given probability distributions. This reduces to function evaluation when, for each pair of inputs, there is only one element of the output distribution with non-zero probability. Assuming the intractability of factoring, Yao shows that every two-party interactive computational problem has a private protocol.

Goldreich, Micali, and Wigderson [44] show how to weaken the intractability assumption from factoring to the existence of any trapdoor permutation. In fact, their protocol solves the following slightly different, but equivalent, problem of "combined oblivious transfer." There are two parties A and B, with private inputs a and b respectively. At the end of the protocol, A is to receive the value $f(a, b)$ without learning anything further about b. Meanwhile, B is to learn nothing at all about A's input a. Trapdoor permutations are needed in this protocol to implement 1-2-OT (via a construction due to Even, Goldreich, and Lempel [39]).

Galil, Haber, and Yung [42] show how to reduce further the complexity assumption from trapdoor permutation to one-way function plus oblivious transfer. As before, one player is the "circuit" constructor, and the other is the "circuit" evaluator. The constructed circuit consists of a number of "gates," each of which enables a single decryption key (output) to be recovered from the knowledge of two decryption keys (inputs). The input decryption keys serve as seeds for a pseudorandom number generator (which can be based on any one-way function [52] [50]) that returns the output decryption key (yet another seed for the generator).

An oblivious circuit evaluation protocol due to Kilian [56] relies only on the existence of a protocol for Oblivious Transfer.

Unlike the previous protocols, a protocol due to Chaum, Damgård, and Van de Graaf [29] for two-party secure computation has both parties contribute to the scrambling of the circuit. This protocol requires a bit commitment scheme with additional nice properties: "blinding" (i.e., the recipient of a committed bit can compute a commitment of any bit xored with the committed bit), and "comparability" (i.e., the recipient of two committed bits can be convinced by the committer that they are equal). Chaum, Damgård, and Van de Graaf show how to achieve such an enhanced bit commitment scheme based on any one of several number-theoretic problems: Jacobi symbol, quadratic residue, or discrete log. For example, a strong bit commitment scheme with blinding and comparability can be based on the discrete log under the DLA. A bit b is unconditionally hidden as $a^b g^r \bmod p$ for prime p, generator g of Z_p^*, $a \in Z_p^*$, random $r \in Z_p^*$. It is then conditionally unforgeable: finding a random r' such that $a^{1-b} g^{r'} = a^b g^r \bmod p$ is equivalent to finding the discrete log of a. To blind a commitment x with a known bit b', the recipient computes $a^{b'} g^{r'} x \bmod p$ for random r'. To compare commitments x and y, the committer reveals the discrete log of $a^{-1} xy$.

Abadi and Feigenbaum [2] present a two-party computation protocol that is similar to the protocol of Chaum, Damgård, and Van de Graaf. The protocol allows one player to keep a circuit hidden unconditionally (except for the number of AND gates), while the other player keeps the input to the circuit hidden conditionally.

4 Multi-Party Cryptographic Secure Distributed Computation

4.1 Main Ideas

In the preceding section, several protocols were described for securely computing any two-input function to which each of two players held one of the inputs, given some "reasonable" intractability assumption. Does this generalize to more than two players?

One tempting approach would be to reduce an n-input function to a series of 2-input functions, e.g., reduce n-ary addition into the pairwise computation of $n - 1$ partial sums. This computes the correct answer, but it risks leaking information to players involved in computing the intermediate results.

In fact, general protocols for conditional (i.e., cryptographic) multi-party computation have been found. Moreover, some of the solutions do rely on reducing the computation to a series of two-party computations, but in a more sophisticated manner.

All of the protocols described in this section follow the three-stage paradigm introduced by Goldreich, Micali, and Wigderson [44]: an input sharing stage, a computation stage, and an output reconstruction stage. The idea is that computation is performed on shares of private inputs, ultimately producing shares of the final answer. These shares are combined in the third stage to reveal only the output to all of the players.

Another technique that is used by several of the protocols is to do "second-order" sharing, i.e., sharing of shares. This idea was first exploited by Galil, Haber, and Yung [42], and is useful for allowing players to prove to one another that their actions are consistent and correct, as well as for enhancing fault tolerance.

Some protocols (beginning with the results of Goldreich, Micali, and Wigderson [44]) that protect against an active adversary use zero-knowledge proofs in the course of the computation. As described in an earlier section, every language in NP has a zero-knowledge proof system [43]. Such a proof system allows anyone to demonstrate membership in the language without leaking any additional information. Consider the set of all legal messages that can be sent at a given moment in the course of a protocol. This is almost always a language in NP, and thus the actual message sent can be followed by a zero-knowledge proof of membership. This "validates" the message without compromising security.

This section is divided into three further subsections. The first describes multi-party computation in the presence of a passive ("gossiping only") adversary. The second considers multi-party computation when the adversary is active, and presents new protocols for this problem as well as modifications of passive-adversary protocols. The third subsection gives multi-party computation protocols that achieve "fairness," a condition which limits the advantage that can be gained by the adversary if faulty players quit the protocol before completion.

4.2 Multi-Party Cryptographic Protocols versus Passive Adversaries

A two-party protocol for any two-input computation, due to Goldreich, Micali, and Wigderson [44], was presented in Section 4. That protocol assumed only the existence of any trapdoor function. In the same paper, multi-party computation is reduced to a succession of two-input computations, in such a way that privacy is maintained.

Using a technique due to Barrington [6], the depth D circuit to be computed can be represented as a straight-line program, of length at most 4^D, whose inputs are permutations in S_5 (the 120 possible permutations on five distinct elements). Each player can then privately convert his input into a single corresponding permutation. The computation reduces to finding the composition of a string of private and public permutations.

In the first stage of the protocol, each player shares its permutation with all of the players. A permutation σ is shared by giving a new permutation σ_i to each player i, where $\sigma_1, \cdots, \sigma_n$ is an otherwise random collection of permutations whose composition is $\sigma_1 \circ \cdots \circ \sigma_n = \sigma$.

In the second stage of the protocol, the straight-line program is performed, from left to right, on the shares of the inputs. As the computation proceeds, each player will hold a share of the partial result obtained thus far.

Composition with a public permutation is immediate. If the straight-line program calls for composing a given permutation σ with a public permutation ϕ to get $\phi \circ \sigma$, then the first player replaces its share σ_1 of σ with the the composition $\phi \circ \sigma_1$.

The composition of two permutations is more difficult. It is not enough for each player to take the composition of its two shares, because of the non-commutativity of composition, i.e., $\sigma \circ \sigma' = (\sigma_1 \circ \cdots \circ \sigma_n) \circ (\sigma'_1 \circ \cdots \circ \sigma'_n) \neq (\sigma_1 \circ \sigma'_1) \circ \cdots \circ (\sigma_n \circ \sigma'_n)$. The composition of two permutations reduces to the composition of $2n$ share permutations, but each player's shares are initially "too far apart."

The solution is to perform a series of "swaps" of neighboring share permutations. A "swap" is a two-party protocol that inputs two permutations ρ and τ, and outputs two otherwise random permutations ρ' and τ' such that $\rho \circ \tau = \tau' \circ \rho'$. Since this is a well-defined two-input random function, a two-party secure computation protocol for swapping can be implemented (using the general techniques from the previous section). This protocol should give only one of the two outputs to each player: ρ' is given only to the player who input ρ, while τ' is given only to the player who input τ. If the swaps are performed purposefully, then one player's two shares are "closer" together after each swap; each player privately finds the composition of its two shares whenever those shares become adjacent. A total of $O(n^2)$ swaps are necessary to maneuver each player's shares into adjacent locations.

Instead of working with straight-line programs and permutations, it is shown in the paper by Galil, Haber, and Yung [42] how to perform private multi-party computation of boolean functions directly on the bits themselves. For private computation, this yields a significant gain in efficiency for round complexity.

In the first stage, each player shares its input bit by giving each other player a one-bit share. The shares are chosen so that their xor yields the input bit, and so that they are otherwise random. Public-key cryptography is used to transmit shares to players. In the second stage, the function is computed by evaluating its corresponding circuit using only the shares of the private inputs. When the evaluation reaches any intermediate wire, each player has a share of the proper value at that wire (for the given circuit and secret inputs). Negation in this stage is immediate (e.g., one player complements its input share). Conjunction relies on iterating the following two-party protocol (as a general secure computation): one player inputs x and y, the second player inputs bit z, and the second player receives the output bit $f((x, y), z) = x \oplus (y \wedge z)$.

Goldreich and Vainish [45] observe that general two-party secure computation is unnecessary for the two preceding protocols. For both permutation swapping and the function $f(x, y, z)$ described above, special-purpose private protocols can implement these protocols under the assumption that Oblivious Transfer is possible.

4.3 Multi-Party Cryptographic Protocols versus Active Adversaries

An active adversary can cause faulty players to deviate from a protocol arbitrarily. If a protocol is to be secure against such an adversary, then the honest players should be able to detect when deviations occur and to take corrective action. Often, the action to be taken is to kick out the offending players and continue the protocol somehow in their absence.

Goldreich, Micali, and Wigderson [44] show how to modify their multi-party private protocol to achieve a multi-party resilient protocol.

In the first stage of the protocol, each player commits to its secret input and to the random bits that it will use during the computation phase. A verifiable secret sharing scheme is used to share the input with all players, and a coin-flipping scheme is used to generate random bits in a way that is reconstructible by the remaining players. After these commitments, any player can be fully "exposed" by the remaining community of honest players.

In the second stage, each message that is sent in the protocol is validated via zero-knowledge proof.

The protocol proceeds otherwise as in the passive adversary case. If any player is caught cheating, then its private input and random tape are reconstructed by the community. For the remainder of the protocol, any messages this player would have sent can be simulated privately by each remaining player.

Galil, Haber, and Yung [42] show how to modify the protocol of Goldreich, Micali, and Wigderson so that the private input and random tape of suspected cheaters are not fully exposed (e.g., to protect an honest victim of a stop failure). The key idea is to use secondary shares (shares of shares) in the first stage of the protocol. When a faulty player is discovered in stage two, then only the primary shares held by that player are exposed (so that the remaining players can adjust their shares accordingly). Whenever a message is needed from the cheating player, the other players compute it, in a secure distributed manner, from the shares they were given in stage one of the faulty player's input and random bits. Not only is the private input kept private, but a simple procedure allows the faulty player to rejoin the protocol at any time.

In the section on two-party secure computation, a protocol due to Chaum, Damgård, and Van de Graaf [29] was mentioned in which both players contributed to the "scrambling" of the circuit (truth table), and then both players evaluated the scrambled circuit at their private inputs. A more general version of this protocol works for any number of players. In the general case, each of n players contributes to the scrambling, and then all n players evaluate the scrambled circuit. This direct approach differs from the other solutions in this subsection, all of which reduce each gate to a collection of $O(n^2)$ two-party subprotocols. An interesting feature of this protocol is that the secrecy of one player's input can be protected unconditionally by this protocol.

4.4 Fairness in Cryptographic Multi-Party Protocols

Another security property of interest, called "fairness," guarantees that the interruption of a protocol confers no special advantage to any parties. The problem of "contract signing," allowing two parties to agree on a contract in a mutually committed fashion, predates and is related to the fairness problem; protocols for contract signing were given by Blum [21]. Fairness itself was originally considered for the basic problem of secret key exchange, introduced and incorrectly solved by Blum [23], and generalized and corrected by Yao [72].

When the secrets are single bits, then no deterministic algorithms are known. A probabilistic solution was found by Luby, Micali, and Rackoff [59]; Vazirani and Vazirani [71] investigated a similar but somewhat weaker notion of exchanging a secret for a valid receipt. Cleve [34] shows how to extend the secret exchange protocol of Luby et al. to allow for a "controlled disclosure" of information.

Yao [73] first defined fairness for two-party computation protocols for boolean functions. Galil, Haber, and Yung [42] describe a means (similar to Blum [23]) for adding fairness to multi-party computation protocols, assuming that trapdoor functions exist. Beaver and Goldwasser [13] present a different way of achieving fairness for boolean functions (similar to Luby et al. [59]), under the weaker assumption that an Oblivious Transfer protocol exists; Goldwasser and Levin [46] extend this technique from boolean values to arbitrary values.

5 Non-Cryptographic Secure Distributed Computation Protocols

5.1 Main Ideas

At first glance, the idea of a secure non-cryptographic protocol for the distributed computation of an arbitrary function seems paradoxical. How can the output of a function be computed by players of unbounded computational power without leaking something, even in only an information theoretical sense, about the private inputs? Indeed, the idea is paradoxical for arbitrary two-input functions computed by two mutually suspicious players, as was previously indicated (e.g., for two-party disjunction). Surprisingly, what is impossible in general for two players becomes possible when there are many players, by substituting secret-sharing, private channels, and "sufficient honesty" for cryptography.

An early result in this area (demonstrating the difference when more than two players participate) was due to Barany and Furedi [4]. They give a 1-private non-cryptographic protocol for an aspect of the multi-player "Mental Poker" problem, i.e., how to deal a deck of cards among mutually untrustworthy players who can only communicate by sending messages. The solution involves all players choosing random permutations of the deck, and sending messages consisting of permutations applied to a player's currently held cards. If player p is to receive the next card, then these messages result in all other players learning the mapping of their current holdings under a permutation π known only to p. One of these other players can then select an unused element in the range of π, and its inverse will be the card next dealt to p. This was extended to any function by Ben-Or, Goldwasser, and Wigderson [19] and by Chaum, Crépeau and Damgård [28].

Most subsequent protocols for non-cryptographic secure computation, and all of the ones given in this section, rely on the secret-sharing method due to Shamir [70] described earlier: To "t-share" a secret s, give each player i a point $(\alpha_i, p(\alpha_i))$ for some predetermined $\alpha_i \neq 0$ (e.g., $\alpha_i = i$), where $p(x)$ is an otherwise random degree t polynomial whose constant term is s. All non-cryptographic computation protocols described in this paper use this method in a "three-stage paradigm" [44]. First, each player shares his secret input with all other players using a Shamir polynomial. Second, all players perform some computation on these shares. Third, the results of this computation are combined to find the actual output.

404

Nice homomorphic properties of Shamir shares (first pointed out by Benaloh [16]) make stage two possible: any linear combination of shares of secrets is itself a share of the linear combination of secrets. For example, if $p(x)$ and $p'(x)$ are otherwise random degree t polynomials with constant terms s and s', then $p(x) + p'(x)$ is an otherwise random degree t polynomial with constant term $s + s'$. Moreover, if player i holds shares $s_i = p(\alpha_i)$ and $s'_i = p'(\alpha_i)$, then $s_i + s'_i$ is his share of the sum. The same can be shown for the product of a secret by a scalar.

A "close call" prevents stage two from being entirely non-interactive, i.e., the product of secret shares is "almost" a share of the product of the secrets. If $p(x)$ and $p'(x)$ are otherwise random degree t polynomials with constant terms s and s', then $p(x)p'(x)$ does have constant term ss', but it is neither degree t (i.e., it is degree $2t$) nor otherwise random (e.g., it cannot be irreducible). Both of these flaws are critical. The larger the degree of the share polynomial, the more players that are needed to recover the secret, so it is essential that the degree not increase with each multiplication. If the share polynomials are not fully random, then information about the secrets may be leaked to fewer than $t + 1$ computationally unbounded agents.

One of the main trends of the non-cryptographic computation protocols described in this paper follow from this "close call." Computation in Stage 2 proceeds directly on shares of secrets, with occasional interaction to "clean up" shares after multiplication.

Another main trend arises from this first one when the adversary is active. If interaction is required for multiplications, then how can the honest players protect themselves against misleading interaction coordinated by an active adversary? The solutions to this problem depend on the level of resilience that is required. If fewer than one-third of the players are faulty, then error-correcting codes can be exploited [19]. For greater resilience, new techniques of verifiable secret sharing [67] can be used. These new techniques can require more interaction among players, even for linear operations.

5.2   Non-cryptographic Protocols versus Passive Adversaries

The first two papers suggesting general non-cryptographic distributed computation [19] [28] both present protocols for t-private computation whenever $n > 2t$. Providing that all players obey the protocols perfectly, no minority of players can pool their knowledge at the end of the protocol to gain further information about the honest players' inputs. In this subsection, we will describe both of these protocols.

Ben-Or, Goldwasser, and Wigderson [19] present a protocol which allows t-private computation among $2t + 1$ players without any cryptographic assumptions. A network of private channels connecting all players is required.

This protocol is based on Shamir's method for secret sharing. Each player i shares its secret input s_i with the other players by giving each other player j a single point $p_i(\alpha_j)$ on an otherwise random degree t polynomial $p_i(x)$ such that $p_i(0) = s_i$. For simplicity, assume that $\alpha_i = i$ for all i, $1 \leq i \leq n$. Any arbitrary linear combination $\lambda(s_1, \cdots, s_n)$ of secret inputs can then be performed by the players without any communication, as described at the start of this section.

Since arbitrary linear computations can be performed directly on t-shares without communication, it suffices to show how multiplication of shared secrets can be performed. For this operation, some communication is needed. If each player multiplies together its shares of the multiplicands, then, analogous to the linear case, each player arrives at a value on a polynomial that passes through the product at the origin. However, this polynomial that "almost-shares" the product is neither random nor of the right degree. It is a degree $2t$ polynomial that has a certain structure, e.g., it cannot be irreducible. The subprotocol for multiplication uses communication to eliminate this structure.

Suppose that the multiplication subprotocol begins with each player i holding the share $p(i)$ of the secret $s = p(0)$ and the share $p'(i)$ of the secret $s' = p'(0)$, where p and p' are otherwise random degree t polynomials. Let B be the n by n (vandermonde) matrix whose (i,j) entry is j^i. Let $Chop_t$ be the n by n matrix whose (i,j) entry is 1 if $1 \le i = j \le t$ and 0 otherwise. The subprotocol is as follows:

1. Each player i privately finds an otherwise random degree t polynomial $r_i(x)$ with constant term zero.

2. Each player i sends to each player j the value $r_i(j)$.

3. Each player i privately finds $v_i = p(i)p'(i) + \sum_{j=1}^{n} r_j(i)$.

4. The players cooperate to give each player i the ith component of $(v_1 v_2 \cdots v_n)B^{-1}Chop_t B$, as follows:

   (a) Each player i privately finds an otherwise random degree t polynomial $q_i(x)$ such that $q_i(0) = v_i$.

   (b) Each player i sends to each player j the value $v_{ij} = q_i(j)$.

   (c) Each player i privately finds $(w_{1i}w_{2i} \cdots w_{ni}) = (v_{1i}v_{2i} \cdots v_{ni})B^{-1}Chop_t B$

   (d) Each player i sends to each player j the value w_{ji}.

   (e) Each player i privately interpolates to find the value at 0 of the degree t curve through $(1, w_{i1}), (2, w_{i2}), \cdots, (n, w_{in})$. This is the desired share at i of the product of s and s'.

Notice that this subprotocol for multiplication first re-randomizes the intermediate degree $2t$ polynomial (by adding random polynomials with zero constant term), and then truncates its degree down to t (by multiplying by $B^{-1}Chop_t B$. Since this truncation step is itself a linear operation, it can be computed privately on shares of the "almost-shares" (step 4c).

Chaum, Crépeau, and Damgård present a protocol that also allows t-private computation among $2t + 1$ players without any cryptographic assumptions, also assuming a network of private channels connecting all players. In fact, this protocol is quite similar to that of Ben-Or, Goldwasser, and Wigderson, although the computation proceeds bit by bit here, on a circuit composed of AND gates and XOR gates.

Assume that each player has some input bits, and that the function to be computed is represented as a circuit composed of AND gates and XOR gates. Each player t-shares his input bits using Shamir polynomials over a finite field that is a power of two (i.e., $GF(2^k)$ for some k such that $2^k > n$). Then XOR in this protocol becomes analogous to addition in the previous protocol, while AND becomes analogous to multiplication.

Each XOR gate can be computed privately without any interaction. Each player simply adds its t-shares of the corresponding inputs, and arrives at a valid t-share of the output. Notice that the new share lies on a polynomial that passes through the XOR at zero only because the underlying field was chosen to be a power of two.

Each AND gate can "almost" be computed by having each player multiply its two t-shares of the corresponding inputs. As before, the resulting values are on a polynomial that passes through zero at the right place, but which is non-random and of twice the desired degree. Chaum, Crépeau, and Damgård have an interesting variation on the interactive "re-randomization" and "truncation" steps for repairing these values, doing both of these steps simultaneously.

After the private multiplication, when each player has an "almost-share" of the AND result, the players proceed as follows. Each player chooses two otherwise random polynomials, one of degree t and one of degree $2t$, such that either both have constant term zero or both have constant term one. Each player distributes shares of both polynomials. Then each player adds its "almost-share" of the AND computation to the sum of the "high-degree shares" it has just received. This yields valid $2t$-shares of an XOR of random values with the AND result. The players reveal these shares and determine what this XOR is. If it is zero, then the XOR of the random values is equal to the AND result; in this case, the sum of the "low-degree shares" for each player is the valid t-share of the AND result. If it is one, then the XOR of the random values is the opposite of the AND result; in this case, one plus the sum of the "low-degree shares" for each player is the valid t-share of the AND result.

5.3 Non-cryptographic Protocols versus Active Adversaries

In this subsection, we will cover four non-cryptographic protocols for distributed computation versus an active adversary, which appeared in the literature as two sets of similar and independent results. The first two [19] [28] achieve t-resilience whenever there are more than $n = 3t$ players, assuming a complete private network of channels. The last two [68] [10] achieve t-resilience when $n > 2t$, but require a broadcast channel in addition to a complete private network, and must allow a small probability of error.

In the same paper that gave the protocol for t-private computation when $n > 2t$ (described in the preceding subsection), Ben-Or, Goldwasser and Wigderson [19] also demonstrated that t-resilient computation was possible for $n > 3t$. The resilient protocol is identical to the private protocol except for changes to the secret-sharing method and to the multiplication stage.

A verifiable secret sharing scheme is used in place of simple secret sharing. Furthermore, instead of having player i receives values at some arbitrary α_i (e.g., $\alpha_i = i$), this protocol requires that each player receive values at a specific nth root of unity.

The importance of this choice for share locations is that the vector of n shares of any secret is a *codeword* of a generalized Reed-Muller error-correcting code [20]. This family of codes can correct a codeword in which up to one-third of the components are incorrect or missing. For our purposes, this means that the $n - t$ honest players can reconstruct a shared secret after t arbitrary shares are corrupted by an active adversary. Furthermore, the correction procedure involves only the computation of $2t$ *linear* combinations of codeword components (called "syndromes"); thus correction may be performed privately on shares of codewords.

The multiplication phase differs from the private protocol in two ways. Re-randomization requires that each player distribute shares of an otherwise random degree $2t$ polynomial with constant term zero. For purposes of resilience, the honest players must be able to prove that the shares that are received are in fact from polynomials of this form. There is a protocol for doing this, based on the fact that degree t polynomials can be so verified, and that a share of an otherwise random degree $2t$ polynomial with zero constant term can be constructed privately from shares of t random degree t polynomials. As pointed out by Beaver [10], though, this extra effort to randomize the higher-degree coefficients is wasted, since these will disappear after truncation anyway.

The second difference is in the truncation step. A subprotocol is inserted to force all players to "validate" their inputs to the degree truncation step. Recall that each player shares its share of the product, so that a linear combination of the first-level shares can be performed privately by each player on the second-level shares. This subprotocol enables the honest players to be sure that these second-level shares are in fact honest shares of the first-level shares.

Chaum, Crépeau and Damgård [28] also give a protocol for t-resilient computation whenever $n > 3t$. There are several important similarities with the protocol of Ben-Or et $al.$ First, this protocol also adapts a privacy-only protocol by overlaying verifiability subprotocols. Second, an important part of both protocols is that players operate on shares of shares of secrets (a technique that was first used in the cryptographic setting by Galil, Haber, and Yung [42]).

There are also some significant differences between the two solutions. The protocol of Chaum, Crépeau, and Damgård relies on an interactive "cut-and-choose" procedure to validate that all of the shares of a secret are consistent with one another, i.e., that all n values lie on a single degree t polynomial. Unlike the protocol of Ben-Or et $al.$, which used error-correcting codes for this purpose, this leaves an exponentially small probability of undetected cheating. In addition, the protocol becomes less efficient in terms of number of rounds and messages required for a given computation.

Since Byzantine Agreement requires that more than two-thirds of the players be honest, these two resilient protocols [19] [28] are the best possible for a network of private channels in the non-cryptographic setting. One natural question is whether the resilience can be increased by adding a broadcast channel. The affirmative answer was supplied independently by T. Rabin and Ben-Or [68], and by Beaver [10], both building on ideas from T. Rabin [67].

At the heart of the protocol is a verifiable secret sharing scheme, due to T. Rabin [67], that can correct up to t errors where $n > 2t$, by giving up on absolute certainty of success. This scheme, which introduces the new concept of "check vectors," has a probability of failure that is exponentially small with respect to a security parameter. This VSS scheme also requires the existence of a broadcast channel.

Because it relies on secondary sharing (shares of shares), Rabin's VSS scheme does not have the same homomorphic properties as Shamir's basic secret-sharing scheme, i.e., shares of linear combinations of secrets cannot be determined through private computation on the original shares. However, with interaction, Rabin shows that there is a t-resilient protocol to give each of $n > 2t$ players the appropriate share of any linear combinations of secrets.

Using the protocol for linear combination, a protocol is given to t-resiliently demonstrate that one secret c is the product of two other secrets a and b. In other words, if a, b, and c have each been shared by Rabin's VSS scheme by one player, and if $c = ab$, then that player can convince the honest players of this fact without revealing anything further about the three secrets. This is done by a type of cut-and-choose protocol. A number of equalities related to the original secrets, of the form $D = (a + R)(b + S)$, are chosen by the verifier; for each such equality, the values R, S, and D are verifiably shared; some of these triples are challenged; those that are challenged are revealed to demonstrate that the corresponding equalities are indeed correct; those triples that are

408

not challenged are combined with the original three secrets to demonstrate that the equalities are indeed related to the original secrets. The probability of undetected cheating is bounded by 2^{-k}, where k is the number of triples that are used by the verifier.

The t-resilient protocol for linear combination also implies a t-resilient protocol for degree truncation. In other words, if each player holds a share of a degree $2t$ polynomial $p(x)$, then this protocol ends with each player holding a corresponding share of a degree t polynomial $\overline{p}(x)$ with the same coefficients for all of its terms. Each output share is in fact a specific linear combination of the input shares: $\overline{p}(\alpha_i) = \sum_{i=i}^{n} \overline{L_i}(\alpha_i)p(\alpha_i)$, where $\overline{L_i}(x)$ is the (publicly computable) degree t truncation of the Lagrange polynomial which is one at α_i and zero at α_j, $j \neq i$.

For re-randomization, each player i can verifiably t-share a random value using a polynomial $p_i(x)$. If each player j now multiplies each received share $p_i(\alpha_j)$ by its own share location α_i, then it obtains a valid $(t + 1)$-share of zero. The sum of all of these $(t + 1)$-shares is a point on an otherwise random degree $t + 1$ polynomial with constant term zero.

Using the t-resilient protocols for product demonstration, re-randomization, and degree truncation, a t-resilient protocol for multiplying two shared secrets is possible. First, each player verifiably t-shares three values: its shares of the two secrets, and the product of its shares of the two secrets. Second, each player t-resiliently demonstrates that the proper product relation holds among these three shared values. Finally, the players perform the t-resilient protocols for re-randomization and truncation.

5.4   Constant-Round Non-cryptographic Computation

One of the open problems in secure distributed computation is whether $O(n)$-resilience is possible in constant rounds and polynomial-sized messages in the non-cryptographic setting. This section presents some of what is currently known about constant-round non-cryptographic computation versus an active adversary.

Bar-Ilan and Beaver [5] demonstrate that the number of rounds of communication needed for secure distributed computation of a function need not be related to the depth of the arithmetic circuit for that function. They show how to compute any NC^1 circuit in constant rounds and polynomial-sized messages.

One of the key ideas in this protocol is the use of a technique, due to Ben-Or and Cleve [18], for mapping an arbitrary arithmetic function to the multiplication of a string of three by three matrices. By a clever self-randomization, this long product of matrices can be securely computed using only two secret matrix multiplications:

$$Y = M_1 M_2 \cdots M_N = R_0 (R_0^{-1} M_1 R_1)(R_1^{-1} M_2 R_2) \cdots (R_{N-1}^{-1} M_N R_N) R_N^{-1}$$

$$= R_0 Z R_N^{-1}.$$

Thus, if the players jointly create and share random invertible matrices $R_0, \cdots, R_N$, they can compute each $S_i = R_{i-1}^{-1} M_i R_i$ in parallel by private computation, public revelation, and interpolation. From these they can find $Z = S_1 S_2 \cdots S_N$, and then two secret multiplication will find $Y = R_0 Z R_N^{-1}$.

When f is not NC^1, the function can still be securely computed in constant rounds, but with messages that are not necessarily polynomial in size. Alternatively, using their technique, an arbitrary function can be securely computed in $D/\log n$ rounds, where D is the circuit depth, using polynomial-sized messages.

Beaver, Feigenbaum, Kilian, and Rogaway [12] show how to achieve non-cryptographic computation of boolean circuits of any size in constant rounds with polynomial-sized messages, by decreasing the resilience to $O(\log n)$ faulty processors.

The method relies on an instance-hiding scheme for any boolean function using $n/\log n + 1$ oracles. This can be converted into an $n + 1$ oracle instance-hiding scheme that is $(\log n)$-private, by replacing each oracle query at x by $n + 1$ oracle queries at $(\log n)$-shares of x.

The method also relies on the constant-round poly-sized computation of functions with log-depth circuits [5]. Creation of appropriate oracle queries, polynomial interpolation, notarized envelope schemes, and majority-voting all have log-depth circuits, and so all are available as subprotocols in this way.

Lastly, Beaver, Micali, and Rogaway [14] show that constant-round poly-sized computation resilient for any faulty minority is achievable assuming that one-way functions exist. The protocol, a generalization of the two-party protocol due to Yao [73] described earlier, uses the one-way function to hide connections between scrambled gates which are all evaluated together. Because of the reliance on one-way functions (along with a broadcast channel and a network of private channels), this is actually a *cryptographic* protocol. However, its relevance is clearer when mentioned along with other constant-round results.

In addition to one-way functions, this protocol requires a broadcast channel and a network of private channels, and allows an exponentially small probability of error. The only use of one-way functions is to guarantee the existence of pseudorandom generators, an equivalence shown by Impagliazzo, Levin and Luby [52] and Håstad [50].

5.5 Hybrid Computation Protocols

Chaum [27] combines both non-cryptographic and cryptographic computation into a single hybrid protocol. The security of this protocol is compromised only if *both* a majority of the players are faulty *and* the underlying cryptographic assumptions are violated.

The key idea is to transform the n-ary function evaluation $f(s_1, \cdots, s_n)$ into an $(n + 1)$-ary function evaluation $f^*(s_1 \oplus r_1, \cdots, s_n \oplus r_n, r_1 \circ \cdots \circ r_n))$. where each player i chooses the random r_i. The computation of f^* is performed using the cryptographic protocol of Chaum, Damgård, and Van de Graaf [29], where all n players jointly determine all messages sent by player $n + 1$. They do this by running a non-cryptographic protocol (e.g., [28]) whenever a message from this "player" is needed.

Recall that the cryptographic protocol of Chaum, Damgård, and Van de Graaf can protect one player's input unconditionally; here this protection is granted to the jointly maintained player. Thus breaking the outer cryptographic protocol reveals only the first n inputs, each of which is a secret xored to a random value. To determine these random values requires breaking the inner non-cryptographic protocols as well. Alternatively, breaking only the inner non-cryptographic protocol reveals only information about the random values, which yields no information about the original secrets.

Huang and Teng [51] give a general multi-party secure computation protocol for which the only non-local computation is repeated instances of a subprotocol for addition ("distributed sum"). This subprotocol requires no cryptographic assumptions; cryptography is only needed to protect against cheating players and to communicate across insecure channels. Their protocol can thus be viewed as a resilient multi-party cryptographic protocol, or as a private multi-party non-cryptographic protocol on a complete private network.

5.6 Non-Cryptographic Computation Protocols on Incomplete Networks

All of the non-cryptographic protocols described in this section thus far have required that the network of private channels connecting the players be complete. It is interesting to consider what weaker assumptions on the private network will still allow secure distributed computation without cryptographic assumptions.

Rabin and Ben-Or [68] show how to perform any secure distributed computation t-resiliently in a network of at least $3t + 1$ players, if the players are joined by a network of private channels with connectivity at least $2t + 1$. This is done by showing how to send a message from any player to any other player through such a network, in expected time equal to the diameter of the network, and with a slight probability of error. This capability can be spliced into previously discussed protocols to achieve the stated result.

Dolev, Dwork, Waarts, and Yung [38] show how to attain the same result without any probability of error. Sending a message from any player to any other player ("perfectly secure message transmission") is divided into two problems: the slightly easier problem of sending a one-time random pad, and the much easier problem of sending the message xored with the one-time pad. The second problem just requires that the encrypted message be sent simultaneously along all $2t + 1$ paths, whereupon the receiver decrypts the majority message. The first problem requires that each bit of random pad be transmitted via an interactive protocol. To transmit one random bit, a three-phase protocol is used in which shares of $nt + 1$ bits and associated checking pieces are sent along different paths from sender to receiver. No amount of cheating by the t faulty players can prevent the sender and receiver from agreeing on one of the $2t + 1$ random bits that is guaranteed to be uncorrupted, or allow the faulty players to learn the value of that bit.

5.7 Non-cryptographic Protocols versus Mobile Adversaries (Virus Model)

Ostrovsky and Yung [65] consider a more powerful adversary that models the behavior of a (detectable) virus in a computer network (with rebootable machines): a different set of up to t players can be under the adversary's control each round. They present a non-cryptographic computation protocol secure against such an adversary where t is some constant fraction of the number of players. One of the features of this protocol is that all of the information held secret by the players must be continually reshared, so that the mobile adversary never gets enough consistent shares of any secret to recover it.

5.8 Parallel Communication-Efficient Protocols

Almost all of the non-cryptographic protocols described in this section rely on algebraic manipulation of Shamir shares of secrets. Franklin and Yung [41] give a general compilation technique for parallelizing such protocols, so that the same computation is performed simultaneously on several sets of inputs. The key idea is to have each player hold a single "multi-share" of a polynomial that hides secrets at many locations (i.e., not just at 0). Linear combinations of multi-shared secrets

411

are multi-shares of the linearly combined secrets, and the algebraic methods for multiplication of ordinary shares can be extended to multi-shares. Using these methods, k sets of inputs can be computed in parallel with the same number of rounds and bits of communication as the sequential case (i.e., at an amortized savings of a multiplicative factor of k in bits of communication), while reducing the number of faulty players tolerated by an additive factor of k (e.g., from t to $t - k + 1$).

This method of parallel computation is the basis of another compilation technique, one which produces a protocol that can detect (but not correct) the tampering of an active adversary with low bit complexity from a protocol secure only against a passive adversary. The protocol is run in parallel on two sets of inputs: the actual inputs, and a random set of inputs constructed jointly by the players. The two computations proceed in lockstep, and the random inputs (together with both outputs) are revealed at the end of the protocol. If the computation on the random inputs is correct, then with high probability the secret computation is also correct.

6 Reductions and Complexity Results

Previous sections have contained descriptions and discussion of what is possible for secure distributed computation. This section will survey what is known about reductions and complexity results for this problem, i.e., what is just as hard as, and what is impossible for, secure distributed computation.

6.1 Reductions Among Primitives

Ostrovsky, Venkatesan, and Yung [63] consider the problem of sufficient conditions for "asymmetric" two-party protocols, i.e., protocols in which one player is polynomially bounded while the other player is computationally unbounded. In particular, they show that Oblivious Transfer is possible, in either direction, assuming the existence of one-way functions. Since Kilian [56] reduces two-party circuit evaluation to Oblivious Transfer, it follows that any asymmetric two-party computation is possible if one-way functions exist.

For the symmetric case, Impagliazzo and Luby [53] show that one-way functions are necessary for bit commitment, and hence necessary for secure computation. They also shows that Oblivious Transfer implies the existence of one-way functions. However, due to a relativization result from Impagliazzo and Rudich [54], it is unlikely that the sufficiency of one-way functions for secure symmetric computation can be shown with current techniques (i.e., a proof that only used one-way functions in a "black-box" manner would prove $P \neq NP$).

Ostrovsky and Yung [64] demonstrate the necessity of a complete network of private channels for implementing private multiparty computation among computationally unbounded players (sufficiency was shown by [19] and [28]).

6.2 Necessary Conditions and Resources

The first complexity results in this field are two impossibility results due to Yao [72], one for a certain type of two-player secret exchange (swapping zeros of trapdoor functions), and the other for the generation of an arbitrarily biased bit by many players.

Ben-Or, Goldwasser, and Wigderson [19] provide matching lower bounds for multiparty non-cryptographic computation on a network of private channels, for both passive and active adversary.

For a passive adversary, there are functions for which there is no t-private protocol among n players when $n \leq 2t$; e.g., the impossibility of a 1-private protocol for two parties to compute the OR of two bits is easy to verify. For an active adversary, there are functions for which there is no t-private protocol among n players when $n \leq 3t$; this follows directly from the lower bound for Byzantine Agreement [58].

For the case of a passive adversary, more is known about when a function does or does not have a t-private protocol among n players when $n \leq 2t$. Chor and Kushilevitz [37] find a "gap" in the maximum level of privacy, in the non-cryptographic zero-error setting for boolean functions. If a function has a t-private protocol, $n \leq 2t$, then it has an $(n-1)$-private protocol. Moreover, every such n-ary function is equivalent to an xor of n single-input functions.

For the case of an active adversary, Cleve [33] has impossibility results for the much simpler problem of computing a single random bit. When at least half of the processors are faulty, it is impossible to compute a random bit in polynomial time with a negligible bias (less than any inverse polynomial in the number of players).

For two player non-cryptographic protocols, a complete characterization of privately computable general functions (i.e., non-boolean) is given independently by Kushilevitz [57] and by Beaver [8]. Bar-Yehuda, Chor, and Kushilevitz [7] generalize to consider two-party "nearly private" computation protocols that leak some information about private inputs to the opposing player. They show that the identity function and the greater-than function can be computed leaking at most $\log n$ bits of information, and that almost all boolean functions can be computed leaking at most $\frac{1}{2}(n - \log n - 3)$ bits of information. They also construct functions for which large gains in round complexity are achievable by leaking small amounts of information; for one artificial example, exponential round complexity can be reduced to two rounds·and linear number of bits of communication by revealing a single bit of information about the inputs.

All of the results in this paper assume that the computation is performed over some finite field. Chor, Gereb-Graus, and Kushilevitz consider the case of private multi-party non-cryptographic computation over countable domains. They show that every n-ary boolean function is either n-private, $\lfloor \frac{n-1}{2} \rfloor$-private but not $\lceil \frac{n}{2} \rceil$-private, or not 1-private. As a surprising special case, private addition over a finite subrange of integers can be n-private, over the positive integers can be at most $\lfloor \frac{n-1}{2} \rfloor$-private, but over all the integers cannot even be 1-private.

Dolev, Dwork, Waarts, and Yung [38] give lower bounds on security versus connectivity for the simpler but related problem of perfectly secure message transmission through an untrusted network. They consider this problem in a setting that generalizes in two ways on the setting considered for the multi-party unconditional protocols in this paper: (1) there are two adversaries, one passive and one active, which control possibly overlapping sets of faulty processors, and which may or may not be able to communicate during execution; and (2) they assume an incomplete network, of known degree. They also show that $3t + 1$ players are necessary to achieve VSS with no probability of error, even assuming a broadcast channel; recall that almost perfect VSS is possible with only $2t + 1$ players [68].

7 Acknowledgments

Thanks to Terry Boult, Joan Feigenbaum, Zvi Galil, Stuart Haber, and Thanasis Tsantilis for helpful suggestions.

413

References

[1] M. Abadi, J. Feigenbaum, and J. Kilian, "On hiding information from an oracle," J. Comput. System Sci. **39** (1989), 21-50.

[2] M. Abadi and J. Feigenbaum, "Secure circuit evaluation: a protocol based on hiding information from an oracle," J. Cryptology **2** (1990), 1-12.

[3] L. Babai and S. Moran, "Arthur-Merlin games: A randomized proof system and a hierarchy of complexity classes," J. Comput. System Sci. **36** (1988), 254-276.

[4] I. Banary and Z. Furedi, "Mental poker with three or more players," Information and Control **59** (1983), 84-93.

[5] J. Bar-Ilan and D. Beaver, "Non-cryptographic fault-tolerant computing in a constant number of rounds of interaction," PODC 1989, 201-209.

[6] D. Barrington, "Bounded-width branching programs recognize exactly those languages in NC^1," J. Comput. System Sci. **38** (1989), 150-164.

[7] R. Bar-Yehuda, B. Chor, and E. Kushilevitz, "Privacy, additional information, and communication," IEEE Structure in Complexity Theory 1990, 55-65.

[8] D. Beaver, "Perfect privacy for two-party protocols," DIMACS Workshop on Distributed Computing and Cryptography, Feigenbaum and Merritt (eds.), AMS, 1990, 65-77.

[9] D. Beaver, "Foundations of secure interactive computing," Crypto 1991.

[10] D. Beaver, "Distributed computations tolerating a faulty minority, and multiparty zero-knowledge proof systems," J. Cryptology, to appear.

[11] D. Beaver and J. Feigenbaum, "Hiding instances in multioracle queries," STACS 1990, 37-48.

[12] D. Beaver, J. Feigenbaum, J. Kilian, and P. Rogaway, "Security with low communication overhead," Crypto 1990.

[13] D. Beaver and S. Goldwasser, "Multiparty computation with faulty majority," IEEE FOCS 1989, 468-473.

[14] D. Beaver, S. Micali, and P. Rogaway, "The round complexity of secure protocols," ACM STOC 1990, 503-513.

[15] M. Bellare, L. Cowen, and S. Goldwasser, "On the structure of secret key exchange protocols," DIMACS Workshop on Distributed Computing and Cryptography, Feigenbaum and Merritt (eds.), AMS, 1990, 79-92.

[16] J. Benaloh (Cohen), "Secret sharing homomorphisms: keeping shares of a secret secret," Crypto '86, 251-260.

[17] J. Benaloh (Cohen) and M. Yung, "Distributing the power of a government to enhance to privacy of voters," PODC 1986, 52-62.

[18] M. Ben-Or and R. Cleve, "Computing algebraic formulas using a constant number of registers," ACM STOC 1988, 254-257.

[19] M. Ben-Or, S. Goldwasser, and A. Wigderson, "Completeness theorems for non-cryptographic fault-tolerant distributed computation," ACM STOC 1988, 1-9.

[20] E. Berlekamp, *Algebraic Coding Theory*, Aegean Park Press, Laguna Hills, CA, 1984.

[21] M. Blum, "Three applications of the Oblivious Transfer: Part I: Coin flipping by telephone; Part II: How to exchange secrets; Part III: How to send certified electronic mail," Department of EECS, University of California, Berkeley, CA, 1981.

[22] M. Blum, "Coin flipping by telephone: a protocol for solving impossible problems," IEEE Computer Conference 1982, 133-137.

[23] M. Blum, "How to exchange (secret) keys," ACM Trans. Comput. Sys. 1 (1983), 175-193.

[24] G. Brassard, D. Chaum, and C. Crépeau, "Minimum disclosure proofs of knowledge," J. Comput. System Sci. 37 (1988) 156-189.

[25] G. Brassard, C. Crépeau, and J. Robert, "Information theoretic reductions among disclosure problems," IEEE FOCS 1986, 168-173.

[26] G. Brassard, C. Crépeau, and M. Yung, "Perfectly concealing computationally convincing interactive proofs in constant rounds," Theoretical Computer Science (to appear).

[27] D. Chaum, "The spymasters double-agent problem: multiparty computations secure unconditionally from minorities and cryptographically from majorities," Crypto 1989, 591-601.

[28] D. Chaum, C. Crépeau, and I. Damgård, "Multiparty unconditionally secure protocols," ACM STOC 1988, 11-19.

[29] D. Chaum, I. Damgård, and J. van de Graaf, "Multiparty computations ensuring privacy of each party's input and correctness of the result," Crypto 1987, 87-119.

[30] B. Chor, M. Gereb-Graus, and E. Kushilevitz, "Private computations over the integers," IEEE FOCS 1990, 335-344.

[31] B. Chor, S. Goldwasser, S. Micali, and B. Awerbuch, "Verifiable secret sharing and achieving simultaneity in the presence of faults," IEEE FOCS 1985, 383-395.

[32] B. Chor and E. Kushilevitz, "A zero-one law for boolean privacy," ACM STOC 1989, 62-72.

[33] R. Cleve, "Limits on the security of coin flips when half the processors are faulty," ACM STOC 1986, 364-369.

[34] R. Cleve, "Controlled gradual disclosure schemes for random bits and their applications," Crypto 1989, 573-588

[35] J. (Benaloh) Cohen and M. Fisher, "A robust and verifiable cryptographically secure election scheme," IEEE FOCS 1985, 372-382.

[36] C. Crépeau, "Equivalence between two flavours of Oblivious Transfer," Crypto 87, 350-354.

[37] C. Crépeau and J. Kilian, "Achieving oblivious transfer using weakened security assumptions," IEEE FOCS 1988, 42-52.

[38] D. Dolev, C. Dwork, O. Waarts, and M. Yung, "Secret Message Transmissions," IEEE FOCS 1990, 36-45.

415

[39] S. Even, O. Goldreich, and A. Lempel, "A randomized protocol for signing contracts," CACM **28** (1985), 637-647.

[40] P. Feldman and S. Micali, "Optimal algorithms for Byzantine agreement," ACM STOC 1988, 148-161.

[41] M. Franklin and M. Yung, "Parallel secure distributed computing," manuscript.

[42] Z. Galil, S. Haber, and M. Yung, "Cryptographic computation: secure fault-tolerant protocols and the public-key model," Crypto 1987, 135-155.

[43] O. Goldreich, S. Micali, and A. Wigderson, "Proofs that yield nothing but their validity and a methodology of cryptographic protocol design," IEEE FOCS 1986, 174-187.

[44] O. Goldreich, S. Micali, and A. Wigderson, "How to play any mental game," ACM STOC 1987, 218-229.

[45] O. Goldreich and R. Vainish, "How to solve any protocol problem – an efficiency improvement," Crypto 1987, 73-86.

[46] S. Goldwasser and L. Levin, "Fair computation of general functions in presence of immoral majority," Crypto 1989, 75-84.

[47] S. Goldwasser and S. Micali, "Probabilistic encryption," J. Comput. System Sci. **28** (1984) 270-299.

[48] S. Goldwasser, S. Micali, and C. Rackoff, "The knowledge complexity of interactive proof systems," SIAM J. Comput. **18** (1989), 186-208.

[49] S. Haber, "Multiparty cryptographic computation: techniques and applications," Ph.D. thesis, Columbia University, 1988.

[50] J. Håstad, "Pseudo-random generators under uniform assumptions," ACM STOC 1990, 395-404.

[51] M. Huang and S. Teng, "Security, verifiability, and universality in distributed computing," J. Algorithms **11** (1990), 492-521.

[52] R. Impagliazzo, L. Levin, and M. Luby, "Pseudorandom number generation from one-way functions," ACM STOC 1989, 12-24.

[53] R. Impagliazzo and M. Luby, "One-way functions are essential for complexity based cryptography," IEEE FOCS 1989, 230-235.

[54] R. Impagliazzo and S. Rudich, "Limits on the provable consequences of one-way permutations," ACM STOC 1989, 44-61.

[55] R. Impagliazzo, and M. Yung, "Direct minimum-knowledge computation," Crypto 1987, 40-51.

[56] J. Kilian, "Founding cryptography on oblivious transfer," ACM STOC 1988, 20-31.

[57] E. Kushilevitz, "Privacy and communication complexity," IEEE FOCS 1989, 416-421.

[58] L. Lamport, R. Shostak, and M. Pease, "The Byzantine generals problem," ACM Trans. on Programming Lang. and Systems (1982), 382-401.

[59] M. Luby, S. Micali, and C. Rackoff, "How to simultaneously exchange a secret bit by flipping a symmetrically-biased coin," IEEE FOCS 1984, 11-21.

[60] S. Micali and P. Rogaway, "Secure computation," Crypto 1991.

416

[61] M. Naor, ""Bit commitment using pseudo-randomness," Crypto 1989, 128-136.

[62] M. Naor, R. Ostrovsky, R. Venkatesan, and M. Yung, "Perfect zero-knowledge arguments for NP can be based on general complexity assumptions," manuscript, 1991.

[63] R. Ostrovsky, R. Venkatesan, and M. Yung, "Fair games against an all-powerful adversary," these proceedings.

[64] R. Ostrovsky and M. Yung, "On necessary conditions for secure distributed computing," DIMACS Workshop on Distributed Computing and Cryptography, Feigenbaum and Merritt (eds.), AMS, 1990, 229-234.

[65] R. Ostrovsky and M. Yung, "Robust computation in the presence of mobile viruses," ACM PODC, 1991, 51-59.

[66] M. Rabin, "How to exchange secrets by oblivious transfer," Tech. Memo TR-81, Aiken Computation Laboratory, Harvard University, 1981.

[67] T. Rabin, "Robust sharing of secrets when the dealer is honest or cheating," M.Sc. Thesis, Hebrew University, 1988.

[68] T. Rabin and M. Ben-Or, "Verifiable secret sharing and multiparty protocols with honest majority," ACM STOC 1989, 73-85.

[69] R. Rivest, A. Shamir, and L. Adleman, "A method for obtaining digital signatures and public key cryptosystems," CACM 21 (1978), 120-126.

[70] A. Shamir, "How to share a secret," CACM 22 (1979), pp. 612-613.

[71] U. Vazirani and V. Vazirani, "Trapdoor pseudo-random number generators, with applications to protocol design," IEEE FOCS 1983, pp. 23-30.

[72] A. Yao, "Protocols for secure computations," IEEE FOCS 1982, 160-164.

[73] A. Yao, "How to generate and exchange secrets," IEEE FOCS 1986, 162-167.

Fair Games Against an All-Powerful Adversary

(EXTENDED ABSTRACT)

Rafail Ostrovsky[*]   Ramarathnam Venkatesan[†]   Moti Yung[‡]

Abstract

Suppose that a weak (i.e., polynomially-bounded) device needs to interact over a clear channel with an infinitely-powerful and adversarial device which he does not trust. Notice that throughout this interaction (game) the infinitely-powerful device can hide information from the weak device using encryption. The weak device, however, is not so fortunate: to keep the game fair, he must hide information from the strong device in the information-theoretic sense. Nevertheless, we show that the weak player can play *any* polynomial length partial-information game (or secure protocol) with the strong player using any one-way function. More specifically, we show that oblivious transfer protocol can be implemented in this model using *any* one-way function and we establish related impossibility results concerning oblivious transfer.

Since many problems fall into the above model (e.g., interactive proofs of [GMR], hiding information from an oracle of [AFK], zero-knowledge arguments with strong verifier [BCC], and two-party partial information games with an infinitely-powerful player [AF, CDV]), our results allow us to simplify and to improve complexity assumptions of a large number of existing protocols (most of which previously required specific assumptions on hardness of various algebraic problems). We also exhibit several practical and theoretical implications of our technique.

---

[*] MIT Laboratory for Computer Science, 545 Technology Square, Cambridge MA 02139. Supported by IBM Graduate Fellowship. Part of this work was done at IBM T.J. Watson Research Center.

[†] Bellcore, 445 South St., Morristown, NJ 07960.

[‡] IBM Research, T.J. Watson Research Center, Yorktown Heights, NY 10598.

1 Introduction

Perfect security of information is cryptography's ultimate goal. This is especially the case when one must interact with an all-powerful adversary, but has to hide from him any information which is not absolutely necessary in order to keep the protocol fair. The difficulty of perfect security is that one can not rely on encryption: absolute security means that even the adversary with an infinitely-powerful resources can not gain any information. This is indeed the requirements for Statistical Zero-knowledge interactive proofs of [GMR], hiding information from an oracle of [AFK], zero-knowledge arguments with strong verifier [BCC], and two-party secure circuit evaluation with perfect security for one side [AF, CDV]. Hence, motivated by this variety of settings, we study in this paper what are the weakest possible assumptions under which all such problems, classified as polynomial length partial-information games against an all-powerful adversary, can be implemented.

Until not long ago, for each partial-information game the answer to our question depended on the specific problem at hand. Fortunately, a very simple protocol called *Oblivious Transfer* (OT) which was suggested by Rabin, was shown to be sufficient for all two-party secure computations (this was put forward in [GMW, Y2] and was shown to be sufficient in [K].) Moreover, the reduction to OT is information-theoretic and holds independently of the power of the players. Informally, OT is the protocol where one player (the Sender) has two bits b_0 and b_1, while the other player (the Receiver) has a selection bit i. After the protocol, it should be the case that the Receiver learns the value of b_i but does not gain any information about b_{1-i}, while the sender does not gain any information about i (the above is one-out-of-two OT defined in [EGL]).

The fact that this simple protocol allows us to achieve all two-party secure computations puts OT in a spot-light: what are the weakest assumptions one can use to implement it? The best previous assumption required a trapdoor one-way permutation [GMW]. An open problem which has been asked was to reduce this assumption [B1, B2, BMR, FMR, GMW, Y1, Y2]. For the case when both players are polynomially bounded, Impagliazzo and Rudich [IR] showed that basing OT solely on any one-way functions is, in some technical sense (i.e., using black-box reduction), as difficult as separating P from NP. However, in numerous cases such as the settings mentioned above, we are not dealing with two polynomially-bounded players. Instead, polynomially-bounded machine is specifically required to deal with an infinitely-powerful adversary and "must" hide information from it, which means that encryption is helpless and therefore this is seemingly hard to do. Indeed, in many settings the best way of doing it was relying on specific algebraic properties of special candidates of one-way functions (e.g., quadratic residues, discrete log), or on a trapdoor function (which is a one-way function with the additional algebraic property that there is a secret associated with it, whose possession enables easy inversion in polynomial-time). For example, given a trapdoor permutation, the OT protocol of [GMW] works when the Sender is infinitely-powerful for an honest receiver (and if cheating is allowed, can be based on a specific algebraic assumption). However, it is not known how the OT protocol (with the strong sender) can be implemented without the trapdoor property. In this paper we show how to get rid of specific algebraic properties and also how the "trapdoor barrier" can be broken. In particular, we show how to perform OT protocol (to and from an all-powerful adversary), based on the availability of *any* one-way function. Moreover, in both directions, we allow the infinitely-powerful device to make arbitrary computations and manage to protect the security of the weak device despite of this fact.

Remark: A NAIVE (AND INCORRECT!) APPROACH: *The following question must be asked: If one of the players is infinitely-powerful, why do we need a trapdoor? Why can't the infinitely-powerful adversary invert a one-way function for the polynomially-bounded machine? The problem, intuitively, it that the infinitely-powerful adversary will then have full information on which inputs he helped the polynomially-bounded player to invert the one-way function. Thus, the question how the polynomially-bounded player can* hide *information from the infinitely-powerful adversary by using only a one-way function is "back to square one". The special game of bit commitment to an all-powerful adversary is a good example to consider in order to realize why the above naive approach is bound to fail.*

Our results allow us to reduce complexity assumptions for a variety of existing protocols (see section 4). In addition, we establish several impossibility results: we show that the two types of OT when the polynomial-time player is either the sender or the receiver, require, both, at least two rounds of interaction, and thus it is impossible to achieve a non-interactive OT. Moreover, we establish that even when dealing with an infinitely-powerful player, we must make complexity-theoretic assumptions. To summarize, we break the "trapdoor barrier" for asymmetric OT, we show a new technique which has both practical and theoretical implications and we establish several impossibility results.

1.1 Definitions

The model we consider for two-party protocols is the standard system of communicating machines [B2, GMR, Ba]. Oblivious Transfer (OT) is a two-party protocol introduced by Rabin [R]. Rabin's OT assumes that S possesses a value x, after the transfer R gets x with probability $\frac{1}{2}$ and it knows whether or not it got it (*equal-opportunity requirement*). A does not know whether B got the value (*oblivious-ness requirement*). A similar notion of 1-2-OT was introduced by [EGL]. In 1-2-OT, player S has two bits b_0 and b_1 and R has a selection bit i. After the transfer, R gets only b_i, while S does not know the value of i. Notice that this could be extended to strings: 1-2-string-OT is similar to 1-2-OT, but S has two strings, instead of two bits. In [Y2] the notion of Combined-OT (or secure circuit evaluation) was proposed: as a common input, S and R are given a poly-size circuit $C(\cdot,\cdot)$. S has a private input x, and R has a private input y. After the protocol, R gets $C(x,y)$, while S gets nothing. The notions of *OT, 1-2-OT, 1-2-string-OT* and *Combined-OT* were all shown to be equivalent to each other [C, BCR, K]. That is, given any one of these protocols, one can implement the other ones. Moreover, the reductions are information-theoretic. Thus, by "OT" we can refer to any one of them. We denote by (STRONG $\xrightarrow{\text{OT}}$ WEAK) an Oblivious Transfer protocol in which polynomially bounded WEAK player is the receiver R

420

(and STRONG player is an infinitely-powerful sender S) and by (WEAK $\xrightarrow{\text{OT}}$ STRONG) an Oblivious Transfer protocol in which polynomially-bounded WEAK player is the sender (while the STRONG player is the receiver).

1.2 Comparison with previous work

Rabin [R] defined the basic OT protocol and developed an implementation for honest parties based on the intractability of factoring. Fischer, Micali and Rackoff [FMR] presented an implementation of OT robust against cheaters, while Even, Goldreich and Lempel [EGL] showed how 1-2-OT implies OT. Relations between various notions of disclosure primitives have been investigated by [BCR, C]. They showed an information theoretic equivalences of the various notions. OT was also shown to be complete for (two- and multi-party) secure distributed circuit evaluation among polynomial-time players [K], and used to implement non-interactive and bounded-interaction zero-knowledge proof systems for NP by Kilian, Micali and Ostrovsky [KMO]. Yao [Y2] used OT to construct secure circuit evaluation (Combined - OT), based on factoring. OT based on the existence of any trapdoor permutation was presented by Goldreich, Micali and Wigderson [GMW]. Thus, secure circuit evaluation was made possible, assuming one-way trapdoor permutations [GMW]. Assuming certain specific algebraic trapdoor functions are secure (e.g., quadratic residues), the construction can be modified to work even if one of the players is infinitely-powerful; in this paper, we substantially reduce the required cryptographic assumption.

We must stress, that our results hold for the insecure communication environment. This should be contrasted with the work of [K, CK] where they assume right from the start that some form of OT already exists, or the non-cryptographic work which assumes that secure channels exist [BGW, CCD, RB]. Instead, we are interested in investigating the required complexity assumptions for achieving OT. Furthermore, we deal with two party games where secure channels do not help.

Since we consider the case when polynomially-bounded player must deal with an infinitely-powerful adversary, let us point out what was done previously, in various settings. In addition to zero-knowledge proof systems of [GMR], this model naturally represents interaction between a usual user and an all-powerful organization (a big brother) which may possess very large (or merely unknown) computational power. One such case is the context of zero-knowledge arguments of [BCC], which assume an all-powerful verifier from which information about the witness to the argument has to be hidden. The same holds for the model of using a powerful oracle to compute a value while keeping the real argument secret [AFK], and the setting of secure circuit computation while keeping one party secure in the information-theoretic sense [CDV, AF]. These works use the hardness of very specific algebraic functions as their underlying assumptions. Here, we reduce complexity assumptions for all these protocols, and we establish related impossibility results.

2 Protocols

2.1 OT with perfect security against a Strong sender

In this protocol an all-powerful STRONG player (the Sender) has a secret random input bit b, which he wants a polynomially-bounded WEAK player (the Receiver) to get only half of the time. The WEAK player does not want the STRONG player to know whether he has gotten the bit or not. We use standard notions of one-way functions and permutations, for example, see [GL, H, ILL]. For clarity, we first describe our OT based on a one-way permutation. The solution based on any one-way function is more complex and we present only the main ideas in this extended abstract.

Let f be a strong one-way permutation, $f : \{0,1\}^n \mapsto \{0,1\}^n$. Let the STRONG player be denoted as S (since he is the Sender in the OT protocol) and the WEAK player be denoted as R (for receiver). In the beginning of the protocol, S is a given a secret input bit b. $B(x,y)$ denotes the dot-product mod 2 of x and y, and all $h_i \in \{0,1\}^n$ are linearly independent. The following OT protocol implements a technique which can be described as gradually "focusing" on a value, while maintaining information-theoretic uncertainty.

$R(0)$: R selects x' at random and computes $x = f(x')$. He keeps both x' and x secret from S.

- **For i from 1 to $(n-1)$ do the following steps:**

  $S(i)$: S selects at random h_i and sends it to R.
  $R(i)$: R sends $c_i := B(h_i, x)$ to S.

$S(n)$: Let x_0, x_1 be the ones which satisfy $\forall i, 1 \le i < n, B(h_i, x_{\{0,1\}}) = c_i$. S flips a random coin j, selects a random string p, $|p| = n$ and sends to R a triple $< p, x_j, v >$, where $v = b \oplus B(p, f^{-1}(x_j))$.

$R(n)$: R checks if for his x, $x = x_j$, and if so, computes $b' = v \oplus B(p, x')$ as the resulting bit he gets from S via an "OT" protocol and outputs (x, b').

Theorem 1 *The above protocol implements OT from an all-powerful (at least probabilistic NP or stronger) player to a probabilistic polynomial-time player, based on any one-way permutation.*

Proof Sketch: Note that the sender S does not have any information weather the receiver R got the bit. To prove the other direction, assume that the above OT protocol is bad and R can predict S's input bit b with probability $\frac{1}{2} + \varepsilon$. Using such R we construct a predicting algorithm $I(y, p)$ which can in expected polynomial time predict $B(f^{-1}(y), p)$ with probability $\frac{1}{2} + \frac{\varepsilon}{2}$ for random p and $y \leftarrow f(x)$, thus contradicting that $B(x, p)$ is a hard-core bit [GL]. The program for $I(y, p)$ is as follows:

- For i from 1 to $(n-1)$ do:

$Step(i)$: Record current configuration of R. Randomly choose the vector h_i and send it to R. If $c_i \neq B(y, h_i)$, discard h_i, reset the R to a configuration before sending h_i and choose another random h_i and repeat until $c_i = B(y, h_i)$.

- At this point, there are exactly two y_0, y_1 which satisfy $\forall i, 1 \leq i < n, B(h_i, y_{\{0,1\}}) = c_i$. Moreover, $y \in \{y_0, y_1\}$. I selects $p \in \{0,1\}^n$ and $j, l \in \{0,1\}$ at random, and sends to R a triple $< p, x_j, l >$. If $x_j = x$ and R outputs b' then I outputs $l \oplus b'$ as the prediction of $B(x, p)$.

Note that the distribution of h_i's is random, since y is and R (for which y is unknown for all $i < n$) can not distinguish if it works in the OT protocol or in the above prediction algorithm. To see that this is so, notice that inductively, for any choice of $h_1, \ldots, h_j$, the distribution of y given $c_1, \ldots, c_j$ and $h_1, \ldots, h_j$ is close to uniform in the subspace defined by the constraints. $\square$

The above protocol works for permutations only and is simulatable (see [GMR]). In case of a general one-way function the size of the range as well as the degeneracy of a given point in the range (only their logarithms are relevant here) are unknown. Never-the-less, it is possible to implement OT based on *any* one-way function:

Theorem 2 *There exists an implementation of OT protocol from an all-powerful (at least probabilistic $P^{\#P}$ or stronger) player to a probabilistic polynomial-time player, given any one-way function.*

Proof sketch: The correctness of the protocol involves an inverting algorithm for the one-way function, which works by guessing the size of the range and the degeneracy of a given point, and "deceives" the "bad" poly-bounded player (receiver) by pruning its computations arising out of incorrect guesses, thereby making the distributions of the conversations indistinguishable for the receiver. Below, we list some of the essential ideas of the protocol:

- S tells R to stop when it isolates exactly two points x_1, x_2 in the range of f, and proves in zero-knowledge that this is the right stopping point.

- S and R restrict the sizes of $|f^{-1}(x_1)| = |f^{-1}(x_2)| = 1$ and S proves in Zero-Knowledge that he did not find out which x_i R holds. (The crucial point is that x_1 and x_2 are never mentioned in the "open" in the protocol — only as part of a Zero-Knowledge proof.

Suppose that R can cheat and guess the bit with probability $\frac{1}{2} + \varepsilon$. The inverting algorithm tries to run this R for all possible stopping points and the logs of the preimage sizes. If the guess was correct, the inverting algorithm can use R to try to invert f for different guesses and *verify* that this was the correct guess. Since there is only polynomial number of guesses, the inverting algorithm will succeed with probability $\frac{\frac{1}{2}+\varepsilon}{poly}$. $\square$

2.2 OT with perfect security against Receiver:
Duality Theorem for OT

In the previous section, we presented the OT scheme with perfect security against the Sender. Here, we present the dual notion of the perfect security of OT against the Receiver. However, instead of developing our protocol "from scratch" , we prove a more general result. In particular, we show that perfect security (for one of the players) can be *reversed:*

Theorem 3 *(Duality Theorem for OT): Given an OT protocol with perfect security for one of the players (i.e. either the Sender or the Receiver) there exist an implementation of OT protocol with perfect security for the other player.*

Proof Sketch: Here, we will show only one direction. The other is similar. We assume that the WEAK player has two bits b_0 and b_1 and he wishes to execute 1-2-OT to the STRONG player and that it is possible for the STRONG player to do (STRONG $\xrightarrow{\text{OT}}$ WEAK). From this assumption, the STRONG player can also *commit* bits by putting them in "envelopes" and can prove properties about this bits [K]. The STRONG player makes four envelopes named $e_1, \cdots, e_4$ satisfying the following: Let pairs $P_0 = \{e_1, e_2\}$ and $P_1 = \{e_3, e_4\}$. The contents of the envelopes in a pair P_b is identical while the contents of the envelopes in P_{1-b} are different for some $b \in \{0, 1\}$. Further there is a label $l(e_i) \in \{0, 1\}$ such that it is distinct for each envelope within a pair. Note that the STRONG player can convince the WEAK player that these properties are satisfied without revealing *any* other property of the envelopes. The WEAK player with random bits b_0, b_1 chooses using (1-2-string STRONG $\xrightarrow{\text{OT}}$ WEAK) the content c_0 of the envelope $e_i \in P_0$ with $l(e_i) = b_0$ and the content c_1 of $e_j \in P_1$ with $l(e_j) = b_1$. Then the WEAK player sends c_0, c_1 to the STRONG player. Note that in the above protocol, for the pair P_i containing equivalent bits, the STRONG player does not gain any information about the WEAK player selection bit. Hence, the bit that WEAK sends to STRONG is lost forever. On the other hand, unless the WEAK player can find out which pair contains equivalent bits, the WEAK player has no idea which bit the STRONG player got as the result of 1-2-OT. □

Remark: For many practical applications we need an OT protocol secure for the Sender. We note that by applying our duality theorem to any protocol which is used in practice and which is information-theoretically secure for the Receiver, (see [GMW], for example) we can now provide an efficient OT protocol which is information-theoretically secure for the Sender — under a general cryptographic assumption. A similar result, achieved independently, was reported to us by Crépeau [C-Pos].

3 Impossibility results

In this section, we provide several impossibility results. First, we show that OT is inherently interactive:

Theorem 4 *It is impossible to implement a non-interactive cryptographic O.T. protocol (and 1-2-O.T. protocol).*

Proof outline: Let us first start with both parties being polynomial-time machines. We apply the procedure for coin-flipping over the phone (due to Blum [B1]). This is specified as following. A and B are exchanging messages, at the end of the process they agree on a result which is either *head* or *tail* each with probability 1/2 (or close to 1/2, less than any inverse polynomial factor away); otherwise the procedure is unfair. We claim that a coin-flipping needs 3 messages.

Suppose A starts the protocol by sending M_1. Then, if one message is enough, in the message space from which this message was drawn, about half of the possible messages lead to *head* and the others to *tail*. By sampling off-line, A can make the protocol unfair. As a result, a fair coin-flipping cannot be executed with one message. Let us next look at the protocol's state after M_1 was sent. Assume it takes two messages. At this point B has to compute and send a message M_2, in his message space about half of the possible messages has to cause the result to be *head* regardless of the first message sent by A (since we assume A cannot bias the outcome). Since the message space is sampleable B can sample a message which gives the result it wants and therefore controls the outcome, a contradiction. Thus, at least three messages are needed. The arguments above do not hold in the third message to be sent by A which depends on the first message (computed by A) and the second message, the actual outcome may have been determined by both messages, and the third message just causes agreement.

Next to show that non-interactive 1-2-OT is impossible. We reduce coin flipping to the existence of 1-2-OT plus one additional message. This implies the result. A sends to B 1-2-OT two strings x and y, which are different, say they have different parity (addition mod 2 of their bits is different). On the first (essential) message of the O.T. protocol the party now also sends a guess of the outcome (0 or 1 parity). After the protocol ends, B convinces A of the result by transmitting the string (of size $k + 1$ say) which he can do either if he got the string or otherwise only by guessing it with probability smaller than $1/2^k$ (inverse exponential). The result computed by both parties is *head* if the guess-of-outcome is correct and *tail* otherwise. This is a coin-flipping. Thus the 1-2-O.T. needs at least two rounds (and cannot be done in one message in a non-interactive fashion). Similar result holds for plain O.T.

Now, for the all-powerful adversary case. R (S) is the powerful (polynomial) participant. Assume 1-2-OT from WEAK to STRONG exists. Then V sends one message based on its internal state which includes the two messages, from correctness, R should get one and only one of them with probability 1/2, but this is trivially impossible. Assume he gets one message, he can then use its power and simulate an internal state of itself which accepts the other message. When S sends to R in an 1-2-OT from STRONG to WEAK, it does not know the internal state of R which may determine the accepted message. However, it is again possible to rely formally on the reduction argument above. A coin-flip which starts by S moving first requires three rounds, as after the first round S can determine the outcome while after the second round R should not know the outcome as well (and about half of its messages should lead to *head*, and half to *tail*). Thus a third message is needed. However a one-message OT (with a random guess by S) followed by opening of the outcome by R gives a coin-flip, which gives the result by contradiction. $\square$

We give certain impossibilities and bounds on resource requirements of partial-information games to contrast with our upper bounds. We note that when both players are polynomially bounded, it was shown that OT protocol implies the existence of a one-way function [IL, BCG]. In contrast, we deal with the case when a poly-bounded player must interact with an infinitely-powerful adversary. Thus, we first ask if complexity assumptions are needed at all for OT. Next, we formally prove that:

Theorem 5 *It is impossible to implement an information-theoretic O.T. protocol.*

Proof Sketch: Let us start with Shamir, Rivest and Adleman's impossibility proof of Mental-Poker [SRA]. They considered the minimal non-trivial scenario of two unfaulty (but curious) players, A and B, who try to deal a hand of one card to each out of a deck of three cards $\{x, y, z\}$. (This scenario can be extended to larger decks and any size of hands). Let the protocol *execution* M be the finite sequence of transmitted messages. M should coordinate the cards drawn by each player (the cards drawn are a function of M and the player's internal computations). The hands should be drawn uniformly at random and be disjoint. They have shown, in a very elegant way, that this is impossible.

Next, we use this proof and the 1-out-of-k-OT construction (which is equivalent to OT) and translate a protocol for Mental-Poker to the existence of an O.T. protocol. Since we have O.T. protocol we can implement a 1-3-O.T. and 1-2-O.T., thus the following procedure will implement a card dealing as above. First A, using a random encoding of the deck (using, say two bit long strings), transfers using 1-out-of-3-O.T. one card to B, the properties of this execution imply that this card is random and it is secure to A. In the second step the encoding of cards is revealed (A decides which string represent what card). As a result B knows what card he has gotten, A does not gain any advantage in guessing the card. Next, B decides on a random encoding of the cards and he transfers one of the two cards which are not in his hand to A, using the 1-out-of-2-O.T. protocol. Then again, the encoding of cards is revealed. As a result, by the O.T. properties, A gets a random card different from B's card and B cannot tell what card was dealt. In addition A does not gain any advantage in knowing what card A has, and the two cards different than his are equally likely to be in B's hand. This procedure achieves a random, secret, and disjoint dealing. This proves that O.T. cannot be implemented perfectly. □

4 Applications

In [GMW] the notion of *(purely) playable* games is introduced. Intuitively, these are the games that can be played by the participants themselves, without invoking any trusted parties. In particular, they ask the following question: *Which partial-information games are (purely) playable?* The work of [Y2, GMW] answers this question for polynomial-time players. We show that:

Theorem 6 *If one-way functions exist, then any two-party polynomial game with an all-powerful adversary is playable.*

The proof uses the techniques of [Y2, GMW] and the availability of OT protocol (in both directions). The details will be given in the final version. The above theorem has many applications which will be given in the final version as well.

Interaction is a very expensive resource, and not always an available one. This motivates the study of how to bound the interaction required by a zero-knowledge proofs as much as possible [BFM, FLS, DY, KMO]. The issue of bounding interaction (to a preprocessing stage) for zero-knowledge proofs was addressed in [KMO]. The complexity assumption used in the efficient procedure of [KMO], required the existence of a *trapdoor permutation*. Utilizing theorem 2, we can eliminate the trapdoor property:

Theorem 7 *After the preprocessing stage consisting of $O(k)$ rounds of interaction between an infinitely-powerful player and a polynomially bounded player, any polynomial number of NP-theorems of any poly-size can be proved non-interactively and in zero-knowledge, based on the existence of any one-way permutation, so that the probability of accepting a false theorem is less than $\frac{1}{2^k}$.*

Our protocols provide us with a general methodology to force one of the players to choose two inputs, such that for one of inputs the player does not have an inverse of a one-way function, while keeping perfect security as for which input the player does not have this information. Using these techniques, in [OY] it is shown that any Statistical Zero-Knowledge protocol for an honest verifier can be compiled into a Statistical Zero-Knowledge protocol for any (even cheating) verifier, based on one-way permutation; this compilation is an important design tool and was originally based on discrete logarithm [BMO2]. In addition, the above technique was used in the proof of characterization of Instance-Hiding Zero-Knowledge proof systems [FO] and to implement strong bit-commitment [NOVY].

Acknowledgments

We would like to thank Mihir Bellare, Gilles Brassard, Claude Crépeau, Oded Goldreich, Shafi Goldwasser, Silvio Micali, Moni Naor, Shimon Even, and Noam Nisan for helpful discussions.

References

[AF]   M. Abadi and J. Feigenbaum. *Simple Protocol for Secure Circuit Computation* STACS 88.

[AFK]   M. Abadi, J. Feigenbaum and J. Kilian. *On Hiding Information from an Oracle* J. Compute. System Sci. 39 (1989) 21-50.

[B1]   Blum M., *Applications of Oblivious Transfer*, Unpublished manuscript.

427

[B2]   Blum, M., *Coin Flipping over the Telephone*, IEEE COMPCON 1982, pp. 133-137.

[BFM]   Blum M., P. Feldman, and S. Micali *Non-Interactive Zero-Knowledge Proof Systems*, STOC 89.

[Ba]   Babai L., *Trading Group Theory For Randomness*, STOC 86.

[BCC]   G. Brassard, D. Chaum and C. Crepeau, *Minimum Disclosure Proofs of Knowledge*, JCSS, v. 37, pp 156-189.

[BCR]   G. Brassard, C. Crépeau and J.-M. Robert, *Information Theoretic Reductions among Disclosure Problems*, FOCS 86 pp. 168-173.

[BCG]   M. Bellare L. Cowen, and S. Goldwasser *The Nature of Key-Exchange*, DIMACS proceedings, Workshop on Distributed Computing and Cryptography, 1991.

[BMR]   D. Beaver, S. Micali and P. Rogaway *The Round Complexity of Secure Protocols* STOC 90.

[BMO2]   Bellare, M., S. Micali and R. Ostrovsky, *The (True) Complexity of Statistical Zero Knowledge* STOC 90.

[BGW]   Ben-Or M., S. Goldwasser and A. Wigderson, *Completeness Theorem for Non-cryptographic Fault-tolerant Distributed Computing*, STOC 88, pp 1-10.

[CCD]   D. Chaum, C. Crepeau and I. Damgard, *Multiparty Unconditionally Secure Protocols*, STOC 88, pp 11-19.

[CDV]   D. Chaum, I. Damgard and J. van-de-Graaf, *Multiparty Computations Ensuring Privacy of each Party's Input and Correctness of the Result*, Crypto 87, pp 87-119.

[C]   C. Crépeau, *Equivalence between Two Flavors of Oblivious Transfer*, Crypto 87.

[C-Pos]   C. Crépeau, Personal communication at Sequences'91, Positano, Italy.

[CK]   C. Crépeau, J. Kilian *Achieving Oblivious Transfer Using Weakened Security Assumptions* , FOCS 88.

[DY]   A. DeSantis and M. Yung, *Cryptographic Applications of the Non-interactive Metaproof and Many-Prover Systems* Crypto 90.

[EGL]   S. Even, O. Goldreich and A. Lempel, *A Randomized Protocol for Signing Contracts*, CACM v. 28, 1985 pp. 637-647.

[FLS]   Feige, U., D. Lapidot and A. Shamir, *Multiple Non-Interactive Zero-Knoweldge Proofs Based on a Single Random String* FOCS 90.

[FO]   F. Feigenbaum and R. Ostrovsky *A Note On Characterization of Instance-Hiding Zero-Knowledge Proof Systems*, manuscript.

[FMR]   Fischer M., S. Micali, C. Rackoff *An Oblivious Transfer Protocol Equivalent to Factoring*, Manuscript.

[GL]   O. Goldreich and L. Levin, *Hard-core Predicate for ANY one-way function*, STOC 89.

[GMW]   O. Goldreich, S. Micali and A. Wigderson, *How to Play any Mental Game* , STOC 87.

[GMR]   S. Goldwasser, S. Micali and C. Rackoff, *The Knowledge Complexity of Interactive Proof-Systems*, STOC 85, pp. 291-304.

[H]   Hastad, J., *Pseudo-Random Generators under Uniform Assumptions*, STOC 90.

[IL]   R. Impagliazzo and M. Luby, *One-way Functions are Essential for Complexity-Based Cryptography* FOCS 89.

[ILL]   R. Impagliazzo, L. Levin, and M. Luby *Pseudo-Random Generation from One-Way Functions* STOC 89.

[IR]   R. Impagliazzo and S. Rudich, *On the Limitations of certain One-Way Permutations* , STOC 89.

[K]   J. Kilian, *Basing Cryptography on Oblivious Transfer* , STOC 1988 pp 20-31.

[KMO]   J. Kilian, S. Micali and R. Ostrovsky *Minimum-Resource Zero-Knowledge Proofs*, FOCS 1989.

[NOVY]   M. Naor, R. Ostrovsky, R. Venkatesan, M. Yung, manuscript in preparation.

[OY]   R. Ostrovsky and M. Yung, manuscript in preparation.

[R]   M., Rabin *How to Exchange Secrets by Oblivious Transfer* TR-81 Aiken Computation Laboratory, Harvard, 1981.

[RB]   T. Rabin and M. Ben-Or, *Verifiable Secret Sharing and Secure Protocols* , STOC 89.

[SRA]   A. Shamir, R. Rivest and L. Adleman, *Mental Poker*, Technical Memo MIT (1979).

[Y1]   A. C. Yao, *Theory and Applications of Trapdoor functions*, FOCS 82.

[Y2]   A. C. Yao, *How to Generate and Exchange Secrets*, FOCS 86.

AN ASYMPTOTIC CODING THEOREM
FOR AUTHENTICATION AND SECRECY

Andrea Sgarro

Dipartimento di Matematica e Informatica
Università di Udine
33100 UDINE (Italy)

and: Dipartimento di Scienze Matematiche
Università di Trieste
34100 TRIESTE (Italy)

Abstract. We give an overview of lower bounds for the impersonation attack in authentication theory. In the frame of a Shannon-theoretic model for authentication codes, we prove an asymptotic coding theorem for the mixed impersonation-decryption attack.

Authentication: a biassed overview.

A Shannon-theoretic frame for authentication theory has been put forward by G. Simmons; cf e.g. [1]. A *multicode* is a finite random triple XYZ (*message, codeword, key*). Under each key (encoding rule), encoding and decoding are assumed to be deterministic. Key and message are independent random variables. Below we give an example of an encoding matrix and of the corresponding authentication matrix χ; in the latter one has $\chi(z,y)=1$ iff key z authenticates codeword y, that is iff there exists a message x which is encoded to y under key z.

	x1	x2	x3			y1	y2	y3	y4
z1	y1	y2	y3		z1	1	1	1	0
z2	y3	y4	y1		z2	1	0	1	1

Examples of multicodes are source codes (Z has one value), ciphers (then Y is rather called the cryptogram), and *authentication codes*. Simmons prescribes a zero-error decoding scheme and so each codeword can appear at most once in each row of the encoding matrix.

Authentication theory begins for good when the attacks are described against which an authentication code must be resistant. *Impersonation attacks* are the simplest possible attacks against an authentication code. In this case the mischievous opposer chooses a codeword y hoping it to be authenticated by the current key Z, which he ignores. The probability of fraud (successful attack) for the enemy's optimal strategy is:

$$P_I = \max_y \; \mathrm{Prob}\{\chi(Z,y)=1\}$$

The most popular lower bound to P_I is *Simmons bound* which involves the mutual information $I(Y;Z)$ (in terms of Shannon entropies $I(Y;Z)$ is defined as $H(X)+H(Y)-H(XY)$):

$$P_I \geq 2^{-I(Y;Z)}$$

In [2] it has been shown that the second side is a bound also to a suboptimal random strategy called the "aping strategy", (which consists in sending a random codeword with the distribution of Y, but independently of Z); this given, it is not surprising that R. Johannesson and this writer were able to strenghten Simmons bound [3]; they obtained the following JS-bound:

$$P_I \geq 2^{-R(0)}$$

R(0) being the rate distortion function at zero distortion for source Z, primary alphabet equal to the key-alphabet, secondary alphabet equal to the codeword-alphabet, and distortion measure χ. (This intrusion of rate-distortion theory is rather surprising.) Unlike Simmons bound, the JS-bound implies the simple and well-known combinatorial bound

$$P_I \geq \frac{|X|}{|Y|}$$

Criteria for equality are the same both for the combinatorial bound and the JS-bound: $\mathrm{Prob}\{\chi(Z,y)=1\}$ must be independent of y [2] (instead, for Simmons' bound independence of y is only a necessary condition for equality).

More sophisticated attacks are *substitution* and *deception*, whose fraud probability are, respectively:

$$P_S = \Sigma_c \Pr\{Y=c\} \ \max_{y \neq c} \Pr\{\chi(Z,y)=1|Y=c\}$$
$$P_D = \max (P_I, P_S)$$

In the case of substitution, the opposer grabs the legal codeword c and replaces it by y; in the case of deception one has to beware both of impersonation and substitution. We are using Massey's definition of deception, which is possibly more palatable to information theorists; in Simmons' more involved game-theoretic setting the right side is just a lower bound to P_D; whichever one's preferences, lower bounds to P_I are *ipso facto* bounds to P_D.

In the literature Simmons bound is often taken as a coding theorem in authentication theory, mutual information playing the role of the "capacity" of the "authentication channel". Since Simmons bound has been strengthened one is tempted to shift that role to R(0). In [2] genuine Shannon-theoretic coding theorems for impersonation codes have been put forward. Rather disappointingly, asymptotically optimal code constructions were inspired by the "naive" combinatorial bound (actually, this is not so surprising: since the tight JS-bound holds with equality iff the combinatorial bound holds with equality, the rather primitive combinatorial bound is a *tight* bound for *good* codes!). In the terminology of information theory Simmons codes are zero-error codes; in [2] we took into examination also the case when a "small" probability of erroneous decoding is allowed: *it turned out that in this case the relevant information measure is good old Shannon entropy, rather than mutual information or rate-distortion functions.* Here we deepen and extend that research in the negligible-decoding-error case by covering the "mixed" attack decryption-impersonation:

$$P_M = \max (P_x, P_I), \text{ with } P_x = \Sigma_y \Pr\{Y=y\} \ \max_x \Pr\{X=x|Y=y\}$$

Other cryptanalytic attacks are key-disclosure given the cryptogram, or given the couple message-cryptogram:

$$P_z = \Sigma_y \Pr\{Y=y\} \ \max_z \Pr\{Z=z|Y=y\}$$
$$P_{z|xy} = \Sigma_{xy} \Pr\{X=x,Y=y\} \ \max_z \Pr\{Z=z|X=x,Y=y\}$$

This allows for a range of 31 pure and mixed attacks, and, hopefully, for more work to be done.

A coding theorem

In the negligible-error case we insist on deterministic decoding, *but we do not insist that it should be deterministically successful!* In other words (as is usual in the Shannon-theoretic approach) we allow for a "small" decoding error-probability P_e.

We formally define our model in the case the mixed attack decryption-impersonation; the changes to make in the other cases are obvious. $\mathcal{X}^n$, $\mathcal{Z}$ and $\mathcal{Y}$ are the message-alphabet, the key-alphabet and the codeword-alphabet, respectively. The random key Z is uniform and independent of the random message X^n of length n.

Let $f: \mathcal{Z} \times \mathcal{X}^n \to \mathcal{Y}$, $g: \mathcal{Z} \times \mathcal{Y} \to \mathcal{X}^n$ be a multi-code (an encoder-decoder pair), and let P_e, P_M be the corresponding probabilities of erroneous decoding and of successful decryption-impersonation attack. (We are no longer requiring that codewords appear at most once in each row of the encoding matrix; functions f and g are arbitrary). We denote by $R_z = n^{-1} \log |\mathcal{Z}|$, $R_y = n^{-1} \log |\mathcal{Y}|$ the key-rate and the codeword-rate, respectively; these parameters measure the "cost" of the code, while P_e, P_M measure its "reliability". We say that a rate-couple (x,y) in the non-negative real quadrant is ε-*achievable* against decryption and impersonation when for any τ ($\tau > 0$) and for all n large enough there are codes with $P_e \leq \varepsilon$, $P_M \leq \varepsilon$, $R_z \leq x + \tau$, $R_y \leq y + \tau$.

The following lemma is an obvious "trick" to convert a zero-error code (and the corresponding construction based, say, on combinatorial designs) into a small-error code; it can be used whenever the probability of erroneous decoding for the former is exponentially small. From now on we assume that the message source is stationary and memoryless ruled by the probability vector P, or rather by its product P^n (generalizations are feasible whenever suitable asymptotic equipartition properties are available). We shall use the standard notion of *typical sequences*: we shortly recall that a sequence (a message, in our case) is typical when the relative frequencies of its letters are "almost" equal to their probabilities; asymptotically, typical sequences become equiprobable, their rate becomes H(P), and their overall probability becomes 1; cf e.g. [4]. By $\mathcal{T}_n$ we denote the set of typical messages of length n. We say that a multicode is typical when the decoder always decodes to the unique typical message having that codeword under that key (this implies that a codeword y which appears at least once in a row of the encoding matrix appears exactly once in correspondence to typical messages).

Lemma. Let a typical multicode be given. Let X^* be a uniform random variable over T_n (messages outside T_n are dropped in the case of X^*); then, with self-explaining notation:

$$P_I = P^*_I \qquad P_S \leq \Pr\{X^n \notin T_n\} + 2^{+n\xi_n} P^*_S$$
$$P_X \leq \Pr\{X^n \notin T_n\} + 2^{+n\xi_n} P^*_X \qquad P_Z \leq \Pr\{X^n \notin T_n\} + 2^{+n\xi_n} P^*_Z$$
$$P_{Z|XY} \leq \Pr\{X^n \notin T_n\} + 2^{+n\xi_n} P^*_{Z|XY} \qquad \text{for suitable } \xi_n, \xi_n \to 0$$

(Proofs follow immediately from the properties of typical sequences).

We give below an example where this lemma is used to obtain the direct (constructive) part of a coding theorem. We stress that in the obvious two-step construction below the key Z and the message X^* are uniform random variables, and the decoding error probability is exactly zero.

Fix δ, $\delta > 0$. We write the key Z as a juxtaposition Z'T ($|Z'| = \lceil 2^{n\delta} \rceil$, $|T| = \lceil 2^{n\delta} \rceil$, $|Z| \cong 2^{n(2\delta)}$); Z' is the subkey for secrecy, T is the subkey for authentication; Z' and T are independent. Recall that $|X^*| = |T_n| \cong 2^{nH(P)}$. In the first step we write an encoding matrix for the subkey Z' by just putting Vigenère tables (modular addition tables; more generally, Latin squares) one on top of the other until the matrix is complete; (in general at the bottom of the matrix we have an incomplete table; actually, for $|Z'| < |X^*|$ we have nothing else but the incomplete table). In the second step each row of the old matrix is repeated $|T|$ times, one for each value of the suffix T; new codewords are obtained from the old ones by simply appending the value of T (the "authentication tag"). Since H(X|Z)=H(X|Z') the value of T is of no help to the decryptor and can as well be discarded by him; instead, the only concern of the impersonator are suffixes (authentication tags): the codeword is authenticated iff the tag is authenticated. With this construction, as easily seen, both P^*_I and P^*_X go exponentially to zero as n goes to infinity. The arbitrariness of δ and the lemma imply that (0,H(P)) is an achievable rate-couple for codes against impersonation and decryption.

434

The converse theorems is equally easy. Once a code with P_e, P_I, $P_x \leq \varepsilon$ is given, one can construct a normal source-code for the joint source ZX^n by appending the value of Z to the old codewords (preliminarily, re-write the values of Z as n-tuples). Then the condition $P_e \leq \varepsilon$ *by itself* allows one to apply the converse part of Shannon's source coding theorem, which, asymptotically, gives for the rate $R_y + R_z$ of the source-code: $R_y + R_z \geq H(P) + H(Z)$, that is $R_y \geq H(P)$ (details are soon filled in).

Taken together, the direct and the converse part yield the following:

Theorem. The region of ε-achievable rate-couples for codes against decryption and impersonation is the quadrant

$$\{R_z, R_y : R_z \geq 0, R_y \geq H(P)\}$$

Remark 1. The theorem implies that the best (most economical) achievable rate-couple is $(0, H(P))$. Note that, even if $R_z = 0$ is achievable, ours is not a zero-keyrate scheme: the achievability of 0 simply means that exponentially many keys will do, however slow the exponential growth.

Remark 2. More to the point with respect to our earlier proposal in [2], we have modified our formalization so as to include key material into the available resources, as we did and we do for codeword material. Imposing that the key be uniformly distributed is not really restrictive in an asymptotic negligible-error scheme, since, for large n's, this situation would approximately hold anyway, provided only that the random key generator be sufficiently regular (memoryless and stationary, say) so as to have asymptotic equipartition properties.

Remark 3. The region given in the theorem is the achievable region also for non-secret codes resistant against impersonation (this was the contents of our former theorem, which, however, used a less perspicuous construction). Symmetrically, it is also the achievable region for codes against decryption without authentication. Actually, the conditions $P_x \leq \varepsilon$, $P_I \leq \varepsilon$ were never used to prove the converse. In the case of impersonation alone, one can simplify our two-step construction by taking Z'=constant and using a single source code for typical messages in the first step. So, what we are doing to achieve resistance against impersonation is simply appending exponentially many ($2^{n\delta}$, δ arbitrarily small) authentication tags to a cipher, or to a source code, respectively. This policy turns out to be asymptotically optimal: subtler constructions, however, are required for finite-length optimality.

Remark 4. Even if the results obtained so far are few and easy (e.g. speed of convergence is not taken into account), a genuine Shannon-theoretic construction of coding theorems is now in the doing. Open problems concern not only the remaining types of attack, but also questions like: is splitting, i.e. probabilistic encoding, asymptotically useful? (In the case of impersonation it is easy to see that splitting never helps; instead, it is well-known that, at least for finite lengths, splitting does help in the case of substitution attacks; the merits of homophonic, i.e. probabilistic, ciphers are also well-known). Does variable-length coding offer advantages?

References.

[1] G. Simmons, "A survey of information authentication", Proceedings of the IEEE, may 1988, 603-620

[2] A. Sgarro, "A Shannon-theoretic coding theorem in authentication theory", Eurocode 90, Udine 1990, to be printed by Springer Verlag in Lecture Notes in Computer Science

[3] R. Johannesson, A. Sgarro, "Strengthening Simmons' bound on impersonation", to be published by IEEE Transactions on Information Theory

[4] I. Csiszár, J. Körner, *Information Theory*, Academic Press, 1981

Automata and Combinatorics on Words

GRAY CODES AND STRONGLY SQUARE-FREE STRINGS

L.J. Cummings
University of Waterloo

Abstract

A *binary Gray code* is a circular list of all 2^n binary strings of length n such that there is a single bit change between adjacent strings. The *coordinate sequence* of a binary Gray code of order n is viewed as a circular string of 2^n integers chosen from $\{1,\ldots,n\}$ so that adjacent strings of the Gray code differ in that bit determined by the corresponding integer in the coordinate sequence. A *deleted coordinate sequence* of a binary Gray code is a linear sequence obtained from the coordinate sequence by deletion of exactly one integer.

A string is *strongly square-free* if it contains no "Abelian square"; i.e., a factor adjacent to a permutation of itself. Every deleted coordinate sequence of a Gray code is a strongly square-free string of length $2^n - 1$ over $\{1,\ldots,n\}$. We further show that Gray codes equivalent to the reflected binary Gray code are characterized by having only deleted coordinate sequences which are maximal strongly square-free strings.

1 Introduction

Definition 1 *A* binary Gray code of order n *is a circular list of all 2^n binary strings of length n such that there is a single bit change between adjacent strings.*

The name "Gray code" is used because a research physicist, Frank Gray, at Bell Laboratories used the lists now called Gray codes in U.S. Patent 2632058 in 1953. The application was error correction in signal transmission by pulse code modulation. Actually, such listings of binary strings are very old. Binary Gray codes yield solutions to the Chinese ring puzzle which is mentioned in Gerolamo Cardano's *De Subtilitate Rerum* in 1550. The puzzle has been known in China since at least the third century A.D.

Definition 2 *The* coordinate sequence *of a binary Gray code of order n is a circular string* $s = a[1] \cdots a[2^n]$ *of integers chosen from $\{1,\ldots,n\}$ so that adjacent strings differ in that bit whose index is the corresponding integer in the coordinate sequence.*

Given any non-empty sequence of integers from $\{1, \ldots, n\}$, a list of binary n-strings is uniquely determined by starting with the constant string of all 0's and changing the entries from 0 to 1 or 1 to 0 according to the integers in the given sequence.

Definition 3 *A deleted coordinate sequence of a binary Gray code is a string obtained from the coordinate sequence of the code by deletion of exactly one integer. The $\underline{ith}$ deleted coordinate sequence of $\mathbf{s} = a[1] \cdots a[2^n]$ is*

$$\hat{\mathbf{s}}_i = a[i+1] \cdots a[2^n]a[1] \cdots a[i-1]$$

where $a[0]$ is empty and $a[2^n + 1] = a[1]$.

Every binary Gray code of order n has at most 2^n distinct deleted coordinate sequences each of length $2^n - 1$.

Definition 4 *The n-cube is the graph whose vertices are the binary strings of length n. Two vertices are adjacent if they differ in exactly one bit. The n-cube is often denoted by Q_n.*

Since the binary strings of length n are vertices of the n-cube, a binary Gray code of order n is a Hamilton cycle in the n-cube; i.e., a path in Q_n which visits each vertex once and only once. The problem of enumerating all Hamilton cycles of Q_n is still unresolved but both upper and lower bounds have been studied for some time [7].

Since the columns of a binary Gray code can be permuted arbitrarily and the rows cyclically, the symmetry group of Q_n is the hyperoctahedral group $C_{2^n} \times S_n$ of order $2^n n!$. The symmetries of the hyperoctahedral group determine an equivalence relation on Q_n and we consider two Gray codes equivalent if one can be obtained from the other by a permutation of the columns or a cyclic permutation of the rows. Note that complementation of the entries of each string by $\bar{0} = 1$ and $\bar{1} = 0$ does not change the coordinate sequence.

E.N. Gilbert [5] gave perhaps the most important characterization of the coordinate sequences of Gray codes: A sequence $a[1], \ldots, a[2^n]$ is the coordinate sequence of a cycle in Q_n if and only if each of its proper substrings contains at least one integer from $\{1, \ldots, n\}$ an odd number of times while $a[1], \ldots, a[2^n]$ itself contains every integer from $\{1, \ldots, n\}$ an even number of times.

In [5] Gilbert included a complete classification of the coordinate sequences for distinct types of cycles in Q_4. He also gave a simple construction of a Hamilton cycle of the n-cube for every $n \geq 1$ as follows: define the coordinate sequence $\tau(k)$ for $k = 1, \ldots, 2^n$ by

$$\tau(k) = \max_{1 \leq d \leq n} \{d : 2^{d-1} \mid k\}. \tag{1}$$

For example, when $n = 4$ we obtain the coordinate sequence

$$1213121412131214.$$

The Gray codes of order n with coordinate sequences (1) are the reflected binary Gray codes, $B_2(n)$. Coordinate sequences of the codes $B_2(n)$ can also be generated by the *rule of succession*. Starting with the n-string of all 0's, this well-known procedure yields the next integer in the coordinate sequence given the previous integer and the current string.

```
procedure succession
  var
    string: array[1..n] of Boolean;
    k,entry: integer;
    begin
      if not odd(k) then
        entry :=1
      else
        entry :=1 + first1(string);
    end;
```

where first1 is a function which returns the index of the first 1 in string.

2 Strongly Square-Free Strings

Definition 5 *A string which consists of two consecutive equal adjacent substrings is called a "square". A string without any substrings which are squares is said to be* square-free.

It has been shown by Ross and Winkleman [9] that over any alphabet of at least three symbols, the set of strings containing "squares" is not context-free.

Definition 6 *A non-empty factor of the form*

$$BB^\sigma = b_1 \cdots b_k b_{\sigma(1)} \cdots b_{\sigma(k)}$$

where σ is a permutation of $\{1, \ldots, k\}$ is called an Abelian square. *A string is* strongly square-free *if it contains no Abelian squares.*

Clearly every strongly square-free string is square-free. Main [6] has shown that for every alphabet with at least 16 elements the set of strings which contain Abelian squares is not context-free.

Proposition 1 *If ω is a strongly square-free string over an alphabet A and $a \in A$ does not appear in ω then $\omega a \omega$ is strongly square-free.*

Proof. Suppose $\omega a \omega$ contains an Abelian square BB^σ. Then, BB^σ cannot be a subword of ω since ω is strongly square-free. Therefore, either a occurs in B or a occurs in B^σ. But since B^σ is a permutation of B, a must appear in *both* B and B^σ, contradicting the implicit assumption that a appears only once in $\omega a \omega$. $\square$

Obviously, Proposition 1 can be generalized considerably, but it is sufficient for our purposes here.

Theorem 1 *Every deleted coordinate sequence of a binary Gray code of order n is a strongly square-free string of length $2^n - 1$ on $\{1, \ldots, n\}$.*

441

Proof. Deleting one "change digit" of the coordinate sequence of an order n Gray code leaves a linear string of length $2^n - 1$. Assume that one such deletion contains an Abelian square

$$BB^\sigma = b_1 \cdots b_k b_{\sigma(1)} \cdots b_{\sigma(k)}.$$

If b_1 is the "change digit" which changes string i into string $i+1$ then after 2k successive changes $b_{\sigma(k)}$ changes string $i + 2k - 1$ back into string i. That is, the strings in position i and $i + 2k$ are identical, contradicting Definition 1. This follows because the bit change effected by any b_i in any binary string is "undone" by the bit change indicated by $b_{\sigma(m)}$ where $\sigma(m) = i$. □

More generally, if any integer sequence from $\{1, \cdots, n\}$ contains an Abelian square BB^σ, where σ is a permutation of $\{1, \ldots, |B|\}$ then regardless of the initial string, the change effected by BB^σ results in the same string before BB^σ was applied. This follows because the strings are binary and each change complements one of the entries. Since one permutation follows another without intervening entries, each string position is changed back to its initial setting.

The square-free sequence 32131232 cannot be a coordinate sequence of any binary Gray code with strings of length 3 since, regardless of the initial string this coordinate sequence yields the initial string after only 6 bit changes. Note that 32131232 begins with an Abelian square.

Dekking [2] has shown that there exist infinite sequences on two symbols in which no four blocks appear together which are permutations of one another. We mention that it is an open question whether there exists an infinite string on 4 symbols which is strongly square-free [3]. Dekking [2] has referred to this question as another four-color problem.

3   The Reflected Binary Gray Code

The most familiar Gray codes are the reflected binary codes which are easily generated recursively by hand. Given the code of order n, immediately below it copy its mirror image. Then extend each string on the right, say, in the original copy by 0 and each string in the mirror image by 1. Starting with the column which has 0 and 1 as entries, a reflected Gray code is generated for every order n.

As already noted, the coordinate sequences for the reflected binary Gray codes can be generated by (1). For our purposes it is convenient to give yet another definition of these coordinate sequences.

For each integer n define the coordinate sequence $\delta(n)$ recursively by:

$$\begin{aligned} \delta'(1) &= 1 \\ \delta'(k+1) &= \delta'(k)(k+1)\delta'(k) \quad for \; k = 1, ..., n-1 \\ \delta(n) &= \delta'(n)n. \end{aligned} \tag{2}$$

442

Here we assume without explicit mention that $\delta(n)$ is a circular string. By induction, the length of $\delta(n)$ is 2^n. Obviously, the definition of $\delta(n)$ shows it is the square, $(\delta'(n-1)n)^2$.

Definition 7 *The reflected binary Gray code of order n, $B_2(n)$, is the Gray code determined by $\delta(n)$ starting with the constant string of all 0's.*

Proposition 2 *For each $n \geq 1, \delta'(n)$ is a strongly square-free string.*

Proof. If $n = 1$ then $\delta'(1) = 1$ is strongly square-free. Assume that $\delta'(n-1)$ is strongly square-free. By definition,

$$\delta'(n) = \delta'(n-1)n\delta'(n-1).$$

Since n does not appear in $\delta'(n-1)$, it follows that any Abelian square in $\delta'(n)$ must be contained in $\delta'(n-1)$, contrary to the induction hypothesis. $\square$

Alternatively, this result can be seen to follow from Proposition 1.

4 Maximal Strongly Square-Free Strings

Definition 8 *A finite string ω over an alphabet A is a* maximal *strongly square-free string if for every $a \in A$, both $a\omega$ and ωa contain Abelian squares.*

Note that $\delta'(n)$ is maximal over $\{1, \ldots, n\}$ but not over $\{1, \ldots, n, n+1\}$ since it is easily extended by $n+1$ in either direction over the larger alphabet.

Although square-free implies strongly square-free, a maximal strongly square-free string need not be a maximal square-free string. A simple example is the string 2131213. It was shown in [4] that the maximal square-free strings s_n on n symbols are obtained recursively by

$$\begin{aligned} s_1 &= \pi(1) \\ s_n &= s_{n-1}\pi(n)s_{n-1} \end{aligned} \tag{3}$$

where π is a permutation of $\{1, \ldots, n\}$.

We remark that if π is the identity then

$$s_n = \delta'(n),$$

is a deleted coordinate sequence of the reflected binary code defined in (2). Since $\delta(n)$ is a square, the following observation applies:

Proposition 3 *If $c = (c_1 \cdots c_n)^2$ is a circular square then its deleted sequences, $\hat{c}_i$, starting at c_1 satisfy*

$$\hat{c}_i = \widehat{c_{i+n}}.$$

Lemma 1 *Every deleted coordinate sequence of $\delta(n)$ is a maximal strongly square-free string.*

Proof. Let $\widehat{\delta_i}$ denote the *ith* deleted coordinate sequence of $\delta(n) = \delta'(n)n$ starting with the first bit of $\delta'(n)n$. Since $\delta(n)$ is a circular square, according to Proposition 3 we have

$$\widehat{\delta_i} = \widehat{\delta_{i+2^{n-1}}}.$$

Thus, to show that each deleted coordinate sequence of $\delta(n)$ is a maximal strongly square-free sequence, we need only consider the strings $a\widehat{\delta_i}$ for $i = 1, \ldots, 2^{n-1}$.

Since $\delta(n)$ is the coordinate sequence of the reflected binary Gray code, each of its deleted coordinate sequences is strongly square-free by Theorem 1.

If $i = 2^{n-1}$, then $\widehat{\delta_i}$ is precisely $\delta'(n)$ which was seen to be strongly square-free in Proposition (2). That $\delta'(n)$ is maximal follows from [4] or directly by an easy inductive argument using Proposition 1.

For $1 \le i < 2^n$ we can write

$$\widehat{\delta_i} = sn\delta'(n-1)np \tag{4}$$

where s is a suffix and p is a prefix of $\delta'(n-1)$ such that $\widehat{\delta_i} = pxs$, where x is the deleted integer.

We consider first two special cases:

If $a = x$ then

$$
\begin{aligned}
a\widehat{\delta_i} &= xsn\delta'(n-1)np \\
&= (xsnp)^2
\end{aligned}
$$

and, of course, any square is an Abelian square.

If $a = n$ then

$$
\begin{aligned}
a\widehat{\delta_i} &= nsn\delta'(n-1)np \\
&= nsnp's'np
\end{aligned}
$$

where p' is that prefix of $\delta'(n-1)$ which is a palidrome of s. Since np' is a permutation of ns, $a\widehat{\delta_i}$ has an Abelian square for each $i = 1, \ldots, 2^{n-1}$ in this case.

In the following we assume that $a \ne x, n$. The proof now requires two further cases:

If $x = 1$ then $a \ne 1$. Thus, each $a\widehat{\delta_i}$ begins with an integer greater than 1. Set $\delta(n) = s[1]s[2] \cdots s[2^n]$. Define

$$f(\widehat{\delta_i}) = (s[2]-1)(s[4]-1)\cdots(s[2^n]-1), \tag{5}$$

where subscripts are taken modulo 2^n when necessary. Equivalently, f deletes the 1's from $\widehat{\delta_i}$ and reduces the remaining integers by 1. It follows from (1) that (5) is $\delta(n-1)$ or a circular

444

permutation of it written linearly. Since $a \neq x, n$ we see that $af(\widehat{\delta_i}) = a\delta(n-1)$ has an Abelian square. Reinserting the 1's and increasing the other integers by 1 yields an Abelian square in $a\widehat{\delta_i}$.

The situation is slightly different if $x > 1$. In this case each $\widehat{\delta_i}$ begins with 1 and (5) deletes an extra 1 leaving a string of length $2^{n-1} - 1$. It now follows that $f(\widehat{\delta_i})$ is a deleted coordinate sequence $\widehat{\delta_i}(n-1)$. By induction, $a\widehat{\delta_i}(n-1)$ has an Abelian square. Reinserting the 1's and increasing the other integers by 1 in $a\widehat{\delta_i}$ yields an Abelian square. $\square$

Theorem 2 *A Gray code Γ of order n is equivalent to $B_2(n)$ if and only if every deleted coordinate sequence of Γ is a maximal strongly square-free string over the alphabet of n symbols.*

Proof. If a Gray code Γ is equivalent to $B_2(n)$ then Γ can be obtained from $B_2(n)$ by permuting its columns and/or a cyclic permutation of the rows. The coordinate sequence $g[1] \cdots g[2^n]$ of Γ is a circular string and is unchanged by a cyclic permutation of the rows of Γ. A permutation σ of the columns of Γ induces the permutation $\sigma(g[1]) \ldots \sigma(g[2^n])$ of the coordinate sequence of Γ. Therefore,

$$\delta(n) = \sigma(g[1]) \cdots \sigma(g[2^n]),$$

for some permutation σ of $\{1, \ldots, n\}$. It follows that any deleted coordinate sequence of $g[1] \cdots g[2^n]$ has the same recursive structure as any deleted coordinate sequence of $\delta'(n)$ in (2). Hence, the proof of Lemma 1 applies.

Conversely, suppose every deleted coordinate sequence of a Gray code Γ is a maximal strongly square-free string and, in particular, a maximal square-free string. Therefore, any deleted coordinate sequence of Γ is of the form s_n as in (3). coordinate sequence of a Gray code has length $2^n - 1$.

Any permutation of $\{1, \ldots n\}$ changes $\delta'(n)$ to a deleted coordinate sequence of a Gray code equivalent to $B_2(n)$. $\square$

References

[1] B. Arazi, *An approach for generating different types of Gray codes*, Information and Control **63**(1984), 1-10.

[2] F. M. Dekking, *Strongly non-repetitive sequences and progression-free sets*, J. Combinatorial Theory, **27**(1979), 181-185.

[3] T.C. Brown, *Is there a sequence on four symbols in which no two adjacent segments are permutations of one another?*, Amer. Math. Monthly **78**(1971), 886-888.

[4] L.J. Cummings, *On the construction of Thue sequences*, Proc. 9 th S-E Conf. on Combinatorics, Graph Theory, and Computing, pp. 235-242.

[5] E.N. Gilbert, *Gray codes and paths on the n-cube*, Bell System Technical Journal **37**(1958), 815-826.

[6] M.G. Main, *Permutations are not context-free: an application of the 'Interchange Lemma'*,Information Processing Letters **15**(1982), 68-71.

[7] M. Mollard, *Un nouvel encadrement du nombre de cycles Hamiltoniens du n-cube*, European J. Combinatorics, **9**(1988), 49-52.

[8] H. Prodinger, *Non-repetitive sequences and Gray code*, Discrete Mathematics **43**(1983), 113-116.

[9] R. Ross, K. Winkelmann, *Repetitive strings are not context-free*, RAIRO Informatique Théorique **16**(1982), 191-199.

A NEW UNAVOIDABLE REGULARITY IN FREE MONOIDS

Aldo de Luca

Dipartimento di Matematica , Università di Roma " La Sapienza"
e Istituto di Cibernetica del CNR, Arco Felice, Napoli

and

Stefano Varricchio

Dipartimento di Matematica Pura ed Applicata, Università degli Studi dell'Aquila
e L.I.T.P., Institut Blaise Pascal, Université Paris 6

1. Introduction and preliminaries

The study of "unavoidable regularities" of very long words over a finite alphabet is a subject of great importance in combinatorics on words both for interest of argument itself and for the many applications in algebra and theoretical computer science.

A famous theorem, discovered by A. I. Shirshov in 1957, states that for any positive integers n and p every sufficiently long word contains a subword which is either a p-power of a non-empty word or n-divisible. A. Restivo and C.Reutenauer [11] proved in 1984 , as an application of the Shirshov theorem, a finiteness condition for semigroups of interest for the Burnside problem.

Another interesting unavoidable regularity may be expressed using bi-ideal sequences, i.e. sequences of words such that each term is both a prefix and a suffix of the next term. In [3] M.Coudrain and M.P.Schützenberger proved that for any positive integer n all sufficiently long words contain a subword which is the n-term of a bi-ideal sequence. As a consequence of that they gave an interesting finiteness condition for finitely generated semigroups.

In this paper we prove some combinatorial properties of uniformly recurrent infinite words which can be expressed in terms of bi-ideal and n-divided sequences.The main result of the paper is the following proposition: for any positive integers n and p , every sufficiently large word in a finite alphabet has a subword which is a p-power or a sesquipower of order n whose canonical factorization is an n-divided sequence. This theorem generalizes both the Shirshov theorem and the theorem of Coudrain and Schützenberger. A proof of this result is in [6] where some further unavoidable regularities are considered. In this paper we shall prove the main result in a more constructive way.

447

Let A be a finite non-empty set , or *alphabet*, and A⁺ (resp. A*) the *free semigroup* (resp. *free monoid*) over A. The elements of A are called *letters* and those of A* *words* . The identity elements of A* is called *empty word* and denoted by Λ. For any word w , |w| denotes its *length* . A word u is a *factor* (or *subword*) of the word w if w∈ A*uA*. For any w ∈ A* , F(w) denotes the set of all its factors. A word w ∈ A* contains as a factor a p-power , p>0, if there exists a word u ∈ A⁺ such that $u^p \in F(w)$. A language L in the alphabet A is any subset of A* . By F(L) we denote the set of all factors of the words of L , i.e. F(L) = ∪ $_{w \in L}$ F(w) . A language is *closed by factors* if L = F(L).

If the alphabet A is totally ordered then one can extend this order to A* by the so-called *lexicographic order* <$_A$, or simply < , defined as: for all u,v ∈ A* , u < v if and only if v ∈ uA* or u = h x ξ , v =h y η , h,ξ,η ∈ A* , x, y ∈ A and x<y. In the following for any positive integer n , $\mathcal{S}_n$ will denote the symmetric group on n objects.

2. Results

In the first part of the section we analyze some combinatorial properties of bi-ideal sequences . Subsequently we prove the existence of an unavoidable regularity which appears in uniformly recurrent infinite words ; from this we derive a theorem which is a generalization of the theorems of Shirshov and Coudrain-Schützenberger.

Definition 2.1. A finite sequence $f_1, ..., f_n$ of words of A* is called a *bi-ideal sequence of order* n if $f_1 \in A^+$, $f_i \in f_{i-1}A^*f_{i-1}$, for i ∈ [2, n]. For any i ∈ [2, n] we denote by g_{i-1} the word such that $f_i = f_{i-1}g_{i-1}f_{i-1}$. The n-th term f_n of a bi-ideal sequence is also called a *sesquipower of order* n .

Let $\{f_i\}_{i=1,...,n}$ be a bi-ideal sequence of order n. We set

$$w_n = f_1$$

and

$$w_{n-i} = f_i g_i , \quad i \in [1, n - 1].$$

From Definition 2.1 for all i ∈ [1, n-1], $f_{i+1} = f_i g_i f_i = w_{n-i} f_i$, so that , by iteration, one has:

$$f_{i+1} = w_{n-i} ... w_n \quad , \text{for all } i \in [0, n-1] \tag{1.1}.$$

Moreover , since $w_i = f_{n-i} g_{n-i}$, i ∈ [1, n - 1] then from Eq.(1.1) one has

$$w_i = w_{i+1} ... w_n g_{n-i} \in w_{i+1} ... w_n A^* \tag{1.2}$$

From Eq. (1.1), $f_n = w_1 w_2 ... w_n$. The n-tuple ($w_1, w_2, ... ,w_n$) is called the *canonical factorization* of f_n.

Conversely, one easily verifies that if $(w_1, w_2, \dots, w_n)$ is an n-tuple of words such that $w_n \neq \Lambda$ and for any $i \in [1, n-1]$, $w_i \in w_{i+1} \dots w_n A^*$ then the sequence of words $f_i = w_{n-i+1} \dots w_n$, $1 \leq i \leq n$, is a bi-ideal sequence of order n whose canonical factorization is just $(w_1, w_2, \dots, w_n)$.

Definition 2.2. A sequence $(u_1, u_2, \dots, u_n)$ of n non empty words of A^* is called a *maximal n-division* of the word $u = u_1 u_2 \dots u_n$ if for any non-trivial permutation σ of $\mathscr{S}_n$ one has $u_1 u_2 \dots u_n > u_{\sigma(1)} u_{\sigma(2)} \dots u_{\sigma(n)}$. If $(u_1, u_2, \dots, u_n)$ is a maximal n-division of u we simply say that $(u_1, u_2, \dots, u_n)$ is *n-divided*. A word u of A^* is called *n-divisible* if u can be factorized as $u = u_1 u_2 \dots u_n$, with $u_i \in A^+$ $(i=1,\dots, n)$ and $(u_1, u_2, \dots, u_n)$ is n-divided.

Proposition 2.1. Let $(w_1, \dots, w_n)$ be the canonical factorization of a sesquipower of order n. The sequence $(w_1, \dots, w_n)$ is n-divided if and only if for all $i \in [1, n-1]$, $w_{i+1} w_i < w_i w_{i+1}$.

Proof. $(\Rightarrow)$. Let us suppose that the sequence $(w_1, \dots, w_n)$ is n-divided and prove that for any $i \in [1, n-1]$ $w_{i+1} w_i < w_i w_{i+1}$. Assume, by contradiction, that an integer $i \in [1, n-1]$ exists for which $w_{i+1} w_i \geq w_i w_{i+1}$. This implies that $w_1 \dots w_{i-1} w_i w_{i+1} \leq w_1 \dots w_{i-1} w_{i+1} w_i$. Moreover since $|w_i w_{i+1}| = |w_{i+1} w_i|$ from a property of the lexicographic order (cf.[10]), it follows that $w_1 \dots w_{i-1} w_i w_{i+1} w_{i+2} \dots w_n \leq w_1 \dots w_{i-1} w_{i+1} w_i w_{i+2} \dots w_n$ which is a contradiction since $(w_1, \dots, w_n)$ is n-divided.

$(\Leftarrow)$ We begin by proving that $w_j w_i < w_i w_j$ for any i, j with $1 \leq i < j \leq n$. If $i = j-1$ the result follows from the hypotheses. Then let us suppose that $i < j-1$. We can write

$$w_i \in w_{j-1} w_j \dots w_n A^*$$

and

$$w_i w_j = w_{j-1} w_j \mu, \ \mu \in A^*.$$

By hypothesis one has

$$w_j w_{j-1} < w_{j-1} w_j \ ;$$

since $i \leq j-2$, we can write

$$w_i = w_{j-1} \lambda, \ \lambda \in A^*.$$

Being $|w_j w_{j-1}| = |w_{j-1} w_j|$ and $w_j w_{j-1} < w_{j-1} w_j$, for a well known property of the lexicographic ordering, one has:

$$w_j w_i = w_j w_{j-1} \lambda < w_{j-1} w_j \mu = w_i w_j .$$

Now we have to prove that for any non-trivial permutation $\sigma \in \mathscr{S}_n$ one has $w_{\sigma(1)} w_{\sigma(2)} \dots w_{\sigma(n)} < w_1 w_2 \dots w_n$. Let us observe that it is possible to reorder the n-block sequence $w_{\sigma(1)} w_{\sigma(2)} \dots w_{\sigma(n)}$ up to $w_1 w_2 \dots w_n$ by a "sorting" algorithm which consists in a finite number of transpositions of

449

consecutive words w_j, w_i with $i < j$. In such a way one can easily reach the result by observing that if $w_j w_i < w_i w_j$ then for any u, v ∈ A* one has $uw_j w_i v < uw_i w_j v$. Q.E.D.

Now we prove the existence of an "unavoidable regularity" which appears in uniformly recurrent words. This regularity includes , as particular cases, the bi-ideal sequences of Coudrain and Schützenberger [3], and the n-divisible words considered by Shirshov in [12, cf. also 10]. To this end we need some preliminary definitions and notations.

Let **Z** be the set of the integers. A two-sided infinite (or bi-infinite) word w in the alphabet A is any map w : **Z** → A . For each n ∈ **Z**, we set $w_n = w(n)$. A word u∈ A⁺ is a finite factor of w if there exist integers i,j ∈ **Z** , i ≤ j , such that u = w_i ....w_j ; the sequence w[i , j] = w_i ....w_j is also called an "occurrence" of u in w . The set of all finite factors of w will be denoted by F(w). The set of all two-sided infinite words over A will be denoted by $A^{\pm\omega}$. For w ∈ $A^{\pm\omega}$, alph(w) will denote the set of all letters of A occurring in w.

Let **N₊** be the set of positive integers. A one-sided (from left to right) infinite word is any map w : **N₊** → A . For each n > 0 the factor w[1, n] = w_1 ....w_n of length n is called the *prefix* of w of length n and will be simply denoted by w[n] . The set of all infinite words w : **N₊** → A will be denoted by A^ω . If w ∈ $A^{\pm\omega}$ one can associate to it the one-sided infinite word w_+ ∈ A^ω defined for all n>0 as $w_+(n) = w(n)$; one has that $F(w_+) \subseteq F(w)$.

The following lemma is a slight generalization of the famous König's lemma. It can be proved by a standard technique based essentially on the "pigeon-hole principle" (cf.[10], Chap.2):

Lemma 2.2. Let L ⊆ A* be an infinite language. Then there exists an infinite word w ∈ $A^{\pm\omega}$ such that F(w) ⊆ F(L).

Definition 2.3. An infinite word w∈ $A^{\pm\omega}$ (resp.w∈ A^ω) is *recurrent* if any factor u∈ F(w) will occur infinitely often in w, i.e. there exist infinitely many pairs of integers (i,j) with i < j and such that u = w[i , j] . An infinite word w∈ $A^{\pm\omega}$ (resp.w∈ A^ω) is *uniformly recurrent* if there exists a map k_w : F(w) → **N** , or simply k , such that: for any u, v ∈ F(w), with |v| ≥ k(u), one has u ∈ F(v).

It is clear that if w∈ $A^{\pm\omega}$ is uniformly recurrent then w_+ ∈ A^ω will be so. For any uniformly recurrent word w we denote by K_w, or simply by K, the map K: **N** → **N**, defined as:

$$K(n) = \max \{k(w) \mid w \in F(w) \cap A^n \}.$$

If v ∈ F(w) and |v| > K(n), then

$$F(w) \cap A^n \subseteq F(v).$$

The function K will be called *recurrence function*. The relevance of uniformly recurrent infinite words is due to the following lemma [8, cf. Theorem 4.1 and Proposition 4.2] which is a much more strong version of Lemma 2.2:

Lemma 2.3. Let L ⊆ A* be an infinite language. There exists an infinite word w∈ $A^{\pm\omega}$ such that:

450

i) w is uniformly recurrent

ii) $F(w) \subseteq F(L)$.

Definition 2.4. An infinite word w$\in A^{\pm\omega}$ (resp.w$\in A^\omega$) is called ω-*power-free* if for any u $\in F(w)$ there exists an integer $p_w(u)$, or simply $p(u)$, such that $u^{p(u)} \notin F(w)$. We shall take as $p(u)$ the minimal integer for which the above condition is verified. Moreover for any n > 0 we pose $P_w(n) = \max\{p(u) \mid u \in F(t) \text{ and } |u| \leq n\}$. If $p(u) = p$ for all u $\in F(w)$ then w is called p-*power-free*. The map $P_w : \mathbf{N} \to \mathbf{N}$, that we simply denote by P, will be called the *power-free function* of w .

We remark that one can easily prove (cf.[6]) that a uniformily recurrent word is ω -power-free if and only if it is not ultimately periodic.

Lemma 2.4. Let x $\in A^\omega$ be a uniformly recurrent ω-power free infinite word over a finite alphabet A. For any n >1, x has a factor which is a sesquipower of order n whose canonical factorization is n-divided.

Proof. We shall prove the statement of the lemma in the case of uniformly recurrent two-sided infinite words . Since by Lemma 2.3 for any uniformly recurrent one-sided infinite word x there exists a uniformly recurrent two-sided infinite word w such that $F(w) \subseteq F(x)$, then also x will satisfy the statement of the lemma. Then we can suppose that x is a uniformly recurrent bi-infinite word on the alphabet A. . We give now a procedure in order to find for any integer n > 0 a sequence of totally ordered finite alphabets Σ_i (i = 1,..., n + 1) and a sequence of bi-infinite uniformly recurrent words x_i $\in \Sigma_i^{\pm\omega}$, with $\Sigma_i = \text{alph}(x_i)$, $x_1 = x$ and for any i , $2 \leq i \leq n + 1$, there exists a monomorphism

$$\delta_i : \Sigma_i{}^* \to \Sigma_{i-1}{}^*,$$

such that

$$\delta_i (x_i) = x_{i-1}.$$

The construction is given inductively. We totally order Σ_1 = alph(x) and extend this order to the lexicographic order of $\Sigma_1{}^*$. Let us suppose that we have constructed the sequence until the i-th step. Let then $x_i \in \Sigma_i^{\pm\omega}$ where $\Sigma_i = \text{alph}(x_i)$. Let $b_i = \min(\Sigma_i)$; one can then construct the set

$$X_i = b_i{}^+(\Sigma_i\backslash b_i)^+.$$

Then consider the set

$$\Lambda_{i+1} = F(x_i) \cap X_i.$$

Since x_i is uniformly recurrent and ω-power-free, there are not in x_i enough long subwords which are either a power of b_i or do not contain b_i as a factor, so that Λ_{i+1} is finite.

One can then consider an alphabet Σ_{i+1} with $\text{card}(\Sigma_{i+1}) = \text{card}(\Lambda_{i+1})$ and a bijection γ_{i+1}: $\Sigma_{i+1} \to \Lambda_{i+1}$. Let $\delta_{i+1} : \Sigma_{i+1}{}^* \to \Sigma_i{}^*$ be the morphism which extends γ_{i+1} . Since Λ_{i+1} is a code with bounded

451

synchronization delay (cf. [1]), it is easy to verify that δ_{i+1} is injective and x_i can be uniquely "parsed" in terms of the elements of Λ_{i+1}. Then there exists a bi-infinite word $x_{i+1} \in \Sigma_{i+1}^*$ such that

$$\delta_{i+1}(x_{i+1}) = x_i .$$

Since, by the inductive hypothesis, x_i is uniformly recurrent and ω-power free, then one can easily derive that so x_{i+1} will be.

Now we define in Σ_{i+1} a total ordering by setting:

$$x <_{\Sigma_{i+1}} y \Leftrightarrow \delta_{i+1}(x) <_{\Sigma_i} \delta_{i+1}(y), \text{ for x, y } \in \Sigma_{i+1}.$$

Moreover for any u, v $\in$ A* one has (cf. [10], Chapter 7)

$$u <_{\Sigma_{i+1}} v \Rightarrow \delta_{i+1}(u) <_{\Sigma_i} \delta_{i+1}(v).$$

Let $w_1, w_2, \ldots, w_{n+1} \in A^*$ be defined as:

$$w_1 = b_1 = \min(\Sigma_1),$$

and for $i = 2, \ldots, n + 1$

$$w_i = \delta_2(\delta_3(\ldots \delta_i(b_i))),$$

with $b_i = \min(\Sigma_i)$.

We prove that $u = w_n w_{n-1} \ldots w_1$ is a sesquipower of order n and its canonical factorization (w_n, w_{n-1}, $\ldots$, w_1) is n-divided. By construction for any $i > 1$ and for any $b \in \Sigma_i$ one has

$$\delta_i(b) = b_{i-1} u_1 , \text{ with } u_1 \in (\Sigma_{i-1})^+,$$

thus, applying the same relation to the first letter of u_1, we obtain

$$\delta_{i-1}(\delta_i(b)) = \delta_{i-1}(b_{i-1}) b_{i-2} u_2, \text{ with } u_2 \in (\Sigma_{i-2})^+.$$

Iterating this procedure one has

$$\delta_2(\ldots \delta_{i-1}(\delta_i(b))) = \delta_2(\ldots \delta_{i-1}(b_{i-1})) \ldots \delta_2(b_2) b_1 \xi_i , \text{ with } \xi_i \in A^+ .$$

Thus, for $b = b_i$, we obtain for any i, $2 \le i \le n + 1$,

$$w_i = w_{i-1} \ldots w_1 \xi_i$$

and this implies that (w_n, w_{n-1}, $\ldots$, w_1) is the canonical factorization of a sesquipower of order n. Now, since $\delta_i(b_i) \in b_{i-1}^+(\Sigma_{i-1} \backslash b_{i-1})^+$, we can write

$$\delta_i(b_i) = b_{i-1}^{r_1} u_1 , \text{ with } r_1 > 0 \text{ and } u_1 \in (\Sigma_{i-1} \backslash b_{i-1})^+$$

and, since $b_{i-1} = \min(\Sigma_{i-1})$, one has

$$b_{i-1}\delta_i(b_i) = b_{i-1}b_{i-1}{}^{r_1} u_1 = b_{i-1}{}^{r_1}b_{i-1} u_1 <_{\Sigma_{i-1}} b_{i-1}{}^{r_1} u_1 b_{i-1} = \delta_i(b_i)b_{i-1},$$

so that

$$b_{i-1}\delta_i(b_i) <_{\Sigma_{i-1}} \delta_i(b_i)b_{i-1}.$$

Since the lexicographic ordering is preserved by δ_i, for any i, we can write

$$w_{i-1}w_i = \delta_2(\dots \delta_{i-1}(b_{i-1}\delta_i(b_i))) <_A \delta_2(\dots \delta_{i-1}(\delta_i(b_i)b_{i-1})) = w_i w_{i-1},$$

for $i = 2,\dots, n$.

By Proposition 2.1, this implies that $(w_n, w_{n-1}, \dots, w_1)$ is n-divided. It remains to prove that $u \in F(x)$. This is a consequence of the fact that $w_{n+1} = \delta_2(\dots \delta_{n+1}(b_{n+1})) \in F(x)$. Since $w_{n+1} = w_n \dots w_1 \xi_{n+1}$, it follows that $w_n \dots w_1 \in F(x)$. Q.E.D.

We observe that Lemma 2.4 was proved by us in [6] with a different technique. The previous proof has the advantage to give an explicit expression for the sesquipower of order n which is n-divided. This procedure is effective if the word x , the recurrence function K and the power-free function P are effectively computable.

A classical and remarkable theorem of Shirshov [12] states that *for any positive integers* n *and* p, *each sufficiently large word over a totally ordered finite alphabet has either a factor which is a p-power of a non-empty word or a factor which is n-divisible*. As a consequence of Lemma 2.4 we can give the following improvement of Shirshov's theorem.

Theorem 2.5. For all n,p,k positive integers there exists an integer $N(n,p,k)$ such that for any totally ordered alphabet A of cardinality k any word $w \in A^*$ such that $|w| \geq N(n,p,k)$ is such that either

    i. $\exists u \neq \Lambda$ such that $u^p \in F(w)$

    ii. $\exists s \in F(w)$ such that s is a sesquipower of order n whose canonical factorization $(w_1, \dots, w_n)$ is n-divided.

Proof. Let A be a totally ordered alphabet of cardinality k . Suppose by contradiction that such an integer $N(n,p,k)$ does not exist; then the set L of the words w not verifying i. and ii. is infinite and closed by factors. By Lemma 2.2 and Lemma 2.3, there exists an infinite uniformly recurrent word x $\in A^\omega$ such that $F(x) \subseteq L$. Since x is p-power-free, by Lemma 2.4 , x has to contain a subword which is a sesquipower of order n whose canonical factorization is n-divided and this is a contradiction. Q.E.D.

We mention that a non standard proof of Shirshov's theorem based on the "uniform recurrence" was recently given by Justin and Pirillo [9].

3. An application to semigroups

The results contained in the previous section, as we have shown in [6] , admit further refinements and generalizations which allows us to find some remarkable finiteness conditions for semigroups. In this section we give a short overview of the most recent results obtained in this area ; in particular we recall some of these which are based on different concepts such as *permutability* and *iteration* .

Definition 3.1. Let S be a semigroup and $s_1, \ldots , s_n$ be a sequence of n elements of S. We say that the sequence $s_1, \ldots , s_n$ is *permutable* (resp. *weakly-permutable*) if there exists a permutation $\sigma \in \mathcal{A}_n$, $\sigma \neq$ id (resp. two permutations $\sigma, \tau \in \mathcal{A}_n, \sigma \neq \tau$) such that $s_1 s_2 \ldots s_n = s_{\sigma(1)} s_{\sigma(2)} \ldots s_{\sigma(n)}$ (resp. $s_{\sigma(1)} s_{\sigma(2)} \ldots s_{\sigma(n)} = s_{\tau(1)} s_{\tau(2)} \ldots s_{\tau(n)}$. A semigroup S is called n-*permutable* (resp. n-*weakly-permutable*) if all sequences of n elements of S are permutable (resp. weakly permutable). S is called *permutable* (resp. *weakly-permutable*) if there exists an integer n > 1 such that S is n-permutable (resp. n-weakly-permutable).

Permutation properties are very important in the study of Burnside's problem as shown by the following theorem due to Restivo and Reutenauer [11].

Theorem 3.1. [Restivo and Reutenauer] Let S be a finitely generated and periodic semigroup. Then S is finite if and only if S is permutable.

The previous theorem does not hold, in general, if one substitutes permutable with weakly-permutable. However in the case of groups the result is true as a consequence of the following deep theorem due to Blyth [2]:

Theorem 3.2. Let G be a group. G is permutable if and only if it is weakly permutable.

Definition 3.2. Let S be a semigroup and n, k two integers such that n > 0 and k > 1. We say that the sequence $s_1, \ldots , s_n$ of n elements of S is k-*iterable* if there exist i, j with $1 \leq i \leq j \leq n$ such that $s_1 \ldots s_n = s_1 \ldots s_{i-1}(s_i \ldots s_j)^k s_{j+1} \ldots s_n$. A semigroup is called (n,k)-*iterable* if all sequences of n elements are k-iterable. S is called k-*iterable* if there exists an integer n > 1 such that S is (n,k)-iterable. We denote by $C_{n,k}$ (resp. C_k) the class of all the (n,k)-iterable (resp. k-iterable) semigroups.

In [4] the following stronger iteration properties, called *iteration properties on the right,* were considered:

Definition 3.3. Let S be a semigroup and n, k two integers such that n > 0 and k > 1. We say that the sequence $s_1, \ldots , s_n$ of n elements of S is k-*iterable on the right* if there exist i, j with $1 \leq i \leq j \leq n$ such that $s_1 \ldots s_j = s_1 \ldots s_{i-1}(s_i \ldots s_j)^k$. A semigroup is called (n,k)-*iterable on the right* if all its sequences of n elements are k-iterable on the right. S is called k-*iterable on the right* if there exists an integer $n \geq 1$ such that S is (n,k)-iterable on the right. We denote by $D_{n,k}$ (resp. D_k) the class of all the (n,k)-iterable on the right (resp. k-iterable on the right) semigroups.

454

Let S be a finitely generated semigroup. It was proved in [4] that D_2 assures the finiteness of S. We proved in [7], as a consequence of Theorems 3.1 and 3.2, that also D_3 implies the finiteness of S.

Theorem 3.3. Let S be a finitely generated semigroup. S is finite if and only if S belongs to the class D_2 or D_3.

In a recent work [5], using techniques both of algebraic and combinatorial nature, we proved the following more general finiteness condition:

Theorem 3.4. Let S be a finitely generated semigroup. S is finite if and only if S belongs to the class C_2 or C_3.

The following theorem, whose proof is given in [6], has been obtained as an application of the results contained in the previous section.

Theorem 3.5. Let S be a finitely generated and periodic semigroup. If there exists a pair of integers (n,k) with $k = 2$ or 3, such that any sequence of n elements of S is either permutable or k-iterable on the right, then S is finite.

We remark that Theorem 3.5 includes the Restivo-Reutenauer theorem and Theorem 3.3 as particular cases. In conclusion a problem which remains open is whether in Theorem 3.5 one can replace the property of iteration on the right with the central iteration.

REFERENCES

[1] J. Berstel and D. Perrin, *"Theory of codes"*, Academic Press, New York, 1985.

[2] R.D. Blyth, *Rewriting products of group elements*, Ph.D. Thesis, 1987, University of Illinois at Urbana-Champain, see also *Rewriting products of group elements, I*, J. of Algebra 116 (1988), 506-521.

[3] M. Coudrain and M.P. Schützenberger, *Une condition de finitude des monoides finiment engendrés*, C.R. Acad. Sc. Paris, Série A, 262 (1966), 1149-1151.

[4] A. de Luca and A. Restivo, *A finiteness condition for finitely generated semigroups*, Semigroup Forum, 28 (1984), 123-134.

[5] A. de Luca and S. Varricchio, *Finiteness and iteration conditions for semigroups*, Theoretical Computer Science, 87 (1991) 315-327.

[6] A. de Luca and S. Varricchio, *Combinatorial properties of uniformly recurrent words and application to semigroups*, International Journal of Algebra and Computation, n.2, 1991.

[7] A. de Luca and S. Varricchio, *A finiteness condition for semigroups generalizing a theorem of Hotzel*, J. of Algebra, 136 (1990) 60-72.

[8] H. Furstenberg, *Poincaré recurrence and number theory*, Bull. Amer. Math. Soc., 5 (1981) 211-234.

[9] J. Justin and G. Pirillo, *Shirshov's theorem and ω-permutability of semigroups*, Advances in Mathematics, 87 (1991) 151-159.

[10] M. Lothaire, *"Combinatorics on words"*, Addison Wesley, Reading Mass. 1983.

[11] A. Restivo and C. Reutenauer, *On the Burnside Problem for semigroups*, J. of Algebra, 89 (1984), 102-104.

[12] A.I. Shirshov, *On certain non associative nil rings and algebraic algebras*, Mat. Sb., 41(1957) 381-394.

THE STAR HEIGHT ONE PROBLEM
FOR IRREDUCIBLE AUTOMATA

R. Montalbano and A. Restivo

Dipartimento di Matematica e Applicazioni, Università di Palermo
Via Archirafi, 34
I - 90123 PALERMO   (ITALY)

1 - INTRODUCTION

The star height of a regular expression is, informally, the maximum number of nested stars in the expression. The star height of a regular language is the minimal star height of a regular expression denoting this language. The notion of star height indicates in a certain sense the "loop complexity" of a regular expression and thus it gives a measure of the complexity of a regular language.

A first question is whether the star height of a regular expression denoting a given language can be always "reduced" below some constant indipendent from the language. A negative answer to this question is given by an important result of Dejean and Schützenberger [3] which states that the star height hierarchy is infinite.

Another problem, which has been raised since the beginning of automata theory, concerns the effective computability of the star height. Since a language has star height zero if and only if it is finite, the first non trivial case is the decidability of star height one. This has been proved by K. Hashiguchi in [5]; an extention of the techniques of this paper provides also the solution of the general problem, see [6]. However, as stressed by D. Perrin [7], "the proof is very difficult to understand and a lot remains to be done to make it a tutorial presentation".

This paper is a contribution in improving the understanding on star height one. We consider a special family of automata, irreducible automata (strongly connected automata in which any state is starting state and accepting state), and the corresponding family of languages, FTR (factorial, transitive and regular) languages. The notions of irreducible automaton and FTR language appear in many interesting applications of automata theory. In particular they appear in the theory of symbolic dynamical systems, where FTR languages provide a description of transitive sofic systems (cfr. [4], [9]). These notions play also an important role in problems concerning the encoding of digital data into a constrained channel; in fact, the system of constraints can be described by the set of available words, which is a FTR language (cfr. [10]).

The main result of this paper (Theorem 4.1.) gives a simple characterization of FTR languages of star height one: *a FTR language has star height one if and only if it is the set of factors of a finitely generated submonoid.* By this characterization we derive a decision procedure of the star height one for FTR languages which is simpler than that given by Hashiguchi. In the last section of the paper we report some open problems and suggestions for further investigations.

2 REGULAR EXPRESSIONS AND THE STAR HEIGHT PROBLEM

In this section we introduce some basic definitions and results concerning regular languages, finite automata and the star height problem.

Let A be a finite alphabet and A^* the free monoid generated by A. The elements of A are called letters, the elements of A^* words and the lenght of a word u is denoted by l(u); 1 is the empty word. A word w is *factor* of a word x if there exist words u, v such that $x = uwv$. For any subset X of A^*, denote by X^* the submonoid generated by X and denote by F(X) the set of factors of words in X. We call *language* any subset of A^*.

A *finite deterministic automaton* is a quadruple $\mathcal{Q} = (Q, \delta, q_0, F)$, where Q is a finite set of states, $q_0 \in Q$ is the initial state, $F \subseteq Q$ is the set of final states and δ is the transition function, $\delta : Q \times A \to Q$.

For any automaton $\mathcal{Q}$ the set $L(\mathcal{Q}) = \{w \in A^* \mid \delta(q_0, w) \in F \}$ is the language recognized by $\mathcal{Q}$.

Given a language L of A^*, the *syntactic congruence modulo L* is defined in A^* by

$$u \sim v \text{ iff } \forall x, y \in A^* \quad xuy \in L \Leftrightarrow xvy \in L.$$

The *syntactic monoid* of L, denoted by Synt(L), is the quotient of A^* by this congruence.

Now let us introduce the class of regular expressions and the one of the languages represented by these expressions, that is the class of regular languages. For each of them we give the definition of star height and report some fundamental results.

The class of **regular expressions** *is defined inductively as follows:*

- 1, Ø *and* a *are regular expressions (for any* a ∈ A).
- *If* E *and* F *are regular expressions,* (E ∪ F), (EF) *and* E^* *are regular expressions.*

For any regular expression E, *we denote by* $|E|$ *the language of* A^* *represented by* E, *which is defined inductively as follows:*

- $|1| = \{1\}, |\emptyset| = \emptyset$ *and* $|a| = \{a\}$ *(for any* $a \in A$*).*
- $|E \cup F| = |E| \cup |F|$, $|EF| = \{vw \in A^* | v \in |E|$ *and* $w \in |F|\}$, *and* $|E^*| = \{1\} \cup \{v_1 v_2 \ldots v_k | k > 0$ *and* $v_i \in |E|, i = 1, \ldots, k\}$.

The star height $h(E)$ **of a regular expression** E *is defined inductively as follows:*

- $h(1) = h(\emptyset) = h(a) = 0$ (*for any* $a \in A$*).*
- $h(E_1 \cup E_2) = h(E_1 E_2) = \max\{h(E_1), h(E_2)\}$.
- $h(E^*) = h(E) + 1$.

The star height $h(L)$ **of a regular language** L *is defined by*

$$h(L) = min \{h(E) | E \text{ is a regular expression denoting } L\}.$$

Remark: For any regular language L, $h(L) = 0$ iff L is finite.

Theorem 2.1.(Dejean, Schützenberger [3]) *For any* $k \geq 0$, *there exists a regular language* L *such that* $h(L) = k$.

In order to study the problem of determining the star height of a regular language, it is convenient to consider a particular class of regular expressions denoting the language: the class of regular expressions in string form defined inducively as follows.

For any regular expression E,

- *If* $h(E) = 0$, *then* E *is in string form iff* $E = w_1 \cup w_2 \cup \ldots \cup w_n$, *where* $n > 0$ *and* w_j *are words (* $j = 1, 2, \ldots, n$*).*
- *If* $h(E) = k > 0$, *then* E *is in string form iff* $E = F_1 \cup F_2 \cup \ldots \cup F_n$, *where* $n > 0$ *and each* F_i *(* $i = 1, \ldots, n$*) is a string of the form:*

$$w_1 H_1^* w_2 H_2^* \ldots w_h H_h^* w_{h+1}, \quad h > 0,$$

where w_m *are words (* $m = 1, \ldots, h+1$*),* H_j *are in string form and* $h(H_j) \leq k - 1$ *(* $j = 1, \ldots, h$*).*

459

We have the following:

Proposition 2.1.(Cohen [2]) *For any regular expression* E, *there exists a regular expression in string form* E_s *such that:*

$$- |E| = |E_s|.$$
$$- h(E) = h(E_s).$$

The problem of determining the star height of a language L is of importance because it gives a measure of the complexity of L; Hashiguchi (1982) presented in [5] an algorithm for deciding whether a regular language is of star height one, or not. We report briefly this result, using a notation of Pin [8].

Let $\mathfrak{X}$ be a finite set of regular languages, and let us consider the operators U and P_n (where n > 0) defined as follows:

$$U(\mathfrak{X}) = \{\text{finite unions of languages of } \mathfrak{X}\}.$$
$$P_n(\mathfrak{X}) = \{\text{products of m languages of } \mathfrak{X} \mid m \leq n\}.$$

We denote by $\mathfrak{X}(\cdot, U, n)$ the following set:

$$\mathfrak{X}(\cdot, U, n) = \{\{1\}\} \cup \{\emptyset\} \cup UP_n(\mathfrak{X}).$$

Remark: Observe that, since $\mathfrak{X}$ is a finite set, also $\mathfrak{X}(\cdot, U, n)$ is finite.

Let L be a regular language, m = card(Synt(L)) and k = 16m (m + 2) (m (m + 2) + 1). Let us consider the following finite set of regular languages:

$$\mathfrak{C}_k = \{\{w\} \mid w \in A^* \text{ and } l(w) \leq k\} \cup \{(w_1 + \dots + w_r)^* \mid w_i \in A^* \text{ and } l(w_i) \leq k\},$$

and let Q_i be the set of states of the minimal automaton recognizing each $L_i \in \mathfrak{C}_k$ (i = 1,..., card($\mathfrak{C}_k$)) and Q the set of states of the minimal automaton of L. We have the following

Theorem (Hashiguchi [5]) *Let* L *and* $\mathfrak{C}_k$ *be as above, and* $h(L) \geq 1$. *Then we have:*

$$h(L) = 1 \Leftrightarrow L \in \mathfrak{C}_k(\cdot, U, n),$$

where $n = p(p + 1)^3 p^2 2p(6p^2 + 1)$, $p = 2^t$ *and* $t = 2^{Card(Q)}(Card(Q_1) + \dots + Card(Q_s))$.

For deciding whether L is of star height one, we decide whether or not it is finite. If L is finite, then h(L) = 0; if h(L) ≠ 0, then we construct $\mathfrak{C}_k(\cdot, U, n)$ (which is a finite set). If L ∈ $\mathfrak{C}_k(\cdot, U, n)$ then h(L) = 1, otherwise h(L) > 1.

3 IRREDUCIBLE AUTOMATA AND FTR LANGUAGES

Since we are interested to the star height one problem for FTR languages, in this section we introduce the notion of irreducible automata and FTR languages, i.e. languages that are Factorial, Transitive and Regular, and give some examples.

These notions appear in the theory of *symbolic dynamical systems* (cfr. [4], [10]), where FTR languages are related to *transitive sofic systems* . They also play an important role in the problem of encoding digital data in a constrained channel. In fact, in this case the system of constraints can be described by the set L of the available words, and time invariance and ergodicity of the encoding process imposes that L is a factorial and transitive language.

A finite deterministic automaton is **irreducible** *if:*

> *- Its state graph is strongly connected;*
> *- All states are starting states and accepting states.*

A language $L \subseteq A^*$ *is*
> - **Factorial** *if* $L = F(L)$;
> - **Transitive** *if* $\forall$ u, v $\in$ L, $\exists$ w $\in A^*$ *such that*   uwv $\in$ L.

The following results, which relate the notions of irreducible automaton and FTR language, point out the theoretical interest of these concepts.

Theorem 3.1.(Fisher [4]) *Every FTR language is recognized by an irreducible automaton and conversely.*

Theorem 3.2.(Beauquier [1]) *For every FTR language there exists a unique minimal irreducible automaton which recognizes the language.*

Example 3.1. Consider the following irreducible automaton over the alphabet A = {a, b, c}:

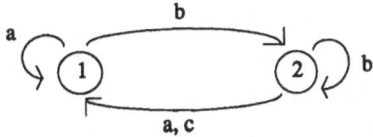

461

The FTR language recognized by this automaton has star height one since it is represented by the regular expression $(c +1) (a + b + bc)^*$.

Example 3.2. Consider the following irreducible automaton over the alphabet $A = \{a, b, c\}$:

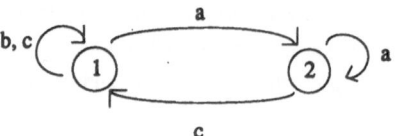

The FTR language L recognized by this automaton may be represented by the regular expression, of star height two, $(b + (a)^*c)^*$. By using the results of next sections we are able to show that does not exist any regular expression of star height one representing L. Thus $h(L) = 2$.

Remark that one can specialize the result of Deiean and Schützenberger (Th. 2.1.) and prove that the star height hierarchy is infinite for the class of FTR languages.

4 FTR LANGUAGES OF STAR HEIGHT ONE

In this section we give a characterization of the FTR languages of star height one. In the next one we deduce from this characterization an algorithm for deciding whether a given FTR language is of star height one. We need the following preliminary lemma.

Lemma 4.1. Let L be a factorial and transitive language; let $L_1, L_2, ..., L_k$ be factorial languages such that $L = L_1 \cup L_2 \cup ... \cup L_k$. Then there exists an integer i, $1 \le i \le k$, such that $L = L_i$.

Proof: The proof is by induction on k.
$k = 2$: if $L = L_1 \cup L_2$, it suffices to prove that either $L_1 \subseteq L_2$ or $L_2 \subseteq L_1$. Suppose, by contradiction, that there exist u, v such that $u \in L_1 / L_2$ and $v \in L_2 / L_1$. Since u, v $\in$ L, there exists w such that uwv $\in$ L (for the transitivity of L). So we have that either uwv $\in L_1$ or uwv $\in L_2$; it follows, for the factoriality of L_1 and L_2, that either $v \in L_1$ or $u \in L_2$ and this is against the hypothesis. Then either $L = L_1$ or $L = L_2$.

$k > 2$: $L = L_1 \cup L_2 \cup ... \cup L_k$. Setting $L' = L_2 \cup ... \cup L_k$, we have $L = L_1 \cup L'$, with L_1 and L' both factorial; then, for the basis of the induction, we have that either $L = L_1$ or $L = L'$. If $L = L_1$ then $i = 1$ and the lemma is proved, otherwise $L = L_2 \cup ... \cup L_k$ and, by the induction hypothesis, there exists an integer i, $2 \leq i \leq k$, such that $L = L_i$. This completes the proof.

Now we are able to prove the following characterization:

Theorem 4.1. *Let* L *be a FTR language.*

$$h(L) = 1 \Leftrightarrow \textit{there exists a finite set } H \textit{ such that } L = F(H^*).$$

Proof: Let us suppose that $h(L) = 1$. Let E be a regular expression in string form denoting L:

$$E = F_1 \cup F_2 \cup ... \cup F_l,$$

where $l > 0$ and each F_i is a string of the form:

$$w_1 H_1^* w_2 H_2^* ... w_h H_h^* w_{h+1},$$

where $h > 0$, $w_m \in A^*$ and $h(H_j) = 0$ $(m = 1, ..., h + 1; j = 1, ..., h)$.
Observe that the latter condition implies that any H_j is a finite set; the proof consists in showing that there exists an integer k such that $E = F(H_k^*)$.
Since L is factorial, we have that $E = F(F_1) \cup ... \cup F(F_l)$. For lemma 4.1. there exists i, $1 \leq i \leq l$, such that $E = F(F_i)$, and then we can write:

$$E = F(w_1 H_1^* w_2 H_2^* ... w_h H_h^* w_{h+1}).$$

The following step is to prove that $E = F(H_1^*) \cup F(H_2^*) \cup ... \cup F(H_h^*) = \cup_j F(H_j^*)$; then the existence of a finite set H such that $E = F(H^*)$ will be a consequence of lemma 4.1. .
Let us set $w_1 H_1^* w_2 H_2^* ... w_h H_h^* w_{h+1} = Y$, so that we can write $E = F(Y)$. Obviously $\cup_j F(H_j^*)$ is included in E; in order to prove the converse inclusion, we will first state that $Y \subseteq \cup_j F(H_j^*)$. Let us consider a word $y \in Y$, and observe that, for the definition of Y, the words $w_1, w_2, ..., w_h, w_{h+1}$ are proper factors of y, and then $l(w_i) < l(y)$, $1 \leq i \leq h+1$.
Since E is transitive and $y \in E$, for any integer n there exist $u_1, u_2, ..., u_n \in A^*$ such that $y u_1 y u_2 ... y u_n y \in E$; this implies that there exist $t, z \in A^*$ and $v_j \in H_j^*$, $1 \leq j \leq h$, such that the word $v = t (y u_1 y u_2 ... y u_n y) z$ admits the following factorization:

$$v = w_1 v_1 w_2 v_2 ... w_h v_h w_{h+1}.$$

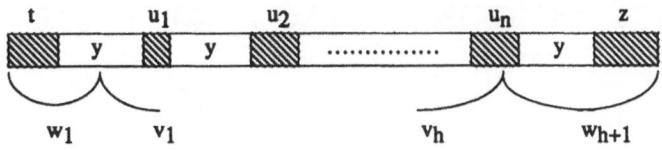

We will see that, for $n \geq 2h$, y is factor of one of $v_1, ..., v_h$.

First we will prove this in a particular case (that is for a particular possible factorization of v) that represents the worst case (this means that in any other possible case y will result *a fortiori* factor of one of $v_1, ..., v_h$).

Let us consider a factorization of v such that any w_i has an overlap, or a common factor, with some occurence of y; in particular let w_i, $2 \leq i \leq h$, have an overlap with two consecutive occurences of y (observe that w_i can not have overlaps with three or more occurences of y, because otherwise one of these would be proper factor of w_i, and this is impossible), and let w_1 and w_{h+1} have an overlap respectively with the first and the latter occurence of y in v (for the same reason they can not have overlaps with two or more occurences of y) . Let us also suppose that any y in v has an overlap with only one w_i.

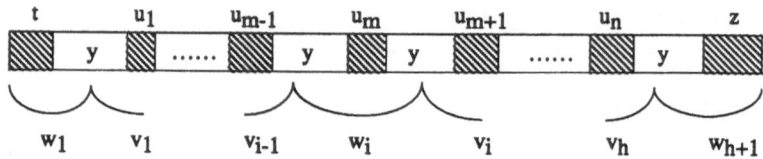

If we denote with N the number of occurences of y in the word v that have an overlap with some w_i, in a factorization of v like the one just described it will be $N = 2h$; in fact, since both w_1 and w_{h+1} have an overlap with one occurence of y, and $w_2, ..., w_h$ have an overlap with two consecutive occurences of y, we have $N = 2 + 2(h - 1) = 2h$. But the total number of occurences of y in v is $n + 1 \geq 2h + 1$, and then there is at least an occurence of y in v that have no overlap with any w_i, that is y is factor of one of $v_1, ..., v_h$.

Example 4.1. Let $E = F(Y)$, where $Y = w_1 H_1^* w_2 H_2^* w_3$.
We have that $n = 2h = 4$, so that for any $y \in Y$ there exist t, z, $u_1,..., u_4 \in A^*$, $v_1 \in H_1^*$, $v_2 \in H_2^*$ such that $v = t (yu_1yu_2yu_3yu_4y) z = w_1v_1w_2v_2w_3$. Two possible factorizations of v are shown in the following figures:

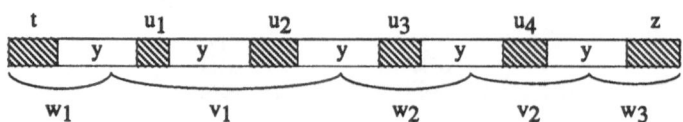

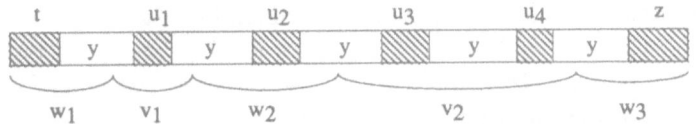

We can see from the figures that in any case y will be factor of one of v_1, v_2.

In any other possible factorization of y in the product $w_1 v_1 w_2 v_2 \cdots w_h v_h w_{h+1}$, the value of N is less than 2h, so that *a fortiori* there exists at least one occurence of y which is factor of one of $v_1, \ldots, v_h$. We can conclude that $y \in U_j F(H_j^*)$, and that $Y \subseteq U_j F(H_j^*)$.

Finally let $z \in E/Y$; since there exists $y \in Y$ such that z is factor of y (because $E = F(Y)$) and $y \in U_j F(H_j^*)$, also $z \in U_j F(H_j^*)$ and then $E \subseteq U_j F(H_j^*)$.

The equality $E = U_j F(H_j^*)$ holds true and so we are in the hypothesis of lemma 4.1.: there exists an integer k, $1 \leq k \leq h$, such that $E = F(H_k^*)$. This completes the first part of the proof.

Now, suppose that $L = F(H^*)$, where H is a finite subset of A^*; in order to state that $h(L) = 1$, we prove that L admits a regular expression in string form of star height one.

If we denote by F the set of factors of the words in H, we have that F is obviously a finite set so that the set

$$U_{u \, \in \, F} U_{v \, \in \, F} (u \, H^* v)$$

is of star height one, because is a finite union of set of star height one. Since it is easy to prove that $L = F(H^*) = U_{u \, \in \, F} U_{v \, \in \, F} (u \, H^* v)$, we have that $h(L) = 1$. This completes the proof.

5 DECIDIBILITY

In this section we give an algorithm for deciding whether a given FTR language is of star height one. Let us first introduce the notion of L-maximal submonoid.

Let L *be a language; a submonoid* M *included in* L *is called* **L-maximal** *if, for any submonoid* N, $M \subseteq N \subseteq L$ *implies* M = N.

Let $\mathfrak{M}(L)$ denote the family of all the L-maximal submonoids.

Theorem 5.1. *For any regular language* L:

        i) $\mathfrak{M}(L)$ *is a finite set;*

        ii) every element of $\mathfrak{M}(L)$ *is a regular set and can be effectively constructed.*

In order to prove this theorem we introduce the notion of stabilizer of a set of states.

Let $\mathcal{Q} = (Q, \delta, q_0, F)$ *be any finite deterministic automaton over the alphabet* A; *the* **stabilizer** *of a subset of states* $S \subseteq Q$ *is the language*

$$\text{Stab}(S) = \{w \in A^* \mid \delta(q, w) \in S, \text{for each } q \in S\}.$$

Remark. Given $\mathcal{Q}$ and $S \subseteq Q$, a finite automaton recognizing Stab(S) can be effectively constructed.

Theorem 5.1. is obtained as a consequence of the following lemma.

Lemma 5.1. *Let* $\mathcal{Q} = (Q, \delta, q_0, F)$ *be an automaton recognizing* L. *For any* $M \in \mathcal{M}(L)$ *there exists a set* $S \subseteq Q$ *such that* M = Stab(S).

Proof: We first prove that for any submonoid $N \subseteq L$, there exists $S \subseteq Q$ such that $N \subseteq$ Stab(S). Set $S = \{ q = \delta(q_0, v) \mid v \in N \}$ and let $u \in N$; we have to prove that $u \in$ Stab(S), that is $\delta(q, u) \in S$ for any q of S. Since $q \in S$, there exists $v \in N$ such that $q = \delta(q_0, v)$; then $\delta(q, u) = \delta(q_0, uv) \in S$, because $uv \in N$. The maximality of M completes the proof.

Example 5.1. Let L be the language recognized by the following irreducible automaton

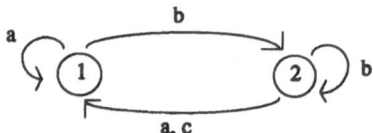

There are two L-maximal submonoids:

$$M_1 = \text{Stab} (\{2\}) = (b + (a + c) a^* b)^*;$$
$$M_2 = \text{Stab}(\{1,2\}) = (a + b + bc)^*.$$

The submonoid $M_3 = \text{Stab}(\{1\}) = (a + b^+(a + c))^*$ is not a L-maximal submonoid, since $M_3 \subseteq M_2$.

The following lemma in [9] is here reported without proof. Let $k = \text{Card}(\text{Synt}(X^*))$, $d = 2(k^2 + 2)$ and $X_d = \{w \in X | l(w) \leq d\}$, then we have:

Lemma 5.2.(Restivo) *Given a regular set X, if there exists a finite subset Y of X such that $F(Y^*) = F(X^*)$, then $F(X_d^*) = F(X^*)$.*

Corollary 5.1. *Given a rational set X, it is decidable whether there exists a finite subset Y of X such that $F(Y^*) = F(X^*)$.*

Theorem 5.2. *Given a FTR language L, it is decidable whether $h(L) = 1$.*

Proof: Let us suppose that $h(L) = 1$. Then, by theorem 4.1., there exists a finite set Z such that $L = F(Z^*)$. Since Z is a submonoid included in L, it is included in a L-maximal submonoid X^*. Then there exists a finite subset Y of X such that $Z \subseteq Y^*$. We have that $Z^* \subseteq Y^* \subseteq X^* \subseteq L$ and that $F(Z^*) = L$; then $F(Y^*) = F(X^*)$. The theorem is then a consequence of corollary 5.1. and of theorem 5.1.

Given a FTR language L, we can decide whether $h(L) = 1$ as follows: first we construct the bases of all L-maximal submonoids, by computing $\text{Stab}(S)$ for any set of states of the irreducible automata recognizing L. Then, for any base X, we verify whether $F(X_d^*) = F(X^*)$. If, for some base X, $F(X_d^*) = F(X^*)$, then by theorem 5.2. we can conclude that $h(L) = 1$; otherwise $h(L) > 1$.

Example 5.1.(continued) Since the base $X = \{a, b, bc\}$ of the L-maximal submonoid M_2 is itself a finite set, then we can conclude that $L = F(X^*)$, and then $h(L) = 1$.

6 OPEN PROBLEMS

In the example of the previous section, in which L is of star height one, we have seen that there exists a L-maximal submonoid that is itself finitely generated; since this is verified in all known examples, we conjecture that this holds true in general:

Conjecture 6.1. *A FTR language L is of star height one if and only if there exists a L-maximal submonoid which is finitely generated.*

A subset $X \subseteq A^*$ is a **minimal factor generator** *if for any $Y \subseteq X$, $F(Y^*) \subseteq F(X^*)$.*

467

Then we have that conjecture 6.1. is based on the following

Conjecture 6.2. *Let* X *be such that:*

- X *is a minimal generating set of* X^*;
- X^* *is a maximal submonoid in* $F(X^*)$.

Then X *is a minimal factor generator.*

Example 6.1. Let $X = \{a^2,\ ab,\ ba,\ b^2,\ aba\}$; X is not a minimal factor generator, since $F(X^*) = F(Y^*)$, with $Y = \{a^2,\ ab,\ ba,\ b^2\}$. This is in accord with conjecture 6.2., because X is a minimal generating set of X^*, but it is not a maximal submonoid in $F(X^*)$, since $F(X^*) = A^*$ is itself a submonoid and then the unique maximal submonoid in $F(X^*)$.

Remark. If the conjecture is true, for deciding whether L is of star height one it suffices to construct the bases of all L-maximal submonoids and to verify whether there exists a finite base.

There exist some questions related to the star height one problem for FTR languages, which arise in the formalization of the encoding process from a free source into a constrained channel (cfr. [10]). In particular, if we recall that a *code* is the base of a free submonoid, the following decision problem remains open:

Problem 6.1. *Given a FTR language* L, *decide whether there exists a finite code* C *such that* $L = F(C^*)$.

A second problem needs some definitions:

Given a set of words X, *a subset* $Y \subseteq X^*$ *is called a* **composition** *of X. If moreover* $F(Y^*) = F(X^*)$, *then* Y *is called a* **rich composition** *of* X.

Problem 6.2. *Given a finite set, decide whether there exists a finite rich composition of it, which is a code.*

Example 6.2. Let $X = \{a^2, a^3, b^2, b^3\}$ and consider the subset of X^*

$$Y = \{a^2, b^2, a^2b^3, a^3b^2, a^3b^3\}.$$

Y is a finite rich composition of X, which is a code.

There exist however set of words which does not admit any finite rich composition which is a code; an example is the set $Z = \{a, bc, ab, c\}$.

Theorem 5.2. states some properties of the family $\mathfrak{M}(L)$; we can consider a new family $\mathfrak{F}\mathfrak{M}(L)$ of submonoids in L, for which some questions arise.

Let L *be a language; a free submonoid* M *included in* L *is called* **L-maximal free** *if, for any free submonoid* N, $M \subseteq N \subseteq L$ *implies* $M = N$.

Problem 6.3. *i) Is* $\mathfrak{F}\mathfrak{M}(L)$ *a finite family ?;*

ii) Are the elements of $\mathfrak{F}\mathfrak{M}(L)$ *regular sets, and how can be constructed ?*

Remark. A L-maximal free submonoid is a submonoid maximal among the free submonoids included in L; there exist submonoids which are L-maximal free, but not L-maximal.

REFERENCES

[1] D. Beauquier, Minimal automaton for a factorial, transitive and rational language, *Report LITP* (1987) 1-12.

[2] R. S. Cohen, Star height of certain families of regular events, *J. Comput. System Sci.* (1970) 281-297.

[3] F. Dejean and M. P. Schutzenberger, On a question of Eggan, *Inform. and Control* 9 (1966) 23-25.

[4] R. Fisher, Sofic systems and graphs, *Monatsh. Math.* 80 (1975) 179-186.

[5] K. Hashiguchi, Regular language of star height one, *Inform. and Control* 53 (1982) 199-210.

[6] K. Hashiguchi, Algorithms for determining relative star height and star height, *Inform. and Comput.* 78 (1987) 124-169.

[7] D. Perrin, Finite automata, in: J. van Leeuwen, ed., *Handbook of Theoretical Computer Science, Vol. B* (North-Holland, Amsterdam, 1990) 1-57.

[8] J.E. Pin, Rational and recognizable languages, in: L.M. Ricciardi, ed., *Lectures in Applied Math. and Informatic*, (Manchester University Press, 1990) 62-104.

[9] A. Restivo, Finitely generated sofic systems, *Theoret. Comput. Sci.* 65 (1989) 265-270.

[10] A. Restivo, Codes and local constraints, *Theoret. Comput. Sci.* 72 (1990) 55-64.

SYNCHRONIZING AUTOMATA

Dominique Perrin,
Institut Blaise Pascal, Paris

ABSTRACT. — *In this paper, we survey some recent results on synchronizing words for automata. We discuss in particular partial solutions of the so-called road coloring problem.*

INTRODUCTION

The notion of a synchronizing word is a natural one for a finite automaton. It is an input sequence such that the state reached at the end is independant of the starting state. This property has important consequences for the applications. It makes in particular the behavior of the automaton resistant against *input errors* since, after an error, a synchronizing word replaces the automaton in the same state as if no error had occurred. It plays also a role in the *identification* of automata, since a synchronizing word can put an unknown automaton in a prescribed state and check some characteristic property of this state.

The variety of names given to synchronizing words is a testimony on the variety of their use : reset sequence, resolving block, magic word appear in particular also in the litterature.

It is interesting that such a simple and useful notion is the subject of several open problems. In this paper, we survey some recent results about a problem kown as the road coloring problem. It was invented by Adler and Weiss and no general solution is yet known.

The paper is divided into three sections. In the first one, we give definitions and some elementary properties of non-synchronizing automata. In the second section, we present some properties of a special class of automata that has been studied by several authors and plays a role in some solutions to the road coloring problem in particular cases. Finally, in the third section, we present a discussion on the road coloring problem itself.

1. SYNCHRONIZING WORDS

Let $\mathcal{A} = (Q, A)$ be a strongly connected finite automaton, where Q is the set of states and A is the alphabet (we do not specify the acceptance conditions). A *synchronizing word* for $\mathcal{A}$ is a word $w \in A^*$ such that $\mathrm{Card}(Q \cdot w) = 1$. Hence, w is synchronizing if all paths labeled w lead to the same state, irrespective of their starting point. The automaton $\mathcal{A}$ itself is said to be *synchronizing* if there exist some synchronizing word. We give below two examples of automata, the first one being synchronizing and the second one being not.

Example 1.1. — The four-states automaton given by Figure 1.1 is synchronizing.

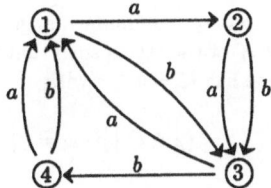

Figure 1.1. A synchronizing automaton.

Indeed, the word $w = bab$ is synchronizing since all paths labeled w lead to state 3.

Example 1.2. — The four-states automaton given by Figure 1. 2. is not synchronizing.

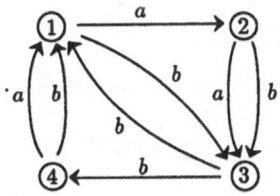

Figure 1.2. A non-synchronizing automaton.

471

Indeed the diagram of Figure 1.3 shows that the family of sets $\{1,3\}$ and $\{2,4\}$ is globally invariant under the action of the symbols a, b.

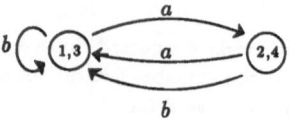

Figure 1.3.

Given a word w, the set $Q \cdot w$ is called its *image*. It is actually the image of the map $q \mapsto q \cdot w$ induced by w on the state set Q. An image of minimal cardinality is called simply a *minimal image*. The *degree* of a strongly connected automaton $\mathcal{A}$, denoted $d(\mathcal{A})$ or d, is the cardinality of the minimal images. The automaton is clearly synchronizing iff its degree is equal to one.

As a dual notion to that of a minimal image, we have the *maximal kernels*. A kernel is a set K of the states such that for some word w in A^* and some state $q \in Q$, one has $K = w^{-1}q$ with

$$w^{-1}q = \{p \in Q \mid p \cdot w = q\}.$$

It is clear that for a word w, the image $Q \cdot w$ is minimal iff for all q in Q, the kernel $w^{-1}q$ is maximal.

Let now E be the adjacency matrix of the underlying graph of the automaton. Thus $E_{p,q}$ is the number of edges from state p to state q. Let w be the left eigenvector of E corresponding to the eigenvalue $\alpha = \text{Card}(A)$ chosen with integer coordinates of g.c.d. equal to 1. For a state p, we call the component w_p of w the *weight* of p. For a set S of states, we denote $w_S = \sum_{p \in S} w_p$. The following property of maximal kernels was already observed by J. Friedman (1990).

PROPOSITION 1.1. — *For any maximal kernel K, one has the equality*

$$w_K = w_Q/d.$$

Proof. — It is enough to prove that all maximal kernels have the same weight. We consider the set of kernels of maximal weight. If K is such a kernel, then so is $u^{-1}K$ for any word u. Indeed, since $wE = \alpha w$, we have $\sum_{a \in A} w_{a^{-1}K} = \alpha w_K$ and thus $w_{a^{-1}K} = w_K$ for all a in A. Thus the family of kernels of maximal weight is invariant under the left action $K \mapsto u^{-1}K$ of A^* on sets of states. Since this action is transitive on the set of maximal kernels and since any kernel with maximal weight is also maximal, this proves that all maximal kernels have the same weight. □

Example 1.2 (continued). — The eigenvector w is

$$w = [2\ 1\ 2\ 1]$$

and the maximal kernels are $\{1,2\}$, $\{3,4\}$, $\{1,4\}$ and $\{2,3\}$ all with weight 3.

An interesting consequence of the preceding result, noted by J. Friedman, is that the degree of the automaton is always a divisor of the integer w_Q, therefore giving a limitation on the possible degree of an automaton, irrespective of the labeling of the graph.

2. SPECIAL AUTOMATA

We describe here the properties, especially with respect to synchronization, of a class of automata called here *special*. They are defined as follows : A strongly connected automaton is called special if there is a letter, called a *special letter*, such that for any pair p, q of states there exist integers $i, j \geq 0$ such that

$$p \cdot a^i = q \cdot a^j.$$

In other words, the automaton is special if there is an integer $k \geq 0$ such that the words a^k brings all states into the ultimate cycle of a.

We may find special automata in two important particular cases. The first one is the case where the special letter is a circular permutation of all states. We observe that the automaton of Example 1.2 is of this type. The second one is the case where for some state q_0, the set of first returns to q_0 is finite. Suppose indeed that the set of first returns to state q_0 is finite. Let q be any state. Then for any letter a, there exists an integer i such that $q \cdot a^i = q_0$ and hence the property with all letter being special.

The automaton of Example 1.1 is of this type. The second type is actualy a property of the graph rather than of the automaton itself since an equivalent definition is that all cycles pass though q_0.

Let now $\mathcal{A} = (Q, A)$ be a special automaton. Let $a \in A$ be a special letter and let $n \geq 1$ be the length of the ultimate cycle of a. We call n the *order* of a. We denote by $0, 1, \ldots, n-1$ the elements of this cycle, in such a way that for $0 \leq i \leq n-1$

$$i \cdot a = i + 1 \qquad \text{mod } n.$$

Since a is special, there are minimal images contained in $\{0, 1, \ldots, n-1\}$. Let I be such an image. Let also K be any maximal kernel and let $J = K \cap \{0, 1, \ldots, n-1\}$.

The following property is well known and easy to verify (see Perrin, 1977 or Pin, 1978).

PROPOSITION 2.1. — *The pair (I, J) is a factorization of the cyclic group $\mathbf{Z}/n$, i.e. any element of $\mathbf{Z}/n$ can be written uniquely as the sum modulo n of an element of I and of an element of J.*

This result has many interesting consequences. One of them is that the degree is a divisor of the order of the special letters. On the other hand, it indicates an interesting connexion between the theory of automata and the study of factorizations of cyclic groups (see Perrin, 1977 for a bibliography).

3. THE ROAD COLORING PROBLEM

If we use colors instead of symbols to label the edges in a finite automaton, we get a new image of a synchronizing word as a sequence of colors leading back to a fixed arrival point irrespective of the starting point. If we further agree to interpret the states of the automaton as cities and the edges as roads connecting them we may interpret a synchronizing word as a prescription allowing a traveller to find his way back home wherever he starts from, even in the case where he would be lost.

This picturesque representation explains the name of *road coloring problem* given by R. Adler and B. Weiss to the following question : is it always possible to label a connected aperiodic graph with all vertices of the same outdegree in such a way that the resulting automaton is synchronizing? It is generally conjectured that the answer to the question is affirmative in all cases under the above hypotheses which are obviously necessary. We recall that the period of a connected directed graph is the g.c.d. of the lengths of its cycles. The graph is said to be *aperiodic* if the period is one.

No solution of the road coloring problem is known presently. We review here the cases that, to our knowledge, have been already solved.

A first result, due to G.-L. O'Brien (1981) gives a solution under the following two restrictive additional hypotheses :

(i) the graph has no multiple edges ;

(ii) there is a cycle of prime length.

The proof uses Proposition 2.1, or rather the weaker corollary that in a special automaton, the degree is a division of the degree of the special letter.

Another result, due to J. Friedman (1990), says that a positive solution to the road coloring problem exists when, with the notation of section 1, the total weight w_Q is prime to the length of the some cycle. The proof uses essentially Propositions 1.1 and 2.1 again.

The last result, due to M.P. Schützenberger and myself (1991) says that the road coloring problem has a solution under the additional hypothesis that all the vertices of the graph except one have indegree one. This hypothesis is actually a condition on the set of first returns to the special state having indegree greater than one. This set is a prefix code and the hypothesis implies that it is finite.

The proof relies on a deep result of C. Reutenauer (1985) on the noncommutative polynomial associated to a code. It says that, when the code is not synchronizing, then the polynomial has a non trivial factorization. The proof then consists in showing that a different labeling of the graph destroys the possibility of a non-trivial factorization of the polynomial, hence making the automaton synchronizing.

BIBLIOGRAPHY

ADLER (R.L.), GOODWIN (L.W.), Weiss (B.). — *Equivalence of topological Markov shifts*, Israel J. Math., t. **27**, 1977, p. 49–63.

BERSTEL (J.), PERRIN (D.). — *Theory of Codes.* — Academic Press, 1985.

FRIEDMAN (J.). — *On the road coloring problem*, Proc. Amer. Math Soc., t. **110**, 1990, p. 1133–35..

O'BRIEN (G.L.). — *The road coloring problem*, Israel J. Math., t. **39**, 1981, p. 145–154.

PERRIN (D.). — *Codes asynchrones*, Bull. Soc. Math. de France, t. **105**, 1977, p. 385–404.

PERRIN (D.), SCHÜTZENBERGER (M.-P.). — *Synchronizing prefix codes and automata, and the road coloring problem*, 1991, submitted for publication.

PIN (J.E.). — *Le problème de la synchronisation. Contribution à l'étude de la conjecture de Cerny*, Thèse, 1978, Université Paris VI.

REUTENAUER (C.). — *Non commutative factorizations of variable length codes*, J. Pure applied Algebra, t. **36**, 1985, p. 167–186.

Author Index